Dictionary of
BIOINFORMATICS
and COMPUTATIONAL
BIOLOGY

Dictionary of
BIOINFORMATICS
and COMPUTATIONAL
BIOLOGY

Edited by

John M. Hancock
Bioinformatics Group
Medical Research Council Mammalian Genetics Unit
Harwell, Oxfordshire, United Kingdom

Marketa J. Zvelebil
Ludwig Institute for Cancer Research
London, United Kingdom

A John Wiley & Sons, Inc., Publication

Published by John Wiley & Sons, Inc., Hoboken, New Jersey.
Published simultaneously in Canada.

For general information on our other products and services please contact our Customer Care Department within the U.S. at 877-762-2974, outside the U.S. at 317-572-3993 or fax 317-572-4002.

Wiley also publishes its books in a variety of electronic formats. Some content that appears in print, however, may not be available in electronic format.

Library of Congress Cataloging-in-Publication Data

Dictionary of bioinformatics and computational biology / edited by John M. Hancock, Marketa J. Zvelebil.
     p. ; cm.
Includes bibliographical references and index.
  ISBN 0-471-43622-4 (cloth : alk. paper)
  1. Bioinformatics—Dictionaries. 2. Computational biology—Dictionaries.
  [DNLM: 1. Computational Biology—Dictionary—English. QU 13 W676 2004] I. Title: Dictionary of bioinformatics and computational biology. II. Hancock, John M. III. Zvelebil, Marketa J.
  QH324.2.W54 2004
  570′.285—2c22

                                2003018952

Printed in the United States of America.

10 9 8 7 6 5 4 3 2 1

Marketa J. Zvelebil
would like to thank
her parents for their continued support throughout—
may it long continue;
Dr. Martin Scurr, without whom she would have long lost her cool;
and last but not least Alice Yang and Qiong Gao
for help in formatting the files.

John M. Hancock
would like to thank his wife, Liz,
for all her support
and his parents for everything they have done.

CONTENTS

LETTER CONTENTS

CONTRIBUTORS

PATRICK ALOY EMBL Heidelberg Heidelberg Germany

ROLF APWEILER EMBL Outstation—Hinxton, European Bioinformatics Institute, Wellcome Trust Genome Campus, Hinxton, Cambridge, United Kingdom

TERRI ATTWOOD School of Biological Sciences, The University of Manchester, Manchester, United Kingdom

JEREMY BAUM Department of Biochemistry, Wolfson Laboratories, Imperial College London, United Kingdom

MARTIN J. BISHOP HGMP Resource Centre, Hinxton, Cambridge, United Kingdom

STUART M. BROWN Research Computing Resource, NYU School of Medicine New York, New York

JAMIE J. CANNONE Cellular and Molecular Biology, The University of Texas at Austin, Austin, Texas

LIZ P. CARPENTER Department of Biological Sciences, Imperial College, London, United Kingdom

JEAN-MICHEL CLAVERIE Information Génétique et Structurale, C.N.R.S.—UMR, Marseille, France

NELLO CRISTIANINI Department of Statistics, University of California, Davis, California

MICHAEL P. CUMMINGS Center for Comparative Molecular Biology and Evolution, Marine Biological Laboratory, Woods Hole, Massachusetts

NIALL DILLON MRC Clinical Sciences Centre, Hammersmith Hospital, London, United Kingdom

ROLAND L. DUNBRACK Institute for Cancer Research Fox Chase Cancer Center, Philadelphia, Pennsylvania

JAMES W. FICKETT Global Director, Bioinformatics, AstraZeneca R&D, Waltham, Massachusetts

ALAN FILIPSKI 648 W. La Donna Drive Tempe Arizona

KATHELEEN GARDINER Eleanor Roosevelt Institute of Cancer Research, Denver, Colorado

DOV S. GREENBAUM 100 York Street, New Haven, Connecticut

RODERIC GUIGÓ Genome Informatics, GRIM, Barcelona, Spain

ROBIN R. GUTELL University of Texas at Austin, Austin, Texas

JOHN M. HANCOCK MRC Mammalian Genetics Unit/UK Mouse Genome Centre Harwell, Oxfordshire, United Kingdom

ANDREW HARRISON Department of Biochemistry and Molecular Biology, University College, London United Kingdom

JAAP HERINGA Bioinformatics Unit, Department of Computer Science, Faculty of Sciences, Vrije Universiteit de Boelelaan, Amsterdam, The Netherlands

A. RUS HOELZEL Department of Biological Sciences, South Road, Durham, United Kingdom

AUSTIN L. HUGHES Department of Biological Sciences, University of South Carolina, Columbia, South Carolina

DAVID T. JONES Department of Computer Science, University College, London, United Kingdom

SUDHIR KUMAR Department of Biology, Arizona State University, Tempe, Arizona

ROMAN A. LASKOWSKI Department of Crystallography, Birkbeck College, London, United Kingdom

ERIC MARTZ Department of Microbiology, University of Massachusetts, Amherst, Massachusetts

MARK MCCARTHY Unit of Metabolic Medicine, St Mary's Hospital, Imperial College School of Medicine, London, United Kingdom

IRMTRAUD M. MEYER Department of Statistics, The Peter Medawar Building for Pathogen Research, University of Oxford, Oxford, United Kingdom

CHRISTINE A. ORENGO Biomolecular Structure and Modelling Unit, Department of Biochemistry and Molecular Biology, University College London, London, United Kingdom

LASZLO PATTHY Biological Research Center, Hungarian Academy of Sciences, Budapest, Hungary

THOMAS D. SCHNEIDER Laboratory of Experimental and Computational Biology CCR, NCI/NIH, Frederick, Maryland

RODGER STADEN Medical Research Council, Laboratory of Molecular Biology, Cambridge, United Kingdom

ROBERT STEVENS Department of Computer Science, The University of Manchester, Manchester, United Kingdom

GUENTER STOESSER EBI-Hinxton, Wellcome Trust Genome Campus, Hinxton, Cambridge, United Kingdom

DENIS THIEFFRY ESIL- GBMA, Universite de la Mediterranee, Campus de Luminy, Marseille, France

STEVEN WILTSHIRE Bioinformatics and Statistical Genetics, University of Oxford, Wellcome Trust Centre for Human Genetics, Oxford, United Kingdom

MARKETA J. ZVELEBIL Ludwig Institute for Cancer Research, London, United Kingdom

PREFACE

The first application of computers to biological data may be traced to 1965, when Margaret Dayhoff published her seminal paper "Computer aids to protein sequence determination" in the *Journal of Theoretical Biology*. The first use of the term "bioinformatics" in scientific literature as represented by PubMed only goes back to 1993. Despite being a relatively new science, the field has come a long way and is now an essential part of any up-to-date undergraduate training in molecular biology and related disciplines.

Since its early days, bioinformatics has diversified and now incorporates subdisciplines ranging from databases and ontologies to the modeling of complex biological systems by way of molecular evolution and protein structure prediction. Its practitioners are also diverse, with principle training in disciplines such as biology, chemistry, physics, mathematics, and computer science, to name but a few. Because of its diversification, bioinformatics has quickly reached the stage at which practitioners can no longer be expected to have expertise in every area.

Our belief that the field deserves a dedicated, broadbased dictionary compelled us to agree to edit this work, which we entitled *Dictionary of Bioinformatics and Computational Biology*. For the purposes of preparing this dictionary, we applied a broad definition of "Bioinformatics and Computational Biology," with an emphasis on the fact that the field is about biology as much as, if not more than, the storage and organization of information.

The aim of this dictionary is to provide clear definitions of the fundamental concepts of bioinformatics and computational biology. The entries were written and edited to enhance the book's utility for newcomers to the field, particularly undergraduate and postgraduate students. Those already working in the field should also find it handy as a source for quick introductions to topics with which they are not too familiar.

We thank our contributors for their efforts in writing the vast majority of the entries in this book, although as editors we accept responsibility for their accuracy. This dictionary is a first attempt to cover the field in this way. As such, it is liable to contain omissions and perhaps even the odd (hopefully minor!) error. We hope it will be successful enough to give us the opportunity to correct these errors and omissions in future editions.

JOHN M. HANCOCK
MARKETA J. ZVELEBIL

March, 2004

A

Ab Initio
Ab Initio Gene Prediction
ABNR
Acceptor Splice Site
Accuracy (of Prediction)
Accuracy Measures
Acrocentric Chromosome
Adaptation
Admixture Mapping
Adopted-Basis Newton–Raphson
    Minimization (ABNR)
Affymetrix GeneChip Oligonucleotide
    Microarray
Affymetrix Probe-Level Analysis
After Sphere
After State
Affine Gap Penalty
AGADIR
Alanine
Alignment—Multiple
Alignment—Pairwise
Alignment of Nucleic Acid Sequences
Alignment of Protein Sequences
Alignment Score
Aliphatic
Allele-Sharing Methods
Allelic Association
Allopatric Evolution
Allopatric Speciation
Alpha Carbon
Alpha Helix
Alu Repeat

AMBER
Amide Bond
Amino Acid
Amino Acid Abbreviations
Amino Acid Composition
Amino Acid Exchange Matrix
Amino Acid Index
Amino Acid Substitution Matrix
Amino-Terminus
Amphipathic
Analog
Ancestral Genome
Ancestral State Reconstruction
Anchor Points
Annotation Transfer
Anomalous Dispersion
APBioNet
Apomorphy
Arginine
Aromatic
Artificial Neural Networks
Asia Pacific Bioinformatics Network
Asparagine
Aspartic Acid
Association Analysis
Asymmetric Unit
Atomic Coordinate File
Attribute
Autapomorphy
Autosome
Axiom

Dictionary of Bioinformatics and Computational Biology. Edited by Hancock and Zvelebil
ISBN 0-471-43622-4 © 2004 John Wiley & Sons, Inc.

AB INITIO

Roland L. Dunbrack and Marketa J. Zvelebil

In quantum mechanics, calculations of physical characteristics of molecules based on first principles such as the Schrödinger equation. In protein structure prediction, calculations made without reference to a known structure homologous to the target to be predicted. In other words, these methods attempt to predict protein structure essentially from first principles (i.e., from physics and chemistry). The main advantage of this type of method is that no homologous structure is required to predict the fold of the target protein. However, the accuracy of ab initio methods is not as high as threading or homology modeling.

There are a number of often-used ab initio methods: lattice folding, FragFOld, Rosetta, and Unres.

Related Websites

Basic ab initio quantum chemistry guide	http://www.chem.swin.edu.au/modules/mod5/
Lattice-folding method	http://www.cs.brandeis.edu/~cs178/classnotes/PFolding/
Rosetta	http://www.bioinfo.rpi.edu/~bystrc/hmmstr/server.php

Further Reading

Hinchliffe A (1995). *Modelling Molecular Structures*. Wiley, New York.

Jones DT (2001). Predicting novel protein folds by using FRAGFOLD. *Proteins* (Suppl 5): 127–132.

Defay T, Cohen FE (1995). Evaluation of current techniques for ab initio protein structure prediction. *Proteins* 23: 431–445.

See also Gene Prediction, Ab Initio.

AB INITIO GENE PREDICTION. *SEE* GENE PREDICTION, AB INITIO.

ABNR. *SEE* ENERGY MINIMIZATION.

ACCEPTOR SPLICE SITE

Thomas D. Schneider

The binding site of the spliceosome on the 3' side of an intron and the 5' side of an exon is called an acceptor splice site. This term is preferred over 3' site because there can be multiple acceptor sites, in which case 3' site is ambiguous. Also, one would

Dictionary of Bioinformatics and Computational Biology. Edited by Hancock and Zvelebil
ISBN 0-471-43622-4 © 2004 John Wiley & Sons, Inc.

have to refer to the 3′ site on the 5′ side of an exon, which is confusing. Mechanistically, an acceptor site defines the beginning of the exon, not the other way around. For example, there are two acceptor sites at the 3′ end of intron 3 of the iduronidase synthetase gene: the normal site (12.7-bit) and a strong (8.9-bit) cryptic acceptor site (Rogan et al., 1998). Calling both of these 3′ sites would be confusing. In a second example, a mutation, *G863A* in the *ABCR* gene, creates a 9.8-bit acceptor site 3 bases downstream from the normal site (Allikmets et al., 1998).

Related Websites

rfs	http://www.lecb.ncifcrf.gov/~toms/paper/rfs/latex/paper.htmlfig.ids
ABCR Mut. G863A	http://www.lecb.ncifcrf.gov/~toms/g863a.html

Further Reading

Allikmets R, et al. (1998). Organization of the ABCR gene: Analysis of promoter and splice junction sequences. *Gene* 215: 111–122. http://www.lecb.ncifcrf.gov/~toms/paper/abcr/.

Rogan PK, et al. (1998). Information analysis of human splice site mutations. *Hum. Mut.* 12: 153–171. http://www.lecb.ncifcrf.gov/~toms/paper/rfs/.

Stephens RM, Schneider TD (1992). Features of spliceosome evolution and function inferred from an analysis of the information at human splice sites. *J. Mol. Biol.* 228: 1124–1136. http://www.lecb.ncifcrf.gov/~toms/paper/splice/.

See also Donor Splice Site, Sequence Walker.

ACCURACY (OF PREDICTION)
David T. Jones

The measurement of the agreement between a predicted structure and the true native conformation of the target protein.

Depending on the type of prediction, a variety of metrics can be defined to measure the accuracy of a prediction experiment. For predictions of *protein* secondary structure (i.e., assigning residues to helix, strand, or coil states), the percentage of residues correctly assigned (the Q_3 score) is the simplest and most widely used metric. For predictions of three-dimensional structure, there are many different metrics of accuracy with a variety of advantages and disadvantages. Probably the most widely used metric is the root-mean-square deviation (RMSD) between the model and the native structure. Unfortunately, small errors in the model, particularly those which accrue from errors in the alignment used to build the model, result in very large RMSD values, and so this metric is less useful for low-quality models. For low-quality models, it is more common to measure prediction accuracy in terms of the percentage of residues which have been correctly aligned when compared with a structural alignment of the target and template proteins, or the percentage of residues which have been correctly positioned to within a certain distance cutoff (e.g., 3 Å).

Related Website

Accuracy estimators	http://predictioncenter.llnl.gov/local/local.html

See also Q index, Room-Mean-Square Deviation, Secondary-Structure Prediction of Protein.

ACCURACY MEASURES. *SEE* ERROR MEASURES.

ACROCENTRIC CHROMOSOME
Katheleen Gardiner

A chromosome with a centromere at or near one end.

Human chromosomes 13–15 and 21 and 22 are acrocentric as are most mouse chromosomes. The short arms contain largely tandem repeats of ribosomal RNA genes and few or no single copy genes.

Further Reading
Miller OJ, Therman E (2000). *Human Chromosomes.* Springer-Verlag, Vienna.
Wagner RP, Maguire MP, Stallings RL (1993). *Chromosomes—A Synthesis.* Wiley, New York.

ADAPTATION
A. Rus Hoelzel

The differential survival of phenotypes based on their relative fitness leads to the adaptation of form to suit a given set of environmental conditions.

Adaptation is the product of the interaction between natural selection and environment. Rapid periods of diversification are referred to as adaptive radiations. Examples include the radiation of a number of groups of higher taxa during the Cenozoic, including birds, mammals, and reptiles. Possible explanations for these events include ecological release (such as adaptation to niches left open following mass extinctions) and invasions (such as founders to island habitats; *but see* Founder Effect).

Among the best researched examples of adaptation are the specializations of Galapagos finch species, all thought to be based on variation derived from a single mainland ancestor. There are at least 14 species that differ in size, coloration, behavior, and, in particular, the shape and size of their beaks. These adaptations were noted by Darwin following his expeditions on the *Beagle* and more recently were researched in considerable detail during an ongoing study initiated in the 1970s (Grant, 1986). Grant and coworkers have shown the importance of environmental change in influencing the survival of individuals with subtly different shaped beaks.

Further Reading
Grant PR (1986). *Ecology and Evolution of Darwin's Finches.* Princeton University Press, Princeton, New Jersey.

See also Evolution, Natural Selection.

ADMIXTURE MAPPING (MAPPING BY ADMIXTURE LINKAGE DISEQUILIBRIUM)
Mark McCarthy and Steven Wiltshire

A potentially powerful method for identifying genes that underlie ethnic differences in disease risk: It focuses on recently admixed populations and detects linkage by testing for association of the disease with ancestry at each typed marker locus.

Several major multifactorial traits (such as diabetes and hypertension) show marked ethnic differences in disease frequency, which may, at least in part, reflect differences in the prevalence of major susceptibility variants. Admixture mapping makes use of populations which have arisen through recent admixture of ancestral populations with widely differing disease prevalences. In such admixed populations, the location of disease susceptibility genes responsible for the prevalence difference between ancestral populations can be revealed by identifying chromosomal regions in which affected individuals show increased representation of the high-prevalence ancestral line. Such analyses bear many similarities to linkage analysis in inbred rodent strains. If the underlying requirements are met, this approach to linkage mapping of multifactorial traits should be more powerful than standard approaches. However, the major limitations to its exploitation to date have been the need to recruit suitable admixed populations and the requirement for a marker map of loci displaying substantial interethnic allele frequency differences (essential for inference of ancestry). Such markers are rapidly emerging from current efforts to develop and validate a genomewide single-nucleotide polymorphism map.

Related Website

ADMIXMAP	http://www.lshtm.ac.uk/eph/eu/GeneticEpidemiologyGroup.htm

Further Reading

Collins-Schramm HE, et al. (2002). Ethnic-difference markers for use in mapping by admixture linkage disequilibrium. *Am. J. Hum. Genet.* 70: 737–750.

McKeigue PM (1998). Mapping genes that underlie ethnic differences in disease risk: Methods for detecting linkage in admixed populations, by conditioning on parental admixture. *Am. J. Hum. Genet.* 63: 241–251.

Shriver MD, et al. (1998). Ethnic-affiliation estimation by use of population-specific DNA markers. *Am. J. Hum. Genet.* 60: 957–964.

Stephens JC, et al. (1994). Mapping by admixture linkage disequilibrium in human populations: Limits and guidelines. *Am. J. Hum. Genet.* 55: 809–824.

See also Genome Scans, Linkage Analysis.

ADOPTED-BASIS NEWTON–RAPHSON MINIMIZATION (ABNR). *SEE* ENERGY MINIMIZATION.

AFFYMETRIX GENECHIP OLIGONUCLEOTIDE MICROARRAY
Stuart M. Brown and Dov S. Greenbaum

Technology for measuring expression levels of large numbers of genes simultaneously.

An oligonucleotide microarray technology, known as *GeneChip* arrays, has been developed by Fodor, Lockhart, and colleagues at Affymetrix. This involves the in situ synthesis of short (25-base) DNA oligonucleotide probes directly on glass slides using photolithography techniques (similar to the methods used for the fabrication of computer chips). This method allows more than 400,000 individual features to be placed on a single 1.28-cm^2 slide. GeneChips produced in a single lot are identical within rigorous quality control standards. In the GeneChip system, a single messenger RNA (mRNA) sample is fluorescently labeled and hybridized to a chip, then scanned. Instead of having entire genes or large gene fragments on a chip, GeneChips typically have tens of oligonucleotides corresponding to portions of a sequence from a single gene. For each oligonucleotide probe that matches the gene sequence, an additional probe is included that contains a single base mismatch. In this way the technology aims to provide a higher degree of specificity than other microarray methodologies.

The entire laboratory process for using the GeneChip system is highly standardized with reagent kits for fluorescent labeling, automated equipment for hybridization and washing, and a fluorescent scanner with integrated software to control data collection and primary analysis. The system also provides indicators of sample integrity, assay execution, and hybridization performance through the assessment of control hybridizations of complementary DNA (cDNA) sequences spiked into the labeling reagents. As a result, Affymetrix arrays produce highly reproducible results when the same RNA is labeled and hybridized to duplicate GeneChips. The Affymetrix microarray software automatically calculates local background values for regions of each chip and provides a quality score for each gene as a flag (Present, Absent, or Marginal) together with each gene's measured signal strength. This allows for the easy identification of genes with expression levels below the reliable detection threshold of the system, an important step in the data analysis process.

Affymetrix GeneChip expression experiments generate large amounts of data, 72 Mbytes for a typical array, so it is critical to establish robust bioinformatics systems for data storage and management.

Related Website

Affymetrix	http://www.affymetrix.com

Further Reading

Fodor SP, et al. (1991). Light-directed, spatially addressable parallel chemical synthesis. *Science* 251: 767–773.

Lockhart DJ, et al. (1996). Expression monitoring by hybridization to high-density oligonucleotide arrays. *Nat. Biotechnol.* 14: 1675–1680.

AFFYMETRIX PROBE-LEVEL ANALYSIS
Stuart M. Brown

Normalization of Affymetrix chip data.

In the Affymetrix GeneChip system, each gene is represented by a set of eleven to twenty 25-base oligonucleotide probes that match positions distributed across the known complementary DNA (cDNA) sequence (a perfect match, or PM, probe). For each probe that matches the cDNA sequence, an additional probe is created with a single base mismatch in the center (MM probe). The Affymetrix system uses the Microarray Analysis Suite (MAS) software to drive a fluorescent scanner that captures data from labeled target cDNA hybridized to each feature (probe sequence) on the GeneChip. The observed intensities of the fluorescent signal show huge variability among the probes in a probe set due to sequence-specific differences in optimal hybridization temperatures as well as variable amounts of cross-hybridization between individual probes and cDNA from other genes. In the MAS version 4 software, the signal for each probe set was calculated as a simple average of the differences between each PM probe and its corresponding MM probe (average difference). In the MAS version 5 software, a median weighted average (Tukey's biweight estimate) of the fluorescent intensity differences between paired PM and MM probes is used to compute a signal for each gene, but MM intensities that are greater than the corresponding PM are discarded so that negative signals are not produced for any genes.

The Affymetrix MAS software also assigns a quality flag (Present, Absent, or Marginal) to each signal based on the overall signal strengths of PM vs. MM probes. In the MAS 4 software, each gene was called Present if a majority of probe pairs had stronger PM than MM signals. In MAS 5 software, this determination is made using a statistical approach. A "discrimination score" of $(PM-MM)/(PM+MM)$ is calculated for each probe pair, which is then compared to a cutoff value (a small positive number between 0 and 0.1) and an overall p value for each probe set is computed using the one-sided Wilcoxon's signed-rank test. By adjusting the cutoff value, it is possible to change the probability that genes with low signal levels (in both PM and MM probes) will be flagged as Absent. In any case, if a probe set has many MM probes with stronger signals than their paired PM probes, then the gene will be flagged as Absent.

Several alternative techniques have been proposed to integrate probe set signals into a number that represents the expression level of a single gene. The model-based expression index (MBEI) developed by Li and Wong (2001) (made available in the free dChip software) uses a model-fitting technique for PM and MM values for a probe set across all chips in an experiment to achieve improved results. Speed and coworkers (2002) have developed a system called Robust Multichip Analysis (RMA) which calculates expression values for each gene using $\log_2$ of quantile normalized, background-adjusted signals from only the PM probes across all of the chips in an experiment. The MBEI and RMA methods reduce the observed variability of low-expressing genes allowing for the reliable detection of differential expression at lower messenger RNA (mRNA) concentrations.

Related Websites

Bioconductor	http://www.bioconductor.org
dCHIP	http://www.biostat.harvard.edu/complab/dchip/

Further Reading

Affymetrix. (1999). *Affymetrix Microarray Suite User Guide*, Version 4. Affymetrix, Santa Clara, CA.

Affymetrix (2001a). *Affymetrix Microarray Suite User Guide*, Version 5. Affymetrix, Santa Clara, CA.

Affymetrix (2001b). *Statistical Algorithms Reference Guide*. Affymetrix, Santa Clara, CA.

Li C, Wong WH (2001). Model-based analysis of oligonucleotide arrays: Expression index computation and outlier detection. *Proc. Natl. Acad. Sci. USA* 98: 31–36.

Liu WM, et al. (2001). Rank-based algorithms for analysis of microarrays. *Proc. SPIE* 4266: 56–67.

See also Microarray Normalization.

AFTER SPHERE. *SEE* AFTER STATE.

AFTER STATE (AFTER·SPHERE)
Thomas D. Schneider

The low-energy state of a molecular machine after it has made a choice while dissipating energy is called its after state. This corresponds to the state of a receiver in a communications system after it has selected a symbol from the incoming message while dissipating the energy of the message symbol. The state can be represented as a sphere in a high-dimensional space.

Further Reading
Schneider TD (1991). Theory of molecular machines. I. Channel capacity of molecular machines. *J. Theor. Biol.* 148: 83–123. http://www.lecb.ncifcrf.gov/~toms/paper/ccmm/

See also Channel Capacity, Gumball Machine, Shannon Sphere, Molecular Machine, Message.

AFFINE GAP PENALTY. *SEE* GAP PENALTY.

AGADIR
Patrick Aloy

AGADIR is an algorithm based on the helix–coil transition to predict the helical behavior of monomeric peptides. It considers short-range interactions under different conditions of pH, temperature, and ionic strength. For the definition of phase space the authors used the helix–coil transition, leading to a two-state model for each residue. When a residue is in a random coil state, it is allowed to explore the whole conformational space. However, when it is in the α-helical state, only the conformational search is restricted to the α-helical angles. The best conformations are then chosen by means of statistical mechanics.

AGADIR was tested on 323 peptides in solution and showed a good correlation between predicted and observed behavior.

Related Website

Agadir-start	http://www.embl-heidelberg.de/Services/serrano/agadir/agadir-start.html

Further Reading
Muñoz V, Serrano L (1994). Elucidating the folding problem of helical peptides using empirical parameters. *Nature Struct. Biol.* 1: 399–409.

ALANINE
Jeremy Baum

Alanine is a small nonpolar amino acid with side chain —CH_3 found in proteins. In sequences, written as Ala or A.

Further Reading
Branden C, Tooze J (1998). *Introduction to Protein Structure*, Second Edition Garland Science, New York.

See also Amino Acid.

ALIGNMENT—MULTIPLE
Jaap Heringa

A multiple sequence alignment is an alignment of two or more sequences. The automatic generation of an accurate multiple alignment is potentially a daunting task. Ideally, one would make use of an in-depth knowledge of the evolutionary and structural relationships within the family, but this information is often lacking or difficult to use. General empirical models of protein evolution (Benner et al., 1992; Dayhoff, 1978; Henikoff and Henikoff, 1992) are widely used instead, but these can be difficult to apply when the sequences are less than 30% identical (Sander and Schneider, 1991). Furthermore, mathematically sound methods for carrying out alignments using these models can be extremely demanding in computer resources for more than a handful of sequences (Carillo and Lipman, 1988). To be able to cope with practical data set sizes, heuristics have been developed that are used for all but the smallest data sets.

The most commonly used heuristic methods are based on the tree-based progressive alignment strategy (Hogeweg and Hesper, 1984; Feng and Doolittle, 1987; Taylor, 1988) with CLUSTAL W (Thompson et al., 1994) being the most widely used implementation. The idea is to establish an initial order for joining the sequences and to follow this order in gradually building up the alignment. Many implementations use an approximation of a phylogenetic tree between the sequences as a guide tree that dictates the alignment order. Although appropriate for many alignment problems, the progressive strategy suffers from its greediness. Errors made in the first alignments during the progressive protocol cannot be corrected later as the remaining sequences are added in. Attempts to minimize such alignment errors have generally been targeted at global sequence weighting (Altschul et al., 1989; Thompson et al., 1994), where the contributions of individual sequences are weighted during the alignment process. However, such global sequence-weighting schemes carry the risk of propagating rather than reducing error when used in progressive multiple-alignment strategies (Heringa, 1999).

The main alternative to progressive alignment is the simultaneous alignment of all the sequences. Two such implementations are available, Multiple Sequence Alignment (MSA) (Lipman et al., 1989) and Divide and Conquer Alignment (DCA) (Stoye et al., 1997). Both methods are based on the Carillo and Lipman (1988) algorithm to limit computations to a small area in the multidimensional search matrix. They nonetheless remain an extremely CPU- and memory-intensive approach, applicable only to about nine sequences of average length for the fastest implementation (DCA). Iterative strategies (Hogeweg and Hesper, 1984; Gotoh, 1996; Notredame and Higgins, 1996; Heringa, 1999, 2002) are an alternative to optimizing multiple alignments by reconsidering and correcting those made during preceding iterations. Although such iterative strategies do not provide any guarantees of finding optimal solutions, they are reasonably robust and much less sensitive to the number of sequences than their simultaneous counterparts.

All of these techniques perform global alignment and match sequences over their full lengths. Problems with this approach can arise when highly dissimilar sequences are compared. In such cases global alignment techniques might fail to recognize highly similar internal regions because these may be overshadowed by dissimilar regions and the high gap penalties normally required to achieve proper global matching. Moreover, many biological sequences are modular and show shuffled domains which can render a global alignment of two complete sequences meaningless. The occurrence of varying numbers of internal sequence repeats (Heringa, 1998) can also severely limit the applicability of global methods. In general, when there is a large difference in the lengths of two sequences to be compared, global alignment routines become unwarranted. To address these problems, Smith and Waterman (1981) early on developed a so-called *local* alignment technique in which the most similar regions in two sequences are selected and aligned. For multiple sequences, the main automatic local alignment methods include the Gibbs sampler (Lawrence et al., 1993), Multiple Expectation-maximization for Motif Elicitation (MEME) (Bailey and Elkan, 1994), and Dialign2 (Morgenstern, 1999). These programs often perform well when there is a clear block of ungapped alignment shared by all of the sequences. However, performance on representative sets of test cases is poor when compared with global methods (Thompson et al., 1999; Notredame et al., 2000). Some recent methods have been developed in which global and local alignment are combined to optimize multiple-sequence alignment (Notredame et al., 2000; Heringa, 2002). Here the popular method T-Coffee (Notredame et al., 2000) and the Praline technique (Heringa, 2002) are robust and sensitive implementations. The accompanying figure shows an example of an alignment of 13 flavodoxin sequences and the sequence of the widely divergent transduction protein cheY.

Related Websites

CLUSTAL W	http://www.ebi.ac.uk/clustalw/
T-Coffee	http://www.ch.embnet.org/software/TCoffee.html

Further Reading

Altschul SF, et al. (1989). Weights for data related by a tree. *J. Mol. Biol.* 207: 647–653.

Bailey TL, Elkan C (1994). Fitting a mixture model by expectation maximization to discover motifs in biopolymers. In *Proceedings of the Second International Conference on Intelligent Systems for Molecular Biology*, AAAI Press, Washington, DC, pp. 28–36.

Benner SA, et al. (1992). Response to Barton's letter: Computer speed and sequence comparison. *Science* 257: 609–610.

Carillo H, Lipman DJ (1988). The multiple sequence alignment problem in biology. *SIAM J. Appl. Math.* 48: 1073–1082.

Dayhoff MO, et al. (1978). A model of evolutionary change in proteins. In *Atlas of Protein Sequence and Structure*, Vol. 5, Suppl. 3, MO Dayhoff, Ed. National Biomedical Research Foundation, Washington, DC, pp. 345–352.

Feng DF, Doolittle RF (1987). Progressive sequence alignment as a prerequisite to correct phylogenetic trees. *J. Mol. Evol.* 21: 112–125.

Gotoh O (1996). Significant improvement in accuracy of multiple protein sequence alignments by iterative refinement as assessed by reference to structural alignments. *J. Mol. Biol.* 264: 823–838.

Henikoff S, Henikoff JG (1992). Amino acid substitution matrices from protein blocks. *Proc. Natl. Acad. Sci. USA* 89: 10,915–10,919.

Heringa J (1998). Detection of internal repeats: How common are they? *Curr. Opin. Struct. Biol.* 8: 338–345.

Heringa J (1999). Two strategies for sequence comparison: Profile-preprocessed and secondary structure–induced multiple alignment. *Comput. Chem.* 23: 341–364.

Heringa J (2002). Local weighting schemes for protein multiple sequence alignment. *Comput. Chem.* 26: 459–477.

Heringa J, Taylor WR (1997). Three-dimensional domain duplication, warping and stealing. *Eur. Opin. Struct. Bio.* 7: 410–421.

Hogeweg P, Hesper B (1984). The alignment of sets of sequences and the construction of phyletic trees: An integrated method. *J. Mol. Evol.* 20: 175–186.

Lawrence CE, et al. (1993). Detecting subtle sequence signals: A Gibbs sampling strategy for multiple alignment. *Science* 262: 208–214.

The colour assignments have been adapted from the defaults in CLUSTALX (Thompson *et al*, 1997) <u>Abstract</u> :

G, P, S, T  H, K, R F, W, Y I, L, M, V

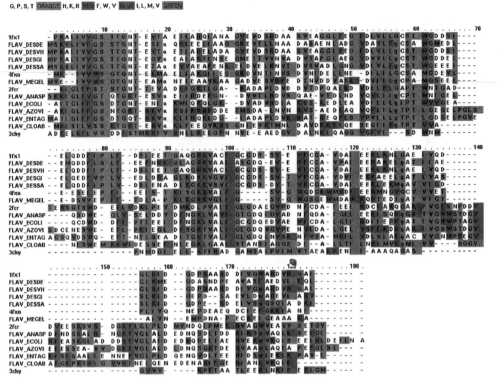

Figure

Lipman DJ, et al. (1989). A tool for multiple sequence alignment. *Proc. Natl. Acad. Sci. USA* 86: 4412–4415.

Morgenstern B (1999). DIALIGN 2: Improvement of the segment-to-segment approach to multiple sequence alignment. *Bioinformatics* 15: 211–218.

Notredame C, Higgins DG (1996). SAGA: Sequence alignment by genetic algorithm. *Nucleic Acids Res.* 24: 1515–1524.

Notredame C, Higgins DG, Heringa J (2000). T-coffee: a novel method for fast and accurate multiple sequence alignment. *J. Mol. Bio.* 307: 205–217.

Sander C, Schneider R (1991). Database of homology derived protein structures and the structural meaning of sequence alignment. *Proteins* 9: 56–68.

Smith TF, Waterman MS (1981). Identification of common molecular subsequences. *J. Mol. Bio.* 147: 195–197.

Stoye J, et al. (1997). DCA: An efficient implementation of the divide-and-conquer approach to simultaneous multiple sequence alignment. *Comput. Appl. Biosci.* 13: 625–626.

Taylor WR (1988). A flexible method to align large numbers of biological sequences. *J. Mol. Evol.* 28: 161–169.

Thompson JD, et al. (1994). CLUSTAL W: Improving the sensitivity of progressive multiple sequence alignment through sequence weighting, positions-specific gap penalties and weight matrix choice. *Nucleic Acids Res.* 22: 4673–4680.

Thompson JD, et al. (1999). A comprehensive comparison of multiple sequence alignment programs. *Nucleic Acids Res.* 27: 2682–2690.

Wang L, Jiang T (1994). On the complexity of multiple sequence alignment. *J. Comput. Bio.* 1: 337–348.

ALIGNMENT—PAIRWISE (DOMAIN ALIGNMENT, REPEATS ALIGNMENT)

Jaap Heringa

Sequence alignment is the most common task in bioinformatics. It involves matching the amino acids or nucleotides of two or more sequences in such a way that their similarity can be determined best. Although many properties of nucleotide or protein sequences can be used to derive a similarity score (e.g., nucleotide or amino acid composition, isoelectric point, or molecular weight), the vast majority of sequence similarity calculations presuppose an alignment between two sequences from which a similarity score is inferred. Ideally, the alignment of two sequences should be in agreement with their evolution, i.e., the patterns of descent as well as molecular structural and functional evolution. Unfortunately, the evolutionary traces are often very difficult to detect; e.g., in divergent evolution of two protein sequences from a common ancestor where amino acid mutations, insertions and deletions of residues, gene duplication, transposed gene segments, repeats, domain structures, and the like can blur the ancestral relationship beyond recognition. The outcome of an alignment operation can be studied by eye for conservation patterns or be given as input to a variety of bioinformatics methods, ranging from sequence database searching to tertiary structure prediction.

Although very many different alignments can be created between two nucleotide or protein sequences, a simplifying assumption is that there has been only a single evolutionary pathway from one sequence to another. In the absence of observed evolutionary traces, the matching of two sequences is regarded as mimicking evolution best when the minimum number of mutations is used to arrive at one sequence from the other. An approximation of this is finding the highest similarity value determined from summing substitution scores along matched residue pairs minus any insertion/deletion penalties. Such an alignment is generally called the optimal alignment.

Unfortunately, testing all possible alignments, including the insertion of a gap at each position of each sequence, is unfeasible. For example, there are about 10^{88} possible alignments of two sequences of 300 amino acids (Waterman, 1989), a number clearly beyond all computing capabilities. However, when introductions of gaps are also assigned scoring values such that they can be treated in the same manner as the mutation of one residue to another, the number of calculations is greatly reduced and becomes readily computable. The technique to calculate the highest scoring or optimal alignment is generally known as the dynamic programming (DP) technique, introduced by Needleman and Wunsch (1970) to the biological community. The DP algorithm is an optimization technique which guarantees, given an amino acid exchange matrix and gap penalty values, that the best scoring or optimal alignment is found.

There are two basic types of sequence alignment: global alignment and local alignment. Global alignment implies the matching of sequences over their complete lengths, whereas with local alignment the sequences are aligned only over the most similar parts of the sequences, carrying the clearest trace of evolutionary relatedness. In cases of distant sequences, global alignment techniques may fail to recognize such highly similar internal regions because these may be overshadowed by dissimilar regions and strong gap penalties required to achieve their proper matching. Moreover, many biological sequences are modular and show shuffled domains, which can render a global alignment of two complete sequences meaningless. For example, a two-domain sequence with domains A and B consecutive in sequence

(AB) will be incorrectly aligned against a sequence with domain organization BA. Also, global alignment of a three-domain sequence ABC against a two-domain sequence AC is likely to position the domain alignments out of register, making it difficult for the alignment method to insert a gap which spans all but the smallest representatives of domain B. The occurrence of varying numbers of internal sequence repeats is known to limit the applicability of global alignment methods, as the necessity to insert gaps spanning excess domains and confusion due to recurring strong motifs is likely to lead to erroneous alignment. In general, when there is a large difference in the lengths of two sequences to be compared, local alignment methods should be included in the analysis.

The first pairwise algorithm for local alignment was developed by Smith and Waterman (1981) as an adaptation of the algorithm of Needleman and Wunsch (1970). The Smith–Waterman technique selects the most similar region in each of two sequences, which are then aligned. Waterman and Eggert (1987) generalized the local alignment routine by devising an algorithm that allows the calculation of a user-defined number of top-scoring local alignments instead of only the optimal local alignment. Various implementations of the Waterman–Eggert technique exist with memory requirements reduced from quadratic to linear, thereby allowing very long sequences to be searched at the expense of only a small increase in computational time (e.g., Huang et al., 1990; Huang and Miller, 1991).

It is not always clear whether the highest scoring (local or global) alignment of two sequences is biologically the most meaningful. There may be biologically plausible alignments that score close to the highest value. For example, alignments of proteins based on their structures or of DNA sequences based on evolutionary changes are often different from their associated optimal sequence alignments. Since for most sequences the true alignment is unknown, methods that either assess the significance of the optimal alignment or provide a few "close" alternatives to the optimal one are of great value. A suboptimal alignment is an alignment whose score lies within the neighborhood of the optimal score. It can therefore be a natural candidate for an alternative to the optimal alignment. Enumeration of suboptimal alignments is not very useful since there are many such alignments. Other approaches that use only partial information about suboptimal alignments are more successful in practice. Vingron and Argos (1990) and Zuker (1991) approached this issue by constructing an algorithm which determines all optimal and suboptimal alignments and depicts them in a dot plot. Reliably aligned regions can be defined as those for which alternative local alignments do not exist.

Pairwise alignments can also be made based on tertiary structures if those are available for a target set of protein sequences. Methods have been devised for optimal superpositioning of two protein structures represented by their main-chain trace (Cα atoms). The matched residues in a so-called structural alignment then correspond to equivalenced Cα atoms in the tertiary structure superpositioning (Taylor and Orengo, 1989). Following the notion that in divergent evolution protein structure is more conserved than sequence, structural alignments are often used as a standard of truth to evaluate the performance of different sequence alignment methods.

Related Websites

malign	http://www.lecb.ncifcrf.gov/toms/delila/malign.html
BLAST	http://www.ncbi.nlm.nih.gov/BLAST/
CLUSTAL W	http://www.ebi.ac.uk/clustalw/

Further Reading

Huang X, Miller W (1991). A time-efficient, linear-space local similarity algorithm. *Adv. Appl. Math.* 12: 337–357.

Huang X, et al. (1990). A space-efficient algorithm for local similarities. *Comput. Appl. Biosci.* 6: 373–381.

Karlin S, Altschul SF (1990). Methods for assessing the statistical significance of molecular sequence features by using general scoring schemes. *Proc. Natl. Acad. Sci. USA* 87: 2264–2268.

Needleman SB, Wunsch CD (1970). A general method applicable to the search for similarities in the amino acid sequence of two proteins. *J. Mol. Biol.* 48: 443–453.

Smith TF, Waterman MS (1981). Identification of common molecular subsequences. *J. Mol. Biol.* 147: 195–197.

Taylor WR, Orengo CA (1989). Protein structure alignment. *J. Mol. Biol.* 208: 1–22.

Vingron M, Argos P (1990). Determination of reliable regions in protein sequence alignments. *Prot. Eng.* 3: 565–569.

Waterman MS (1989). Sequence alignments. In *Mathematical Methods for DNA Sequences, MS* Waterman, Ed. CRC Press, Boca Raton, FL.

Waterman MS, Eggert M (1987). A new algorithm for best subsequences alignment with applications to the tRNA-rRNA comparisons. *J. Mol Biol.* 197: 723–728.

Zuker M (1991). Suboptimal sequence alignment in molecular biology. Alignment with error analysis. *J. Mol. Biol.* 221, 403–420.

ALIGNMENT OF NUCLEIC ACID SEQUENCES. *SEE* ALIGNMENT—MULTIPLE, ALIGNMENT—PAIRWISE.

ALIGNMENT OF PROTEIN SEQUENCES. *SEE* ALIGNMENT—MULTIPLE, ALIGNMENT—PAIRWISE.

ALIGNMENT SCORE
Laszlo Patthy

Alignment of the sequences of homologous genes or proteins requires that the nucleotides (or amino acids) from "equivalent positions" (i.e., of common ancestry) are brought into vertical register. The procedures that attempt to align sites of common ancestry are referred to as sequence alignment procedures.

The correct alignment of two sequences is the one in which only sites of common ancestry are aligned, and all sites of common ancestry are aligned. The true alignment is most likely to be found as the one in which matches are maximized and mismatches and gaps are minimized. This may be achieved by using some type of scoring system that rewards similarity and penalizes dissimilarity and gaps. The alignment score is calculated as the sum of the scores assigned to the different types of matches, mismatches, and gaps. Most commonly used sequence alignment programs are based on the Needleman–Wunsch algorithm, which identifies the best alignment as the one with maximum alignment score among all possible alignments.

Further Reading

Needleman SB, Wunsch CD (1970). A general method applicable to the search for similarities in the amino acid sequence of two proteins. *J. Mol. Biol.* 48: 443–53.

ALIPHATIC

Roman A. Laskowski

Acyclic or cyclic, saturated or unsaturated carbon compounds, excluding aromatic compounds.

In proteins, the amino acid residues leucine, isoleucine, and valine have aliphatic side chains.

Related Websites

Amino acid repository	http://www.imb-jena.de/IMAGE_AA.html
Review of amino acids	http://wbiomed.curtin.edu.au/teach/biochem/tutorials/AAs/AA.html

See also Amino Acid, Aromatic, Side Chain.

ALLELE-SHARING METHODS (NONPARAMETRIC LINKAGE ANALYSIS)

Mark McCarthy and Steven Wiltshire

Basis for one type of linkage analysis methodology, typically employed in the analysis of multifactorial traits, which relies on the detection of an excess of allele sharing among related individuals who share a phenotype of interest (or, equivalently, a deficit of allele sharing among related individuals with differing phenotypes).

The simplest and most frequently used substrates for the application of allele-sharing methods are collections of affected sibling pairs (or sibpairs). The expectation, under the null hypothesis, is that any pair of full siblings selected at random will, on average, share 50% of their genomes identity by descent and that a large collection of such sibpairs will show approximately 50% sharing at all chromosomal locations. If, however, sibpairs are not selected at random but instead are ascertained on the basis of similarity for some phenotype of interest (e.g., sharing a disease known to have some inherited component), then the location of genes contributing to variation in that phenotype can be revealed through detecting regions where those affected sibpairs share significantly more than 50% identity by descent. In other words, when relatives are correlated for phenotype, they will tend to show correlations of genotype around variants contributing to phenotype development. Some of the earliest programs for allele-sharing linkage analysis relied on sharing defined as identity by state, but current methods focus more appropriately on identity-by-descent sharing. A variety of methods have been developed to extend this methodology to include other combinations of relatives, the analysis of quantitative as well as qualitative traits, and relative pairs with dissimilar as well as similar phenotypes. Allele-sharing methods are, essentially, equivalent to nonparametric linkage analysis.

The first example of a genome scan for a multifactorial trait using allele-sharing methods was published by Davies et al. (1994) using approximately 100 sibpair families ascertained for type 1 diabetes. This scan revealed evidence for linkage in several chromosomal regions, although the largest signal by far (as expected) fell in the histocompatibility locus antigen (HLA) region of chromosome 6. Wiltshire (2001) describe a scan of over 500 sibpair families ascertained for type 2 diabetes which illustrates some of the methodological issues associated with this approach.

Related Websites
Programs for allele-sharing analyses:

GENEHUNTER	http://www-genome.wi.mit.edu/ftp/distribution/software/genehunter
MERLIN	http://www.sph.umich.edu/csg/abecasis/Merlin/reference.html
SOLAR	http://www.sfbr.org/sfbr/public/software/solar/index.html
SPLINK	www-gene.cimr.cam.ac.uk/clayton/software
ASPEX	ftp://lahmed.stanford.edu/pub/aspex/

Further Reading

Abecasis GR, et al. (2002). Merlin—Rapid analysis of dense genetic maps using sparse gene flow trees. *Nat. Gene.t* 30: 97–101.

Davies JL, et al. (1994). A genome-wide search for human type 1 diabetes susceptibility genes. *Nature* 371: 130–136.

Kruglyak L, et al. (1996). Parametric and non-parametric linkage analysis: A unified multipoint approach. *Am. J. Hum. Genet.* 58: 1347–1363.

Lander ES, Schork NJ (1994). Genetic dissection of complex traits. *Science* 265: 2037–2048.

Ott J (1999). *Analysis of Human Genetic Linkage*, 3rd ed. Johns Hopkins University Press, Baltimore, MD, pp. 272–296.

Risch N, Zhang H (1995). Extreme discordant sib pairs for mapping quantitative trait loci in humans. *Science* 268: 1584–1589.

Terwilliger J, Ott J (1994). *Handbook of Human Genetic Linkage*. Johns Hopkins University Press, Baltimore, MD.

Wiltshire S (2001). A genome-wide scan for loci predisposing to type 2 diabetes in a UK population (The Diabetes (UK) Warren 2 Repository): Analysis of 573 pedigrees provides independent replication of a susceptibility locus on chromosome 1q. *Am. J. Hum. Gene.* 69: 553–569.

See also Haseman–Elston Regression Method, Linkage Analysis, Multifactorial Trait.

ALLELIC ASSOCIATION
Mark McCarthy and Steven Wiltshire

Nonindependent distribution of alleles at different variant loci.

In the most general sense, allelic association refers to any observed association, whatever its basis. However, most associations of interest (at least from the gene-mapping point of view) are those between loci that are linked, in which case *allelic association* equates with *linkage disequilibrium* (in the stricter definition of that term).

Allelic association is the basis of association analysis (either case-control or family-based association) where the aim is to demonstrate allelic association between a marker locus (of known location and genotyped) and a disease susceptibility variant (the position of which is unknown and which is represented in the analysis by the disease phenotype).

Examples: The histocompatibility locus antigen (HLA) region on chromosome 6 comprises a number of highly polymorphic loci and displays strong linkage disequilibrium. Consequently, alleles at neighboring HLA loci are in very strong allelic association in all studied populations. Since this region is known to harbor disease susceptibility genes (particularly for diseases associated with disturbed immune regulation), there are many examples of strong associations between certain HLA alleles and a variety of disease states. For example, possession of the *HLA-B27* allele has long been known to be associated with ankylosing spondylitis, and individuals who carry *DR3* and/or *DR4* (now termed *HLA-DQA1*0501-DQB1*0201/DQA1*0301-DQB1*0302*) alleles are at increased risk of type 1 diabetes.

Related Website
Repository for planned haplotype map data:

| HapMap | http://www.ncbi.nlm.nih.gov/SNP/HapMap/index.html |

Further Reading
Cardon LR, Bell JI (2001). Association study designs for complex disease. *Nat. Rev. Genet.* 2: 91–99.

Herr M, et al. (2000). Evaluation of fine mapping strategies for a multifactorial disease locus: Systematic linkage and association analysis of IDDM1 in the HLA region on chromosome 6p21. *Hum. Mol. Genet.* 9: 1291–1301.

Jeffreys AJ, et al. (2001). Intensely punctate meiotic recombination in the class II region of the major histocompatibility complex. *Nat. Genet.* 29: 217–222.

Lonjou C, et al. (1999). Allelic association between marker loci. *Proc. Natl. Acad. Sci. USA* 96: 1621–1626.

Owen MJ, McGuffin P (1993). Association and linkage: Complementary strategies for complex disorders. *J. Med. Genet.* 30: 638–639.

Sham P (1998). *Statistics in Human Genetics*. Arnold, London, pp. 145–186.

See also Association Analysis, Family-Based Association Analysis, Haplotype, Linkage Disequilibrium, Phase.

ALLOPATRIC EVOLUTION (ALLOPATRIC SPECIATION)
A. Rus Hoelzel

Allopatric evolution occurs when populations of the same species diverge genetically in geographically isolated locations (having nonoverlapping geographic ranges).

Such differentiation, together with the development of pre- or postzygotic reproductive isolation, can result in allopatric speciation. In general terms this is the most widely accepted mechanism for speciation in animal species, although there is some controversy about the relative importance of natural selection and genetic drift as factors generating divergence in allopatry. Obvious examples of allopatric evolution include differentiation between populations on different continents, different oceans, or separated by physical barriers to migration (e.g., a mountain range or river).

The generation of natural biogeographical barriers to gene flow over time is referred to as a vicariance event, such as the incursion of a glacier, and this can lead to differentiation in allopatry, sometimes followed by reconvergence (e.g., after a glacier recedes). Populations may also be established in allopatry by long-range dispersal in some cases.

Further Reading
Hartl DL (2000). *A Primer of Population Genetics*. Sinauer Associates, Sunderland, MA.

See also Evolution, Sympatric Evolution.

ALLOPATRIC SPECIATION. *SEE* ALLOPATRIC EVOLUTION.

ALPHA CARBON. *SEE* Cα (C-ALPHA).

ALPHA HELIX
Roman A. Laskowski

The most common regular secondary structure in globular proteins. It is characterized by the helical path traced by the protein's backbone, a complete turn of the helix taking 3.6 residues and involving a translation of 5.41 Å. The rodlike structure of the helix is maintained by hydrogen bonds from the backbone carbonyl oxygen of each residue, i, in the helix to the backbone =NH of residue $i+4$.

In proteins, most alpha helices are right handed; that is, when viewed down the axis of the helix, the backbone traces a clockwise path. The right-handed form is more stable than the left-handed form as the latter involves steric clashes between side chains and backbone.

Certain types of amino acid (residues), such as alanine, arginine, and leucine, favor formation of alpha helices, whereas others tend to occur less often in this type of secondary structure, the most significant being proline, which breaks the hydrogen-bonding pattern necessary for alpha-helix formation and stability.

Other types of helix found in globular proteins, albeit very rarely, are the 3_{10} and *pi* helix, which differ in that their hydrogen bonds are to residues $i+3$ and $i+5$, respectively.

Further Reading
Branden C, Tooze J (1991). *Introduction to Protein Structure*. Garland Science, New York.
Lesk AM (2001). *Introduction to Protein Architecture*. Oxford University Press, Oxford, UK.

See also Backbone Models, Globular, Amino acid Secondary Structure of Protein, Side Chain.

ALU REPEAT
Katheleen Gardiner, Dov S. Greenbaum

The most common family of short interspersed repeated sequences (SINEs) in the primate genome.

Alu repeats are ~300 bp in length and are present in over 500,000 copies in the human genome, comprising about 5% of it. Identity among copies ranges from 70% to 100%. Alu repeats are believed to be degenerate copies of the 7SL RNA, which encodes the RNA component of the signal recognition particle, and originally to have been derived by reverse transcription followed by integration into the genomic DNA. While the gene is transcribed with an internal promoter, it requires the enzymatic activity of reverse transcriptase, encoded elsewhere in the genome, to move. It is thought that these repeats multiplied to their present high-copy-number fairly recently in evolution. They can be classified as either transposable elements or mobile pseudogenes.

Alu repeats are typically found within introns and intergenic regions, but examples are known where Alu repeats form parts of coding or regulatory sequences. Their average density is around one per 6 kb, with the highest frequency in the GC-rich R bands.

Transposition of Alu repeats can cause mutations within genes, while their presence can cause misalignments during meiosis. Because of their frequency within genomes, database-searching algorithms offer to mask them and not consider them within a sequence similarity query, as their inclusion could result in false positives.

Further Reading
Batzer MA, Deininger PL (2002). Alu repeats and human genomic diversity. *Nat. Rev. Genet.* 3: 370–379.
Cooper DN (1999). *Human Gene Evolution.* Bios Scientific Publishers, Oxford, UK, pp. 265–285.
Ridley M (1996). *Evolution.* Blackwell Science, Cambridge, MA, pp. 265–276.
Rowold DJ, Herrera RJ (2000). Alu elements and the human genome. *Genetics* 108: 57–72.
Schmid CW, Jelinek WR (1982) The Alu family of dispersed repetitive sequences. *Science* 216: 1065–1070.

See also Chromosome Band, SINE.

AMBER
Roland L. Dunbrack

A molecular dynamics simulation and modeling computer program developed by the late Peter Kollman and colleagues at the University of California at San Francisco. AMBER (Assisted Model Building Using Energy Refinement) uses an empirical molecular mechanics potential energy function for studying the structures and dynamics of proteins, DNA, and other molecules.

Related Website

AMBER	http://www.amber.ucsf.edu/amber/amber.html

Further Reading
Cornell WD, et al. (1995). A second generation force field for the simulation of proteins, nucleic acids, and organic molecules. *J. Am. Chem. Soc.* 117: 5179–5197.

AMIDE BOND (PEPTIDE BOND)
Roman A. Laskowski

The covalent bond joining two amino acids, between the carboxylic (—COOH) group of one and the amino group (—NH$_2$) of the other, to form a peptide.

The bond has a partial double-bond character, and so the atoms shown below tend to lie in a plane and act as a rigid unit (see Figure 2 in Amino Acid).

Because of this, the dihedral angle ω(omega), defined about the bond, tends to be close to 180°.

Further Reading
Branden C, Tooze J (1991). *Introduction to Protein Structure.* Garland Science, New York.
Lesk AM (2001). *Introduction to Protein Architecture.* Oxford University Press, Oxford.

See also Dihedral Angle, Amino Acid, Peptide.

AMINO ACID (RESIDUE)
Roman A. Laskowski and Jeremy Baum

An organic compound containing one or more amino groups ($-NH_2$) and one or more carboxyl groups ($-COOH$). The 20 alpha-amino acids that are the component molecules of proteins each have a central carbon atom, called the C-alpha (C alpha, $C\alpha$), to which are bonded an amino group (NH_2), a carboxyl group (COOH), a hydrogen atom, and a side chain group R. The different amino acids are distinguished by the side chain substituent R (see Figure 1 on page 22) and the stereochemistry of the alpha carbon.

The sequence of amino acids that make up each protein is encoded in the DNA using the genetic code wherein each amino acid is represented by a codon of 3 bases in the DNA sequence.

Except for glycine, the central $C\alpha$ is asymmetric and can adopt one of two isomers, identified as L and D. In proteins, amino acids predominantly adopt the L configuration.

When amino acids polymerize, they do so in linear chains with neighboring residues connected together by (approximately) planar peptide bonds. (See Figure 2 on page 22.) The basic structure of an amino acid in a polymer is shown in the figure. They are then often referred to as amino acid residues. The $-NH-CH-CO-$ component in common to all residues is called the backbone.

Related Websites

Amino acid nomenclature	http://www.chem.qmw.ac.uk/iupac/AminoAcid
Amino acid information	http://prowl.rockefeller.edu/aainfo/contents.htm
Amino acid information	http://www.imb-jena.de/IMAGE_AA.html
Amino acid viewer	http://info.bio.cmu.edu/Courses/BiochemMols/AAViewer/AAVFrameset.htm

See also $C\alpha$, Codons, Genetic Code, Sequence of Proteins, Side Chain.

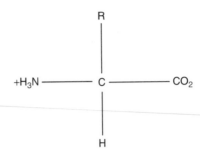

Figure 1

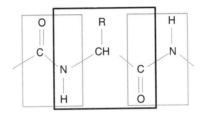

Figure 2

AMINO ACID ABBREVIATIONS. *SEE* IUPAC-IUB CODES.

AMINO ACID COMPOSITION
Jeremy Baum

The relative abundance of the 20 amino acids (usually) in a protein. This measure can be used to make predictions of the protein class, as, e.g., integral membrane proteins tend to have more hydrophobic residues than do cytoplasmic proteins.

See also Amino Acid.

AMINO ACID EXCHANGE MATRIX (LOG-ODDS SCORE, PAM MATRIX)
Jaap Heringa

An amino acid exchange matrix is a 20×20 matrix which contains probabilities for each possible mutation between 20 amino acids.

The matrix is symmetric, so that a mutation from amino acid X into Y is assigned an identical probability as the mutation Y into X. The matrix diagonal contains the odds for self-conservation. Amino acid exchange matrices constitute a model for protein evolution and are a prerequisite for sequence alignment methods. For example, dynamic programming techniques rely on an amino acid exchange weight matrix and gap penalty values. The optimal, or highest scoring, alignment of two sequences is evaluated by the pairwise amino acid substitution scores summed over all matched positions less the penalties arising from each gap in the alignment. For alignment of similar sequences (>35% in sequence identity), the scoring system utilized is not critical (Feng et al., 1985). In more divergent comparisons with residue identity fractions in the so-called twilight zone (15% to 25%) (Doolittle, 1987), different scoring regimens can lead to dramatically deviating alignments. Many different substitution matrices have been devised over more than three decades, each trying to approximate divergent evolution in order to optimize the signal-to-noise ratio in the detection of homologies among sequences. A combination of physicochemical characteristics of amino acids can be used to derive a substitution matrix which then basically contains pairwise amino acid similarity values. Other data from which residue exchange matrices have been computed include sequence alignments, structure-based alignments, and common sequence motifs.

Fitch (1966) constructed the first nonidentity residue exchange weights matrix. He used the minimum number of nucleotide base changes for each amino acid substitution. Values of 0, 1, 2, and 3, required to substitute one residue with another, were converted to similarity values 4, 2, 1, and 0, respectively. The currently most widely used amino acid exchange matrices are the PAM 250 matrix (Dayhoff, 1978) and the BLOSUM 62 matrix (Henikoff and Henikoff, 1992), members of the PAM and BLOSUM series, respectively.

Dayhoff et al. (1978) derived the classical PAM 250 matrix from an evolutionary model for residue substitutions. Sequences from 72 protein families for which similarities were high enough (a fraction of 85% or more identical residues) were compared by eye to yield accurate multiple alignments. The amino acid substitutions observed in these matches were then tabulated and converted to mutational probabilities according to 1% accepted point mutations (one amino acid changed out of 100). This so-called PAM 1 matrix is converted into a PAM 250 matrix by 250 self-multiplications, but this number can be varied to yield matrices associated with greater or smaller genetic distances. The most widely used substitution table is the PAM 250 log-odds matrix, where each PAM-250 matrix element C is converted by $10 \times \log(C)$. Jones et al. (1992) repeated the work of Dayhoff et al. (1978, 1983) and constructed a PAM 250 matrix over a database about 18 times larger (23,000 sequences compared to about 1300 in Dayhoff et al., 1978). Gonnet et al. (1992) performed an exhaustive matching of a database of 1.7×10^6 residues where sequences were pregrouped using a special tree formalism. They also derived a PAM 250–based substitution matrix and suggested a gap penalty regimen exponentially related to the gap length.

Henikoff and Henikoff (1992) used the PROSITE sequence motif database (Hofmann et al., 1999) to construct the BLOCKS database (Henikoff et al 2000) with about 2000 multiple subsequence alignments associated with the conserved PROSITE motifs. From the latter database, they constructed the so-called BLOSUM series of exchange matrices. For example, the alignment blocks showing pairwise sequence identities of ≤62% were used to construct the scoring matrix BLOSUM 62. Very similar sequence groups were down-weighted by taking the average value of their contributions to the matrix.

The currently most widely used matrix is BLOSUM 62, which has relatively high diagonal values as compared to the PAM 250 matrix and so is a more conservative matrix. This has an appreciable effect on the pairwise identity values, the most popular way of identifying sequence relationships, as the BLOSUM 62–based identity scores are dramatically higher as compared to the PAM 250 matrix, particularly between divergent sequences. However, "softer" matrices than the BLOSUM 62 matrix, such as the higher PAM series for the Dayhoff or the Gonnet matrices or the BLOSUM 50 matrix, are particularly useful in global alignment, as they are more suitable for aligning divergent sequences. They can also be useful in local alignment searches, aiding the recognition of distant homologous relationships.

It is evident that constructing a single amino acid exchange matrix to represent all localized exchange patterns in protein structures is a gross oversimplification and might lead to error in specific protein families. Attempts to group local characteristics of protein structure and to construct specific exchange matrices for each of those groups have included secondary structure (Eisenberg et al. 1991; Mehta et al., 1995) and solvent accessibility (Thompson and Goldstein, 1996). Recent attempts have included adapting the amino acid exchange matrix to a local family in order to optimize the representation of the family and thus enhance searching for distant members (Koshi and Goldstein, 1998).

Related Websites

PAM	http://www.cmbi.kun.nl/bioinf/tools/pam.shtml
BLOSUM	http://www.md.huji.ac.il/courses/bioinfo01/exercises/ex6/blosum.html
Dayhoff	http://www.cryst.bbk.ac.uk/PPS2/course/section2/matrixtab.html

Further Reading

Dayhoff MO, et al. (1978). A model of evolutionary change in proteins. In *Atlas of Protein Sequence and Structure*, Vol. 5, Suppl. 3, MO Dayhoff, Ed. National Biomedical Research Foundation, Washington, DC, pp. 345–352.

Dayhoff MO, et al. (1983). Establishing homologies in protein sequences. *Methods Enzymol.* 91: 524–545.

Doolittle RF (1987). *Of URFS and ORFS. A Primer on How to Analyze Derived Amino Acid Sequences.* University Science Books, Mill Valley, CA.

Eisenberg D, et al. (1991). Secondary structure-based profiles: Use of structure-conserving scoring tables in searching protein sequence databases for structural similarities. *Proteins* 10: 229–239.

Feng DF, et al. (1985). Aligning amino acid sequences: Comparison of commonly used methods. *J. Mol. Evol.* 25: 351–360.

Fitch W (1966). An improved method of testing for evolutionary homology. *J. Mol. Biol.* 16: 9–16.

Gonnet GH, et al. (1992). Exhaustive matching of the entire protein sequence database. *Science* 256: 1443–1445.

Henikoff S, Henikoff JG (1992). Amino acid substitution matrices from protein blocks. *Proc. Natl. Acad. Sci. USA* 89: 10,915–10,919.

Henikoff S, Henikoff JG (1993). Performance evaluation of amino acid substitution matrices. *Proteins Struct. Func. Genet.* 17: 49–61.

Hofmann K, et al. (1999). The PROSITE database, its status in 1999. *Nucleic Acids Res.* 27: 215–219.

Jones DT, et al. (1992). The rapid generation of mutation matrices from protein sequences. *Comput. Appl. Biosci.* 8: 275–282.

Koshi JM, Goldstein RA (1998). Models of natural mutations including site heterogeneity. *Proteins* 32: 289–295.

Mehta PK, et al. (1995). A simple and fast approach to prediction of protein secondary structure from multiply aligned sequences with accuracy above 70%. *Prot. Sci.* 4: 2517–2525.

Thompson MJ, Goldstein RA (1996). Constructing amino acid residue substitution classes maximally indicative of local protein structure. *Proteins* 25: 28–37.

See also Dayhoff Amino Acid Substitution Matrix.

AMINO ACID INDEX
Jeremy Baum

The properties of amino acids are often used in predicting the properties of the proteins in which they occur. These properties (e.g., size, polarity, weight, hydrophobicity) must first be quantified. Many workers have independently determined numerical scales (indices) for amino acids based on chosen properties. Further "indices" exist that quantify the substitution of one amino acid for another during protein evolution. A database of these indices exists.

Related Websites

| aaindex | http://www.genome.ad.jp/dbget/aaindex.html |

AMINO ACID SUBSTITUTION MATRIX. *SEE* AMINO ACID EXCHANGE MATRIX.

AMINO-TERMINUS. *SEE* N-TERMINUS.

AMPHIPATHIC (AMPHIPHILIC)
Roman A. Laskowski

A molecule having both a hydrophilic ("water-loving" or polar) and a hydrophobic ("water-hating" or nonpolar) end. Examples are the phospholipids that form cell membranes. These have polar head groups, which form the outer surface of the membrane, and two hydrophobic hydrocarbon tails, which extend into the interior.

In proteins amphipathicity maybe be observed in surface α-helices and β-strands. Residues (amino acids) on the surface of proteins tend to be polar as they are in contact with the surrounding solvent, whereas those on the inside of the protein tend to be hydrophobic.

Thus an amphipathic helix is one which lies on or close to the protein's surface and hence has one side consisting largely of polar residues, facing out into the solvent, and the opposite side consisting largely of hydrophobic residues, facing the protein's hydrophobic core. Because of the helix's periodicity of 3.6 residues per turn, this gives rise to a characteristic pattern of polar and hydrophobic residues in the protein's sequence. For example, the pattern of hydrophobic residue might be of the form: i, $i+$, $i+4$, $i+7$, with polar residues in between.

Similarly, an amphipathic strand has one side hydrophobic and the other side polar. From the geometry of strand residues, this gives rise to a simple pattern of alternating hydrophobic and polar residues.

Further Reading
Branden C, Tooze J (1991). *Introduction to Protein Structure*. Garland Science, New York.
Lesk AM (2001). *Introduction to Protein Architecture*. Oxford University Press, Oxford.

See also Amino Acid, Hydrophobicity, Polar, Sequence of Proteins.

ANALOG
Dov S. Greenbaum

Genes or proteins that are similar in some way but show no signs of any common ancestry.

Examples of analogs are structural analogs—those proteins that share the same fold—and functional analogs—proteins that share the same function.

Analogs are interesting in that they are the exception to the rule in annotation transfer, such that a protein which shares no evolutionary history and does not have any similar structure can still have the same function of its analog. Structural similarities are thought to be the result of convergence to a favorable fold, although an extremely distant divergence cannot be ruled out.

Determining the difference in sequences between analogs and distant homologs is important; evolutionary theory can only be used to analyze proteins that are similar through descent from a common ancestor and not due to random mutations that result in either similar fold or function. Conversely, a protein that is misidentified as an analog rather than a distant homolog may not be analyzed in its proper evolutionary context. Either error will be deleterious to a comparative genomic study.

Further Reading
Fitch W (1970). Distinguishing homologous from analogous proteins. *Syst. Zool.* 19: 99–113.

ANCESTRAL GENOME (CENANCESTOR, LAST UNIVERSAL COMMON ANCESTOR)
Dov S. Greenbaum

The common ancestor of all extant living organisms.

It is assumed that all life on Earth arose from a common ancestral organism, represented by the root of the universal phylogenetic tree. Originally it was thought that this simple organism, with minimal metabolism, was similar to a prokaryote. Given that all kingdoms of life are thought to arise from this single organism, the image of an early prokaryote is no longer in vogue. We have yet to arrive at a consistent view of this earliest organism.

One possibility is suggested by the genetic annealing model proposed by Woese, the father of the tripartite kingdom division. This proposes that the universal ancestor was not a single organism but rather a diverse, loosely knit community of primitive cells that shared innovative proteins. Eventually this community split into three distinct lineages to become the three separate kingdoms. The genetic code would have been translated loosely at first, but as the cells developed, a more complex and accurate code would have also developed.

Other possibilities include a totipotent ancestor, more complex than many present-day organisms, that contained most of the systems which it then transferred to evolutionarily later life forms. Another alternative is that there are lost branches in the phylogenetic tree that may have also contributed to the present kingdoms prior to becoming extinct.

It is thought that the universal ancestor, or cenancestor, living more than 3.5 billion years ago, had DNA, transcription, translation, and operons but no nucleus.

Related Website

| Woese | http://www.life.uiuc.edu/micro/woese.html |

Further Reading

Snel B, et al. (2002). Genomes in flux: The evolution of archaeal and proteobacterial gene content. *Genome Res.* 12: 17–25.

Woese CR (1998). The Universal Ancestor. *Proc Natl Acad Sci USA* 95: 6854–6859.

ANCESTRAL STATE RECONSTRUCTION
Sudhir Kumar and Alan Filipski

The process of inferring the nucleotide or amino acid state of a specified site of a (unobservable) sequence for an internal node in a phylogenetic tree.

Maximum parsimony and Bayesian inference methods (using maximum-likelihood or distance-based approaches) are available for ancestral state reconstruction. Parsimony methods work reasonably well when the sequences are closely related, so that the expected number of substitutions per site is small, and when branches are similar in length. Bayesian methods are used to compute the posterior probability of any given set of ancestral states, and the ancestral state with the highest posterior probability is chosen to be the best estimate.

Further Reading

Maddison WP, Maddison DR (1992). *MacClade: Analysis of Phylogeny and Character Evolution*. Sinauer Associates, Sunderland, MA.

Nei M, Kumar S (2000). *Molecular Evolution and Phylogenetics*. Oxford University Press, Oxford, UK.

Yang Z (1997). PAML: A program package for phylogenetic analysis by maximum likelihood. *Comput. Appl. Biosci.* 13: 555–556.

Yang Z, et al (1995). A new method of inference of ancestral nucleotide and amino acid sequences. *Genetics* 141: 1641–1650.

See also Maximum-Parsimony Principle, Bayesian Phylogenetic Analysis.

ANCHOR POINTS
Roland L. Dunbrack

In protein loop modeling, usually one or two residues from the parent structure before and after a loop that is being constructed to fit onto the template structure. These residues are usually kept fixed, and the constructed loop must be connected to them to produce a viable model for the loop. Therefore, if a five-residue loop is being modeled with a two-residue anchor point at either end, a fragment of nine residues long will have to be found from which the two residues at either end "fit" well onto the parent structure.

See also Loop Prediction/Modeling.

ANNOTATION TRANSFER
Dov S. Greenbaum

The transfer of information on the known function or other property of a gene or protein to another gene or protein based on predefined criteria.

Annotation transfer is often termed *guilt by association*. That is, given some knowledge of some other gene or protein that has some similarity to an unknown gene, it is possible to transfer the known annotation to the novel gene. For example, in the case of homologous proteins, it is often assumed that proteins with a high degree of sequence similarity share the same function. Another example could be that proteins that interact with each other are assumed to have similar functions; thus one can transfer annotation information from one gene to its interaction partner.

One has to be careful in utilizing annotation transfer, as it may not always be the case that the two proteins have similar function, even though they may share an interaction, localization, expression, or sequence similarity. One pitfall to be avoided is the incorporation of potentially misleading information resulting from an annotation transfer into a database, as this will only propagate the error, when additional annotation transfers are made from the formerly novel protein.

Further Reading
Hegyi H, Gerstein M. (2001). Annotation transfer for genomics: Measuring functional divergence in multi-domain proteins. *Genome Res.* 11: 1632–1640.

ANOMALOUS DISPERSION. *SEE* MULTIPLE ANOMALOUS DISPERSION PHASING.

APBIONET. *SEE* ASIA PACIFIC BIOINFORMATICS NETWORK.

APOMORPHY
A. Rus Hoelzel

A derived or newly evolved state of an evolutionary character.

As a hypothesis about the pattern of evolution among operational taxonomic units, a phylogenetic reconstruction can inform us about the relationship between character states. We can identify ancestral and descendant states and therefore identify primitive and derived states. An apomorphy (meaning "from shape" in Greek) is a derived state (shown as filled circles in the figure).

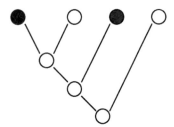

Figure   Illustration of apomorphy. The two black circles represent examples of apomorphy.

Changes have occurred compared to the common ancestor, and these are now represented by the derived, or apomorphic, states. In the figure, there are two apomorphies, each derived independently from the ancestral state, which is an example of homoplasy.

Further Reading
Maddison DR, Maddison WP (2001). *MacClade 4: Analysis of Phylogeny and Character Evolution*. Sinauer Associates, Sunderland, MA.

See also Autapomorphy, Plesiomorphy, Synapomorphy.

ARGININE
Jeremy Baum

Arginine is a positively charged amino acid with side chain $—(CH_2)_3NHC(NH_2)_2$ found in proteins. In sequences, written as Arg or R.

Further Reading
Branden C, Tooze J (1991). Introduction to Protein Structure. Garland Science, New York.

See also Amino Acid.

AROMATIC
Roman A. Laskowski

An aromatic molecule, or compound, is one that contains a planar or near-planar closed ring with particularly strong stability. The enhanced stability results from the overlap of *p*-orbital electrons to constitute a delocalized π molecular orbital system. Huckel's rule for identifying aromatic cyclic rings states that the total number of electrons in the π system must equal $4n+2$, where n is any integer.

The archetypal aromatic molecule is benzene, which satisfies the Huckel rule with $n=1$.

In proteins, the amino acid residues phenylalanine, tyrosine, tryptophan, and histidine have aromatic side chains. Interactions between aromatic side chains in the cores of protein structures often play an important part in conferring stability on the structure as a whole.

Related Website

Atlas of Protein Side-Chain Interactions	http://www.biochem.ucl.ac.uk/bsm/sidechains/index.html

Further Reading
Burley SK, Petsko GA (1985). Aromatic-aromatic interaction: A mechanism of protein structure stabilization. *Science* 229: 23–28.

Singh J, Thornton, JM (1985). The interaction between phenylalanine rings in proteins. *FEBS Lett.* 191: 1–6.

Singh J, Thornton JM (1992). *Atlas of Protein Side-Chain Interactions*, Vols. I and II. IRL Press, Oxford.

See also Amino Acid, Side Chain.

ARTIFICIAL NEURAL NETWORKS. *SEE* NEURAL NETWORK.

ASIA PACIFIC BIOINFORMATICS NETWORK (APBIONET)
John M. Hancock and Martin J. Bishop

APBioNet is an organization dedicated to the promotion of bioinformatics in the Asia-Pacific region.

The organization was founded in 1998. Its four main areas of activity are the development of bioinformatics network infrastructure; the exchange of data and information; the development of training programs, workshops, and symposia; and the encouragement of collaboration. It has organizational members in Australia, Canada, China, Hong Kong, India, Japan, South Korea, Malaysia, Russia, Singapore, and the United States.

Related Website

APBioNet	http://www.apbionet.org/

Further Reading
Sugawara H, Miyazaki S (1998). Towards the Asia-Pacific Bioinformatics Network. *Pac. Symp. Biocomput.* 1998: 759–764.

See also EMBnet.

ASPARAGINE
Jeremy Baum

Asparagine is a polar amino acid with side chain —CH_2CONH_2 found in proteins. In sequences, written as Asn or N.

Further Reading
Branden C, Tooze J (1991). *Introduction to Protein Structure*. Garland Science, New York.

See also Amino Acid, Polar.

ASPARTIC ACID
Jeremy Baum

Aspartic acid is a polar (often negatively charged) amino acid with side-chain —CH_2CO_2H found in proteins. In sequences, written as Asp or D.

Further Reading
Branden C, Tooze J (1991). *Introduction to Protein Structure*. Garland Science, New York.

See also Amino Acid, Polar.

ASSOCIATION ANALYSIS (LINKAGE DISEQUILIBRIUM ANALYSIS)
Mark McCarthy and Steven Wiltshire

Genetic analysis which aims to detect associations between alleles at different loci and a central tool in the identification and characterization of disease susceptibility genes.

In this context, the simplest structure is a case–control analysis, where a comparison is made of allele (or haplotype) frequencies between a sample of cases (individuals with the disease or phenotype of interest and therefore enriched for disease susceptibility alleles) and controls (either a population-based sample or individuals selected not to have the disease). The interest here is in identifying associations which reflect linkage disequilibrium between the typed marker and disease (and hence with disease susceptibility variants). Such associations indicate either that the typed marker is the disease susceptibility variant or that it is in linkage disequilibrium (LD) with it (and, in turn, given the limited genomic extent of linkage disequilibrium in most populations, that the disease susceptibility variant must be nearby). Since there is concern that case–control studies may generate false positives due to latent population stratification (i.e., reveal associations which do not reflect linkage disequilibrium), family-based association methods are often used. While association analysis has hitherto largely been restricted to candidate gene studies, there is increasing interest in conducting genome scans for association (analogous to those for linkage). However, the technical and analytical demands of this remain substantial, not least because the limited extent of linkage disequilibrium in human populations means that a very large number of markers (at least 100,000) would need to be typed to achieve genomewide coverage.

Examples: Altshuler and colleagues (2000) combined a variety of association analysis methods to demonstrate that the *Pro12Ala* variant in the *PPARG* gene is associated with type 2 diabetes. Dahlman and colleagues (2002) sought to replicate a previously reported association between a variant in the 3'UTR region of the interleukin 12 p40 gene (*IL12B*) and type 1 diabetes: Despite typing a much larger data set, they were unable to find evidence to support the previous association.

Further Reading

Altshuler D, et al. (2000). The common PPARgamma Pro12Ala polymorphism is associated with decreased risk of type 2 diabetes. *Nat. Genet.* 26: 76–80.

Cardon LR, Bell JI (2001). Association study designs for complex disease. *Nat. Rev. Genet.* 2: 91–99.

Dahlman I, et al. (2002). Parameters for reliable results in genetic association studies in common disease. *Nat. Genet.* 30: 149–150.

Hästbacka J et al. (1993). Linkage disequilibrium mapping in isolated founder populations: diastrophic dysplasia in Finland. *Nat Genet* 2: 204–211.

Hirschhorn JN, et al. (2002). A comprehensive review of genetic association studies. *Genet. Med.* 4: 45–61.

Lander ES, Schork NJ (1994). Genetic dissection of complex traits. *Science* 265: 2037–2048.

Risch N (2001). Implications of multilocus inheritance for gene-disease association studies. *Theor. Popul. Biol.* 60: 215–220.

Risch N, Merikangas K (1996). The future of genetic studies of complex human diseases. *Science* 273: 1516–1517.

See also Allelic Association, Family-Based Association Analysis, Linkage Disequilibrium, Multifactorial Trait.

ASYMMETRIC UNIT
Liz P. Carpenter

The asymmetric unit of a crystal is the smallest building block from which a crystal can be created by applying the crystallographic symmetry operators followed by translation by multiples of the unit cell vectors. The asymmetric unit may contain one or several copies of the molecules under study.

See also Crystal, Macromolecular, Space Group, Unit Cell, X-Ray Crystallography for Structure Determination.

ATOMIC COORDINATE FILE (PDB FILE)
Eric Martz

A data file containing a list of the coordinates in three-dimensional space for a group of atoms. Typically the atoms constitute a molecule or a complex of molecules. An atomic coordinate file is required in order to look at the three-dimensional structure of a molecule using visualization software. Sources of coordinates, in order of decreasing reliability, are empirical (X-ray crystallography or nuclear magnetic resonance), homology modeling, or ab initio theoretical modeling.

There are over a dozen popular formats for atomic coordinate files. One of the most popular for macromolecules is the original one used by the Protein Data Bank, commonly called a *PDB file*. The Protein Data Bank has recently adopted a new standard format, the macromolecular crystallographic information format, or mmCIF, but some popular software remains unable to process this newer format. Therefore the PDB format continues to be supported at the Protein Data Bank as well as by most other sources of atomic coordinate files. When atomic coordinate files are transmitted through the Internet, their formats are identified with MIME types.

Related Website

Protein Data Bank	http://www.rcsb.org/pdb/

See also MIME Types; Visualization, Molecular; Protein Data Bank.

Eric Martz is grateful for help from Eric Francoeur, Peter Murray-Rust, Byron Rubin, and Henry Rzepa.

ATTRIBUTE. *SEE* FEATURE.

AUTAPOMORPHY
A. Rus Hoelzel

A unique derived state of an evolutionary character.

As a hypothesis about the pattern of evolution among operational taxonomic units, a phylogenetic reconstruction can inform us about the relationship between character states. We can identify ancestral and descendant states and therefore identify primitive and derived states. An apomorphy (meaning "from shape" in Greek) is a derived state (shown as the filled circle in the figure).

A unique derived state (as shown in the figure) is known as an autapomorphy (*aut* means "alone").

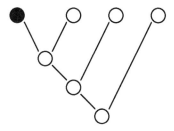

Figure Illustration of autapomorphy. The black circle is in a derived state not shared by any other OTU.

Further Reading

Maddison DR, Maddison WP (2001). *MacClade 4: Analysis of Phylogeny and Character Evolution*. Sinauer Associates, Sunderland, MA.

See also Apomorphy, Plesiomorphy, Synapomorphy.

AUTOSOME
Katheleen Gardiner

A chromosome other than a sex chromosome.

Humans have 22 pairs of autosomes. The gene content of each chromosome in an autosome pair is identical.

Further Reading

Miller OJ, Therman E (2000). *Human Chromosomes*. Springer, Vienna.

Wagner RP, Maguire MP, Stallings RL (1993). *Chromosomes—A Synthesis*. Wiley-Liss, New York.

AXIOM
Robert Stevens

An axiom is a statement asserted into a logical system without proof.

Some knowledge representation languages, such as Description Logics, make use of axioms to make statements adding information to ontologies. For example, I can assert that the class nucleic acid has two children, DNA and RNA. A disjointness axiom can then be made stating that DNA and RNA are disjoint, that is, it is not possible to be both a DNA and an RNA; that is, the classes do not overlap. Some systems include other axiom types that can assert the equivalence of two categories, that a set of subcategories covers the meaning of a supercategory exhaustively, or additional necessary properties of a category or individual.

B

BAC
Backbone
Backbone-Dependent Rotamer Library
Backbone Models
Backpropagation Networks
Bacterial Artificial Chromosome
Ball-and-Stick Models
BAMBE
Base-Call Confidence Values
Base Caller
Base Composition
Base Pair
Bayesian Analysis
Bayesian Network
Bayesian Phylogenetic Analysis
Before State, Before Sphere
Belief Network
Beta Barrel
Beta Breakers
Beta Sheet
Beta Strand
Binary Relation

Binding Site
Binding Site Symmetry
BioChip
Bioinformatics
Bioinformatics.org
Birth-and-Death Evolution
Bit
BLAST
BLASTX
BLAT
Block
Blocks
BLOSUM Matrix
Boltzmann Factor
BP
Bootstrap Analysis
Born Solvation Energy, Generalized
Bottleneck
Box
Bragg's Law
Branch-Length Estimation

Dictionary of Bioinformatics and Computational Biology. Edited by Hancock and Zvelebil
ISBN 0-471-43622-4   © 2004 John Wiley & Sons, Inc.

BAC (BACTERIAL ARTIFICIAL CHROMOSOME)
Rodger Staden

Bacterial artificial chromosome used to clone segments of DNA up to 200 kb.

These are possibly the most common unit of DNA sequence currently making up draft genome sequences. The term is therefore widely applied to individual components of a draft sequence.

BAC. *SEE* VECTOR.

BACKBONE
Roman A. Laskowski

When amino acids link together to form a peptide or protein chain, the atoms that constitute the continuous link running the length of the chain are referred to as the backbone atoms. Each amino acid contributes its—N—Cα—C—atoms to this backbone. Also included as part of the backbone is the carbonyl oxygen attached to the backbone carbon, C—, and any hydrogens attached to these backbone atoms. All other atoms are termed side-chain atoms. For most of the amino acids, the side-chain atoms spring off from the main-chain Cα, the exception being glycine, which has no side chain (other than a single hydrogen atom), and proline, whose side chain links back onto its main-chain nitrogen.

Further Reading

Branden C, Tooze J (1991). *Introduction to Protein Structure*. Garland Science, New York.

Lesk AM (2001). *Introduction to Protein Architecture*. Oxford University Press, Oxford, UK.

See also Amino Acid, Glycine, Main Chain, Side Chain.

BACKBONE-DEPENDENT ROTAMER LIBRARY. *SEE* ROTAMER LIBRARY.

BACKBONE MODELS
Eric Martz

Simplified depictions of proteins or nucleic acids in three dimensions that enable the polymer chain structures to be seen. (See illustration at Models, Molecular.) Lines are drawn between the positions of alpha-carbon atoms (for proteins) or phosphorus atoms (for nucleic acids), thus allowing the backbone chains to be visualized. These lines do not lie in the positions of any of the covalent bonds. Another type of backbone model is a line following the covalent bonds of the main-chain atoms, rendering the polypeptide as polyglycine.

Dictionary of Bioinformatics and Computational Biology. Edited by Hancock and Zvelebil
ISBN 0-471-43622-4 © 2004 John Wiley & Sons, Inc.

For nucleic acids, an alternative backbone consists of lines connecting the centers of the pentose rings. Most macromolecular visualization software packages can display backbone models. Sometimes there is an option to smooth the backbone trace; smoothed backbones may be rendered as wires, ribbons, or schematic models. In the early 1970s, before computers capable of displaying backbone models were readily available, physical backbone models were made of metal wire. One apparatus for constructing wire backbone models was Byron's Bender (see Related Websites, Further Reading). It was popular in the 1970s and early 1980s and was then superceded for most purposes by computer visualization methods.

Further Reading
Martz E, Francoeur E. *History of Visualization of Biological Macromolecules,* http://www.umass.edu/microbio/rasmol/history.htm.

Rubin B, Richardson JS (1972). The simple construction of protein alpha-carbon models. *Biopolymers* 11: 2381–2385.

Rubin, B (1985). Macromolecule backbone models. *Methods Enzymol.* 115: 391–397.

See also Models, Molecular; Schematic (Ribbon, Cartoon) Model; Visualization, Molecular.
   Eric Martz is grateful for help from Eric Francoeur, Peter Murray-Rust, Byron Rubin, and Henry Rzepa.

BACKPROPAGATION NETWORKS. *SEE* NEURAL NETWORK.

BACTERIAL ARTIFICIAL CHROMOSOME. *SEE* BAC.

BALL-AND-STICK MODELS
Eric Martz

Three-dimensional molecular models in which atoms are represented by balls and covalent bonds by sticks connecting the balls. Also called (in the United Kingdom) *ball-and-spoke models.* (See illustration at Models, Molecular.) Ball-and-stick models can be traced back to John Dalton in the early nineteenth century. Because they are so detailed, ball-and-stick models are generally more useful for small molecules than for macromolecules. For macromolecules, hydrogen atoms are often omitted to simplify the model. Ball-and-stick models originated in physical models and later were implemented in computer visualization programs. One of the early and widely used packages is ORTEP (see Related Websites).

In order to see details beneath the outer surface of the model, the balls are typically considerably smaller than the van der Waals radii of the atoms. In RasMol and its derivatives (see Visualization, Molecular), the balls have a fixed radius of 0.45 Å, and the cylindrical sticks have a radius of 0.15 Å. Some software offers scaled ball-and-stick models, in which the balls, while still much smaller than the van der Waals radii, vary in size in proportion to the van der Waals radii. When Kendrew and co-workers solved the structure of myoglobin at atomic resolution in the early 1960s, they first built a wire-frame model. Shortly thereafter, they built ball-and-spoke models, 29 of which were sold to interested research groups in the late 1960s (see Related Websites). Due to the complexity of proteins, it is difficult to discern major structural features from a ball-and-stick model. Therefore, physical models of entire proteins or protein domains were later more commonly backbone or schematic.

Related Website

ORTEP—Oak Ridge Thermal Ellipsoid Plot Program for Crystal Structure Illustrations	http://www.ornl.gov/ortep/ortep.html

Further Reading

Brode W, Boord CE (1932). Molecular models in the elementary organic laboratory I. *J. Chem. Ed.* 9: 1774–1782.

Corey RB, Pauling L (1953). Molecular models of amino acids, peptides and proteins. *Rev. Sci. Instrum.* 24: 621–627.

Kendrew JC, Dickerson RE, Strandberg RG, Hart RG, Davies DR, Phillips DC, Shore VC (1960) Structure of myoglobin. A three-dimensional Fourier synthesis at 2 Å resolution. *Nature* 185: 422–427.

Martz E, Francoeur E. *History of Visualization of Biological Macromolecules.* http://www.umass.edu/microbio/rasmol/history.htm.

See also Models, Molecular; Visualization, Molecular; Wire-Frame Models.

Eric Martz is grateful for help from Eric Francoeur, Peter Murray-Rust, Byron Rubin, and Henry Rzepa.

BAMBE (BAYESIAN ANALYSIS IN MOLECULAR BIOLOGY AND EVOLUTION)
Michael P. Cummings

Program for phylogenetic analysis of nucleotide sequence data using a Bayesian approach.

A Metropolis–Hastings algorithm for Markov chain Monte Carlo (MCMC) is used to sample model space through a process of parameter modification proposal and acceptance/rejection steps (also called cycles or generations). After the process becomes stationary, the frequency with which parameter values are visited in the process represents an estimate of their underlying posterior probability. Model parameters include substitution rates and specific taxa partitions.

Several commonly used likelihood models are available, as are choices for starting tree of the Markov Chain, including user defined, UPGMA (unweighted pair-group method with arithmetic mean), neighbor joining neighbor-joining and random. An accessory program, Summarize, is used to process the output file of the tree topology information and reports on topological features.

The programs are written in C++ and are available as source code and binaries for some systems.

Related Website

BAMBE	http://www.mathcs.duq.edu/larget/bambe.html

Further Reading

Larget B, Simon D (1999). Markov chain Monte Carlo algorithms for the Bayesian analysis of phylogenetic trees. *Mol. Biol. Evol.* 16: 750–759.

Mau B, et al. (1999). Bayesian phylogenetic inference via Markov Chain Monte Carlo methods. *Biometrics* 55: 1–12.

Newton MA, et al. (1999). Markov chain Monte Carlo for the Bayesian analysis of evolutionary trees from aligned molecular sequences. *Stat. Mol. Biol. Genet. IMS Lect. Notes-Monogr. Ser.* 33: 143–162.

See also MrBayes.

BASE-CALL CONFIDENCE VALUES
Rodger Staden

Numerical values assigned to each base in a sequence reading to predict its reliability.

The values are assigned by analysis of the fluorescence signals from which the base calls were derived. The most widely used scale of values is named after the program PHRED, which defines the confidence $C = -10 \log(P_{error})$ where P_{error} is the probability that the base call is erroneous.

BASE CALLER
Rodger Staden

A computer program which interprets the DNA sequence traces from a sequencing instrument to determine its sequence.

See also DNA Sequencing Trace, PHRED.

BASE COMPOSITION (GC COMPOSITION, GC RICHNESS)
Katheleen Gardiner

The percent of G + C or A + T bases in a DNA sequence or genome.

For example, the average base composition of the human genome is 38% GC. This is not a uniform value, however, and is related to chromosome bands. G bands are consistently AT rich, ~34% to 40% GC. R bands are variable, with some segments, especially many telomeric regions, >60% GC.

Further Reading

Nekrutenko A, Li WH (2000). Assessment of compositional heterogeneity within and between eukaryotic genomes. *Genome Res.* 10: 1986–1995.

See also Chromosome Band, Isochore.

BASE PAIR (BP)

Basic unit of DNA structure.

DNA molecules consist of two polymers of DNA bases (adenine, guanine, cytosine and thymine). The two strands are held together by hydrogen bonds between pairs of bases, forming structures known as base pairs. In the standard Watson-Crick structure of DNA only two types of base pair are permitted (adenine-thymine and guanine-cytosine pairs), constraining the relationship between the sequences of the two strands and allowing accurate replication of DNA.

The term is also used as a measure of the length of DNA sequences. Lengths in the thousands of bases are measured in kilobases (kb) and lengths of millions of bases in megabases (Mb). Lengths of single stranded DNA or RNA molecules may be expressed in nucleotides (nt).

Further Reading

Lewin B (2000). *Genes VII*. Oxford University Press, Oxford, UK.

Watson JD, Crick FHC (1953). Molecular structure of nucleic acids; a structure for deoxyribose nucleic acid. *Nature* 171: 737–738.

See also DNA Sequence, DNA Replication.

BAYESIAN ANALYSIS (BAYES' THEOREM)
John M. Hancock

A statistical method of estimating the probability of an observation taking account of some prior knowledge or expectation.

For example, we might ask what is the probability of a particular model M given a particular set of data D. By Bayes' theorem (named after Thomas Bayes, 1702–1761), the probability of M given D is

$$P(M|D) = \frac{P(D|M)P(M)}{P(D)} = P(M)\frac{P(D|M)}{P(D)}$$

Thus, if we can estimate or guess the probabilities of M and D arising by chance or some other, known process and the probability of obtaining D if M is true, we can estimate the likelihood (posterior probability) of M being true. The probability $P(M)$ is known as the prior probability or prior; $P(D|M)$ is known as the likelihood.

This basic principle is widely applicable in bioinformatics from phylogeny estimation to microarray data analysis.

Further Reading

Baldi P, Brunak S (2001). *Bioinformatics: The Machine Learning Approach (Adaptive Computation and Machine Learning)*. MIT Press, Cambridge, MA.

Townsend JP, Hartl DL (2002). Bayesian analysis of gene expression levels: Statistical quantification of relative mRNA level across multiple strains or treatments. *Genome Biol.* 3: research0071.1–0071.16.

See also Bayesian Phylogenetic Analysis.

BAYESIAN NETWORK (BELIEF NETWORK, CAUSAL NETWORK, KNOWLEDGE MAP, PROBABILISTIC NETWORK)
John M. Hancock

A modeling tool that combines directed acyclic graphs with Bayesian probability.

The figure shows a simple example of a Bayesian network, which consists of a causal graph combined with an underlying probability distribution. Each node of the network in the figure corresponds to a variable and the edges represent causality

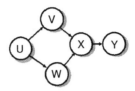

Figure

between these events, which is directional. The other elements of a Bayesian network are the probability distributions [e.g., $P(Y|X)$, $P(X|V,W)$] associated with each node. With this information the network can model probabilities of complex causal relationships.

Bayesian networks are widely used modeling tools. Techniques also exist for inferring or estimating network parameters. Such an approach is applicable, e.g., to modeling gene networks from microarray data.

Further Reading
Friedman N, et al. (2000). Using Bayesian networks to analyze expression data. *J. Comput. Biol.* 7: 601–620.

Neapolitan RE (2003). *Learning Bayesian Network*. Prentice-Hall, Englewood Cliffs, NJ.

See also Bayesian Analysis, Directed Acyclic Graph.

BAYESIAN PHYLOGENETIC ANALYSIS
Sudhir Kumar and Alan Filipski

A probabilistic method based on Bayes' rule. It is often used to infer phylogenetic trees and estimate evolutionary parameters.

Given a substitution model, an assumed prior distribution for the parameters, and a set of sequences, Bayes' rule can be used to compute a posterior distribution for the unknown quantities (e.g., tree topology, branch lengths, ancestral states, divergence times). Estimates for these quantities are obtained by sampling from these posterior distributions. Directly applying Bayes' rule to compute the posterior distributions for parameters of interest is often not practical because of the need to evaluate high-dimensional integrals. This difficulty is often addressed by using Monte Carlo Markov chain (MCMC) methods. MCMC provides a computationally practical method for sampling from the posterior distribution of the parameters (e.g., topology, rates, or branch lengths) being estimated.

Further Reading
Felsenstein J (1993). *PHYLIP: Phylogenetic Inference Package*. University of Washington, Seattle, WA.

Huelsenbeck JP, Ronquist F (2001). MRBAYES: Bayesian inference of phylogenetic trees. *Bioinformatics* 17: 754–755.

Murphy WJ, et al. (2001). Resolution of the early placental mammal radiation using Bayesian phylogenetics. *Science* 294: 2348–2351.

Yang Z (1997). PAML: A program package for phylogenetic analysis by maximum likelihood. *Comput. Appl. Biosci.* 13: 555–556. Yang Z, Rannala B (1997). Bayesian phylogenetic inference using DNA sequences: A Markov Chain Monte Carlo Method. *Mol. Biol. Evol.* 14: 717–724.

See also Markov Chain, Monte Carlo Simulations.

BEFORE STATE, BEFORE SPHERE
Thomas D. Schneider

The high energy state of a molecular machine before it makes a choice is called its before state. This corresponds to the state of a receiver in a communications system before it has selected a symbol from the incoming message. The state can be represented as a sphere in a high dimensional space.

Further Reading
Schneider TD (1991). Theory of molecular machines. I. Channel capacity of molecular machines. *J. Theor. Biol.*, 148: 83–123. http://www.lecb.ncifcrf.gov/~toms/paper/ccmm/.

See also Channel Capacity, Gumball Machine, Shannon sphere.

BELIEF NETWORK. *SEE* BAYESIAN NETWORK.

BETA BARREL
Roman A. Laskowski

A beta sheet that curves around to close up into a single, cylindrical structure. Barrels most commonly consist of purely parallel or antiparallel sheets of six or eight beta strands.

Related Websites

3Dee database	http://www.compbio.dundee.ac.uk/3Dee/
CATH server	http://www.biochem.ucl.ac.uk/bsm/cath_new

Further Reading
Buchanan SK, Smith BS, Venkatramani L, Xia D, Esser L, Palnitkar M, Chakraborty R, van der Helm D, Deisenhofer J (1999). Crystal structure of the outer membrane active transporter FepA from *Escherichia coli. Nature Struct. Biol.* 6: 56–63.

See also Beta Sheet, Beta Strand.

BETA BREAKERS
Patrick Aloy

Beta breakers are amino acid residues that have been found to disrupt and terminate beta strands. They were first split into two categories by Chou and Fasman: *breakers* (Lys, Ser, His, Asn, Pro) and *strong breakers* (Glu), depending on the frequency in which these residues were found in extended structures. More recent analyses performed on larger data sets of protein structures have changed the classification, the breakers now being Gly, Lys, and Ser and the strong breakers Asp, Glu, and Pro. It is worth noting that the structural parameters empirically derived do not always agree with those determined experimentally for individual proteins.

Further Reading
Chou PY, Fasman GD (1974). Prediction of protein conformation. *Biochemistry* 13: 222–245.
Munoz V, Serrano L (1994). Intrinsic secondary structure propensities of the amino acids, using statistical phi-psi matrices: Comparison with experimental scales. *Proteins* 20: 301–311.

BETA SHEET
Roman A. Laskowski

The second most commonly found regular secondary structure in globular proteins (after the alpha helix). The sheet is formed from a number of beta strands lying side by side and held together by hydrogen bonds between the backbone NH and CO groups of adjacent strands. The backbone of each strand is in a fully extended conformation. The sheet can be made from beta strands running parallel to one another (i.e., in the same N-terminal to C-terminal sense), antiparallel, or a mixture of the two, the most common being purely antiparallel sheets.

Further Reading
Branden C, Tooze J (1991). *Introduction to Protein Structure*. Garland Science, New York.
Lesk AM (2001). *Introduction to Protein Architecture*. Oxford University Press, Oxford.

See also Alpha Helix, Backbone, Beta Strand, C-Terminus, Globular, N-Terminus, Secondary Structure of Protein.

BETA STRAND
Roman A. Laskowski

Part of the polypeptide chain of a protein which, together with other beta strands, forms a beta sheet. The backbone in a beta strand is almost fully extended (in contrast to the alpha-helix structure, in which the backbone is tightly coiled).

Further Reading
Branden C, Tooze J (1991). *Introduction to Protein Structure*. Garland Science, New York.
Lesk AM (2001). *Introduction to Protein Architecture*. Oxford University Press, Oxford, UK.

See also Alpha Helix, Backbone, Beta Sheet, Polypeptide.

BINARY RELATION
Robert Stevens

A relation between exactly two objects or concepts.

Most knowledge representation languages are based on binary relationships. Binary relationships are easy to visualize as trees or graphs. [Systems of binary relationships are sometimes said to consist of object–attribute–value (OAV) triples, where the attribute represents the relationship between the object and its value.]

BINDING SITE
Thomas D. Schneider

A binding site is a place on a nucleic acid that a recognizer (protein or macromolecular complex) binds. A classic example is the set of binding sites for the bacteriophage lambda repressor (cI) protein on DNA (Ptashne et al., 1980). These happen to be the same as the binding sites for the lambda cro protein.

Related Websites

Hawaii figure	http://www.lecb.ncifcrf.gov/toms/gallery/hawaii.fig1.gif
Hawaii	http://www.lecb.ncifcrf.gov/toms/paper/hawaii/

Further Reading

Ptashne M, et al. (1980). How the λ repressor and *cro* work. *Cell*, 19: 1–11.

Shaner MC (1993). Sequence logos: A powerful, yet simple, tool. *Proceedings of the Twenty-Sixth Annual Hawaii International Conference on System Sciences*, Vol. 1: *Architecture and Biotechnology Computing*, TN Mudge, V Milutinovic, L Hunter, Eds. IEEE Computer Society Press, Los Alamitos, CA, pp. 813–821. http://www.lecb.ncifcrf.gov/\ ~toms/paper/hawaii/.

See also Binding Site Symmetry, Sequence Logo, Sequence Walker.

BINDING SITE SYMMETRY
Thomas D. Schneider

Binding sites on nucleic acids have three kinds of symmetry:

Asymmetric: All sites on RNA and probably most if not all sites on DNA bound by a single polypeptide will be asymmetric. Examples: RNA, splice sites (Figure 1); DNA, T7 RNA polymerase (Figure 2).

Symmetric: Sites on DNA bound by a dimeric protein usually (there are exceptions) have a twofold dyad axis of symmetry. This means that there is a line passing through the DNA perpendicular to its long axis, about which a 180° rotation will bring the DNA helix phosphates back into register with their original positions. There are two places that the dyad axis can be set:

Odd symmetric: The axis is on a single base so that the site contains an odd number of bases. Examples: lambda cI and cro and lambda O (Figure 3 on page 46).

Even symmetric: The axis is between two bases, so that the site contains an even number of bases. Examples: 434 cI and cro, ArgR, CRP, TrpR, FNR, LexA (Figure 2).

See also Binding Site, Sequence Logo, Symmetry Paradox.

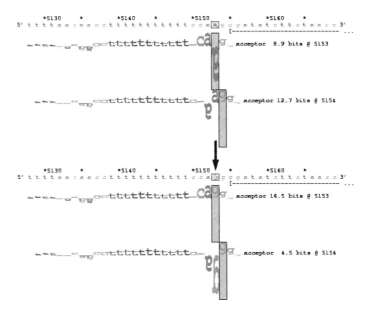

Figure 1 Asymmetric binding site symmetry.

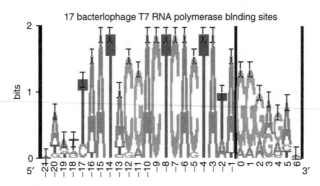

Figure 2   Asymmetric binding site symmetry.

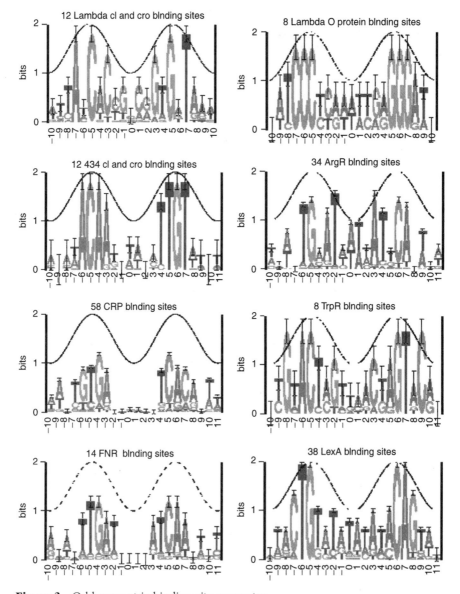

Figure 3   Odd symmetric binding site symmetry.

BIOCHIP. *SEE* MICROARRAY.

BIOINFORMATICS (COMPUTATIONAL BIOLOGY)
John M. Hancock

In general terms, the application of computers and computational techniques to biological data.

The field covers a wide range of applications from the databasing of fundamental data sets such as protein and DNA sequences, and even laboratory processes, to sophisticated analyses such as evolutionary modeling, to the modeling of protein structures and cellular networks. Areas related to bioinformatics are neuroinformatics, the modeling of nervous systems, and medical informatics, the application of computational techniques to medical data sets. While the boundaries of these related disciplines are reasonably clearly drawn, it is not unambiguously clear whether they form part of bioinformatics, but they are not included in this dictionary. Systems biology, the modeling of biological systems in general, is included here as a subdiscipline of bioinformatics.

Bioinformatics may be regarded as a synonym for *computational biology*. It is the more often used synonym in the United Kingdom and Europe, whereas computational biology is more commonly used in the United States, although these differences are not exclusive. A case could be made for defining bioinformatics more narrowly in terms of the computational storage and manipulation (but not analysis) of biological information and computational biology as a more biology-oriented discipline aimed at learning new knowledge about biological systems (see www.bisti.nih.gov/CompuBioDef.pdf). However, given the difficulty of drawing this distinction at the margins, particularly as researchers may move seamlessly between the two areas, we have not adopted it here.

Bioinformatics draws on a range of disciplines, including biochemistry, molecular biology, genomics, molecular evolution, computer science, and mathematics.

Related Websites

Definition of bioinformatics	http://bioinformatics.org/faq/ #definitionOfBioinformaticsTight
Definition of computational biology	http://bioinformatics.org/faq/#definitionOfCompbiol

Further Reading
Luscombe NM, et al. (2001). What is bioinformatics? A proposed definition and overview of the field. *Methods Inf. Med.* 40: 346–58.

BIOINFORMATICS.ORG
John M. Hancock and Martin J. Bishop

An organization dedicated to freedom and openness in bioinformatics.

The organization, which is nonprofit, provides resources for research, software development, and education about bioinformatics. The website provides project hosting for software development projects, access to major online databases, a suite of online sequence analysis tools, and a repository of molecular biology programs for Linux. It also provides bioinformatics educational material in the form of frequently asked questions (FAQ).

Related Website

| Bioinformatics.org | http://bioinformatics.org/ |

See also Open-Source Bioinformatics Organizations.

BIRTH-AND-DEATH EVOLUTION
Austin L. Hughes

A process of long-term evolution within multigene families whereby family membership changes gradually over time as new genes are added by gene duplication while other genes are lost by gene loss or inactivation.

This process was first described in certain families of vertebrate immune system genes, but it is now known to occur in a wide variety of gene families, including such highly conserved genes as histones. The birth-and-death model contrasts with that of concerted evolution in that it does not lead to homogenization of genes within a species. However, because the birth-and-death process causes a turnover of genes over evolutionary time, in families subject to this process there will be few if any orthologous relationships among the genes of distantly related species.

Further Reading

Hughes AL, Nei M (1989). Evolution of the major histocompatibility complex: Independent origin of nonclassical class I genes in different groups of mammals. *Mol. Biol. Evol.* 6: 559–579.

Nei M, et al. (1997). Evolution by the birth-and-death process in multigene families of the vertebrate immune system. *Proc. Natl. Acad. Sci. USA* 94: 7799–7806.

Piontkivska H, et al. (2002). Purifying selection and birth-and-death evolution in the histone H4 gene family. *Mol. Biol. Evol.* 19: 689–697.

See also Concerted Evolution, Gene Loss and Inactivation, Orthologous Genes.

BIT
Thomas D. Schneider

A bit is a binary digit, or the amount of information required to distinguish between two equally likely possibilities or choices. If I tell you that a coin is "heads," then you learn one bit of information. It's like a knife slice between the possibilities. Likewise, if a protein picks one of the four bases, then it makes a 2-bit choice. For eight things it takes 3 bits. In simple cases the number of bits is the log base 2 of the number of choices or messages M:

$$\text{Bits} = \log_2 M$$

Claude Shannon figured out how to compute the average information when the choices are not equally likely. The reason for using this measure is that when two communication systems are independent, the number of bits is additive. The log is the only mathematical measure that has this property. Both averaging and additivity are important for sequence logos and sequence walkers.

Even in the early days of computers and information theory people recognized that there were already two definitions of bit and that nothing could be done about it. The most common definition is "binary digit," usually a 0 or a 1 in a computer. This definition allows only for two integer values. The definition that Shannon came up with is an average number of bits that describes an entire communication message (or, in molecular biology, a set of aligned protein sequences or nucleic acid binding sites). This latter definition allows for real numbers. Fortunately the two definitions can be distinguished by context.

The idea that DNA can carry 2 bits per base goes back a long way. It was implied by Seeman et al.'s (1976) famous paper that the major groove of DNA can support up to 2 bits of sequence conservation, while the minor groove can only support 1 bit, but practical application of this idea to molecular biology only came when it was discovered that RepA binding sites are strangely anomalous in this regard (Papp et al., 1993). More recent experiments by Lyakhov et al. (2001) have confirmed this prediction.

A byte is a binary string consisting of 8 bits.

Related Websites

Primer	http://www.lecb.ncifcrf.gov/~toms/paper/primer
Hawaii	http://www.lecb.ncifcrf.gov/~toms/paper/hawaii
Nano2	http://www.lecb.ncifcrf.gov/~toms/paper/nano2

Further Reading

Lyakhov IG, et al. (2001). The P1 phage replication protein RepA contacts an otherwise inaccessible thymine N3 proton by DNA distortion or base flipping. *Nucleic Acid Res.* 29: 4892–4900. http://www.lecb.ncifcrf.gov/~toms/paper/repan3/.

Papp PP, et al, (1993). Information analysis of sequences that bind the replication initiator RepA. *J. Mol. Biol.* 233: 219–230.

Schneider TD (1994). Sequence logos, machine/channel capacity, Maxwell's demon, and molecular computers: A review of the theory of molecular machines. *Nanotechnology.* 5: 1–18, http://www.lecb.ncifcrf.gov/~toms/paper/nano2/.

Seeman NC, et al. (1976). Sequence-specific recognition of double helical nucleic acids by proteins. *Proc. Natl Acad. Sci. USA* 73: 804–808.

See also Binding site, Information, Message, Nit, Shannon, Sequence Walker Sequence Logo.

BLAST
Jaap Heringa

Suite of programs facilitating rapid searching of nucleic acid and protein databases.

BLAST (Basic Local Alignment Search Tool; Altschul et al., 1990) is a fast heuristic homology search algorithm which comprises five basic routines to search with a query sequence against a sequence database, including all combinations of nucleotide and protein sequences: (i) BLASTP compares an amino acid query sequence against a protein sequence database; (ii) BLASTN compares a nucleotide query sequence against a nucleotide sequence database; (iii) BLASTX compares the six-frame conceptual protein translation products of a nucleotide query sequence against a protein sequence database; (iv) TBLASTN compares a protein query sequence against a nucleotide sequence database translated in six reading frames; and (v) TBLASTX compares the six-frame translations of a nucleotide query sequence against the six-frame translations of a nucleotide sequence database.

The BLAST suite is the most widely used technique for sequence database searching that maintains sensitivity based on an exhaustive statistical analysis of ungapped alignments (Karlin and Altschul, 1990). The basic idea behind BLAST is the generation of all tripeptides from a query sequence and for each of those the derivation of a table of tripeptides deemed similar, the number of which is only a fraction of the total number possible. The BLAST program quickly scans a database of protein sequences for ungapped regions showing high similarity, which are called high-scoring segment pairs (HSPs), using the tables of similar peptides. The initial search is done for a word of length W that scores at least the threshold value T when compared to the query using a substitution matrix. Word hits are then extended in either direction in an attempt to generate an alignment with a score exceeding the threshold of S and as far as the cumulative alignment score can be increased. Extension of the word hits in each direction are halted when (i) the cumulative alignment score falls off by a quantity X from its maximum achieved value; (ii) the cumulative score goes to zero or below due to the accumulation of one or more negative-scoring residue alignments; or (iii) the end of either sequence is reached. The T parameter is the most important for the speed and sensitivity of the search resulting in the HSPs. A maximal-scoring segment pair (MSP) is defined as the highest scoring of all possible segment pairs produced from two sequences.

The BLAST algorithm provides a rigorous statistical framework (Karlin and Altschul, 1990) based on the extreme-value theorem to estimate the statistical significance of tentative homologs. The E value given for each database sequence found indicates the randomly expected number of sequences with an alignment score equal to or greater than that of the sequence considered. Only if the value is lower than the user-selectable threshold (E parameter) will the hit will be reported to the user.

The original BLAST program only detects local alignments without gaps and therefore might miss some significant similarities. A more recent version of the BLAST algorithm, coined Gapped BLAST, is able to insert gaps into the alignments, leading to increased sensitivity (Altschul et al., 1997). The original statistical framework for ungapped alignments is also employed to assess the significance of the gapped alignments, although no mathematical proof for this is available yet. However, computer simulations have indicated that the theory probably applies to gapped alignments as well (Altschul and Gish, 1996).

The most recent development for the BLAST engine is position-specific iterated BLAST (PSI-BLAST) (Altschul et al., 1997), which exploits increased sensitivity offered by multiple alignments and derived profiles in an iterative fashion. To optimize meaningful searching, query sequences are first scanned for the presence of so-called low-complexity regions (Wooton and Federhen, 1996), i.e., regions with a biased composition likely to lead to spurious hits, which are excluded from alignment. The program then initially operates on a single query sequence by performing a gapped BLAST search. Then, the program takes the significant local alignments found, constructs a multiple alignment, and abstracts a position-specific scoring matrix (PSSM) from this alignment. This is a type of profile which is used to rescan the database in a subsequent round to find more homologous sequences. The scenario is iterated until the user decides to stop or the search has converged; i.e., no more significantly scoring sequences can be found in subsequent iterations. The web server for PSI-BLAST, located at http://www.ncbi.nlm.nih.gov/BLAST, enables the user to specify at each iteration round which sequences should be included in the profile, while by default all sequences are included that score below a user-set expectation value (*E* value). However, the user needs to activate every subsequent iteration. An alternative to the PSI-BLAST web server is a stand-alone version of the program, downloadable from the aforementioned Word Wide Web address, which allows the user to specify beforehand the desired number of iterations. Although a consistent and very powerful tool, a limitation of the PSI-BLAST engine is the way in which the PSSM is generated, as this is essentially a simple stacking of the found local regions onto the query sequence used as a template (*N*-to-1 alignment), without keeping track of cross similarities between the added regions. As there are also no safeguards to control the number of sequences added to the PSSM at each iterative step (all sequences having an *E* value lower than a preset threshold are selected), it is clear that erroneous alignments are likely to progressively drive the method into inclusion of false positives (profile wander). Also, the various BLAST programs are not generally useful for motif searching, for which specialized software has been developed

Related Website

| NCBI BLAST Home Page | http://www.ncbi.nlm.nih.gov/BLAST/ |

Further Reading

Altschul SF, Gish W (1996). Local alignment statistics. *Methods Enzymol.* 266: 460–480.

Altschul SF, et al. (1990). Basic local alignment search tool. *J. Mol. Biol.* 215: 403–410.

Altschul SF, et al. (1997). Gapped BLAST and PSI-BLAST: A new generation of protein database search programs. *Nucleic Acids. Res.* 25: 3389–3402.

Karlin S, Altschul SF (1990). Methods for assessing the statistical significance of molecular sequence features by using general scoring schemes. *Proc. Natl. Acad. Sci. USA* 87: 2264–2268.

Wooton JC, Federhen S (1996). Analysis of compositionally biased regions in sequence databases. *Methods Enzymol.* 266: 554–571.

BLASTX
Roderic Guigó

A program of the BLAST suite for sequence comparison, BLASTX compares the translation of the nucleotide query sequence to a protein database.

Because BLASTX translates the query sequence in all six reading frames and provides combined significance statistics for hits to different frames, it is particularly useful when the reading frame of the query sequence is unknown or it contains errors that may lead to frame shifts or other coding errors. Thus a BLASTX search is often the first analysis performed when analyzing an anonymous genomic sequence. Sequence conservation with known proteins may indicate the existence (and location) of coding exons on the genomic sequence.

BLASTX is particularly useful when analyzing predicted open reading frames (ORFs) in bacterial genomes or expressed sequence tags (ESTs). It is also useful for analyzing short anonymous eukaryotic sequences, although BLASTX searches do not resolve exon splice boundaries well. For large sequences (such as eukaryotic chromosomes), however, the BLASTX search may become computationally prohibitive.

Related Website

| NCBI BLAST | http://www.ncbi.nlm.nih.gov/BLAST/ |

Further Reading
Gish W, States D (1993). Identification of protein coding regions by database similarity search. *Nature Genet.* 3: 266–272.

See also Alignment Multiple, Alignment-Pairwise, BLAST, Homology-Based Gene Prediction.

BLAT (BLAST-LIKE ALIGNMENT TOOL)
John M. Hancock

Program to rapidly align protein or nucleic acid sequences to a genome.

BLAT uses a heuristic similar to that used by BLAST to carry out rapid alignment of input sequences (protein or nucleic acid) to a genomic sequence. The program makes use of a catalogue of 11-mer oligonucleotide sequences that is small enough to be held in memory, making searches rapid. The program identifies matches between 11-mers in the probe sequence and 11-mers in the genomes and retrieves the genomic region likely to match the input sequence. It then aligns the sequence to the region and links together nearby alignments into a longer alignment. Finally it attempts to find small exons that might have been missed and attempts to adjust gap boundaries to correspond to splice sites. For protein searches BLAT uses 4-mers rather than 11-mers. The program is designed to find matches of better than 95% identity and length of 40 or more bases, or for proteins >80% similarity and length 20 or more amino acids.

Related Website

| BLAT | http://genome.ucsc.edu/cgi-bin/hgBlat |

Further Reading
Kent WJ (2002). BLAT—the BLAST-like alignment tool. *Genome Res.* 12: 656–664.

See also BLAST.

BLOCK
Terri Attwood

An ungapped local alignment derived from a conserved region (motif) of an aligned protein family and used to build a characteristic signature of family membership. Within blocks, sequence segments are clustered to reduce multiple contributions to residue frequencies from groups of highly similar or identical sequences. This is important for deep motifs (with tens or hundreds of sequences), which are often dominated by numerous identical or near-identical sequences, reflecting the innate bias of the source sequence databases. Each cluster is treated as a single segment, each of which is assigned a score giving a measure of its relatedness—the higher the score, the more dissimilar the segment is from others in the block. In practice, blocks are seldom used in isolation, but, rather, multiple blocks are generally used to characterize particular families. The power of this approach derives principally from the fact that the more blocks matched by a query sequence, the greater the confidence there is that the sequence belongs to the given family, provided the blocks are matched in the correct order and have appropriate distances between them. Although diagnostically powerful, care must be taken to assess the reliability of matches; particular caution should be used in the interpretation of single-block matches, which are usually spurious. Blocks are the basis of the Blocks database, which is derived from sequence families encoded in the InterPro and PRINTS resources. They are also the source of data for the BLOSUM series of substitution matrices, which are a commonly used alternative to the PAM series.

Related Websites

| Blocks search | http://blocks.fhcrc.org/blocks/blocks_search.html |
| Block maker | http://blocks.fhcrc.org/blockmkr/make_blocks.html |

Further Reading

Henikoff JG, Greene EA, Pietrokovski S, Henikoff S (2000). Increased coverage of protein families with the blocks database servers. *Nucleic Acids Res.* 28(1): 228–230.

Henikoff S, Henikoff JG (1991). Automated assembly of protein blocks for database searching. *Nucleic Acids Res.* 19: 6565–6572.

Henikoff S, Henikoff JG (1994). Position-based sequence weights. *J. Mol. Biol.* 243(4): 574–578.

BLOCKS

Terri Attwood

A database in which protein families are characterized by groups of calibrated blocks (motifs) which together form signatures of family membership. Blocks are generated automatically by looking for the most highly conserved regions within sequence families. Automatic block detection is based on two separate approaches. The first is based on the identification of conserved residue triplets (which need not be contiguous in space)—this method is guaranteed to find motifs, even in random sequences. The second is a more robust statistical approach based on *Gibbs sampling,* a strategy that picks random positions along all but one of the sequences within a family and then aligns the remaining sequence for best fit with the others. This procedure is iterated until the score is maximized. Blocks identified by both methods are encoded as ungapped local alignments and calibrated against SWISS-PROT to obtain a measure of the likelihood of a chance match. Two scores are noted for each block: The first denotes the level at which 99.5% of matches are true negatives; the second is the median value of the true positive scores, which is normalized by multiplying by 1000 and dividing by the 99.5% score in order for the performance of individual blocks to be meaningfully compared. The median standardized score for known true positive matches is termed "strength."

By virtue of using multiple motifs, Blocks is a diagnostically more reliable resource than those that exploit individual motifs (e.g., PROSITE). When searching the database, the scoring scheme may influence the quality of individual block matches (i.e., can sometimes lead to overprediction), but the more blocks matched, the greater the likelihood that the match has not arisen by chance. Consequently, probability values are calculated for multiple hits. Two versions of Blocks are made available from the Fred Hutchinson Cancer Research Center, Seattle, Washington. The first is derived from sequence families embodied in InterPro; the second is based explicitly on motifs encoded in PRINTS and is termed the PRINTS database in Blocks format. Because the databases are generated automatically, family annotation is not provided directly, but links are made to the relevant annotation in InterPro and PRINTS.

Related Websites

Blocks	http://www.blocks.fhcrc.org/
About Blocks	http://blocks.fhcrc.org/blocks/help/about_blocks.html
Blocks search	http://blocks.fhcrc.org/blocks/blocks_search.html

Further Reading

Henikoff JG, Greene EA, Pietrokovski S, Henikoff S (2000). Increased coverage of protein families with the blocks database servers. *Nucleic Acids Res.* 28(1): 228–230.

Henikoff S, Henikoff JG (1991). Automated assembly of protein blocks for database searching. *Nucleic Acids Res.* 19: 6565–6572.

Henikoff S, Henikoff JG (1994). Protein family classification based on searching a database of blocks. *Genomics* 19(1): 97–107.

See also InterPro, Motif, PRINTS, PROSITE, SWISS-PROT.

BLOSUM MATRIX
Laszlo Patthy

Whereas Dayhoff amino acid substitution matrices are based on substitution rates derived from global alignments of protein sequences that are at least 85% identical, the BLOSUM (BLOcks amino acid Substitution Matrix) matrices are derived from local alignments of conserved blocks of aligned amino acid sequence segments of more distantly related proteins. The primary difference is that in the case of distantly related proteins the Dayhoff matrices make predictions based on observations of closely related sequences, whereas the BLOSUM approach makes direct observations on blocks of distantly related proteins. The matrices derived from a database of blocks in which sequence segments are identical at 45% and 80% of aligned residues are referred to as BLOSUM 45, BLOSUM 80, etc.

The BLOSUM matrices show some consistent differences when compared with Dayhoff matrices. Since the blocks were derived from the most highly conserved regions of proteins, the differences between BLOSUM and Dayhoff matrices arise from the more significant structural and functional constraint on conserved regions. The differences between BLOSUM and Dayhoff matrices are primarily due to the fact that the most variable, surface-exposed regions of proteins (loops, β-turns) are underrepresented and the highly conserved regions (secondary structure elements) that form the conserved core of protein folds are overrepresented in BLOSUM matrices. The relatively weak conservation of polar residues in the Dayhoff matrices is due to the fact that in the case of surface residues it is the hydrophilicity rather than the actual residue that is conserved.

BLOSUM matrices are used in the alignment of amino acid sequences to calculate alignment scores and in sequence similarity searches and are especially useful for the detection of distant homologies.

Related Website

BLOCKS	http://www.blocks.fhcrc.org/

Further Reading
Henikoff S, Henikoff TG (1992). Amino acid substitution matrix from protein blocks. *Proc. Natl. Acad. Sci. USA.* 89: 10915–10919.

See also Alignment Score, Amino Acid Exchange Matrix, Dayhoff Amino Acid Substitution Matrix, Global Alignment, Local Alignment, Sequence Similarity.

BOLTZMANN FACTOR
Roman A. Laskowski

In any system of molecules at equilibrium, the number possessing an energy E is proportional to the Boltzmann factor $\exp^{(-E/kT)}$, where k is Boltzmann's constant and T is the temperature of the system.

It can be used in various dynamic studies of proteins.

Further Reading
Dill KA, Bromberg S (2003). *Molecular Driving Forces: Statistical Thermodynamics in Chemistry and Biology.* Garland Science, New York.

BP. *SEE* BASE PAIR.

BOOTSTRAP ANALYSIS
Sudhir Kumar and Alan Filipski

A commonly used, computation-intensive approach for calculating reliability of inferred phylogenies and the variance of evolutionary parameter estimates.

In bootstrap analysis for sequence data, the original sequence data set is used to generate a large number (hundreds to thousands) of replicates by pseudorandomly sampling sites from the original data set with replacement. Each replicate is a complete data set containing the same number of sites and taxa as the original data set. These replicates are then subject to the same analysis, and the estimates obtained are used to generate confidence intervals, variances, or other measures of the robustness of the initial inference. For instance, in tree making, each pseudorandom replicate is used to generate a phylogeny under a given criterion and the percent frequency of occurrence of the dichotomy of taxa induced by a branch in the original tree is referred to as the bootstrap value for that branch.

Related Website

SEQBOOT: program from the PHYLIP package that carries out bootstrapping on a sequence alignment	http://www.din.or.jp/~qph/research/phylip/seqboot.html

Further Reading

Dopazo J (1994). Estimating errors and confidence intervals for branch lengths in phylogenetic trees by a bootstrap approach. *J. Mol. Evol.* 38: 300–304.

Efron B (1982). *The Jackknife, the Bootstrap, and Other Resampling Plans*. Society for Industrial and Applied Mathematics, Philadelphia, PA.

Efron B, Tibshirani R (1993). *An Introduction to the Bootstrap*. Chapman & Hall, New York.

Felsenstein J (1985). Confidence limits on phylogenies: An approach using the bootstrap. *Evolution* 39: 783–791.

Kumar S, et al. (2001). MEGA2: Molecular evolutionary genetics analysis software. *Bioinformatics* 17: 1244–1245.

Nei M, Kumar S (2000). *Molecular Evolution and Phylogenetics*. Oxford University Press, Oxford, UK.

Swofford DL (1998). *PAUP*: Phylogenetic Analysis Using Parsimony (and Other Methods)*. Sinauer Associates, Sunderland, MA.

Zhaxybayeva O, Gogarten JP. (2002). Bootstrap, Bayesian probability and maximum likelihood mapping: Exploring new tools for comparative genome analyses. *BMC Genomics* 3: 4.

See also Jackknife, Phylogenetic Reconstruction.

BORN SOLVATION ENERGY, GENERALIZED
Roland L. Dunbrack

An approximate solution of the Poisson equation for calculation of solvation and interaction energies of ionic and polar molecules in solution. The calculation of the generalized Born solvation energy is much faster than numerical solution of the Poisson or Poisson–Boltzmann equations and can be used in molecular dynamics simulations with programs such as CHARMM and AMBER.

Further Reading

Bashford D, Case DA (2000). Generalized Born models of macromolecular solvation effects. *Annu. Rev. Phys. Chem.* 51: 129–52.

Still WC, et al. (1991). Semianalytical treatment of solvation for molecular mechanics and dynamics. *J. Am. Chem. Soc.* 112: 6127–6129.

See also AMBER, CHARMM, Solvation Free Energy.

BOTTLENECK. *SEE* POPULATION BOTTLENECK.

BOX
Thomas D. Schneider

Commonly used term for a region of sequence with a particular function.

A sequence logo of a binding site will often reveal that there is significant sequence conservation outside such a "box." The term "*core*" is sometimes used to acknowledge this, but sequence logos reveal that the division is arbitrary. A better usage is to replace this concept with *binding site* for nucleic acids or "*motif*" for proteins. For example, in a paper by Margulies and Kaguni (1996), the authors use the conventional model that DnaA binds to nine bases and they call the sites boxes. However, in the paper they demonstrate that there are effects of the sequence outside the box.

Further Reading

Margulies C, Kaguni JM (1996). Ordered and sequential binding of DnaA protein to oriC, the chromosomal origin of *Escherichia coli*. *J. Biol. Chem.* 271: 17035–17040.

BRAGG'S LAW. *SEE* DIFFRACTION OF X-RAYS, X-RAY CRYSTALLOGRAPHY FOR STRUCTURE DETERMINATION.

BRANCH-LENGTH ESTIMATION
Sudhir Kumar and Alan Filipski

The process of estimating branch lengths in a phylogenetic tree.

Given a set of sequences, their pairwise distances, and a phylogenetic tree, the branch lengths can be estimated by an ordinary least-squares method, which chooses branch lengths so as to minimize the sum of squared differences between observed distances and patristic distances. (The *patristic distance* between two sequences is the sum of branch lengths connecting them.) A computationally efficient analytical formula to obtain this estimate was developed by Rzhetsky and Nei (1993).

Alternatively, branch lengths may be estimated using the maximum-likelihood method, in which the probability of observing the given sequences is maximized. Under the maximum-parsimony model, algorithms are available to compute branch lengths by reconstructing the ancestral states and comparing the ancestral and descendant sequences for a given branch. However, this latter method can lead to underestimation because of multiple substitutions at individual sites.

Further Reading

Felsenstein J (1981). Evolutionary trees from DNA sequences: A maximum likelihood approach. *J. Mol. Evol.* 17: 368–376.

Kumar S, et al, (2001). *MEGA2*: Molecular evolutionary genetics analysis software. *Bioinformatics* 17: 1244–1245.

Maddison WP, Maddison DR (1992). *MacClade: Analysis of Phylogeny and Character Evolution*. Sinauer Associates, Sunderland, MA.

Nei M, Kumar S (2000). *Molecular Evolution and Phylogenetics*. Oxford University Press, Oxford, UK.

Rzhetsky A, Nei M (1993). Theoretical foundation of the minimum-evolution method of phylogenetic inference. *Mol. Biol. Evol.* 10: 1073–1095.

Swofford DL (1998). *PAUP*: Phylogenetic Analysis Using Parsimony (and Other Methods)*. Sinauer Associates, Sunderland, MA.

See also Phylogenetic Reconstruction.

C

Cα (C-alpha)
Candidate Gene
Carboxy-Terminus
Car–Parinello Simulations
CASP
Catalytic Triad
Category
CDS
Cenancestor
Centimorgan
Centromere
Channel Capacity
Character
CHARMM
Chimeric DNA Sequence
Chou–Fasman Prediction Method
Chromatin
Chromosomal Deletion
Chromosomal Inversion
Chromosomal Translocation
Chromosome
Chromosome Band
CINEMA
Cis-Regulatory Module
Clade
Classification
Classification in Machine Learning
Classifier
CLUSTAL
CLUSTALY, CLUSTALX
Cluster
Cluster Analysis
Clusters of Orthologous Groups (COG)
Clustering
CluSTr
CNS
Code, Coding
Coding Region (CDS)
Coding Region Prediction

Coding Statistic
Coding Theory
Codon
Codon Usage Bias
Coevolution
Coevolution of Protein Residues
Cofactor
COG
Coil
Coiled Coil
Coincidental Evolution
Comparative Gene Prediction
Comparative Genomics
Comparative Modeling
Complement
Complexity
Complexity Regularization
Complex Trait
Components of Variance
Composer
Composite Regulatory Element
Computational Biology
Concept
Conceptual Graph
Concerted Evolution
Conformation
Conformational Analysis
Conformational Energy
Congen
Conjugate Gradient Minimization
Connectionist Networks
Consed
Consensus Sequence
Consensus Tree
Conservation
ConsInspector
Contact Map
Contig
Contig Mapping

Dictionary of Bioinformatics and Computational Biology. Edited by Hancock and Zvelebil
ISBN 0-471-43622-4   © 2004 John Wiley & Sons, Inc.

Cα (C-ALPHA)
Roman A. Laskowski

The central carbon atom, Cα, common to all amino acids to which is attached an amino group (NH_2), a carboxyl group (COOH), a hydrogen atom, and a side chain. Each amino acid is distinguished from every other by its specific side chain.

Further Reading

Branden C, Tooze J (1991). *Introduction to Protein Structure*. Garland Science, New York.

Lesk AM (2001). *Introduction to Protein Architecture*. Oxford University Press, Oxford, UK.

See also Amino Acid, Side Chain.

CANDIDATE GENE
Mark McCarthy and Steven Wiltshire

In the context of disease–gene mapping, used to denote a gene with a strong prior claim for involvement in determining trait susceptibility, usually based on a perceived match between the known (or presumed) function of the gene and/or its product and the biology of the disease under study.

By extension, the term is often used (as in "candidate-gene-based analysis") to represent a particular strategy for susceptibility gene identification, which focuses on association analysis of biological candidates. This is often contrasted with approaches which rely on an unbiased genomewide analysis (such as genome scans for linkage). Since, for many diseases, the basic pathophysiological mechanisms are unclear, the candidate approach has obvious limitations: Most obviously, it will be unlikely to uncover susceptibility variants when they lie in pathways not previously suspected of a role in disease involvement. Increasingly, the boundaries between the "candidate–gene" and "genomewide" approaches are becoming blurred as many gene-mapping efforts start from an initial genome scan, followed by detailed examination of all the positional candidates within linked regions of interest. That is, they seek to identify the strongest candidates based on both chromosomal position and biology.

Examples: Multifactorial trait susceptibility loci which were initially identified through biological candidacy include the histocompatibility locus antigen (HLA) region and the insulin gene (in type 1 diabetes) and factor V Leiden (in deep-vein thrombosis)

Related Websites

Online Mendelian Inheritance in Man	http://www.ncbi.nih.gov/Omim/
Genecanvas (candidate genes for cardiovascular disease)	http://genecanvas.idf.inserm.fr/main.asp.htm

Further Reading

Collins FS (1995). Positional cloning moves from the perditional to traditional. *Nat. Genet.* 9: 347–350.

Dictionary of Bioinformatics and Computational Biology. Edited by Hancock and Zvelebil
ISBN 0-471-43622-4   © 2004 John Wiley & Sons, Inc.

Halushka MK, et al. (1999). Patterns of single-nucleotide polymorphisms in candidate genes for blood-pressure homeostasis. *Nat. Genet.* 22: 239–247.

Hirschhorn JN, et al. (2002). A comprehensive review of genetic association studies. *Genet. Med.* 4: 45–61.

Tabor HK, et al. (2002). Candidate-gene approaches for studying complex genetic traits: Practical considerations. *Nat. Rev. Genet.* 3: 1–7.

See also Genome Scans, Linkage, Positional Candidate Approach.

CARBOXY-TERMINUS. *SEE* C-TERMINUS.

CAR–PARINELLO SIMULATIONS
Roland L. Dunbrack

A method for ab initio molecular dynamics calculations based on density-functional theory developed by Roberto Car and Michele Parinello. The theory has allowed for molecular dynamics simulations of much larger systems than was possible previously.

Further Reading
Car R, Parinello M (1985). *Phys. Rev. Lett.* 55, 2471–2474.

Teter MP, et al. (1992). Iterative minimization techniques for ab initio total-energy calculations—Molecular dynamics and conjugate gradients. *Rev. Modern Phys.* 64: 1045–1097.

See also Ab Initio, Molecular Dynamics Simulations.

CASP
David T. Jones

CASP stands for Critical Assessment in Structure Prediction—an ongoing international experiment run every two years to assess the state-of-the-art in protein structure prediction by blind testing.

One problem with benchmarking protein structure prediction methods is that, if predictions are made on protein whose structure is already known, then it is hard to be certain that prior knowledge of the correct answer did not influence the prediction. Published benchmarking results for prediction methods might not therefore be representative of the results that might be expected on proteins of entirely unknown structure.

To tackle this problem, John Moult and colleagues initiated an international experiment to evaluate the accuracy of protein structure prediction methods by blind testing. The first CASP experiment was in 1994, and since then one has been run every two years. In CASP experiments, crystallographers and nuclear magnetic resonance (NMR) spectroscopists release the sequences of as-yet unpublished protein structures to the prediction community. Each research group attempts to predict the structures of these proteins and these predictions are e-mailed to the CASP evaluators before the structures are made available to the public. At the end of the prediction season, all predictions are collated and evaluated by a small group of independent assessors. In this way, the current state-of-the-art in protein structure prediction can be determined without any chance of bias or cheating. CASP has proven to be an extremely important development in the field of protein structure prediction and has not only allowed progress in the field to be accurately estimated but also greatly stimulated interest in the field itself.

Related Website

| CASP | http://predictioncenter.llnl.gov/ |

Further Reading

Fischer D, Barret C, Bryson K, Elofsson A, Godzik A, Jones D, Karplus KJ, Kelley LA, Maccallum RM, Pawowski K, Rost B, Rychlewski L, Sternberg MJ (1999). CAFASP-1: Critical assessment of fully automated structure prediction methods. *Proteins: Struc. Funct. Genet.* 3(Suppl): 209–217.

CATALYTIC TRIAD
Roman A. Laskowski

A group of three amino acid residues in an enzyme structure responsible for its catalytic activity. The three residues may be far apart in the amino acid sequence of the protein yet in the final, folded structure come together in a specific conformation in the active site to perform the enzyme's function on its substrate.

The best-known catalytic triad is the Ser–His–Asp triad of the serine proteinases and lipases, first identified in 1969. Its role is to cleave a given peptide substrate at a specific peptide bond. Specificity is governed by the substrate residue that fits into the P subsite, or specificity pocket, immediately adjacent to the scissile bond. For example, in the serine proteinases trypsin, the peptide bond that is cut is the one downstream from an arginine or lysine residue, either of these fitting neatly into the specificity pocket.

The enzymes that use the Ser–His–Asp triad are a ubiquitous group responsible for a range of physiological responses such as the onset of blood clotting and digestion as well as playing a major role in the tissue destruction associated with arthritis, pancreatitis, and pulmonary emphysema.

A number of different protein families possess the Ser–His–Asp triad and, although their overall folds and topologies may differ substantially, the three-dimensional conformation of the triad itself is remarkably well conserved, illustrating the importance of the precise three-dimensional arrangement of these residues for catalytic activity to take place. What is more, because the protein structures differ so greatly, it is thought the triad may have been arrived at independently as a consequence of convergent evolution.

Related Website

| Catalytic triad of subtilisin | http://www.biochem.ucl.ac.uk/cgi-bin/wallace/enzyme/getECPAGE.pl?ecnum=3.4.21.62.0021 |

Further Reading

Barth A, Wahab M, Brandt W, Frost K (1993). Classification of serine proteases derived from steric comparisons of their active sites. *Drug Design Discovery* 10: 297–317.

Blow DM (1990). More of the catalytic triad. *Nature* 343: 694–695.

Wallace AC, Laskowski RA, Thornton JM (1996). Derivation of 3D coordinate templates for searching structural databases: Application to Ser–His–Asp catalytic triads in the serine proteinases and lipases. *Prot. Sci.* 5: 1001–1013.

See also Conformation, Fold, Peptide, Peptide Bond, Residue, Sequence of Proteins.

CATEGORY
Robert Stevens

Types of things, which are defined through either a collection of shared properties (an intentional definition) or enumeration (an extensional definition).

Country, *person*, and *bacterium* are all examples of categories. Categories can be thought of as having conditions that are necessary for an individual to be a member of that category. For example, a *bachelor* is a man who has never married (strictly speaking). The conditions for category membership are of two kinds:

1. *Necessary condition*: A condition that must be satisfied for an individual to be a member of that category. For example, a protein enzyme is a polymer of amino acids, but not all polymers of amino acids are protein enzymes.
2. *Sufficient condition*: A condition that must be satisfied for an individual to be a member of that category; in addition, fulfilling this condition is enough for category membership. For example, a protein enzyme is a kind of protein that catalyzes a chemical reaction. A protein enzyme must do this and what is more doing this is sufficient for a thing to be a protein enzyme.

Categories that have such sufficiency conditions are known as defined. It is not possible to determine sufficiency conditions for all categories. Such categories are known as natural kinds. It is often true that no matter how many conditions are made or how carefully they are constructed, a member of the category will not fulfill those conditions. For example, an elephant without tusks or a trunk and that is colored pink is still an elephant.

CDS. *SEE* CODING REGION.

CENANCESTOR. *SEE* ANCESTRAL GENOME.

CENTIMORGAN
Katheleen Gardiner

Unit of distance between two genes in a genetic linkage map; based on recombination frequency.

One centimorgan equals a 1% recombination frequency. The correlation between centimorgan and physical distance in base pairs is not uniform within or between organisms. Where recombination frequency is high, 1 cM corresponds to fewer base pairs. In the human genome, 1 cM averages 1 Mb; in mouse, 1 cM averages 2 Mb.

Further Reading

Liu B-H (1997). *Statistical Genomics: Linkage, Mapping and QTL Analysis*. CRC Press, Boca Raton, FL.

Ott J (1999). *Analysis of Human Genetic Linkage*. Johns Hopkins University Press, Baltimore, MD.

Yu A, et al. (2001). Comparison of human genetic and sequence-based physical maps. *Nature* 409: 951–953.

CENTROMERE (PRIMARY CONSTRICTION)
Katheleen Gardiner

The chromosomal region containing the kinetochore, where the spindle fibers attach during meiosis and mitosis. A centromere is essential for stable chromosome maintenance and for separation of the chromatids to daughter cells during cell division. In metaphase chromosomes, where the DNA is highly condensed, the centromere is seen as a narrowed or constricted region. The centromere defines the two chromosome arms, the chromatin (DNA plus associated proteins) above or below the centromere.

Relative centromere location categorizes a chromosome as acrocentric, the centromere at or near one end producing one very short and one long arm; metacentric, the centromere near the middle of the chromosome and two arms of approximately equal size; or submetacentric, the centromere between the middle and one end, producing one long and one short arm. In human chromosomes, the short arm is referred to as the p arm and the long as the q arm.

Human centromeres are comprised of a large block of a complex family of repetitive DNA, called alphoid sequences. The basic alphoid sequences of ~170 bp are similar among chromosomes. They are arranged into tandem arrays and successively larger blocks of arrays to generate segments ranging from several hundred kilobases to several megabases in size that are chromosome specific.

Further Reading

Miller OJ, Therman E (2000). *Human Chromosomes*. Springer, Vienna.

Schueler MG, Higgens AW, Rudd MK, Gustashaw K, Willard HF (2001). Genomic and genetic definition of a functional human centromere. *Science* 294: 109–115.

Scriver CR, Beaudet AL, Sly WS, Valle D, Eds. (2002). *The Metabolic and Molecular Basis of Inherited Disease*. McGraw-Hill, New York.

Sullivan BA, Schwartz S, Willard HF (1996). Centromeres of human chromosomes. *Environ. Mol. Mutagen.* 28: 182–191.

Wagner RP, Maguire MP, Stallings RL (1993). *Chromosomes—A synthesis*. Wiley-Liss, New York.

See also Kinetochore, Satellite DNA.

CHANNEL CAPACITY (CHANNEL CAPACITY THEOREM)
Thomas D. Schneider

The maximum information, in bits per second, that a communications channel can handle is the channel capacity:

$$C = W \log_2 \left(\frac{P}{N} + 1 \right) \text{ bits/sec}$$

where W is the bandwidth (cycles per second, or hertz), P is the received power (joules per second), and N is the noise (joules per second). Shannon derived this formula by realizing that each received message can be represented as a sphere in a high-dimensional space. The maximum number of messages is determined by the diameter of these spheres and the available space. The diameter of the spheres is determined by the noise, and the available space is a sphere determined by the total power and the noise. Shannon realized that by dividing the volume of the larger sphere by the volume of the smaller message spheres, one would obtain the maximum number of messages. The logarithm (base 2) of this number is the channel capacity. In the formula, the signal-to-noise ratio is P/N. Shannon's channel capacity theorem states that if one attempts to transmit information at a rate R greater than C only at best C bits will the information be received. On the other hand, if $R \leq C$, then one may have as few errors as desired so long as the channel is properly coded.

Further Reading
Shannon CE (1948). A mathematical theory of communication. *Bell Syst. Tech. J.* 27: 379–423, 623–656. http://cm.bell-labs.com/cm/ms/what/shannonday/paper.html.
Shannon CE (1949). Communication in the presence of noise. *Proc. IRE* 37: 10–21.

See also Message, Molecular Machine Capacity, Shannon Sphere.

CHARACTER (SITE)
Sudhir Kumar and Alan Filipski

A feature, or an attribute, of an organism or genetic sequence.
For instance, a backbone is an attribute of animals, which takes a value of 1 for vertebrates and 0 for all other animals. In molecular sequence alignment, each column (position or site in the sequence) corresponds to a character. The observed value of a character is called a character state. In molecular sequence analysis, a character can take either 4 (for DNA or RNA) or 20 (for amino acid) possible states or can be absent (indel) for a particular sequence in an alignment.

Further Reading
Schuh, RT (2000). *Biological Systematics: Principles and Applications.* Cornell University Press, Ithaca, NY.

See also Sequence Alignment.

CHARMM
Roland L. Dunbrack

CHARMM (Chemistry at HARvard Macromolecular Mechanics) is a molecular dynamics simulation and modeling computer program developed by Martin Karplus and colleagues at Harvard. It uses an empirical molecular mechanics potential energy function for studying the structures and dynamics of proteins, DNA, and other molecules.

Related Websites

Alexander MacKerell (CHARMM parameterization)	http://www.pharmacy.umaryland.edu/faculty/amackere/
CHARMM Development and Information	http://www.scripps.edu/brooks/charmm_docs/charmm.html

Further Reading

Brooks BR, et al. (1983). CHARMM: A program for macromolecular energy, minimization, and dynamics calculations. *J. Comput. Chem.* 4: 187–217.

Karplus M, McCammon JA (2002). Molecular dynamics simulations of biomolecules. *Nat. Struct. Biol.* 9: 646–652.

MacKerell AD Jr, et al. (1998). All-atom empirical potential for molecular modeling and dynamics studies of proteins. *J. Phys. Chem.* B102: 3586–3616.

See also Modeling, Molecular.

CHIMERIC DNA SEQUENCE
Rodger Staden

A contiguous DNA sequence containing segments from noncontiguous sources. Both physical mapping clones and sequencing clones can be chimeric, hence leading to incorrect maps and assemblies.

CHOU–FASMAN PREDICTION METHOD
Patrick Aloy

The Chou–Fasman method is used to predict protein secondary structure from sequence information. It was one of the first automated implementations and is a "first-generation" structure prediction method. It is based on single residue preferences derived from a very limited database (15 proteins). According to these preferences, to adopt a given conformation, residues were classified as strong formers, formers, weak formers, indifferent formers, breakers, and strong breakers.

The method uses empirical rules for predicting the initiation and termination of helical and beta regions in proteins (i.e., four helix formers out of six residues; three beta formers out of five) and then extends these segments in both directions until a tetra-peptide of breakers is found.

The authors claimed a three-state-per-residue accuracy of 77% when evaluated on the same set used to derive the parameters. Ulterior analyses on realistic benchmarks have shown the accuracy of the Chou–Fasman method to be around 48%.

Related Website

Chou–Fasman program	http://fasta.bioch.virginia.edu/o_fasta/chofas.htm

Further Reading

Chou PY, Fasman GD (1974). Prediction of protein conformation. *Biochemistry* 13: 222–245.

See also Secondary Structure Prediction of Protein.

CHROMATIN
Katheleen Gardiner

DNA within a chromosome complexed with proteins.

Euchromatin, comprising most of the chromosome material, contains transcribed genes, both active and inactive, and is in an expanded conformation during the interphase. Heterochromatin, within and adjacent to centromeres, contains repetitive DNA and remains condensed during the interphase.

Further Reading

Miller OJ, Therman E (2000). *Human Chromosomes.* Springer, Vienna.

Schueler MG, et al. (2001). Genomic and genetic definition of a functional human centromere. *Science* 294: 109–115.

Wagner RP, et al. (1993). *Chromosomes—A Synthesis.* Wiley-Liss, New York.

CHROMOSOMAL DELETION
Katheleen Gardiner

Breakage and loss of a segment of a chromosome.

A deletion may be telomeric, where material is lost from the end of a chromosome, or internal. Large deletions (megabase in size) can be detected as a loss of or as a change in the chromosome band pattern. Some specific deletions are associated with human genetic disease, for example Charcot Marie Tooth (chromosome 4) and DiGeorge syndrome (chromosome 22).

Further Reading

Borgaonkar DS (1994). *Chromosomal Variation in Man.* Wiley-Liss, New York.

Miller OJ, Therman E (2000). *Human Chromosomes.* Springer, Vienna.

Scriver CR, Beaudet AL, Sly WS, Valle D, Eds. (2002). *The Metabolic and Molecular Basis of Inherited Disease.* McGraw-Hill, New York.

Wagner RP, Maguire MP, Stallings RL (1993). *Chromosomes—A Synthesis.* Wiley-Liss, New York.

CHROMOSOMAL INVERSION
Katheleen Gardiner

Breakage within a chromosome at two positions, followed by rejoining of the ends after the internal segment reverses orientation.

Even if no material is lost, gene order is changed. A pericentric inversion occurs when the two breakpoints are located on opposite sides of the centromere, a paracentric inversion when they are located on the same side of the centromere.

Further Reading

Borgaonkar DS (1994). *Chromosomal Variation in Man.* Wiley-Liss, New York.

Miller OJ, Therman E (2000). *Human Chromosomes.* Springer, Vienna.

Scriver CR, et al., Eds. (1995). *The Metabolic and Molecular Basis of Inherited Disease,* McGraw-Hill, New York.

Wagner RP, et al. (1993). *Chromosomes—A Synthesis.* Wiley-Liss, New York.

CHROMOSOMAL TRANSLOCATION
Katheleen Gardiner

Breakage of two nonhomologous chromosomes with exchange and rejoining of broken fragments.

A reciprocal translocation occurs when there is no loss of genetic material and each product contains a single centromere. Chromosome translocations can be identified by an abnormal karyotype, where the chromosome number is correct, but two chromosomes show abnormal sizes and banding patterns. Specific translocations are associated with some forms of malignancy.

Further Reading

Borgaonkar DS (1994). *Chromosomal Variation in Man*. Wiley-Liss, New York.

Miller OJ, Therman E (2000). *Human Chromosomes*. Springer, Vienna.

Scriver CR et al., Eds. (1995) *The Metabolic and Molecular Basis of Inherited Disease*, Mc-Graw-Hill, New York.

Wagner RP, et al. (1993). *Chromosomes—A Synthesis*. Wiley-Liss, New York.

CHROMOSOME
Katheleen Gardiner

A DNA molecule containing genes in an ordered linear sequence and complexed with protein.

In prokaryotes, one or more circular chromosomes generally contain the complete set of genes of the organism. In eukaryotes, the complete set of genes is divided among multiple linear chromosomes. The number, size, and gene content of each chromosome are species specific and invariant among individuals of a species.

Structural features describing eukaryotic chromosomes include the centromere, the telomere, and the long and short arms.

The DNA of the human genome is divided among 23 pairs of chromosomes, the autosomes, numbered in order of decreasing size from 1 to 22, plus the sex chromosomes, X and Y.

Individual chromosomes in a cell can be identified by a light/dark band pattern produced by various staining procedures. The number, size, and chromosome band pattern of an organism define its karyotype.

Further Reading

Miller OJ, Therman E (2000). *Human Chromosomes*. Springer, Vienna.

Wagner RP, et al. (1993). *Chromosomes—A Synthesis*. Wiley-Liss, New York.

See also Centromere, Chromosome Band, Telomere.

CHROMOSOME BAND
Katheleen Gardiner

Metaphase human chromosomes can be stained with a variety of reagents to produce a banding pattern visible under the light microscope.

Banding pattern plus chromosome size usually allows the unambiguous identification of individual normal chromosomes. For human chromosomes, a total of 450 bands is commonly achieved; 850 bands can be identified in longer, prophase chromosomes. Bands are numbered from the centromere.

When chromosomes are treated with trypsin or hot salt solutions and stained with Giemsa, darkly staining regions are referred to as G bands, or Giemsa dark bands. R bands are produced when chromosomes are first heated in a phosphate buffer and then stained with Giemsa. The pattern obtained is the reverse (hence R) of the G-band pattern. T bands are the very bright R bands, resistant to heat denaturation, found most often but not exclusively at the telomeres. Q bands are obtained by staining with quinacrine, with bright bands corresponding to those seen with Giemsa. C banding highlights centromeres by treatment with alkali and controlled hydrolysis.

The exact molecular basis of chromosomal band patterns is not clear; however, there is thought to be a relationship with regional base composition. G bands are regions that are high in AT content. R bands are variable in base composition, but the brightest R bands are the regions that are highest in GC content and include most telomeric regions of human chromosomes. G bands are also gene poor, LINE rich, Alu poor, CpG island poor, and late replicating. Opposite features are found in a subset of R bands.

Further Reading

Holmquist GP (1992). Chromosome bands, their chromatin flavors and their functional features. *Am. J. Hum. Genet.* 51: 17–37.

Miller OJ, Therman E (2000). *Human Chromosomes*. Springer, Vienna.

Wagner RP, et al. (1993). *Chromosomes—A Synthesis*. Wiley-Liss, New York.

Scriver CR, et al., Eds. (1995). *The Metabolic and Molecular Basis of Inherited Disease*, McGraw-Hill, New York.

Standing Committee on Human Cytogenetic Nomenclature (1985). *An International System for Human Cytogenetic Nomenclature*. Karger, Basel, Switzerland.

CINEMA
Martin J. Bishop and John M. Hancock

A color interactive editor for multiple alignments. The program allows visualization and manipulation of both protein and DNA sequences.

CINEMA is a Java applet. It allows construction of multiple alignments either by cut and paste or accessing sequences or alignments from databases. Alignments are edited by dragging segments of sequences relative to one another to add or remove gaps using the mouse. The applet uses an extensible architecture with a set of pluglets that allow additional features to be added. Pluglets available via the CINEMA website allow viewing of protein structures, six-frame translation of DNA sequences, production of dot plots, alignment using CLUSTALW, and BLAST database searching.

Related Website

CINEMA	http://bioinf.man.ac.uk/dbbrowser/CINEMA2.1/

Further Reading

Parry-Smith DJ, et al. (1997). CINEMA—A novel Colour INteractive Editor for Multiple Alignments. *Gene* 211: GC45–GC56.

See also BLAST, CLUSTAL.

CIS-REGULATORY MODULE (COMPOSITE REGULATORY ELEMENT, CRM)
John M. Hancock

Functionally indivisible sequence elements consisting of a number of transcription factor binding sites.

Although these structures are still largely ill-defined, their distinguishing feature appears to be the multiple occurrence of a relatively small number of transcription factor binding sites. COMPEL is a database of this class of element.

Gene Regulatory Sequence Analysis	http://compel.bionet.nsc.ru/new/index.html

Further Reading
Yuh C-H, Davidson EH. (1996). Modular cis-regulatory organization of Endo16, a gut-specific gene of the sea urchin embryo. *Development* 122: 1069–1082.

See also Regulatory Region Prediction.

CLADE
Sudhir Kumar and Alan Filipski

A monophyletic group of taxa, i.e., one in which all species descend from a single ancestor and such that the group includes all descendants of that ancestor.

In a rooted phylogenetic tree, it can refer to a subtree consisting of a given node and all of its descendants.

If a group of taxa fails to be monophyletic, it may be paraphyletic. A group of taxa is called paraphyletic if it contains the most recent common ancestor of all members but fails to include all descendants of that ancestor. An example of a paraphyletic group is the common group "reptile," which fails to be monophyletic because it does not include birds (descendants of the most recent common ancestor of "reptiles").

A non-monophyletic group comprising more than one independent clade is called polyphyletic. A set of taxa may be both paraphyletic and monophyletic.

Further Reading
Schuh, RT (2000). *Biological Systematics: Principles and Applications*. Cornell University Press, Ithaca, NY.

CLASSIFICATION
Robert Stevens

1. A grouping of concepts into useful categories or classes not necessarily in a hierarchical fashion. For most purposes, a "class" is the same as a category. In biology, nucleic acid, protein, and gene would all be classes to which other concepts (classes or individuals) may belong. Often classes are arranged into a hierarchy of progressively more specific subclasses, but the primary nature of classification is the grouping of concepts.
2. The process of arranging individuals into classes or classes into a hierarchy of sub- and superclasses based on their definitions, often with the help of software (a reasoner or classifier) using formal logical criteria.

CLASSIFICATION IN MACHINE LEARNING (DISCRIMINANT ANALYSIS)
Nello Cristianini

The statistical and computational task of assigning data points to one of k classes (for finite k) specified in advance.

This process is also known as discriminant analysis. The function that maps data into classes is called a "classifier" or a "discriminant function." Particularly important is the special case of binary classification ($k = 2$), to which the general case $k > 2$ (multiclass) is often reduced.

The task of inferring a classifier function from a labeled data set is one of the standard problems in machine learning and in statistics.

Discriminant analysis is also used for determining which variables are more relevant in forming the decision function and hence for obtaining information about the "reasons" for a certain classification rule.

Common approaches include linear Fisher discriminant analysis and nonlinear methods such as neural networks or support vector machines.

Further Peading

Cristianini N, Shawe-Taylor J (2000). *An Introduction to Support Vector Machines.* Cambridge University Press, Cambridge, UK.

Duda RO, Hart PE, Stork DG (2001). *Pattern Classification.* Wiley, New York.

See also Clustering, Label, Machine Learning, Neural Network, Regression Analysis, Support Vector Machines.

CLASSIFIER (REASONER)
Robert Stevens

A mechanism, usually a software program, for performing classification.

See also Classification.

CLUSTAL (CLUSTALY, CLUSTALX)
Jaap Heringa

Higgins and Sharp (1988) constructed the first implementation of the fast and widely used method CLUSTAL for multiple-sequence alignment, which was especially designed for use on small workstations. The method follows the heuristic of the tree-based progressive alignment protocol, which implies repeated use of a pairwise alignment algorithm to construct the multiple alignment in $N - 1$ steps (with N the number of sequences), following the order dictated by a guide tree. Speed of the early version of CLUSTAL was obtained during the pairwise alignments of the sequences through the Wilbur and Lipman (1983, 1984) algorithm. From the resulting pairwise similarities, a tree was constructed using the UPGMA clustering criterion, after which the sequences were aligned following the branching order of this tree. For the comparison of groups of sequences, Higgins and Sharp (1988) used consensus sequences to represent aligned subgroups of sequences and also employed the Wilbur–Lipman technique to match these. Since its early inception, the CLUSTAL package has been subjected to a number of revision cycles. Higgins et al. (1992) implemented an updated version, CLUSTAL V, in which the memory-efficient dynamic programming routine of Myers and Miller (1988) is used, enabling the alignment of large sets of sequences using little memory. Further, two alignment positions, each from a different alignment, are compared in CLUSTAL V using the average alignment similarity score of Corpet (1988). The largely extended version CLUSTAL W (Thompson et al., 1994) uses the alternative neighbor-joining (NJ) algorithm (Saitou and Nei, 1987), which is widely used in phylogenetic analysis, to construct a guide tree. Sequence blocks are represented by a profile, in which the individual sequences are additionally weighted according to the branch lengths in the NJ tree. Further carefully crafted heuristics to optimize the exploitation of sequence information include (i) local gap penalties, (ii) automatic selection of the amino acid substitution matrix, (iii) automatic gap penalty adjustment, and (iv) a mechanism to delay the alignment of sequences that appear to be distant at the time they are considered. The method CLUSTAL W does not provide the possibility to iterate the procedure as do several other multiple-sequence alignment algorithms (e.g., Hogeweg and Hesper, 1984; Corpet, 1988; Gotoh, 1996; Heringa, 1999, 2002). An integrated user interface for the CLUSTAL W method has been implemented in CLUSTAL X (Thompson et al., 1997), which is freely available and includes accessory programs for tree depiction. The CLUSTAL W method is the most popular method for multiple-sequence alignment and is reasonably robust for a wide variety of alignment cases.

Related Websites

CLUSTAL W online help	http://www-igbmc.u-strasbg.fr/BioInfo/ClustalW
General description	http://www-igbmc.u-strasbg.fr/BioInfo/ClustalW/clustalw.html
Download	ftp://ftp-igbmc.u-strasbg.fr/pub/ClustalW
CLUSTAL X help	http://www-igbmc.u-strasbg.fr/BioInfo/ClustalX
CLUSTAL X download	ftp://ftp-igbmc.u-strasbg.fr/pub/ClustalX

Further Reading

Corpet, F (1988). Multiple sequence alignment with hierarchical clustering. *Nucl. Acids Res.* 16: 10881–10890.

Gotoh O (1996). Significant improvement in accuracy of multiple protein sequence alignments by iterative refinement as assessed by reference to structural alignments. *J. Mol. Biol* 264: 823–838.

Heringa J (1999). Two strategies for sequence comparison: Profile-preprocessed and secondary structure-induced multiple alignment. *Comp. Chem.* 23: 341–364.

Heringa J (2002). Local weighting schemes for protein multiple sequence alignment. *Comput. Chem.* 26: 459–477.

Higgins DG, Sharp PM (1988). CLUSTAL: A package for performing multiple sequence alignment on a microcomputer. *Gene* 73: 237–244.

Higgins DG, et al. (1992). CLUSTAL V: Improved software for multiple sequence alignment. *Comput. Appl. Biosci.* 8: 189–191.

Hogeweg P, Hesper B (1984). The alignment of sets of sequences and the construction of phyletic trees: An integrated method. *J. Mol. Evol.* 20: 175–186.

Myers EW, Miller W (1988). Optimal alignment in linear space. *Comput. Appl. Biosci.* 4: 11–17.

Saitou N, Nei M (1987). The neighbor-joining method: A new method for reconstructing phylogenetic trees. *Mol. Biol. Evol.* 4: 406–425.

Thompson JD, et al. (1994). CLUSTAL W: Improving the sensitivity of progressive multiple sequence alignment through sequence weighting, positions-specific gap penalties and weight matrix choice. *Nucleic Acids Res.* 22: 4673–4680.

Thompson JD, et al. (1997). The ClustalX windows interface: Flexible strategies for multiple sequence alignment aided by quality analysis tools. *Nucleic Acids Res.* 25: 4876–4882.

Wilbur WJ, Lipman DJ (1983). Rapid similarity searches of nucleic acid and protein data banks. *Proc. Natl. Acad. Sci. USA* 80: 726–730.

Wilbur WJ, Lipman DJ (1984). The context dependent comparison of biological sequences. *SIAM J. Appl. Math.* 44: 557–567.

CLUSTALY, CLUSTALX. *SEE* CLUSTAL.

CLUSTER
Dov S. Greenbaum

A group of related sequences, genes, or gene products.

The relatedness of a cluster is established through the use of different clustering methods and distance metrics; their distance from a seed sequence or gene determines a member's membership within a cluster. Clustering methods, many extracted from other disciplines, include hierarchical, self-organizing maps (SOMs), self-organizing tree algorithms (SOTAs), or *K*-means clustering. Distance methods include Euclidean distance or Pearson correlations. Clustering has become a basic tool of bioinformatics in the study of microarray and other high-throughput data.

Clustering has many uses. Most commonly it associates genes with other genes based on a specific feature such as messenger RNA (mRNA) expression. Based on this clustering, i.e., that genes are coregulated because they have the same expression patterns, one can assign function to novel proteins based on the functions of other proteins in the cluster: guilt by association.

Clustering can be unsupervised, i.e., there is no additional information regarding the genes, or supervised, where information extraneous to the data being clustered is used to help cluster.

Further Reading
Eisen, MB, et al. (1998). Cluster analysis and display of genome-wide expression patterns. *Proc. Natl. Acad. Sci. USA* 95: 14863–14868.

Michaels GS, et al. (1998). Cluster analysis and data visualization of large-scale gene expression data. *Pac. Symp. Biocomput.* 1998: 42–53.

CLUSTER ANALYSIS. *SEE* CLUSTERING.

CLUSTERS OF ORTHOLOGOUS GROUPS (COG)
Dov S. Greenbaum

The NCBI database Clusters of Orthologous Groups (COGs) of proteins contains groups of proteins clustered together on the basis of their purported evolutionary similarity.

Using an all-against-all sequence comparison of all the sequenced genomes, the authors found and clustered together all proteins that are more similar to each of the proteins in the group than any of the other proteins in their respective genomes. This technique was used to account for both slow and rapid evolution of proteins. The major disadvantage in this method is the clustering of proteins that may have only minimal similarity. Each cluster contains at least one protein from archea, prokaryotes, and eukaryotes.

Clusters in the COG database can be selected on the basis of one or any of the following criteria: function, text, species, or groups of species. Clusters are named within the database on the basis of known functions (similar or otherwise) of the proteins in the group. These functions also include putative and predicted. Tools are also provided to search through the database using a gene/protein name, text string, or organism.

The COG database provides three distinct forms of information for the user: (i) functional and structural (two- and/or three-dimensional structures) annotation of the proteins in the COG; (ii) multiple alignments of the proteins in the COG, allowing the user to identify and describe conserved regions in proteins of similar function; and (iii) phylogenetic patterns of each cluster, allowing for the determination of whether a given organism is present or absent from a cluster, thus possibly identifying missing genes in a particular pathway.

The database also provides the user with a tool to assign new proteins to a COG. COGnitor compares the inputted sequence (FASTA, or flat-file format) with all the underlying sequence information in the database and provides a suggested cluster based on these sequences.

Additional information is provided for many of the clusters, including (i) classification schemes (e.g., transport classification or enzyme commission numbers), (ii) gene names, (iii) basis for the cluster (i.e., motif, experimental data, operon structure, similarity based on PSI-BLAST), (iv) domains contained in the clustered proteins, (v) COG structure information (e.g., the existence of subclusters), (vi) protein notes, (vii) background information, (viii) predictions, (ix) references, and (x) modified or new protein sequences.

Related Website

COG	http://www.ncbi.nlm.nih.gov/COG/

Further Reading

Tatusov RL, Koonin EV, Lipman DJ (1997). A genomic perspective on protein families. *Science* 278: 631–637.

Tatusov RL, et al. (2001). The COG database: New developments in phylogenetic classification of proteins from complete genomes. *Nucleic Acids Res.* 29: 22–28.

See also Ortholog.

CLUSTERING (CLUSTER ANALYSIS)
Nello Cristianini

The statistical and computational problem of partitioning a set of data into "clusters" of similar items, or, in other words, of assigning each data item to one of a finite set of classes not known a priori.

No external information about class membership is provided, hence making this an instance of an unsupervised learning algorithm. Often no assumption is made about the number of classes existing in the data set.

This methodology is often used in exploratory data analysis and visualization of complex multivariate data sets and is equally a topic of statistics and machine learning.

Clustering is done on the basis of a similarity measure (or a distance) between data items which needs to be provided to the algorithm. The criteria with which performance should be measured are less standardized than in other forms of machine learning (see Error Measures), but overall the goal is typically to produce a clustering of the data that maximizes some notion of within-cluster similarity and minimizes the between-cluster similarity (over all possible partitions of the data).

Since examining all possible organizations of the data into clusters is computationally too expensive, a number of algorithms have been devised to find "reasonable" clusterings without having to examine all configurations. Typical methods include hierarchical clustering techniques that proceed by a series of successive mergers. At the beginning there are as many clusters as data items; then the most similar items are grouped and these initial groups are merged according to their similarity; and so on. Eventually, this method would produce a single cluster, so that some stopping criterion needs to be specified.

Another popular heuristic is the one known as k-means clustering. In such a method, a set of k prototypes (or centers or means) is found, and each point is assigned to the class of the nearest prototype. The method is iterative: First some partitioning into k classes is found (possibly by choosing random "seeds" as temporary centers). Then the means of the k classes so defined are computed. Then each data point is assigned to the nearest center, so finding another partitioning, that has new means, and so on, iterating until convergence. Notice that the final solution is dependent on the initial choice of centers.

See also Microarray, Phylogenetic Reconstruction.

CLUSTR
Rolf Apweiler

The CluSTr (Clusters of SWISS-PROT and TrEMBL proteins) database is a protein sequence cluster database which offers an automatic classification of SWISS-PROT and TrEMBL proteins into groups of related proteins based on pairwise comparisons between protein sequences.

CluSTr is a protein sequence cluster database which automatically classifies SWISS-PROT and TrEMBL proteins into groups of related proteins based on pairwise comparisons between protein sequences. The CluSTr methodology uses the Smith–Waterman algorithm and Z-score statistics that allows the data to be updated incrementally avoiding time-consuming recalculations. CluSTr provides a hierarchical organization of protein clusters by performing analysis at different levels of sequence similarity.

Related Website

CluSTr	http://www.ebi.ac.uk/clustr/

Further Reading
Kriventseva EV, Biswas M, Apweiler R (2001). Clustering and analysis of protein families. *Curr. Opin. Struct. Biol.* 11: 334–339.
Kriventseva EV, Fleischmann W, Zdobnov EM, Apweiler R, CluSTr (2001). A database of clusters of SWISS-PROT + TrEMBL proteins. *Nucleic Acids Res.* 29: 33–36.

See also Clusters of Orthologous Groups.

CNS
Roland L. Dunbrack

A software suite for determination and refinement of X-ray crystallographic and nuclear magnetic resonance structures developed by Axel Brünger and colleagues at Yale and Stanford. CNS (Crystallography and NMR System) has a hierarchical structure with an HTML interface, a structure determination language, and low-level source code that allow for new algorithms to be easily integrated.

Related Website

CNS	http://cns.csb.yale.edu

Further Reading
Brunger AT, et al. (1998). Crystallography & NMR system: A new software suite for macromolecular structure determination. *Acta Crystallogr. D Biol. Crystallogr.* 54 (Pt 5): 905–921.
Brunger AT, Adams PD (2002). Molecular dynamics applied to X-ray structure refinement. *Acc. Chem. Res.* 35: 404–412.

See also NMR, X-Ray Crystallography for Structure Determination.

CODE, CODING. *SEE* CODING THEORY.

CODING REGION (CDS)
Roderic Guigó

Region in genomic DNA or complementary DNA (cDNA) that codes for protein—either a fraction of the protein (coding exon) or the whole protein.

The term denotes only the regions in genomic DNA or cDNA that end up translated into amino acid sequences. That is, in spliced genes, the coding regions are the coding fraction of coding exons and do not include untranslated exons or the untranslated segments of the coding exons.

CDS (coding sequences) is a synonym for *coding region*, and it is a feature key in the DDBJ/EMBL/GenBank nucleotide sequence data banks. It is defined as "sequence of nucleotides that corresponds with the sequence of amino acids in a protein (location includes stop codon); feature includes amino acid conceptual translation." The CDS feature is used to obtain automatic amino acid translations of the sequences in the DNA databases, such as DAD from DDBJ, TrEMBL from EMBL, or GenPept from GenBank.

Related Websites

DDBJ/EMBL/GenBank feature table definition	http://www.ebi.ac.uk/embl/Documentation/ FT_definitions/feature_table.html#components
TrEMBL	http://www.ebi.ac.uk/trembl/
GenPept	ftp://ftp.ncifcrf.gov/pub/genpept/

Further Reading

Guigó R (1998). Assembling genes from predicted exons in linear time with dynamic programming. *J. Comp. Biol.* 5: 681–702.

See also Coding Region Prediction.

CODING REGION PREDICTION
Roderic Guigó

Identification of regions in genomic DNA or complementary DNA (cDNA) sequences that code for proteins.

Protein coding regions exhibit characteristic sequence composition bias. The bias is a consequence of the unequal representation of amino acids in real proteins and the unequal usage of synonymous codons (see Codon Bias). The bias can be measured and used to discriminate protein coding regions from noncoding ones.

Protein coding regions are usually delimited by sequence patterns (translation initiation and termination codons in prokaryotic organisms, acceptor and donor splice sites, in addition, in eukaryotes) that may also contribute to their identification. The identification of protein coding regions is the first, essential step once the genome sequence of an organism has been obtained—even if only partially.

See also Coding Region; Coding Statistic; Codon Usage Bias; Gene Prediction, Ab Initio.

CODING STATISTIC (CODING MEASURE, CODING POTENTIAL, SEARCH BY CONTENT)
Roderic Guigó

A coding statistic can be defined as a function that computes a real number related to the likelihood that a given DNA sequence codes for a protein (or a fragment of a protein).

Protein coding regions exhibit characteristic DNA sequence composition bias, which is absent on noncoding regions. The bias is a consequence of the uneven usage of (1) the amino acids in real proteins and (2) synonymous codons (see Codon Bias). Coding statistics measure this bias and thus help to discriminate between protein coding and noncoding regions.

Most coding statistics measure directly or indirectly either codon (or dicodon) usage bias, base compositional bias between codon positions, or periodicity (correlation) in base occurrence (or a mixture of them all). Since the early 1980s, a great number of coding statistics have been published in the literature [see, e.g., Gelfand, (1995) and references therein]. Fickett and Tung (1992) showed that all these measures reduce essentially to a few independent ones: the Fourier spectrum of the DNA sequence, the length of the longest open reading frames (ORFs), the number of runs (repeats) of a single base or of any set of bases, and in-phase hexamer frequencies.

Guigó (1998) distinguishes between measures dependent on a model of coding DNA and measures independent of such a model. The model of coding DNA is usually probabilistic, i.e., the probability distribution of codons or dicodons in coding sequences, or more complex Markov models. To estimate the model, a set of previously known coding sequences from the genome of the species under consideration is required. This model is often very species specific. Under the model one can compute the probability of a DNA sequence given that the sequence codes for a protein. The probability of the DNA sequence can also be computed under the alternative noncoding model. Often the log-likelihood ratio of these two probabilities is taken as the coding measure. Examples of model-dependent coding measures are codon usage (Staden and McLachlan, 1982), amino acid usage (McCaldon and Argos, 1988), codon preference (Gribskov et al., 1984), hexamer usage (Claverie et al., 1990), codon prototype (Mural et al., 1991), and heterogeneous Markov models (Borodovsky and McIninch, 1993).

In contrast, model-independent measures do not require a set of previously known coding sequences from the species genome under study. They capture intrinsic bias in coding regions, which is very general, but they do not measure the direction of this bias, which is very species specific. Examples of model-independent measures are position asymmetry (Fickett, 1982), periodic asymmetry (Konopka, 1990), average mutual information (Herzel and Grosse, 1995), and the Fourier spectrum (Silverman and Linsker, 1986). Model-independent coding measures are useful when no previous coding sequences are known for a given genome. Because the signal they produce is weaker than that produced by model-dependent measures, usually longer sequences are required to obtain discrimination. This limits their utility mostly to prokaryotic genomes where genes are continuous ORFs.

Typically, coding statistics are computed on a sliding window along the query genomic sequence. This generates a profile in which peaks tend to correspond to coding regions and valleys to noncoding ones. Nowadays, coding statistics are used within ab initio gene prediction programs that resolve the limits between peaks and valleys at legal splice junctions. Fifth-order Markov models are among the most popular coding statistics used within gene prediction programs.

Two popular coding region identification programs are TestCode (Fickett, 1982) and GRAIL (Uberbacher and Mural, 1991).

Further Reading

Borodovsky M, McIninch J (1993). GenMark: Parallel gene recognition for both DNA strands. *Comput. Chem.* 17: 123–134.

Claverie J-M, et al. (1990). *k*-Tuple frequency analysis: From intron/exon discrimination to T-cell epitope mapping. *Methods Enzymol.* 183: 237–252.

Fickett JW (1982). Recognition of protein coding regions in DNA sequences. *Nucleic Acids Res.* 10: 5303–5318.

Fickett JW, Tung C-S (1992). Assessment of protein coding measures. *Nucleic Acids Res.* 20: 6441–6450.

Gelfand MS (1995). Prediction of function in DNA sequence analysis. *J. Comput. Biol.* 1: 87–115.

Gribskov M, et al. (1984). The codon preference plot: Graphic analysis of protein coding sequences and prediction of gene expression. *Nucleic Acids Res.* 12: 539–549.

Guigó R (1998). Assembling genes from predicted exons in linear time with dynamic programming. *J. Comput. Biol.* 5: 681–702.

Guigó R (1999). DNA composition, codon usage and exon prediction. In *Genetic Databases*. M Bishop, Ed. Academic Press, New York, pp. 53–80.

Herzel H, Grosse I (1995). Measuring correlations in symbol sequences. *Phys. A*, 216: 518–542.

Konopka AK (1990). Towards mapping functional domains in indiscriminantly sequenced nucleic acids: A computational approach. In *Structure and Methods:* Vol. 6: *Human Genome Initiative and DNA Recombination*, R Sarma, M Sarma, Ed. Adenine, Guilderland, NY, pp. 113–125.

Konopka AK (1999). Theoretical molecular biology. In *Encyclopedia of Molecular Biology and Molecular Medicine*, RA Meyers, Ed. Wiley, New York, pp. 37–53.

McCaldon P, Argos P (1988). Oligopeptide biases in protein sequences and their use in predicting protein coding regions in nucleotide sequences. *Proteins* 4: 99–122.

Mural RJ, et al. (1991). Pattern recognition in DNA sequences: The intron-exon junction problem. In *Proceedings of the First International Conference on Electrophoresis, Supercomputing and the Human Genome*, CC Cantor, HA Lim, Eds. World Scientific, Singapore, pp. 164–172.

Silverman BD, Linsker R (1986). A measure of DNA periodicity. *J. Theor. Biol.* 118: 295–300.

Staden R, McLachlan AD (1982). Codon preference and its use in identifying protein coding regions in long DNA sequences. *Nucleic Acids Res.* 10: 141–156.

Uberbacher EC, Mural RJ (1991). Locating protein-coding regions in human DNA sequences by a multiple sensor-neural network approach. *Proc. Natl. Acad. Sci. USA* 88: 11261–11265.

See also Coding Region; Coding Region Prediction; Codon Usage Bias; Gene Prediction, Ab Initio; GENSCAN.

CODING THEORY (CODE, CODING)
Thomas D. Schneider

Coding is the transformation of a message into a form suitable for transmission over a communications line.

This protects the message from noise. Since messages can be represented by points in a high-dimensional space (see Message), the coding corresponds to the placement of the messages relative to each other. When a message has been received, it has been distorted by noise, and in the high-dimensional space the noise distorts the initial transmitted message point in all directions evenly. The final result is that each received message is represented by a sphere. Picking a code corresponds to figuring out how the spheres should be placed relative to each other so that they are distinguishable. This situation can be represented by a gumball machine. Shannon's famous work on information theory was frustrating in the sense that he proved that codes exist that can reduce error rates to as low as one may desire, but he did not say how this could be accomplished. Fortunately a large effort by many people established many kinds of communications codes with the result that we now have many means of clear communications, such as CDs, MP3, DVD, the Internet, and digital wireless phones. One of the most famous coding theorists was Hamming. An example of a simple code that protects a message against error is the parity bit. There are many codes in molecular biology besides the genetic code since every molecular machine has its own code.

Further Reading
Shannon CE (1949). Communication in the presence of noise. *Proc. IRE*, 37: 10–21.
Hamming RW (1986). *Coding and Information Theory*. Prentice-Hall, Englewood Cliffs, NJ.

See also Gumball Machine, Hamming, Message, Noise, Shannon Sphere.

CODON
Niall Dillon

Three-base unit of the genetic code which specifies the incorporation of a specific amino acid into a protein during messenger RNA (mRNA) translation.

Within a gene, the DNA coding strand consists of a series of three-base units which are read in a 5′ to 3′ direction in the transcribed RNA and specify the amino acid sequence of the protein encoded by that gene. The genetic code is degenerate with most of the 20 amino acids represented by several codons. Degeneracy is highest in the third base. Initiation and termination of translation are specified by initiator and terminator codons. Codon usage shows significant variation between phyla.

Further Reading
Lewin B (2000). *Genes VII*. Oxford University Press, Oxford, UK.

See also Genetic Code.

CODON USAGE BIAS
Roderic Guigó

Nonrandomness in the use of synonymous codons in an organism.

In all genetic codes there are 64 codons encoding about 20 amino acids (sometimes one more or a few less) and a translation termination signal. Often, therefore, more than one codon encodes the same amino acid or the termination signal. Codons that encode the same amino acid are called synonymous codons. Most synonymous codons differ by only one base at their 3′ end position. Usage of synonymous codons is nonrandom. This nonrandomness is called *codon usage bias*. The bias is different in different genomes, which have different "preferred" synonyms for a given amino acid.

Codon usage bias depends mostly on mutational bias; in other words codon usage reflects mostly the background base composition of the genome. In some prokaryotic organisms, it appears to correlate also with the relative abundance of alternative transfer RNA (tRNA) isoacceptors (Ikemura, 1985). In general, highly expressed genes tend to show higher codon usage bias.

There are a number of measures of codon usage bias for a given gene. The codon adaptation index (CAI; Sharp and Li, 1987) is one such measure. CAI is a measurement of the relative adaptiveness of the codon usage of a gene toward the codon usage of highly expressed genes.

Related Websites

Codon usage database	http://www.kazusa.or.jp/codon/
Genetic codes resource at NCBI	http://www.ncbi.nlm.nih.gov/htbin-post/Taxonomy/ wprintgc?mode=c
Analysis of codon usage	http://artedi.ebc.uu.se/course/UGSBR/codonusage1.html

| Correspondence analysis of codon usage | http://www.molbiol.ox.ac.uk/cu/ |
| Codon usage indices | http://www.molbiol.ox.ac.uk/cu/Indices.html |

Further Reading

Ikemura T (1985). Codon usage and tRNA content in unicellular and multicellular organisms. *Mol. Biol. Evol.* 2: 13–34.

Sharp PM, Li W-H (1987). The codon adaptation index a measure of directional synonymous codon usage bias, and its potential applications. *Nucleic Acids Res.* 15: 1281–1295.

Sharp PM, et al. (1993). Codon usage: Mutational bias, translational selection or both? *Biochem. Soc. Trans.* 21: 835–841.

See also Coding Region, Coding Region Prediction, Coding Statistic.

COEVOLUTION (MOLECULAR COEVOLUTION)
Laszlo Patthy

Coevolution is a reciprocally induced inherited change in a biological entity (species, organ, cell, organelle, gene, protein, biosynthetic compound) in response to an inherited change in another with which it interacts. Protein–protein coevolution and protein–DNA coevolution are the phenomena describing the coevolution of molecules involved in protein–protein and DNA–protein interactions, respectively. Coevolution of protein residues describes the coevolution of pairs of amino acid residues that interact within the same protein. Coevolution is reflected in cladograms of the interacting entities: the cladograms of coevolving entities are congruent.

Further Reading

Dover GA, Flavell RB (1984). Molecular coevolution: DNA divergence and the maintenance of function. *Cell* 38(3): 622–623.

See also Coevolution of Protein Residues, Protein–DNA Coevolution, Protein–Protein Coevolution.

COEVOLUTION OF PROTEIN RESIDUES
Laszlo Patthy

In general, coevolution is a reciprocally induced inherited change in a biological entity in response to an inherited change in another with which it interacts.

The three-dimensional structure of a protein is maintained by several types of noncovalent interactions (short-range repulsions, electrostatic forces, van der Waals interactions, hydrogen bonds, hydrophobic interactions) and some covalent interactions (disulfide bonds) between the side chain or peptide backbone atoms of the protein. There is a strong tendency for the correlated evolution of interacting residues of proteins. For example, in the case of extracellular proteins, pairs of cysteines forming disulfide bonds are usually replaced or acquired "simultaneously" (on an evolutionary time scale). Such correlated evolution of pairs of cysteines involved in disulfide bonds frequently permits the correct prediction of the disulfide bond pattern of homologous proteins.

Bioinformatic tools have been developed for the identification of coevolving protein residues in aligned protein sequences. The identification of protein sites undergoing correlated evolution is of major interest for structure prediction since these pairs tend to be adjacent in the three-dimensional structure of proteins.

Further Reading

Pollock DD, et al. (1999). Coevolving protein residues: Maximum likelihood identification and relationship to structure. *J. Mol. Biol.* 287: 187–198.

Pritchard L, et al. (2001). Evaluation of a novel method for the identification of coevolving protein residues. *Protein Eng.* 14: 549–555.

Shindyalov IN, et al. (1994). Can three-dimensional contacts in protein structures be predicted by analysis of correlated mutations? *Protein Eng.* 7: 349–358.

Tuff P, Darlu P (2000). Exploring a phylogenetic approach for the detection of correlated substitutions in proteins. *Mol. Biol. Evol.* 17: 1753–1759.

See also Coevolution.

COFACTOR
Jeremy Baum

Cofactors are small organic or inorganic molecules that are bound to enzymes and are active in the catalytic process. Examples include flavin mononucleotide (FMN), flavin adenine dinucleotide (FAD), nicotinamide adenine dinucleotide (NAD), thiamine pyrophosphate (ThP or TPP), coenzyme A, biotin, tetrahydrofolate, vitamin B_{12}, heme, chlorophyll, and a variety of iron–sulfur clusters. With the help of cofactors, the enzymes can catalyze chemistry not available with the limited range of amino acid side chains.

COG. *SEE* CLUSTERS OF ORTHOLOGOUS GROUPS.

COIL (RANDOM COIL)
Roman A. Laskowski

Irregular, unstructured regions of a protein's backbone, as distinct from the regular region characterized by specific patterns of main-chain hydrogen bonds. Not to be confused with the loop region of a protein.

Further Reading

Branden C, Tooze J (1991). *Introduction to Protein Structure*. Garland Science, New York.

Lesk AM (2001). *Introduction to Protein Architecture*. Oxford University Press, Oxford, UK.

See also Loop, Main Chain, Secondary Structure of Protein.

COILED COIL
Patrick Aloy

Coiled coils are elongated helical structures that contain a highly specific heptad repeat that conforms to a specific sequence pattern. The first and fourth positions within this pattern are hydrophobic while the remaining positions show preferences for polar residues.

Coiled coils form intimately associated bundles of long alpha helices and are usually associated with specific functions. Coiled-coil segments should be removed from protein sequences before performing database searches as they can produce misleading results due to their repetitive pattern.

Many programs have been developed to predict coiled-coil segments in protein sequences, COILS being one of the most widely used.

The COILS program was developed by Lupas et al. (1991) to predict coiled-coil regions from protein sequences. It assigns a per-residue coiled-coil probability based on the comparison of its flanking sequences with sequences of known coiled-coil proteins.

Related Website

| COILS | http://www.russell.embl-heidelberg.de/cgi-bin/coils-svr.pl |

Further Reading

Lupas A, Van Dyke M, Stock J (1991). Predicting coiled coils from protein sequences. *Science* 252: 1162–1164.

See also Alpha Helix, Hydrophobic Scale, Polar.

COINCIDENTAL EVOLUTION. *SEE* CONCERTED EVOLUTION.

COMPARATIVE GENE PREDICTION. *SEE* GENE PREDICTION, COMPARATIVE.

COMPARATIVE GENOMICS
Jean-Michel Claverie

A research area and the set of bioinformatic approaches dealing with the global comparison of genome information in terms of their structure and gene content.

The information gathered from whole-genome comparisons may include:

- Conservation of sequences across species (e.g., from highly conserved genes to "ORFans")
- Identification of orthologous and paralogous genes
- Conservation of noncoding regulatory regions
- Conservation of local gene order (colinearity, synteny)
- Coevolution patterns of genes (e.g., subset of genes simultaneously retained or lost in genomes)
- Identification of gene fusion/splitting events (e.g., the Rosetta stone method)Evidence of lateral gene transfer

Comparative genomic approaches have been used to address a number of very diverse problems such as estimating the number of human genes, identifying and locating human genes, proposing new whole-species phylogenetic classification, predicting functional links between genes from their coevolution patterns or fusion (Rosetta stone) events, recognizing lateral gene transfer, or identifying specific adaptation to exotic life-styles. More information is extracted about each individual sequence as new complete genome sequences become available. Among the most significant discoveries resulting from comparative genomics are (i) the large fraction of genes of unknown function and (ii) the fact that every genome, even from small parasitic bacteria, contains a proportion of genes without homologs in any other species.

Related Websites

High-Quality Automated and Manual Annotation of Microbial Proteomes (HAMAP)	http://www.expasy.org/sprot/hamap/
Clusters of Orthologous Groups	http://www.ncbi.nlm.nih.gov/COG/
Human Genome Project Comparative and Functional Genomics page	http://www.ornl.gov/hgmis/research/function.html

Further Reading

Cambillau C, Claverie JM (2000). Structural and genomic correlates of hyperthermostability. *J. Biol. Chem.* 275: 32383–32386.

Mallon AM, et al. (2000). Comparative genome sequence analysis of the Bpa/Str region in mouse and man. *Genome Res.* 10: 758–775.

Marcotte EM, et al. (1999). Detecting protein function and protein-protein interactions from genome sequences. *Science* 285: 751–753.

Pellegrini M, et al. (1999). Assigning protein functions by comparative genome analysis: Protein phylogenetic profiles. *Proc. Natl. Acad. Sci. USA.* 96: 4285–4288.

Roest Crollius H, et al. (2000). Characterization and repeat analysis of the compact genome of the freshwater pufferfish *Tetraodon nigroviridis*. *Nat. Genet.* 25: 235–238.

Snel B, et al. (1999). Genome phylogeny based on gene content. *Nat. Genet.* 21: 108–110.

Wilson MD, et al. (2001). Comparative analysis of the gene-dense ACHE/TFR2 region on human chromosome 7q22 with the orthologous region on mouse chromosome 5. *Nucleic Acids Res.* 29: 1352–1365.

Worning P, et al. (2000). Structural analysis of DNA sequence: Evidence for lateral gene transfer in *Thermotoga maritima*. *Nucleic Acids Res.* 28: 706–709.

COMPARATIVE MODELING (HOMOLOGY MODELING)

Terri Attwood and Roland L. Dunbrack

The prediction of a three-dimensional protein structure from a protein's amino acid sequence based on an alignment to a homologous amino acid sequence from a protein with an experimentally solved three-dimensional structure. Such models are relatively accurate and form a rational basis for explaining experimental observations, redesigning proteins to change their performance, and rational drug design.

General steps involved in modeling, after the identification of a known three-dimensional structure of a homolog of the protein being modeled, are:

1. Alignment of target protein to template protein sequence
2. Transfer of coordinates of aligned atomic positions
3. Building of side chains using rotamer libraries

4. Building of INDELS (insertion/deletions) or variable (loop) regions

5. Local and global energy minimization

6. Assessing correctness of fold using programs such as ProCheck or PROSAII

There are a number of approaches to modeling, including fragment-based modeling, averaged template modeling, and single-template modeling. The accuracy of a modeled structure depends on the accuracy of the alignment and the amount and quality of biological knowledge of the target proteins and is influenced by the evolutionary distance separating the sequences.

An alternative strategy is based on deriving distance constraints from the known structure and using these to build a model by simulated annealing or distance geometry methods.

Related Websites

Swiss-Model	http://www.expasy.ch/swissmod/SWISS-MODEL.html
Quanta and InsightII	http://www.msi.com/
ESyPred3D	http://www.fundp.ac.be/urbm/bioinfo/esypred/
Modeller	http://guitar.rockefeller.edu/modeller/modeller.html

Further Reading

Greer J (1990). Comparative modeling methods: Application to the family of the mammalian serine proteases. *Proteins* 7: 317–334.

Sali A (1995). Modelling mutations and homologous proteins. *Curr. Opin. Biotechnol.* 6: 437–451.

Sali A, Blundell TL (1993). Comparative protein modelling by satisfaction of spatial restraints. *J. Mol. Biol.* 234: 779–815.

See also Alignment—Multiple, Alignment—Pairwise, Homology, Indel, Modeller.

COMPLEMENT

John M. Hancock

As the DNA double helix is made up of two base-paired strands, the sequences of the two strands are not identical but have a strict relationship to one another. For example, a sequence 5′-ACCGTTGACCTC-3′ on one strand pairs with the sequence 3′-TGGCAACTGGAG-5′. This second sequence is the known as the complement of the first. Converting a sequence to its complement is a simple operation and may be required, for example, if a sequence has been read in the 3′ to >5′ direction rather than in the conventional direction.

See also DNA Sequence, Reverse Complement.

COMPLEXITY

Thomas D. Schneider

Like *specificity*, the term *complexity* appears in many scientific papers, but it is not always well defined. When one comes across a proposed use in the literature, one can unveil this difficulty by asking: How would I measure this complexity? What are the units of complexity? An option is to use Shannon's information measure or explain why Shannon's measure does not cover what you are interested in measuring, then give a precise, practical definition. Some commonly used measures of the complexity of a system, such as a genetic network, are the numbers of components or interactions in the system. (See also Kolmogorov Complexity; Li and Vitányi, 1997.)

Further Reading

Li M, Vitányi P (1997). *An Introduction to Kolmogorov Complexity and Its Applications*, 2nd ed. Springer-Verlag, New York.

COMPLEXITY REGULARIZATION. *SEE* MODEL SELECTION.

COMPLEX TRAIT. *SEE* MULTIFACTORIAL TRAIT.

COMPONENTS OF VARIANCE. *SEE* VARIANCE COMPONENTS.

COMPOSER
Roland L. Dunbrack

A program for comparative modeling of protein structure developed by Tom Blundell and colleagues (1993). Composer searches for multiple protein structures homologous to the target sequence, aligns these structures, and identifies residues that occupy equivalent positions in all the structures. These residues are used to build a framework upon which loop modeling and side-chain prediction are performed.

Related Website

| Tripos | http://www.tripos.com/sciTech/inSilicoDisc/bioInformatics/composer.html |

Further Reading

Srinivasan N, Blundell TL (1993). An evaluation of the performance of an automated procedure for comparative modelling of protein tertiary structure. *Protein Eng.* 6: 501–512.

Topham CM, et al. (1990). An assessment of COMPOSER: A rule-based approach to modelling protein structure. *Biochem. Soc. Symp.* 57: 1–9.

See also Comparative Modeling, Loop Prediction/Modeling, Side-Chain Prediction.

COMPOSITE REGULATORY ELEMENT. *SEE*
CIS-REGULATORY MODULE.

COMPUTATIONAL BIOLOGY. *SEE* BIOINFORMATICS.

CONCEPT
Robert Stevens

The International Organization for Standardization (ISO) defines *concept* as a unit of thought.

In practice in computer science we are referring to the encoding of that unit of thought within a system. The word "concept" is used ambiguously to refer either to categories of things, e.g., country, or the individual things themselves, e.g., England. On the whole it is better to avoid the use of "concept" and use the terms *category* and *individual* as defined elsewhere.

CONCEPTUAL GRAPH
Robert Stevens

A graphical representation of logic originated by John Sowa based on the "existential grants" graphs proposed by the American philosopher Charles Stanley Pierce.

Subsets of the formalism are often used as an easily understandable notation for representing concepts. Computer implementations for various subsets are available and there is an active user group, but since the full formalism is capable of expressing not just first order but also higher order logics, full-proof procedures are computationally intractable. Excellent information is available online and there is an active user community.

Related Websites

Sowa	http://www.bestweb.net/sowa/
CG	http://www.cs.uah.edu/delugach/CG/

CONCERTED EVOLUTION (COINCIDENTAL EVOLUTION, MOLECULAR DRIVE)
Dov S. Greenbaum and Austin L. Hughes

A mode of long-term evolution within a multigene family whereby members of the gene family become homogenized over evolutionary time.

This results in greater sequence similarity within a species than between species when comparing members of gene families or repetitive sequence families such as satellite DNA. Concerted evolution is thought to be the result of DNA replication, recombination, and repair mechanisms, e.g., gene conversion and unequal crossing over. It requires homogenization—the horizontal transfer of a gene within the genome—and fixation—the spread of the new sequence combination to all (or most) individuals of the species.

Concerted evolution is thought to help spread advantageous mutations throughout a genome. Conversely, though, it erases prior evidence of divergence and the history of gene duplication.

The most striking example of concerted evolution involves ribosomal RNA (rRNA) genes. The rRNA genes constitute a large gene family in most organisms (approximately 400 members in a typical mammalian genome). Yet rRNA genes are typically identical or nearly so within species.

Further Reading

Arnheim N, et al. (1980). Molecular evidence for genetic exchanges among ribosomal genes on nonhomologous chromosomes in man and apes. *Proc. Natl. Acad. Sci. USA* 77: 7323–7327.

Dover G (1982). Molecular drive: A cohesive mode of species evolution. *Nature* 299: 111–117.

Liao D (1999). Concerted evolution: Molecular mechanism and biological implications. *Am. J. Hum. Genet.* 64: 24–30.

See also Birth-and-Death Evolution.

CONFORMATION
Roman A. Laskowski

A three-dimensional arrangement of atoms making up a particular molecule or part thereof. Different conformations of the same molecule are those involving rearrangements of atoms that do not break any covalent bonds or alter the chirality of any of the atoms. Thus, two rotamers of a given protein side chain are different conformations, whereas the D- and L-isomers of an amino acid are not (they correspond to different configurations).

Further Reading

Branden C, Tooze J (1991). *Introduction to Protein Structure*. Garland Science, New York.

See also Rotamer, Side Chain, Structure—3D Classification.

CONFORMATIONAL ANALYSIS
Roland L. Dunbrack

The process of determining the most likely conformation of a molecule based on its covalent geometry, stereochemistry, and internal degrees of freedom. This is a term long used in organic chemistry to denote analysis of molecular conformations and the dependence of chemical and physical properties (reactivity, density, melting point, etc.) on these conformations. The same term can be applied to analysis of bioorganic molecules such as proteins and DNA. Conformational analysis has been used in protein chemistry to explain the Ramachandran distribution of backbone dihedral angles and side-chain rotamer preferences.

Related Website

Protein Sidechain Conformational Analysis	http://www.fccc.edu/research/labs/dunbrack/confanalysis.html

Further Reading

Chakrabarti P, Pal D (2001). The interrelationships of side-chain and main-chain conformations in proteins. *Prog. Biophys. Mol. Biol.* 76: 1–102.

Dunbrack RL Jr, Karplus M (1994). Conformational analysis of the backbone-dependent rotamer preferences of protein sidechains. *Nature Struct. Biol.* 1: 334–340.

Ramachandran GN, Sasisekharan V (1968). Conformations of polypeptides and proteins. *Adv. Prot. Chem.* 23: 283.

See also Ramachandran Plot.

CONFORMATIONAL ENERGY
Roland L. Dunbrack

Potential energy of a flexible molecule dependent on the structure or conformation. In macromolecular modeling programs such as CHARMM and AMBER, this energy is usually calculated as a function of bond lengths, bond angles, dihedral angles, and nonbonded interatomic interactions. For reasonably small molecules, this energy can also be calculated with quantum mechanics packages.

Related Websites

Alexander MacKerell (CHARMM parameterization)	http://www.pharmacy.umaryland.edu/faculty/amackere/
AMBER	http://www.amber.ucsf.edu/amber/amber.html

Further Reading
Cornell WD, et al. (1995). A second generation force field for the simulation of proteins, nucleic acids, and organic molecules. *J. Am. Chem. Soc.* 117: 5179–5197.

MacKerell AD Jr, et al. (1998). All-atom empirical potential for molecular modeling and dynamics studies of proteins. *J. Phys. Chem.* B102: 3586–3616.

See also AMBER, CHARMM, Conformation.

CONGEN
Roland L. Dunbrack

A program for loop prediction in the process of comparative modeling, developed by Robert Bruccoleri and colleagues. Congen uses a grid search on the backbone dihedral angles in populated regions of the Ramachandran map followed by energy minimization with the CHARMM force field.

Related Website

Congenomics	http://www.congenomics.com/

Further Reading
Bruccoleri RE, Karplus M (1987). Prediction of the folding of short polypeptide segments by uniform conformational sampling. *Biopolymers* 26: 137–168.

Li H, et al. (1997). Homology modeling using simulated annealing of restrained molecular dynamics and conformational search calculations with CONGEN: Application in predicting the three-dimensional structure of murine homeodomain Msx-1. *Protein Sci.* 6: 956–970.

See also CHARMM, Comparative Modeling, Loop Prediction/Modeling, Ramachandran Plot.

CONJUGATE GRADIENT MINIMIZATION. *SEE* ENERGY MINIMIZATION.

CONNECTIONIST NETWORKS. *SEE* NEURAL NETWORK.

CONSED
Rodger Staden

A widely used contig editor

Related Website

PHRAP	http://www.phrap.org/

Further Reading
Gordon D, et al. (1998). Consed: A graphical tool for sequence finishing. *Genome Res.* 8: 195–202.

CONSENSUS SEQUENCE
Dov S. Greenbaum, Thomas D. Schneider, and Rodger Staden

A consensus sequence is a string of nucleotides or amino acids which best represents a set of multiply aligned sequences.

The consensus sequence is decided by a selection procedure designed to determine which residue or nucleotide is placed at each individual position, usually that which is found at the specific position most often.

Consensus sequences are used to determine and annotate DNA sequences, for example, in the case of a TATA sequence in a promoter region, as well as to functionally classify previously unknown peptides such as where a DNA-binding consensus sequence defines a sequence as DNA binding and possibly a transcription factor.

The simplest form of a consensus sequence is created by picking the most frequent character at some position in a set of aligned DNA, RNA, or protein sequences such as binding sites. Where more than one character type occurs at a position, the consensus can be represented by the most frequent character or by a symbol denoting ambiguity (e.g., IUPAC) depending on the algorithm being employed. For protein sequences the symbol can denote membership of an amino acid family, e.g., the hydrophobics. For DNA sequence assembly projects the algorithm can optionally use the base-call confidence values in conjunction with the frequencies.

A disadvantage of using consensus sequences is that the process of creating a consensus destroys the frequency information and can lead to errors in interpreting sequences. For example, suppose a position in a binding site has 75% A. The consensus would be A. Later, after having forgotten the origin of the consensus while trying to make a prediction, one would be wrong 25% of the time. If this was done over all the positions of a binding site, most predicted sites could be wrong. For example, in Rogan and Schneider (1995) a case is shown where a patient was misdiagnosed because a consensus sequence was used to interpret a sequence change in a splice junction. Figure 2 of Schneider (1997b) shows a Fis binding site that had been missed because it did not fit a consensus model. One can entirely replace this concept with sequence logos and sequence walkers.

Related Websites

Phform	http://saturn.med.nyu.edu/searching/thc/phform.html
Scanprosite	http://us.expasy.org/tools/scanprosite/
The Consensus Sequence Hall of Fame	http://www.lecb.ncifcrf.gov/~toms/consensus.html
Pitfalls	http://www.lecb.ncifcrf.gov/~toms/pitfalls.html
Colonsplice	http://www.lecb.ncifcrf.gov/~toms/paper/colonsplice/
Walker	http://www.lecb.ncifcrf.gov/~toms/paper/walker/walker.htmlfigr21

Further Reading

Aitken A. (1999). Protein consensus sequence motifs. *Mol. Biotechnol.* 12: 241–253.

Bork P, Koonin EV. (1996) Protein sequence motifs. *Curr. Opin. Struct. Biol.* 6: 366–376.

Keith JM, et al. (2002) A simulated annealing algorithm for finding consensus sequences. *Bioinformatics* 18: 1494–1499.

Rogan PK, Schneider TD (1995). Using information content and base frequencies to distinguish mutations from genetic polymorphisms in splice junction recognition sites. *Hum. Mut.* 6: 74–76. http://www.lecb.ncifcrf.gov/~toms/paper/colonsplice/.

Schneider TD (1997a). Information content of individual genetic sequences. *J. Theor. Biol.,* 189: 427–441. http://www.lecb.ncifcrf.gov/~toms/paper/ri/.

Schneider TD (1997b). Sequence walkers: A graphical method to display how binding proteins interact with DNA or RNA sequences. *Nucleic Acids Res.* 25: 4408–4415. http://www.lecb.ncifcrf.gov/~toms/paper/walker/. Erratum: *Nucleic Acis Res.* 26: 1135, 1998.

Staden, R. (1984). Computer methods to locate signals in nucleic acid sequences. *Nucleic Acid Res.* 12: 505–519.

Staden R (1988). Methods to define and locate patterns of motifs in sequences. *Comput. Applic. Biosci.* 4: 53–60.

Staden R (1989). Methods for calculating the probabilities of finding patterns in sequences. *Comput. Applic. Biosci.* 5: 89–96.

See also Box, Binding Site, Complexity, Core Consensus, Score, Sequence Logo, Sequence Walker.

CONSENSUS TREE (MAJORITY-RULE CONSENSUS TREE, STRICT CONSENSUS TREE, SUPERTREE)

Sudhir Kumar and Alan Filipski

A single tree summarizing phylogenetic relationships from multiple individual phylogenies.

Several individual phylogenetic trees may need to be combined into a single tree representing a consensus phylogeny when a tree-building method generates more than one topology. Also, trees from different genes may be combined into a consensus tree. Currently, algorithms exist to build consensus trees where each individual tree contains the same set of taxa. When the individual trees have only partial overlaps in taxa representation, the trees produced are referred to as supertrees, which are much harder to reconstruct.

A strict consensus tree is one in which groupings of taxa that appear in all individual trees are shown. A majority-rule consensus tree is one in which groupings of taxa that appear in 50% or more individual trees are shown. Consensus trees typically have multifurcations representing disagreements among the original trees.

Further Reading

Felsenstein J (1993). *PHYLIP: Phylogenetic Inference Package*. University of Washington, Seattle, WA.

Nei M, Kumar S (2000). *Molecular Evolution and Phylogenetics*. Oxford University Press, Oxford, UK.

Schuh RT (2000). *Biological Systematics: Principles and Applications*. Cornell University Press, Ithaca, NY.

Swofford DL (1998). *PAUP*: Phylogenetic Analysis Using Parsimony (and Other Methods)*. Sinauer Associates, Sunderland, MA.

Wilkinson M (1996). Majority-rule reduced consensus trees and their use in bootstrapping. *Mol. Biol. Evol.* 13: 437–444.

See also Phylogenetic Tree.

CONSERVATION
Dov S. Greenbaum

The presence of a sequence, or part thereof, in two or more genomes.

When two genes are present in two different genomes, as defined by some threshold of sequence similarity, they are termed *conserved*. There are different degrees of conservation, the higher the sequence similarity, the more conserved the two genes are, indicating the conservation of more elements (nucleotides, amino acids) between the molecular species.

Additionally, while two genes may have different sequences incorporating different amino acids at a specific position, if those amino acids are similar in their chemical properties, i.e., hydrophobic, and as such do not significantly alter the structure or function of a protein, those two genes are termed conserved. In this case what is conserved is not sequence but (potential) function. Conservation can also be observed at other levels, e.g., structure and gene organization.

Use of the term conservation implies homology between the species being considered. Surprisingly, the degree of biological sequence conservation is neatly given in bits of information. One can envision that eventually all forms of biological conservation could be measured this way.

Further Reading

Graur D, Li W-H (1999). *Fundamentals of Molecular Evolution*. Sinauer Associates, Sunderland, MA.

See also Bits, Conservation, Information, Sequence Logo.

CONSINSPECTOR
John M. Hancock and Martin J. Bishop

A program for predicting protein binding sites in DNA by finding matches to consensus sequences.

The program allows one or more sequences in a sequence file to be scanned for matches for a number of selected consensuses, produced by the program ConsInd. The program first searches for matches to the best conserved bases in the consensus. The matched sequences are then aligned to the consensus after anchoring the two sequences at these sites. The quality of match is assessed by randomly shuffling the test sequence and comparing the similarities to the consensus to random matches found in the shuffled sequences. A quality value is given as

$$Ci\ score\ \frac{sim(real) - meansim(random)}{stdsim(random)}$$

where *sim(real)* is the similarity of the real sequence to the consensus, *meansim(random)* is the mean similarity of randomly generated sites, and *stdsim(random)* is the standard deviation of the random similarities.

Related Website

ConsInspector	http://www.gsf.de/biodv/consinspector.html

Further Reading

Frech K, et al. (1993). Computer-assisted prediction, classification, and delimitation of protein binding sites in nucleic acids. *Nucleic Acids Res.* 21: 1655–1664.

Frech K, et al. (1997). ConsInspector 3.0: New library and enhanced functionality. *Comp. Appl. Biosci.* 13: 109–110.

See also GenomeInspector, MatInspector, ModelInspector, PromoterInspector.

CONTACT MAP
Andrew Harrison and Christine A. Orengo

A contact map is a two-dimensional matrix which is used to capture information on all residues which are in contact within a protein structure. The axes of the matrix are associated with each residue position in the three-dimensional structure, and cells within the matrix are shaded when the associated residues are in contact. Various criteria are used to define a residue contact. For example, the $C\beta$ atoms of two residues should be within 10 Å distance from each other. Alternatively, any atom from the two residues should be within 8 Å from each other. Depending on the fold of the protein, characteristic patterns are seen within the matrix. Contacting residues in secondary structures give rise to a thickening of the central diagonal while antiparallel secondary structures give rise to lines orthogonal to the central diagonal and parallel secondary structures lines which are parallel.

A distance plot is a related two-dimensional representation of a protein, devised by Phillips in the 1970s. In this the matrix cells are shaded depending on the distances between any two residues in the protein. Difference distance plots can be used to show changes occurring in a protein structure, e.g., on ligand binding.

See also Fold, Secondary Structure of Protein.

CONTIG
Rodger Staden

An ordered set of overlapping segments of DNA from which a contiguous sequence can be derived. Confusingly, it is presently also used as a shorthand for a single sequence without gaps. Originally (Staden, 1980) a contig was defined to be an ordered set of overlapping sequences from which a contiguous sequence can be derived. Later the meaning was broadened for use in contig mapping to be a set of overlapping clones. The newest usage (as a shorthand for a contiguous sequence) causes great confusion because the objects being defined are so closely related but very different: One is a set, the other a single sequence.

Further Reading

Staden R (1980). A new computer method for the storage and manipulation of DNA gel reading data. *Nucleic Acids Res.* 8: 3673–3694.

See also Contig Mapping.

CONTIG MAPPING (PHYSICAL MAPPING)
Rodger Staden

The process of arranging a set of overlapping clones into their correct order along the genome.

CONTINUOUS TRAIT. *SEE* QUANTITATIVE TRAIT.

CONTINUUM ELECTROSTATICS
Roland L. Dunbrack

A method for modeling electrostatic interactions in aqueous solution and other media without explicit modeling of solvent molecules. Approximate solvation and interaction energies can be obtained from numerical solution of the Poisson equation or the Poisson–Boltzmann equation when salts are present.

CONVERGENCE
Dov S. Greenbaum

Convergence refers to independent sequences attaining similar traits over evolutionary time.

Functional convergence refers to different protein folds that can converge and all attain a similar function.

Structural convergence refers to differing sequences attaining a similar three-dimensional structures. A probable example of this are alpha–beta barrels, which frequently reoccur during evolution. This is thought to their stability resulting in the structure being energetically favored.

Intuitively, there are six criteria to determine whether sequences have converged: similarity of DNA sequence, protein sequence, structure, enzyme–substrate interactions, catalytic mechanism, and the same segment in the protein responsible for the protein function.

Conversely, *divergence* refers to proteins that may have had common ancestors but have diverged over time.

Related Website

Convergent evolution of protein structures	http://pps98.man.poznan.pl/assignments/projects/debono/title.html

Further Reading
Holbrook JJ, Ingram VA (1973). Ionic properties of an essential histidine residue in pig heart lactate dehydrogenase. *Biochem. J.* 131: 729–738.

COORDINATE SYSTEM OF SEQUENCES
Thomas D. Schneider

A coordinate system is the numbering system of a nucleic acid or protein sequence.

Coordinate systems in primary databases such as GenBank and PIR are usually 1-to-n, where n is the length of the sequence, so they are not recorded in the database. In the DELILA system, one can extract sequence fragments from a larger database. If one does two extractions, then one can go slightly crazy trying to match up sequence coordinates if the numbering of the new sequence is still 1-to-n. The DELILA system handles all continuous coordinate systems, both linear and circular, as described in LIBDEF, the definition of the DELILA database system. For example, on a circular sequence running from 1 to 100, the DELILA instruction "get from 10 to 90 direction -;" will give a coordinate system that runs from 10 down to 1 and then continues from 100 down to 90.

Unfortunately there are many examples in the literature of nucleic acid coordinate systems without a zero coordinate. A zero base is useful when one is identifying the locations of sequence walkers: The location of the predicted binding site is the zero base of the walker (the vertical rectangle). Without a zero base, it would tricky to determine the positions of bases in a sequence walker. With a zero base it is quite natural. Insertion or deletions will make holes or extra parts of a coordinate system. The DELILA system cannot handle these (yet). In the meantime, the sequences are renumbered to create a continuous coordinate system.

Related Website

libdef	http://www.lecb.ncifcrf.gov/~toms/libdef.html

Further Reading
Schneider TD, et al. (1982). A design for computer nucleic-acid sequence storage, retrieval and manipulation. *Nucleic Acids Res.* 10: 3013–3024.

CORE CONSENSUS
Thomas D. Schneider

A core consensus is the strongly conserved portion of a binding site, found by creating a consensus sequence. It is an arbitrary definition, as can be seen from the examples in the sequence logo gallery (see Figure 3 under Binding Site Symmetry). The sequence conservation (measured in bits of information) often follows the cosine waves that represent the twist of B-form DNA. This has been explained by noting that a protein bouncing in and out from DNA must evolve contacts (Schneider, 2001). It is easier to evolve DNA contacts that are close to the protein than those that are further around the helix. Because the sequence conservation varies continuously, any cutoff or "core" is arbitrary. This concept can be replaced by using sequence logos and sequence walkers.

Related Website

| oxyr | http://www.lecb.ncifcrf.gov/~toms/paper/oxyr/ |

Further Reading

Schneider TD (1996). Strong minor groove base conservation in sequence logos implies DNA distortion or base flipping during replication and transcription initiation. *Nucleic Acid Res.* 29: 4881–4891. http://www.lecb.ncifcrf.gov/~toms/paper/baseflip/

See also Bit, Information, Consensus Sequence, Sequence Logo, Sequence Walker.

COVARIATION ANALYSIS

Jamie J. Cannone and Robin R. Gutell

Technique for predicting RNA structure making use of evolutionary conservation of base-pairing interactions.

While comparative sequence analysis is based on the simple proposition that molecules with the same function will have similar secondary and tertiary structures, covariation analysis, a subset of comparative sequence analysis, identifies base pairs that occur at the same positions in the RNA sequence in all of the RNA sequences in the data set. Covariation analysis searches for positions that have the same pattern of variation in an alignment of sequences (see the accompanying figure). The most recent implementation of this method usually base pairs any two positions with the same pattern of variation, regardless of the types of base pairs. While most of the base-pair types that are identified exchange between G:C, A:U, and G:U, covariation analysis has also identified exchanges between noncanonical base pairs.

For example, the following four sets of base-pair exchanges show covariation: (1) A:U ↔ U:A ↔ G:C ↔ C:G, (2) G:U ↔ A:C, (3) U:U ↔ C:C, and (4) A:A ↔ G:G. By searching for these coordinated positional variations in a well-aligned collection of sequences, key elements of an RNA molecule's core structure can be elucidated. Earlier covariation methods searched specifically for helices composed of canonical base pairs. Improvements in covariation algorithms and an ever-growing collection of sequences make comprehensive searches that consider all base-pairing types in a context-independent manner possible. Due to the requirement that the two base-paired positions have the same pattern of variation, covariation analysis will only identify a subset of the total number of base pairs that are in common to different sequences; other comparative methods must be employed to detect the remainder.

Our confidence in the prediction of a base pair with covariation analysis is directly proportional to the dependence of the two "paired" positions. Positions that vary independently of one another are less likely to form a base pair that can be predicted with covariation analysis. In contrast, a greater extent of simultaneous variation at the two paired positions could indicate that the two positions are dependent on one another, and thus we are more confident in base pairs predicted with covariation analysis. One of the family of methods that measures the dependence/independence between two positions that are proposed to be base paired is the chi-square statistic, which gauges the types of base pairs and their frequencies (see Phylogenetic Events Analysis). The accuracy of the base-pair predictions with covariation analysis is very high: Approximately 97% to 98% of the 16S and 23S ribosomal RNA (rRNA) base pairs predicted with covariation analysis are present in the high-resolution crystal structures.

An important facet of covariation analysis is that the current covariation methods are comparing all positions in a sequence alignment, independently of previous predictions, structural context, and principles of RNA structure. In practice, the majority of covariations involve two single nucleotides exchanging to maintain a canonical base pair and are arranged into standard secondary-structure helices. Thus, covariation analyses have independently determined the two most fundamental principles in RNA structure: the Watson–Crick base-pairing relationship and the formation of helices from the antiparallel and consecutive arrangement of these base pairs. In addition to these achievements, a significant number of examples of both covariation between canonical (A:U, G:C, and G:U) and noncanonical base pairs and covariation between noncanonical base-pair types have been predicted for the rRNAs and proven correct by the ribosomal subunit crystal structures. Likewise, examples of tertiary base pairs that are not part of larger helices, both short and long range, have been predicted and shown to exist.

Related Website

Comparative RNA Web (CRW)	http://www.rna.icmb.utexas.edu/

See also Prediction, Phylogenetic Events Analysis, RNA Structure.

A

Sequence 1:	UAGCGAA	nnnnnnn	AUCGCUU
Sequence 2:	UAACAAG	nnnnnnn	GUUGUUU
Sequence 3:	CAGCAGG	nnnnnnn	GCUGCUC
Sequence 4:	CAGGAGG	nnnnnnn	GCUCCUC
Sequence 5:	UAAGAAA	nnnnnnn	AUUCUUU
Position		11111	1111122
Numbers:	1234567	8901234	5678901

B
1:21
U:U, C:C
2:20
A:U
3:19
G:C, A:U
4:18
C:G, G:C
5:17
G:C, A:U
6:16
A:U, G:C
7:15
A:A, G:G

Figure Examples of covariation. A. Schematic alignment. Five sequences are shown from 5′ at left to 3′ at right. Black and red lines above the alignment show base-pairing. Nucleotide position numbers appear in blue at the bottom of the alignment. B. Summary of covariations from the alignment in panel A. Position numbers are followed by the observed base-pair types for the seven base-pairs.

CPG ISLAND
Katheleen Gardiner

A 200–2000-bp segment of genomic DNA with base composition >60% GC and CpG dinucleotide frequency 0.6 observed/expected.

These characteristics stand out in the mammalian genome where the average base composition is ~38% GC and the average CpG frequency is 0.2–0.25 observed/expected. CpG islands are found more often in GC-rich R bands and are most often unmethylated, located at or near the 5' ends of genes and associated with promoters.

Further Reading
Antequera F, Bird A (1993). Number of CpG islands and genes in human and mouse. *Proc. Natl. Acad. Sci.* 90: 11,995–11,999.

Ioshikhes IP, Zhang MQ (2000). Large-scale human promoter mapping using CpG islands. *Nat. Genet.* 26: 61–63.

Takai D, Jones PA (2002). Comprehensive analysis of CpG islands in human chromosomes 21 and 22. *Proc. Natl. Acad. Sci.* 99: 3740–3745.

See also Promoter.

CRM. *SEE* CIS-REGULATORY MODULE.

CROSS-VALIDATION (*K*-FOLD VALIDATION, LEAVE ONE OUT)
Nello Cristianini

A method for estimating the generalization performance of a machine learning algorithm.

This is done by randomly dividing the data into k mutually exclusive subsets (the "folds") of approximately equal size. The algorithm is trained and tested k times: Each time it is trained on the data set minus a fold and tested on that fold (called also the held-out set). The performance estimate is the average performance for the k folds. When as many folds as data points are used, the method reduces to the procedure known as the leave-one-out, or jackknife, method.

Further Reading
Mitchell T (1997). *Machine Learning*. McGraw Hill, New York.

See also Bootstrap Analysis, Error Measures, Jackknife.

CRYO-BUFFER
Liz P. Carpenter

A cryo-buffer is a solution which will form a glass when the temperature is rapidly dropped to below its freezing point, instead of forming ice, as is observed with many solutions.

Cryo-buffers are used in X-ray crystallography to soak crystals of macromolecules prior to freezing at −173°C (100 K). Many crystals survive longer in the X-ray beam at −173°C than at room temperature or 4°C. If, however, ice forms inside the crystal during freezing, this will destroy the internal structure of the crystal as the ice expands and diffraction will be lost. If a crystal is grown or soaked in a suitable cryo-buffer solution before it is frozen, then an amorphous glass is formed instead of ice crystals and the ordered structure of the crystal is preserved. Cryo-buffers have to be carefully selected so that they do not destroy the crystal themselves. Glycerol, certain alcohols, low-molecular-weight polyethylene glycols, and sugars, often at concentrations of 20% to 30%, have been found to be suitable cryo-buffers for a variety of macromolecular crystals. Certain oils can also be used.

Related Websites

Flash Cooling: A Practical Guide by Gitay Kryger	http://www.rose.brandeis.edu/PRLab/Crystallizations/cool/
Oxford cryosystems tutorial guides	http://www.oxfordcryosystems.co.uk/cryo/addons/guide/index.html
Hakon Hopes Cryonotes	http://www-structure.llnl.gov/Xray/cryo-notes/Cryonotes.html

See also Cryo-Cooling of Protein Crystals; Crystal, Macromolecular; Diffraction of X Rays, Unit Cell, X-Ray Crystallography for Structure Determination.

CRYO-COOLING OF PROTEIN CRYSTALS
Liz P. Carpenter

Macromolecular crystals are used in X-ray diffraction experiments for the solution of the three-dimensional structure of the molecule. Crystals are often unstable in a beam of X rays, so that over time the resolution of the diffraction pattern will become worse and the data may become unusable. To overcome this problem, crystals are now routinely stored in liquid nitrogen or in gaseous nitrogen gas at 100 K ($-173°C$).

Ice formation in the crystals when they are stored will destroy the structure of the crystals since the volume increases as the water freezes. If, however, the crystal is grown or soaked in a cryo-buffer (see Cryo-Buffer), then the crystal should survive freezing and will last appreciably longer in the X-ray beam.

Related Websites

Flash Cooling: A Practical Guide by Gitay Kryger	http://www.rose.brandeis.edu/PRLab/Crystallizations/cool/
Oxford cryosystems tutorial guides	http://www.oxfordcryosystems.co.uk/cryo/addons/guide/index.html
Hakon Hopes Cryonotes	http://www-structure.llnl.gov/Xray/cryo-notes/Cryonotes.html

Further Reading
Garman EF, Schneider TR (1997). Macromolecular cryocrystallography. *J. Appl. Cryst.* 30: 211–237.

Watenpaugh K (1991). Macromolecular crystallography at cryogenic temperatures. *Curr. Opin. Struct. Biol.* 1: 1012–1015.

See also Cryo-Buffer; Crystal, Macromolecular; X-Ray Crystallography for Structure Determination.

CRYSTAL, MACROMOLECULAR
Liz P. Carpenter

A crystal is an ordered three-dimensional array of molecules necessary to solve the three-dimensional structure of a macromolecule by X-ray crystallography.

The molecule or molecules in a crystal could be a small molecule such as a drug or enzyme substrate; a macromolecule such as a protein, DNA, or RNA; a dimer or multimer of a single macromolecule; a complex of a protein bound to a drug; a complex of a protein bound to DNA or RNA; or several proteins forming a complex. The crystals used for macromolecular crystallography typically have dimensions of between 5 μm (10^{-6} m) and a few millimeters on each edge.

Each crystal consists of a repeat unit called a unit cell which is duplicated many thousands of times. In order for the crystal to form, each copy of the unit cell must be very nearly identical. Each unit cell is a parallelepiped with edges of length a, b, and c and angles α, β, and γ between the edges. The angle α is between the b and c edges, β is the angle between the a and c edges, and γ is the angle between the a and b edges. The entire crystal can be constructed by translations of an integer number of unit cell lengths along the direction of these cell edges.

The accompanying figure shows the crystallographic unit cell with cell dimensions a, b, and c and angles α, β, and γ.

The simplest type of crystallographic unit cell, called a triclinic cell, occurs when $a \neq b \neq c$ and $\alpha \neq \beta \neq \gamma$. The α, β, and γ angles can be 90° or 120° and two or all of the unit cell lengths can be the same, leading to higher levels of symmetry within the crystal.

Generally, one or more copies of the molecule or molecules under study are contained within the unit cell. The crystal is held together by chemical bonds between adjacent molecules in adjacent unit cells. These bonds can be salt bridges, hydrogen bonds, hydrophobic interactions, and occasionally covalent bonds. The regions of crystals where there are interactions between adjacent molecules as a result of the formation of the crystal are called crystal contacts. Interactions seen in these regions are often the result of the crystal structure and may not be significant for the function of the molecule.

See also Space Group, Unit Cell, X-Ray Crystallography for Structure Determination.

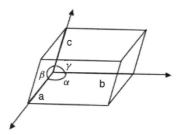

Figure Crystallographic unit cell.

CRYSTAL SYMMETRY. *SEE* SPACE GROUP.

CRYSTALLIZATION OF MACROMOLECULES
Liz P. Carpenter

In order to solve the three-dimensional structure of a protein by X-ray crystallography, it is necessary to obtain well-ordered crystals of the molecule or complex. This is achieved by purifying milligram quantities of the molecule in solution, concentrating the solution to near saturation and then adjusting the conditions until crystals are obtained. Important parameters that have to be optimized to obtain crystals include the amount of the precipitant (salts, alcohols, and various polyethylene glycols), the pH, the buffer, additives (metal ions, heavy atoms, organic molecules, detergents, inhibitors), the temperature, the physical setup of the experiment, and the macromolecule concentration. Two recent advances in the area of crystallization has been the introduction of sparse matrix screening with commercial kits consisting of diverse conditions that have given crystals with other macromolecules and the introduction of robotics to produce crystallization trays with nanoliter volume in the drops. Crystallization remains the limiting step, however, in the solution of macromolecular structures by X-ray crystallography.

Related Websites

Hampton	http://www.hamptonresearch.com/
Emerald	http://www.decode.com/emeraldbiostructures
Enrico Stura's *Protein Crystallisation: Theory and Practice*	http://www.bioc.rice.edu/~berry/papers/ crystallization/crystallization.html
Crystallization lecture by Airlie McCoy	http://perch.cimr.cam.ac.uk/Course/Crystals/intro.html

Further Reading
Bergfors T (Ed.) (1999). *Protein Crystallization Strategies, Techniques, and Tips. A Laboratory Manual.* International University Line, La Jolla, CA.

CRYSTALLOGRAPHIC SYMMETRY. *SEE* SPACE GROUP.

C-TERMINUS
Roman A. Laskowski

In a polypeptide chain the C-terminus (carboxy-terminus) is the end residue that has a free carboxyl group (COOH). The other end is referred to as the N-terminus. The direction of the chain is defined as running from the N- to the C-terminus. C-terminal means of or relating to the C-terminus.

Further Reading
Branden C, Tooze J (1991). *Introduction to Protein Structure*. Garland Science, New York.

See also N-Terminus, Polypeptide, Amino Acid.

C VALUE. *SEE* GENOME SIZE.

C-VALUE PARADOX
Katheleen Gardiner

The paradox that the genome size (*C* values) of a species does not correlate with its relative evolutionary complexity or gene number.

For example, the *C* value of human and other mammals is ~3 billion bp and for birds it is 1−2 billion, while for many fish it is ~6 billion and for many amphibians it is 20−80 billion. It is unlikely that a greater number of genes is required to produce an amphibian than a mammal. It is assumed that the excess genome size is due to repetitive DNA sequences or other sequences of nonprotein coding, nonregulatory function.

Further Reading
Ridley M (1996). *Evolution*. Blackwell Science, Cambridge, MA.

See also Repeat Sequence.

CYSTEINE
Jeremy Baum

Cysteine is a nonpolar amino acid with side chain $-CH_2SH$ found in proteins. In sequences, written as Cys or C.

Further Reading
Branden C, Tooze J (1991). *Introduction to Protein Structure*. Garland Science, New York.

See also Amino Acid.

CYSTINE. *SEE* DISULFIDE BRIDGE.

D

DAG
DAGEdit
DALI
DAML+OIL
Data Integration
Data Mining
Dayhoff Amino Acid Substitution Matrix
dbEST
dbSNP
dbSTS
DDBJ
Dead-End Elimination Algorithm
Decision Tree
Degree of Genetic Determination
DELILA
DELILA Instructions
Dependent Variable
Description Logic
Descriptor
Diagnostic Performance
Diagnostic Power
Dielectric Constant
Diffraction of X Rays
Dihedral Angle
Dinucleotide Frequency
DIP
Directed Acyclic Graph
DISCOVER

Discriminant Analysis
Discriminating Power
Disulfide Bridge
DL
DNA Array
DNA Databank of Japan,
DNA−Protein Coevolution
DNA Replication
DNA Sequence
DNA Sequencing
DNA Sequencing Artifact
DNA Sequencing Error
DNA Sequencing Trace
DNA Structure
DnaSP
DOCK
Docking
Domain
Domain Alignment
Domain Family
Donor Splice Site
Dot Matrix
Dot Plot
Dotter
Double Stranding
Downstream
Dynamic Programming

Dictionary of Bioinformatics and Computational Biology. Edited by Hancock and Zvelebil
ISBN 0-471-43622-4   © 2004 John Wiley & Sons, Inc.

DAG. *SEE* DIRECTED ACYCLIC GRAPH.

DAGEDIT
Robert Stevens

DAGEdit is a Java application to browse, query, and edit a collection of concepts arranged in a directed acyclic graph (DAG).

It has been developed to allow the Gene Ontology editorial team to maintain the Gene Ontology but can be used for any similarly structured ontology. It can load and save ontologies in a variety of formats, including the Gene Ontology Consortium's own text format, an RDF-based format, and a relational database–based format. It provides facilities for adding synonyms, textual definitions for each concept, and cross-references to the sources of those definitions. The most current version of DAGEdit can be downloaded from the publicly accessible source repository at http://sourceforge.net/projects/geneontology.

DALI
John M. Hancock and Martin J. Bishop

DALI aligns protein structures in three dimensions.

The program aligns structures by representing them as two-dimensional matrices of residue–residue distances. It then attempts to superimpose these matrices optimally by breaking the main matrix into submatrices of a defined size and screening for the best match of these submatrices between proteins. Overlapping matched submatrices are then combined to produce an optimal alignment for the entire protein. By optimizing a number of alignments in parallel, the algorithm is able to account for internal repeats.

The program is made available via a network server that can deliver results of either pairwise alignments or alignment of an input protein structure to the Protein Data Bank (PDB) database by email or over the World Wide Web.

Related Website

DALI	http://www.ebi.ac.uk/dali/

Further Reading

Holm L, Sander C (1993). Protein structure comparison by alignment of distance matrices. *J. Mol. Biol.* 233: 123–138.

See also VAST.

DAML+OIL
Robert Stevens

DAML+OIL is a language based on description logic and is used to formally describe an ontology.

Its syntax has much in common with frame-based languages and has a clean and well-defined semantics. DAML+OIL can be used to represent simple, hand-crafted

Dictionary of Bioinformatics and Computational Biology. Edited by Hancock and Zvelebil
ISBN 0-471-43622-4   © 2004 John Wiley & Sons, Inc.

asserted taxonomies of classes. In addition, classes can be described in terms of their properties by relating classes to other concepts within the ontology. DAML+OIL has a mapping to an expressive description logic so that a reasoner can be used to check the logical consistency of the ontology and to infer the subsumption hierarchy implied by the class descriptions.

DATA INTEGRATION
Dov S. Greenbaum

The process and technology of linking together databases, often containing different classes of data, to enable detection of patterns across data sets.

One of the main goals of bioinformatics is to determine the function of every gene and protein. In many cases, though, it is difficult and impractical to directly determine the function of a specific protein. Additionally, there is a lot of information being produced by high-throughput experiments that do not directly apply to function but are associated with protein function. These diverse data can be integrated and, through integration, a researcher may have enough information to create a plausible annotation.

The paradigm of data integration results from the idea that many independent facts can be combined and used to determine the function or some other feature of a novel gene through a computational framework such as a Bayesian net.

Given the heterogenous nature of the data and the methods used to store the data, this still remains a nontrivial problem.

Further Reading
Marcotte E, Date S (2001). Exploiting big biology: Integrating large-scale biological data for function inference. *Brief Bioinform*. 2: 363–374.
Stein LD (2003). Integrating biological databases. *Nature Rev. Genet.* 4: 337–345.

DATA MINING. *SEE* PATTERN ANALYSIS.

DAYHOFF AMINO ACID SUBSTITUTION MATRIX (PAM MATRIX, PERCENT ACCEPTED MUTATION MATRIX)
Laszlo Patthy

Amino acid substitution matrices describe the probabilities and patterns displayed by nonsynonymous mutations of nucleotide sequences during evolution. Amino acid substitution probabilities in proteins were analyzed by Dayhoff (1978), who tabulated nonsynonymous mutations observed in several different groups of global alignments of closely related protein sequences that are at least 85% identical. Mutation data matrices, or percent accepted mutation (PAM) matrices, commonly known as Dayhoff matrices, were derived from the observed patterns of nonsynonymous substitutions, and matrices for greater evolutionary distances were extrapolated from those for lesser ones.

The observed mutational patterns have two distinct aspects: the resistance of an amino acid to change and the pattern observed when it is changed. The data collected on a large number of protein families have shown that the relative mutabilities of the different amino acids show striking differences: On the average, Asn, Ser, Asp, and Glu are most mutable, whereas Trp, Cys, Tyr, and Phe are the least mutable. The relative immutability of cysteine can be interpreted as a reflection of the fact that it has several unique, indispensable functions that no other amino acid side chain can mimic (e.g., it is the only amino acid that can form disulfide bonds). The low mutability of Trp, Tyr, and Phe may be explained by the importance of these hydrophobic residues in protein folding.

The distribution of accepted amino acid replacement mutations suggests that the major cause of preferences in amino acid substitutions is that the new amino acid must function in a way similar to the old one. Accordingly, conservative changes to chemically and physically similar amino acids are more likely to be accepted than radical changes to chemically dissimilar amino acids. Some of the key properties of an amino acid residue that determine its role and replaceability are size, shape, polarity, electric charge, and ability to form salt bridges, hydrophobic bonds, hydrogen bonds, and disulfide bonds.

In general, chemically similar amino acids tend to replace one another: within the aliphatic group (Met, Ile, Leu, Val); within the aromatic group (Phe, Tyr, Trp); within the basic group (Arg, Lys); within the acid, acid–amide group (Asn, Asp, Glu, Gln); within the group of hydroxylic amino acids (Ser, Thr); etc. Cysteine practically stands alone. Glycine–alanine interchanges seem to be driven by selection for small side chains, proline–alanine interchanges by selection for small aliphatic side chains, etc.

Dayhoff amino acid substitution matrices are used in the alignment of amino acid sequences to calculate alignment scores and in sequence similarity searches.

Further Reading
Dayhoff MO (1978). Survey of new data and computer methods of analysis. In *Atlas of Protein Sequence and Structure*, Vol. 5, Suppl. 3, MO Dayhoff, Ed. National Biomedical Research Foundation, Washington, DC, pp. 1–8.

Dayhoff MO, et al. (1978). A model of evolutionary change in proteins. In *Atlas of Protein Sequence and Structure*, Vol. 5, Suppl. 3. National Biomedical Research Foundation, Washington, DC. Dayhoff MO, et al. (1983). Establishing homologies in protein sequences. *Methods Enzymol.* 91: 524–545.

George DG, et al. (1990). Mutation data matrix and its uses. *Methods Enzymol.* 183: 333–351.

See also Alignment of Protein Sequences, Alignment Score, Amino Acid Exchange Matrix, BLOSUM Matrix, Global Alignment, Nonsynonymous Mutation, Sequence Similarity Search.

DBEST
Guenter Stoesser

A database containing sequence data and mapping information on expressed sequence tags (ESTs) from various organisms.

ESTs are typically short (about 300–500 bp), single-pass sequence reads from complementary DNA (cDNA) (mRNA) and are tags of expression for a given cDNA library. ESTs have applications in the discovery of new genes, mapping of genomes, and identification of coding regions in genomic sequences. Sequence data from dbEST are incorporated into the EST division of DDBJ/EMBL/GenBank. dbEST is maintained at the National Center for Biotechnology Information (NCBI), Bethesda, Maryland.

Related Website

| dbEST | http://www.ncbi.nlm.nih.gov/dbEST/ |

Further Reading
Adams MD, et al. (1991). Complementary DNA sequencing: Expressed sequence tags and human genome project. *Science* 252: 1651–1656.
Boguski MS (1995). The turning point in genome research. *Trends Biochem. Sci.* 20: 295–296.
Boguski MS, et al. (1993). dbEST-database for expressed sequence tags. *Nat. Genet.* 4: 332–333.
Schuler GD (1997). Pieces of the puzzle: Expressed sequence tags and the catalog of human genes. *J. Mol. Med.* 75: 694–698.

DBSNP

Guenter Stoesser

A central repository for genetic variations.

The vast majority (>99%) of variations described in dbSNP are single-nNucleotide polymorphisms (SNPs); most of the rest are small insertions and deletions. A SNP (pronounced "snip") is a small genetic change, or variation that can occur within an organism's DNA.sequence. SNPs occur roughly every 100–300 bp in human chromosome sequences.

dbSNP entries include the sequence information around the polymorphism, the specific experimental conditions necessary to perform an experiment, descriptions of the population containing the variation, and frequency information by population or individual genotype. SNP data facilitate large-scale association genetics studies for associating sequence variations with heritable phenotypes. Furthermore, SNPs facilitate research in functional and pharmaco-genomics, population genetics, evolutionary biology, positional cloning, and physical mapping. As an integrated part of the National Center for Biotechnology Information (NCBI), the contents of dbSNP are cross-linked to records in GenBank, LocusLink, the human genome sequence, and PubMed. The result sets from queries in any of these resources point back to the relevant records in dbSNP. The BLAST algorithm compares a query sequence against all flanking sequence records in dbSNP. dbSNP is collaborating to develop data exchange protocols with other public variation and mutation databases, such as HGBASE (human genic biallelic sequences). dbSNP is maintained at NCBI, Bethesda, Maryland.

Related Website

| dbSNP | http://www.ncbi.nlm.nih.gov/SNP |

Further Reading
Osier MV, et al. (2001). ALFRED: An allele frequency database for diverse populations and DNA polymorphisms—An update. *Nucleic Acids Res.* 29: 317–319.
Sherry ST, et al. (1999). dbSNPDatabase for single nucleotide polymorphisms and other classes of minor genetic variation. *Genome Res.* 9: 677–679.
Sherry ST, et al. (2001). dbSNP: The NCBI database of genetic variation. *Nucleic Acids Res.* 29: 308–311.

DBSTS
Guenter Stoesser

A database of sequence-tagged sites (STSs)—short genomic landmark sequences which are used to create high-resolution physical maps of large genomes and to build scaffolds for organizing large-scale genome sequencing.

All STS sequences are incorporated into the STS division of DDBJ/EMBL/GenBank. dbSTS is maintained at the National Center for Biotechnology Information (NCBI), Bethesda, Maryland.

Related Website

dbSTS	http://www.ncbi.nlm.nih.gov/dbSTS/

Further Reading

Benson D, et al. (1998). GenBank. *Nucleic Acids Res.* 26: 1–7.

Olson M, et al. (1989). A common language for physical mapping of the human genome. *Science* 245: 1434–1435.

DDBJ (DNA DATABANK OF JAPAN)
Guenter Stoesser

The DNA Databank of Japan (DDBJ) is a member of the tripartide International Nucleotide Sequence Database Collaboration DDBJ/EMBL/GenBank, together with GenBank (United States) and the EMBL nucleotide sequence database.

Collaboratively, the three organizations collect all publicly available nucleotide sequences and exchange sequence data and biological annotations on a daily basis. Major data sources are individual research groups, large-scale genome sequencing centers, and the Japanese Patent Office (JPO). Sophisticated submission systems are available (Sakura). Specialized databases provided by DDBJ include the Genome Information Broker (GIB) for complete genome sequence data and the Human Genomics Studio (HGS). The database is maintained at the Center for Information Biology (CIB), a division of the National Institute of Genetics (NIG) in Mishima, Japan. Additional to operating DDBJ, CIB's mission is to carry out research in information biology.

Related Websites

Center for Information Biology	http://www.cib.nig.ac.jp
DDBJ	http://www.ddbj.nig.ac.jp
DDBJ submission systems and related information	http://www.ddbj.nig.ac.jp/sub-e.html
Genome Information Broker (GIB)	http://gib.genes.nig.ac.jp
Human Genomics Studio	http://studio.nig.ac.jp
Data analysis	http://www.ddbj.nig.ac.jp/analyses-e.html

Further Reading

Tateno Y, Gojobori T. (1997). DNA Data Bank of Japan in the age of information biology. *Nucleic Acids Res.* 25: 14–17.

Tateno Y, et al. (2002a). DNA Data Bank of Japan (DDBJ) for genome scale research in life science. *Nucleic Acids Res.* 30: 27–30.

Tateno Y, et al. (2002b). DNA Data Bank of Japan (DDBJ) in collaboration with mass sequencing teams. *Nucleic Acids Res.* 28: 24–26.

DEAD-END ELIMINATION ALGORITHM
Roland L. Dunbrack

An algorithm for side-chain placement onto protein backbones that eliminates rotamers for some side chains that cannot be part of the global minimum-energy configuration, originally developed by Johan Desmet and colleagues. This method can be used for any search problem that can be expressed as a sum of single-side-chain terms and pairwise interactions. Goldstein's improvement on the original dead-end elimination (DEE) algorithm can be expressed as follows. If the total energy for all side chains is expressed as the sum of singlet and pairwise energies,

$$E = \sum_{i=1}^{N} E_{bb}(r_i) + \sum_{i=1}^{N-1} \sum_{j>i}^{N} E_{sc}(r_i, r_j)$$

then a rotamer r_i can be eliminated from the search if there is another rotamer s_i for the same side chain that satisfies the following equation:

$$E_{bb}(r_i) - E_{bb}(s_i) + \sum_{j=1, j\neq i}^{N} \min_{r_j}\{E_{sc}(r_i, r_j) - E_{sc}(s_i, r_j)\} > 0$$

In words, rotamer r_i of residue i can be eliminated from the search if another rotamer of residue i, s_i, always has a lower interaction energy with all other side chains regardless of which rotamer is chosen for the other side chains. More powerful versions have been developed that eliminate certain pairs of rotamers from the search. DEE-based methods have also proved very useful in protein design, where there is variation of residue type as well as conformation at each position of the protein.

The algorithm can also be applied to other situations where the energy function can be defined in terms of pairwise interactions, such as sequence alignment.

Related Website

Canonical loop modeling using dead end	http://antibody.bath.ac.uk/index.html

Further Reading

De Maeyer M, et al. (2000). The dead-end elimination theorem: Mathematical aspects, implementation, optimizations, evaluation, and performance. *Methods Mol. Biol.* 143: 265–304.

Desmet J, De Maeyer M, Hazes B, Lasters I (1992). The dead-end elimination theorem and its use in protein sidechain positioning. *Nature* 356: 539–542.

Goldstein RF (1994). Efficient rotamer elimination applied to protein side-chains and related spin glasses. *Biophys. J* 66: 1335–1340.

Looger LL, Hellinga HW (2001). Generalized dead-end elimination algorithms make large-scale protein side-chain structure prediction tractable: Implications for protein design and structural genomics. *J. Mol. Biol.* 307: 429–445.

See also Rotamer.

DECISION TREE
John M. Hancock

A graphical representation of a procedure for classifying or evaluating an item of interest.

A decision tree can be used to classify observations according to predetermined rules, e.g., classifying a set of microarray data into normal or diseased or into different types of cancer with similar or identical pathology.

A more sophisticated use is to use decision trees as machine learning tools by learning the content of the nodes and the structure of the graph. This can permit supervised learning of rules that can distinguish samples, such as different classes of cancerous cells.

Related Website

Introduction	http://www.aaai.org/AITopics/html/trees.html

Further Reading

Zhang H, et al. (2003). Cell and tumor classification using gene expression data: Construction of forests. *Proc. Natl. Acad. Sci. USA.* 100: 4168–4172.

DEGREE OF GENETIC DETERMINATION. *SEE* HERITABILITY.

DELILA
Thomas D. Schneider

DELILA stands for DEoxyribonucleic acid LIbrary LAnguage.

It is a language for extracting DNA fragments from a large collection of sequences, invented around 1980. The idea is that there is a large database containing all the sequences one would like, which we call a "library." One would like a particular subset of these sequences, so one writes up some instructions and gives them to the librarian, Delila, which returns a "book" containing just the sequences one wants for a particular analysis. So "Delila" also stands for the program that does the extraction. Since it is easier to manipulate Delila instructions than to edit DNA sequences, one makes fewer mistakes in generating one's data set for analysis, and they are trivial to correct. Also, a number of programs create instructions, which provides a powerful means of sequence manipulation. One of Delila's strengths is that it can handle any continuous coordinate system. The "Delila system" refers to a set of programs that use these sequence subsets for molecular information theory analysis of binding sites and proteins. Delila is capable of making sequence mutations, which can be displayed graphically along with sequence walkers.

A complete definition for the language is available (LIBDEF), although not all of it is implemented. There are also tutorials on building Delila libraries and using Delila instructions. A Web-based Delila server is available.

Related Websites

NCBI	http://www.ncbi.nlm.nih.gov/
GenBank	http://www.ncbi.nih.gov/Genbank/GenbankOverview.html
DELILA	http://www.lecb.ncifcrf.gov/~toms/delila/delila.html
LIBDEF	http://www.lecb.ncifcrf.gov/~toms/libdef.html
DELILA server	http://www.lecb.ncifcrf.gov/~toms/delilaserver.html

Further Reading
Schneider TD, et al. (1982). A design for computer nucleic-acid sequence storage, retrieval and manipulation. *Nucleic Acids Res.* 10: 3013–3024.
Schroeder JL, Blattner FR (1982). Formal description of a DNA oriented computer language. *Nucleic Acids Res.* 10: 69–84.

See also Binding Sites, Coordinate System, Molecular Information Theory, Sequence Walker.

DELILA INSTRUCTIONS
Thomas D. Schneider

Delila instructions are a set of detailed instructions for obtaining specific nucleic acid sequences from a sequence database. The instructions are written in the computer language Delila. There is a tutorial on using Delila instructions available online.

Related Website

DELILA instructions	http://www.lecb.ncifcrf.gov/~toms/delilainstructions.html

Further Reading
Schneider TD, et al. (1982). A design for computer nucleic-acid sequence storage, retrieval and manipulation. *Nucleic Acids Res.* 10: 3013–3024.

See also DELILA.

DEPENDENT VARIABLE. *SEE* LABEL.

DESCRIPTION LOGIC (DL)
Robert Stevens

Description logics (DLs) are a form of knowledge representation language describing categories, individuals, and the relationships amongst them in a logical formalism.

"They are sub-sets of standard predicate logic. They attempt to give a unified logical basis to the various well known traditions of knowledge representation: frame-based systems, semantic networks" (Description Logic home page, see Related Website below). They are derived from work on the KL-ONE family of knowledge representation languages. Different DLs have different levels of expressiveness. All have some operations for defining new concepts from existing concepts and the relations among them. The operators available almost always include conjunction and existential quantification and may include negation, disjunction, universal quantification, and other logical operations on concepts on relations. Due to the logical formalism of the DL, ontologies expressed in a DL can be submitted to a reasoning application. The reasoner can check the logical consistency of the descriptions (concept satisfiability) and infer a subsumption hierarchy (taxonomy) from the description (subsumption reasoning). "Description logic systems have been used for building a variety of applications including conceptual modeling, information integration, query mechanisms, view maintenance, software management systems, planning systems, configuration systems and natural language understanding" (Description Logic home page).

Related Website

Patrick Lambrix, Description Logic	http://www.ida.liu.se/labs/iislab/people/DL/

DESCRIPTOR. *SEE* FEATURE.

DIAGNOSTIC PERFORMANCE. *SEE* DIAGNOSTIC POWER.

DIAGNOSTIC POWER (DIAGNOSTIC PERFORMANCE, DISCRIMINATING POWER)
Terri Attwood

A measure of the ability of a discriminator to identify true matches and to distinguish them from false matches. A discriminator is a mathematical abstraction (e.g., a regular expression, profile, fingerprint) of a conserved motif or set of motifs used to search either an individual query sequence or a full database for the occurrence of that same or similar motif(s).

True-positive matches to a discriminator are true family members that are correctly diagnosed; true-negatives are genuine nonfamily members that are not matched by the discriminator; false-positive matches are nonfamily members that are incorrectly matched; and false-negative matches are true family members that fail to be diagnosed.

The goal of sequence analysis methods is to improve diagnostic performance, to capture all (or most) true-positive family members, to include no (or few) false-positive family members, and hence to minimize or preclude false negatives. Ideally, there should be perfect separation between true-positive and true-negative results. In practice, however, the populations of each overlap, and scoring thresholds are often used to find the appropriate balance between the ability of the discriminator to capture the majority of true-positive matches with the inclusion of the minimum of false-positives. Given the diagnostic limitations of most sequence analysis methods, it is good practice to use several approaches and to evaluate the reliability of any diagnosis by seeking a consensus.

Related Website

Diagnostic performance illustration	http://www.bioinf.man.ac.uk/dbbrowser/bioactivity/curves.gif

Further Reading
Attwood TK, Parry-Smith DJ (1999). *Introduction to Bioinformatics*. Addison Wesley Longman, Harlow, Essex, United Kingdom.

See also Fingerprint, Motif, Profile, Regular Expression.

DIELECTRIC CONSTANT
Jeremy Baum

The dielectric constant of a medium is a measure of the shielding effect that the medium has on electric fields. It is a scalar quantity, usually denoted by ε or, in the case of a vacuum, by ε_0. It is a macroscopic quantity indicative of the polarizability of the medium.

See also Electrostatic Potential.

DIFFRACTION OF X RAYS
Liz P. Carpenter

X rays are a type of electromagnetic radiation. All waves have three properties: wavelength, amplitude, and phase. The wavelength (λ) is the distance over which the wave repeats itself. X rays have wavelengths between 0.1×10^{-10} and 100×10^{-10} m (0.5–2 Å, where 1 Å $= 10^{-10}$ m). They lie between ultraviolet light and gamma rays on the electromagnetic spectrum. The X rays used for solving the structures of macromolecules generally lie in the range 0.5–2×10^{-10} m. The lengths of atomic bonds are in this range, so X rays are ideally suited to the study of molecular structure.

When two waves are traveling in the same direction, they can be added together. When the sum has a larger amplitude than the original waves, it is called constructive interference. For example, when the waves have the same phase, amplitude, and wavelength, their amplitudes are summed to give a single wave with twice the amplitude. When, however, the waves are exactly 180° out of phase, the waves cancel each other out in a process called *destructive interference*. When a beam of X rays hits a crystal, a diffraction pattern is observed because in some directions the right conditions are obtained to get constructive interference.

When an X ray encounters an electron, the electron will oscillate at the same frequency as the incident wave. These oscillating electrons then emit radiation at the same wavelength as the incident wave in any direction. The ordered structure of a crystal means that there are many planes of atoms within the crystal. The electrons in parallel planes of atoms will scatter X rays. Constructive interference will be observed if two waves have traveled the same distance or if one wave has traveled an integer number of wavelengths more than the other. When the angle between the incident beam and the crystal plane (θ) and the distance separating the two planes are correct, the path difference between the two waves will be an integer number of wavelengths and constructive interference will occur, giving rise to diffraction spots. This is described in Bragg's law (see the accompanying figure on page 118):

$$2d \sin \theta = n\lambda$$

Each diffraction spot will be the sum of all the waves diffracted in that direction by all the electrons in the crystal. Only when Bragg's law is obeyed will there be significant diffraction: This gives rise to a series of intense spots surrounded by areas where the diffraction sums to zero. Each diffraction spot will have an intensity (I) associated with it which is proportional to the square of the amplitude of the wave (F) traveling in that direction. Each spot is referred to as a reflection and has an associated Miller index h,k,l which relates to the set of planes that produced the reflection:

$$|I(h, k, l)| = k|F(h, k, l)|^2$$

where $|F|$ is the amplitude of the **F**, the structure factor, which is a vector with both an amplitude $|F|$ and phase α. The structure factor amplitude is the ratio of the amplitude of the radiation scattered by the contents of the unit cell to the amplitude of the radiation scattered by a point electron.

In order to obtain an entire data set from a crystal of a macromolecule, the crystal is placed in a beam of X rays and an X-ray detector [an image plate or charge-coupled device (CCD) camera] is placed on the other side of the crystal to record the diffraction pattern. The crystal is exposed to X rays and rotated by, say 1°; then the shutter is closed, the diffraction pattern is recorded digitally from the detector, and the detector is blanked. This process is repeated until the crystal has been rotated by up to 180° in 1° images. The images are then processed to produce a list of diffraction spots with indices h,k,l with their associated intensities $I(h,k,l)$.

The electron density ρ at any point in the unit cell (x,y,z) can be calculated from the following equation:

$$\rho(x, y, z) = \frac{1}{V} \sum_{h} \sum_{k} \sum_{l} |F(h, k, l)| \cos 2\pi [hx + ky + lz - \alpha(h, k, l)]$$

where $\rho(x, y, z)$ is the electron density at point x,y,z in the unit cell.

V is the volume of the unit cell.

h,k,l are the indices of the diffraction spots.

$|F(h, k, l)|$ is the amplitude of the wave giving rise to the h,k,l diffraction spot.

$\alpha(h, k, l)$ is the phase of the wave giving rise to the diffraction spot h,k,l.

The only unknown from this equation is the phase of the wave as it hits the detector. The fact that this information is not easily derived from a single data set and the difficulty in obtaining this information are known as the "phase problem" in X-ray crystallography.

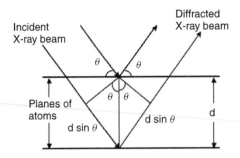

Figure   Bragg's law indicates when constructive interference will give rise to diffraction from a crystal.

Related Websites

Bernard Rupp's teaching	http://www-structure.llnl.gov/Xray/101index.html
Cambridge Structural Medicine course	http://perch.cimr.cam.ac.uk/course.html

See also Crystal, Macromolecular; Crystallization of Macromolecules; Electron Density Map; Heavy-Atom Derivative; Phase Problem; Refinement; *R*-Factor; Space Group; X-Ray Crystallography for Structure Determination and related Further Readings.

DIHEDRAL ANGLE (TORSION ANGLE)
Roman A. Laskowski

Mathematically, a dihedral angle is the angle between two planes. It is zero if the planes are parallel and can take values up to $90°$.

In chemistry, a dihedral angle can be defined for any chain of four covalently bonded atoms: A—B—C—D. The angle is between the plane containing atoms A, B, and C and that containing atoms B, C, and D. It represents the relative rotation of bonds A—B and C—D about the bond B—C.

By convention, dihedral angle values range from $-180°$ to $+180°$. If, when viewed along the bond B—C, the bond A—B eclipses bond C—D, the value of the dihedral angle is $0°$. If the bonds are not eclipsed, the absolute value of the dihedral angle is the angle through which the bond A—B has to be rotated about B—C to achieve eclipse, the sign being positive if the rotation is clockwise and negative if it is anti-clockwise.

In proteins, dihedral angles are defined for the backbone and side-chain atoms. The three backbone dihedral angles and their defining atoms are phi, (C_{i-1}—N_i—$C\alpha_i$—C_i), psi (N_i—$C\alpha_i$—C_i—N_{i+1}), and omega ($C\alpha_i$—C_i—N_{i+1}—$C\alpha_{i+1}$). The phi and psi dihedral angles determine the local secondary structure of the protein chain and are commonly depicted on a Ramachandran plot. The omega dihedral angle is defined about the peptide bond. Because of the partial double-bond nature of this bond, it forms a planar or close to planar unit. The value of omega, therefore, will be close to either omega $= 180°$ (for the highly favored trans conformation) or omega $= 0°$ (for the fairly rare cis conformation).

Side-chain dihedral angles in proteins are designated by chi$_n$, where n is the number of the bond counting out from the central Cα atom. Steric hindrance often limits the ranges the values can take, the favored conformations being known as rotamers.

It is also possible to define a "virtual" dihedral angle for any set of four atoms A, B, C, and D where there is no physical bond between atoms B and C. See, e.g., the kappa and zeta virtual dihedral angles.

See also Amide Bond, Backbone, Kappa Virtual Dihedral Angle, Ramachandran Plot, Rotamer, Secondary Structure of Protein, Side Chain, Zeta Virtual Dihedral Angle.

DINUCLEOTIDE FREQUENCY
Katheleen Gardiner

Frequency at which any two bases are found adjacent on the same strand in a DNA sequence.

The four bases, A, G, C, and T, permit 16 dinucleotides (AA, AT, AC, AG, etc.). In the mammalian genome, the frequency of the CG dinucleotide is statistically low; it is present at only 1/5–1/4 of the frequency expected from the average genome base composition of ~38% GC. This CpG suppression is related to DNA methylation. CpG dinucleotides are enriched in CpG islands.

Further Reading

Nekrutenko A, Li WH (2000). Assessment of compositional heterogeneity within and between eukaryotic genomes. *Genome Res.* 10: 1986–1995.

Swartz M, Trautner TA, Kornberg A (1962). Enzymatic synthesis of DNA: Further studies on nearest neighbor base sequences in DNA. *J. Biol. Chem.* 237: 1961–1967.

See also CpG Island.

DIP
Patrick Aloy

The Database of Interacting Proteins (DIP) catalogs and classifies experimentally determined protein–protein interactions. It combines information from very different sources, such as two-hybrid experiments, large-scale studies on protein complexes, or literature searches, to create a single, consistent, and comprehensive set of protein–protein interactions.

To date, DIP contains information about 17,838 interactions involving 6812 proteins from 110 different organisms. All the interactions in the DIP are manually curated by both human experts and intelligent knowledge-based computer systems.

The database offers several searching possibilities (e.g., ID, BLAST, motif) and graphic tools to display the retrieved interactions.

Related Website

DIP	http://dip.doe-mbi.ucla.edu

Further Reading

Xenarios I, et al. (2002). DIP: The Database of Interacting Proteins. A research tool for studying cellular networks of protein interactions. *Nucleic Acids Res.* 30: 303–305.

DIRECTED ACYCLIC GRAPH
Robert Stevens

A directed acyclic graph (DAG) is a form of multihierarchy of nodes and arcs.

The nodes form the terms or concepts in the ontology and the arcs the relationships. Any one concept can have many parents and many children. More than one kind of relationship can be used. The "directed" nature of the graph comes from the directional nature of the relationships: A is a parent of B is directed. In other words, the graph has a top and a bottom. The gene ontology is referred to as a DAG (see the accompanying figure).

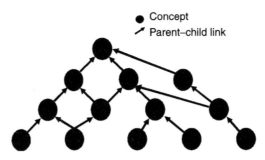

Figure   Organization of concepts in a directed acyclic graph.

DISCOVER
Roland L. Dunbrack

A molecular dynamics simulation and modeling computer program developed by Arnold Hagler and colleagues. DISCOVER uses an empirical molecular mechanics potential energy function for studying the structures and dynamics of proteins, DNA, and other molecules.

Related Website

Accelrys	http://www.accelrys.com/insight/discover.html

Further Reading
Dauber-Osguthorpe P, et al. (1988). Structure and energetics of ligand binding to proteins: *Escherichia coli* dihydrofolate reductase-trimethoprim, a drug-receptor system. *Proteins* 4: 31–47.

Ewig CS, et al. (2001). Derivation of class II force fields. VIII. Derivation of a general quantum mechanical force field for organic compounds. *J. Comput. Chem.* 22: 1782–1800.

See also Molecular Mechanics.

DISCRIMINANT ANALYSIS. *SEE* CLASSIFICATION.

DISCRIMINATING POWER. *SEE* DIAGNOSTIC POWER.

DISULFIDE BRIDGE
Jeremy Baum

A disulfide bridge is formed when the side chains of two cysteine residues interact to form a covalent S—S bond. With the exception of bonded cofactors, these are the only covalent bonds occurring in proteins that deviate from the linear polymer primary structure. Such structures typically occur in secreted (i.e., extracellular) proteins.

See also Cofactor, Cysteine, Side Chain.

DL. *SEE* DESCRIPTION LOGIC.

DNA ARRAY. *SEE* MICROARRAY.

DNA DATABANK OF JAPAN. *SEE* DDBJ.

DNA–PROTEIN COEVOLUTION
Laszlo Patthy

Coevolution is a reciprocally induced inherited change in a biological entity (species, organ, cell, organelle, gene, protein, biosynthetic compound) in response to an inherited change in another with which it interacts.

Protein–DNA interactions essential for some biological function (e.g., binding of RNA polymerase II and transcription factors to specific sites in the promoter region and binding of a hormone receptor to the appropriate hormone response element) are ensured by specific contacts between a DNA sequence motif and the DNA-binding protein. During evolution gene regulatory networks have expanded through the expansion of the families of nuclear hormone receptors, transcription factors, and the coevolution of the DNA-binding proteins with the DNA sequence motifs to which they bind.

For example, the different steroid hormone–receptor systems have evolved through the coevolution of the three interacting partners (the steroid hormone, their protein receptors, and their target DNA sites). Phylogenetic analyses of the steroid receptor systems indicate that the full complement of mammalian steroid receptors (androgen receptor, progesterone receptor, glucocorticoid receptor, mineralocorticoid receptor) evolved from an ancestral steroid receptor through a series of gene duplications. Coevolution of the novel hormone receptor (with novel ligand specificity) and the target DNA binding sites has led to an increased complexity of gene regulatory networks.

Further Reading

Thornton JW (2001). Evolution of vertebrate steroid receptors from an ancestral estrogen receptor by ligand exploitation and serial genome expansions. *Proc. Natl. Acad. Sci. USA* 98: 5671–5676.

Umesono K, Evans RM (1989). Determinants of target gene specificity for steroid/thyroid hormone receptors. *Cell* 57: 1139–1146.

See also Coevolution.

DNA REPLICATION
John M. Hancock

Bacteria typically have a single replication origin and a single terminus, making the genome a single replicon. Eukaryotic genomes have numerous origins, termini, and replicons. Errors in DNA replication at microsatellites and cryptically simple regions can result in replication slippage (slipped-strand mispairing; slippage) and insertions and deletions of repetitive motifs. Replication errors at single bases can result in point mutations. Strand breakage during replication can induce recombination (gene conversion) between different copies of genes or even between paralogs.

DNA replication from a replication origin is bidirectional (proceeds in both directions) and takes place at a structure known as the replication fork. Because DNA synthesis can only take place in the $5' \rightarrow 3'$ direction, DNA synthesis on one strand of the replication fork (the leading strand) is continuous, while on the other strand (the lagging strand) it is discontinuous, proceeding by way of short fragments known as Okazaki fragments, which are primed by RNA and subsequently joined.

Further Reading
Lewin B (2000). *Genes VII.* Oxford University Press, Oxford, UK.

See also Microsatellite, Recombination, Replication Origin, Simple DNA Sequence.

DNA SEQUENCE
John M. Hancock and Rodger Staden

The order of the bases, denoted A (adenine),C (cytosine),G (guanine), and T (thymine), along a segment of DNA.

DNA sequences have a directionality resulting from the chemical structure of DNA. By convention, a DNA sequence is written in the $5' \rightarrow 3'$ direction, reflecting its direction of replication. Upper- or lowercase characters may be used, although lowercase characters are less ambiguous for human reading. Letters other than the four typical ones may be used to represent uncertainty in a sequence.

See also IUPAC-IUB Codes.

DNA SEQUENCING
Rodger Staden

The process of discovering the order of the bases in a DNA sequence.

See also Genome Sequencing.

DNA SEQUENCING ARTIFACT
Rodger Staden

Sequencing artifacts can lead to sequencing errors. Skilled practitioners can spot the presence of artifacts by visual inspection of sequence traces, but it is not always possible to know the correct sequence. A DNA sequencing "compression" occurs when, during electrophoresis, a secondary structure in the fragments causes them to move anomalously in the sequencing gel. This can be seen by a stretching of the trace peaks immediately after the compression but causes base-calling errors. Compressions are often strand specific, and so their presence can be revealed by making sure the whole sequence is covered by data from both strands. A DNA sequencing "stop" occurs when local sequence properties cause the DNA polymerase to terminate prematurely, so producing an anomalously strong signal for fragments of a particular length.

See also DNA Sequencing Error, DNA Sequencing Trace.

DNA SEQUENCING ERROR
Rodger Staden

Individual sequence readings can contain incorrectly called bases, missed bases, and extra bases (overcalls). The reliability of any base in a reading depends on its position along the sequence and the local sequence composition. Ideally each base call should have an associated confidence value. Finished sequences are normally derived from more than one clone and from readings off both strands or by using more than one sequencing chemistry. These combinations should overcome most sequencing artifacts and resolve the majority of problems. At a higher level, it is possible that readings have been assembled in the wrong order, and so, although the sequence is accurate, it is from another part of the genome. Such problems can be discovered by comparing a restriction enzyme digest of the artificial chromosome containing the target DNA with that predicted from the finished sequence. However, this will not reveal chimeric sequences.

See also Chimeric DNA Sequence, Finishing, Finishing Criteria, DNA Sequencing Artifacts.

DNA SEQUENCING TRACE
Rodger Staden

In DNA sequencing the chains of nucleotides are labeled with fluorescent dyes. The intensity of fluorescence measured as DNA fragments are electrophoresed in a sequencing instrument is known as a chromatogram or, more commonly, a trace. The traces are processed to produce the DNA sequence.

DNA STRUCTURE. *SEE* BASE PAIR.

DNASP
Michael P. Cummings

A program for analysis of nucleotide sequence data that includes a comprehensive set of population genetic analyses.

The program estimates numerous population genetic parameters with associated sampling variance, including analyses of polymorphism, divergence, codon usage bias, linkage disequilibrium, gene flow, gene conversion, recombination, and population size. The program also allows for hypothesis tests for departure from neutral theory expectations with and without an outgroup. In addition, the program can perform coalescent simulations.

The program accommodates several input and output formats (e.g., NEXUS, FASTA, PHYLIP, NBRF/PIR, and MEGA), has a graphical user interface, and is available as a Windows-executable file.

Related Website

DnaSP	http://www.ub.es/dnasp/

Further Reading

Rozas J, Rozas R (1999). DnaSP version 3: An integrated program for molecular population genetics and molecular evolution analysis. *Bioinformatics* 15: 174–175.

DOCK
Marketa J. Zvelebil

The DOCK program explores ways in which two molecules, such as a ligand and a protein, can fit together.

DOCK is one of the main efforts in automating the docking process. It is the long-term project of Tack Kuntz and colleagues. DOCK uses shape complementarity to search for molecules that can match the shape of the receptor site. In addition, it takes into account chemical interactions. The program generates many possible orientations (and, more recently, conformations) of a putative ligand within a user-selected region of a receptor structure (protein). These orientations are then subsequently scored using several schemes designed to measure steric and/or chemical complementarity of the receptor–ligand complex.

In the first instance, the program DOCK uses molecular coordinates to generate a molecular surface of the protein with the Connolly molecular surface program which describes the site features. Only the surface for the designated active/binding sites is generated. Subsequently, spheres which are defined by the shape of cavities within the surface are generated and placed into the active site. The centers of the spheres act as the potential location (sites) for the ligand atoms. The ligand atoms are then matched with the sphere centers to determine possible ligand orientations. Many orientations are generated for each ligand. Lastly, for each ligand the orientations are scored based on shape, electrostatic potential, and force field potential.

Related Website

DOCK	http://dock.compbio.ucsf.edu/

Further Reading

Kuntz ID (1992). Structure-based strategies for drug design and discovery. *Science* 257: 1078–1082.

Kuntz ID, Blaney JM, Oatley SJ, Langridge R, Ferrin TE (1982). A geometric approach to macromolecule-ligand interactions. *J. Mol. Biol.* 161: 269–288.

Kuntz ID, Meng EC, Shoichet BK (1994). Structure-based molecular design. *Acc. Chem. Res.* 27(5): 117–123.

Meng EC, Gschwend DA, Blaney JM, Kuntz ID (1993). Orientational sampling and rigid-body minimization in molecular docking. *Proteins* 17(3): 266–278.

Meng EC, Shoichet BK, Kuntz ID (1992). Automated docking with grid-based energy evaluation. *J. Comp. Chem.* 13: 505–524.

Shoichet BK, Bodian DL, Kuntz ID (1992). Molecular docking using shape descriptors. *J. Comp. Chem.* 13(3): 380–397.

DOCKING
Marketa J. Zvelebil

Docking simply attempts to predict the structure of the intermolecular complex formed between the protein and a ligand.

Usually the enzyme structure is considered to be static and the ligand is "docked" into the binding pocket, and its suitability is evaluated by a potential energy function, often semi-empirical in nature. Most docking methods generate a large number of likely ligands, and as a result they have to have a means of scoring each structure to select the most suitable one. Consequently, docking programs deal mainly with the generation and evaluation of possible structures of intermolecular complexes. In conformational flexible docking, where, at least, the ligand is not constrained to be rigid, conformational degrees of freedom have to be taken into account. Monte Carlo methods in combination with stimulated annealing can be used to explore the conformational space of ligands.

Another method deals with building ligand molecules "de novo" inside a binding site, introducing fragments that complement the site to optimize intermolecular interactions.

Further Reading

Camacho CJ, Vajda S (2002). Protein-protein association kinetics and protein docking. *Curr. Opin. Struct. Biol.* 12(1): 36–40.

Lengauer T, Rarey M (1996). Computational methods for biomolecular docking. *Curr. Opin. Struct. Biol.* 6(3): 402–406.

See also DOCK.

DOMAIN. *SEE* PROTEIN DOMAIN.

DOMAIN ALIGNMENT. *SEE* ALIGNMENT OF PROTEIN SEQUENCE.

DOMAIN FAMILY
Terri Attwood

A structurally and functionally diverse collection of proteins that share a common region of sequence that mediates a particular structural or functional role.

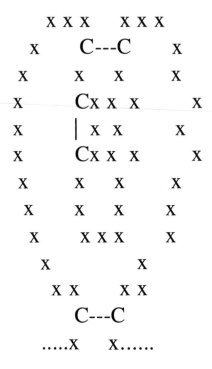

Figure

Domain families are a fundamentally different concept from the one normally held for protein families (gene families), which are usually structurally and functionally discrete and share similarity along the entire length of their sequences. The distinction between domain families and gene families is not just a semantic one, because different processes underpin their evolution, with different functional consequences.

In this context, domains largely refer to protein *modules*. Modules are autonomous folding units believed to have arisen largely as a result of genetic shuffling mechanisms—examples include kringle domains (named after the shape of a Danish pastry), which are structural units found throughout the blood-clotting and fibrinolytic proteins; the WW module (characterized by two conserved tryptophan residues, hence its name), which is found in a number of disparate proteins, including dystrophin, the product encoded by the gene responsible for Duchenne muscular dystrophy; and the apple domain, a 90-residue domain stabilized by three disulfide bonds (shown schematically in the accompanying figure), multiple tandem repeats of which are found in the N-termini of the related plasma serine proteases plasma kallikrein and coagulation factor XI.

Modules are contiguous in sequence and are often used as building blocks to confer a variety of complex functions on a parent protein, either via multiple combinations of the same module or by combinations of different modules to form protein mosaics.

SMART is an example of a domain family database.

Related Website

| SMART | http://smart.embl-heidelberg.de/help/smart_about.shtml |

Further Reading

Apic G, Gough J, Teichmann SA (2001). An insight into domain combinations. *Bioinformatics* 17(Suppl. 1): S83–S89.

Dengler U, Siddiqui AS, Barton GJ (2001). Protein structural domains: Analysis of the 3Dee domains database. *Proteins* 42: 332–344.

Siddiqui AS, Dengler U, Barton GJ (2001). 3Dee: A database of protein structural domains. *Bioinformatics* 17: 200–201.

See also Gene Family, Protein Family.

DONOR SPLICE SITE
Thomas D. Schneider

The binding site of the spliceosome on the 5′ side of an intron and the 3′ side of an exon is called a donor splice site. This term is preferred over 5′ site because there can be multiple donor sites, in which case 5′ site is ambiguous. Also, one would have to refer to the 5′ site on the 3′ side of an exon, which is confusing. Mechanistically, a donor site defines the end of the exon.

See also Acceptor Splice Site, Sequence Walker.

DOT MATRIX. *SEE* DOT PLOT.

DOT PLOT (DOT MATRIX)
Jaap Heringa

A way to represent all possible alignments of two sequences by comparing them in a two-dimensional matrix.

In a dot plot one sequence is written out vertically with its amino acids (or nucleotides) forming the matrix rows, while the other sequence forms the columns. Each intersection of a matrix row with a column represents the comparison of associated amino acids in the two sequences such that all possible local alignments can be found along diagonals paralleling the major matrix diagonal. Overall similarity is discernible by piecing together local subdiagonals through insertions and deletions. The simplest way of expressing similarity of matched amino acid pairs is to place a dot in a matrix cell whenever they are identical. Such matrices are therefore often referred to as dot matrices and were important early on as the major means to visualize the relationship of two sequences (Fitch, 1966; Gibbs and McIntyre, 1970).

More biological insight is normally obtained by using more varying amino acid similarity values than binary identity values. Most dot-plot methods therefore refine the crude way of only showing sequence identities by using amino acid exchange odds for each possible exchange. These odds are normally incorporated in a symmetrical 20 × 20 *amino acid exchange matrix* in which each value approximates the evolutionary likelihood of a mutation from one amino acid into another.

To increase the signal-to-noise ratio for dot plots, McLachlan (1971, 1972, 1983) first developed filtering techniques. He devised "double matching probabilities" to estimate the significance of regions showing high similarity. Such regions were identified by using windows of fixed lengths which were effectively slid over the two sequences to compare all possible stretches of, e.g., five matched residue pairs. McLachlan's initial program, CMPSEQ (McLachlan, 1971), was elaborated by Staden (1982), who devised the widely used dot-matrix program DIAGON. He added output filtering and also allowed different amino acid similarity scoring systems.

Important biological issues include the choice of the amino acid similarity scores to use as well as the adopted length of the windows in the dot-matrix analysis. Argos (1987) included physicochemical amino acid parameters in the calculation of the similarity values. He used these together with windows of different lengths that were tested simultaneously in order to be less dependent on the actual choice of an individual window length.

Further Reading

Argos P (1987). A sensitive procedure to compare amino acid sequences. *J. Mol. Biol.* 193: 385–396.

Fitch W (1966). An improved method of testing for evolutionary homology. *J. Mol. Biol.* 16: 9–16.

Gibbs AJ, McIntyre GA (1970). The diagram, a method for comparing sequences. Its use with amino acid and nucleotide sequences. *Eur. J. Biochem.* 16: 1–11.

McLachlan AD (1971). Tests for comparing related amino acid sequences: Cytochrome c and cytochrome c551. *J. Mol. Biol.* 61: 409–424.

McLachlan AD (1972). Repeating sequences and gene duplications in proteins. *J. Mol. Biol.* 72: 417–437.

McLachlan AD (1983). Analysis of gene duplication repeats in the myosin rod. *J. Mol. Biol.* 169: 15–30.

Staden R (1982). An interactive graphics program for comparing and aligning nucleic acid and amino acid sequences. *Nucleic Acids Res.* 10: 2951–2961.

DOTTER
John M. Hancock and Martin J. Bishop

A popular implementation of the dot-plot method of sequence comparison.

The program accepts input files in FASTA format and can compare DNA or protein sequences or a DNA sequence with a protein sequence. The program can handle sequences of any length and can be run in the batch mode for long sequences. Otherwise it can be run interactively. It allows maximum and minimum cutoffs to be set to reduce noise, visualization of alignments giving rise to features in the plot, zooming in, displaying multiple dot plots simultaneously, and color coding features within the sequence.

Related Website

Dotter	http://www.cgr.ki.se/cgr/groups/sonnhammer/Dotter.html

Further Reading

Sonnhammer ELL, Durbin R (1995). A dot-matrix program with dynamic threshold control suited for genomic DNA and protein sequence analysis. *Gene* 167: GC1-10.

See also Dot Plot, STADEN.

DOUBLE STRANDING
Rodger Staden

Obtaining data from both strands of the DNA. Some sequencing artifacts are only revealed by data from one strand of the DNA so double stranding is required in order to reduce errors in the finished sequence.

Figure 1 shows data from a partially completed sequencing project. The data are still divided into two contigs, depicted by the rulers at the bottom. Above each ruler the long blue, red, and yellow horizontal lines represent the DNA fragments/inserts/templates from which readings have been taken. The extent of each reading is denoted by the short arrows colored light blue (forward readings) and orange (reverse readings). The red template indicates a consistent read pair within a contig. The yellow-colored template means a read pair between contigs, hence showing their relative order and orientation, and that primer walks on this template would eventually join the contigs.

Figure 2 shows a section of a contig with low reading coverage. The display includes the contig positions, reading orientations (negative numbers at the left), names, sequences with confidence as grey scales, and consensus sequence with confidence values. Padding characters introduced to align the readings are shown as asterisks.

Figure 3 shows the traces for three overlapping DNA sequence readings. For each trace the display includes the reading orientation, name, base positions, base calls,

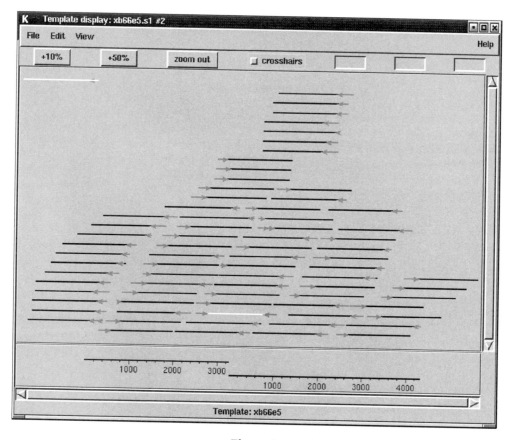

Figure 1

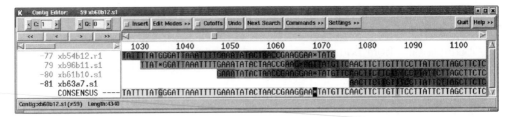

Figure 2

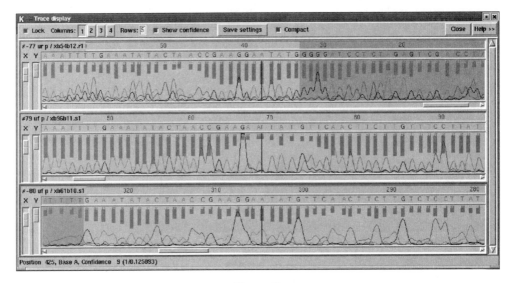

Figure 3

confidence values, and fluorescence amplitudes. The shaded regions at the ends of the top and bottom traces indicate segments that are not used in the consensus calculation because they are from the sequencing vector or of low quality. The inverted histogram shows the confidence values for each base call. Just to the left of the vertical cursor is a position where the top and bottom readings show AGGATA and the middle one AGAATTA. The traces are needed to help resolve the disagreement.

See also Primer Walking.

DOWNSTREAM
Niall Dillon

Describes a sequence located distal to a specific point in the direction of transcription (i.e., in a 3′ direction on the strand being transcribed).

DYNAMIC PROGRAMMING
Jaap Heringa

Optimization technique to calculate the highest scoring or optimal alignment between two protein or nucleotide sequences.

The physicist Richard Bellman first conceived of dynamic programming (DP) (Bellman, 1957) and published a number of papers and a book on the topic between 1957 and 1975. Then Needleman and Wunsch (1970) introduced the technique to the biological community.

A DP algorithm is guaranteed to yield the maximally scoring alignment of a pair of nucleotide or protein sequences, given an appropriate nucleotide or amino acid exchange matrix and gap penalty values. The algorithm operates in two steps. First a search matrix is set up in the same way as a dot matrix with one sequence displayed horizontally and the other vertically. The matrix is traversed from the upper left to the lower right. Each cell $[i, j]$ in the matrix receives as a score the value composed of the maximum value of the scores in row $i - 1$ and column $j - 1$ (with subtraction of the proper gap penalty values) added to the exchange value of the associated matched residue pair of cell $[i, j]$. Cell $[i, j]$ therefore contains the maximum score of all possible alignments of the two subsequences up to cell $[i, j]$. Writing the above in a more explicit form gives

$$S[i,j] = s[i,j] + \text{Max} \begin{Bmatrix} S[i-1, j-1] \\ \max_{1<x<i}(S[i-x, j-1] - \text{gp}(x-1)) \\ \max_{1<y<j}(S[i-1, j-y] - \text{gp}(y-1)) \end{Bmatrix} \qquad (1)$$

where $S[i,j]$ is the alignment score for the first subsequence from 1 to i and the second subsequence from 1 to j, Max denotes the maximum value of the three arguments between brackets, $s[i,j]$ is the substitution value for the amino acid exchange associated with cell$[i,j]$, $\text{gp}(x-1)$ is the nonnegative penalty value for a gap of length $x-1$, and $\max_{1<x<i}$ represents the maximum value of all argument values over the range $1, \ldots, i$.

The choice of proper gap penalty values is closely connected to the residue exchange values used in the analysis. When the search matrix is traversed, the highest scoring matrix cell is selected from the bottom row or the rightmost column, and this score is guaranteed to be the optimal alignment score. The second step of a dynamic programming algorithm is usually called the trace-back step in which the actual optimal alignment is reconstructed from the matrix cell containing the highest alignment score.

Classical Needleman–Wunsch dynamic programming algorithms use a two-dimensional search matrix, so that the algorithmic speed and storage requirements are both of the order $N \times M$ when two sequences consisting of N and M amino acids in length are matched. The large computer memory requirements of Needleman–Wunsch algorithms are due to the trace-back step, where the matches of the optimal alignment are reconstructed. Furthermore, the amount of computation required makes the dynamic programming technique unfeasible for a query sequence search against a large sequence database on personal computers.

Gotoh (1986, 1987) devised a dynamic programming algorithm that drastically decreased the storage requirements from order N^2 to order N (assuming that two sequences each N amino acids in length are matched) while keeping speed at an order of N^2. Myers and Miller (1988) constructed an even more memory efficient linear space algorithm based on the Gotoh approach and on a trace-back strategy proposed by Hirschberg (1975), which is only slightly slower.

Smith and Waterman (1981) developed the so-called *local alignment* technique in which the most similar regions in two sequences are selected and aligned first. To get from global to local dynamic programming, an important prerequisite is that the amino acid exchange values used must include negative values. Any score in the search matrix lower than zero is to be set to zero. Formula (1) is then changed to

$$S[i,j] = \text{Max} \begin{cases} L^{s(i,j)} + S[i-1, j-1] \\ L^{s(i,j)} + \max_{1<x<i}(S[i-x, j-1] - L^p(x-1)) \\ L^{s(i,j)} + \max_{1<y<j}(S[i-1, j-y] - L^p(y-1)) \\ 0 \end{cases} \qquad (2)$$

where Max is now the maximum of four terms. A consequence of this scenario is that the final highest alignment score value does not have to be in the last row or column as in global alignment routines but can be anywhere in the search matrix. The local alignment algorithm relies on dissimilar subsequences producing negative scores which are subsequently discarded by placing zero values in the associated submatrix cells. An arbitrary issue in using the algorithm is deciding the zero cutoff relative to the 20×20 residue exchange weights matrix.

Waterman and Eggert (1987) generalized the local alignment routine by devising an alignment routine which allows the calculation of a user-defined number of top-scoring local alignments instead of only the optimal local alignment. The obtained local alignments do not intersect; i.e., they have no matched amino acid pair in common. If during the procedure an alignment is encountered that intersects with any of the top-scoring alignments listed thus far, the highest scoring of the conflicting pair is retained in the top list. Huang et al. (1990) developed an implementation of the technique in which they reduced the memory requirements from order N^2 to order N, thereby allowing very long sequences to be searched at the expense of only a small increase in computational time. Another popular version of the same technique is LALIGN (Huang and Miller, 1991), which is part of the popular FASTA package (Pearson and Lipman, 1988).

Further Reading

Bellman RE (1957). *Dynamic Programming*. Princeton University Press, Princeton, NJ.

Gotoh O (1986). Alignment of three biological sequences with an efficient traceback procedure. *J. Theor. Biol.* 121: 327–337.

Gotoh O (1987). Pattern matching of biological sequences with limited storage. *Comput. Appl. Biosci.* 3: 17–20.

Hirschberg DS (1975). A linear space algorithm for computing longest common subsequences. *Commun. Assoc. Comput. Mach.* 18: 341–343.

Huang X, Miller W (1991). A time-efficient, linear-space local similarity algorithm. *Adv. Appl. Math.* 12: 337–357.

Huang X, et al. (1990). A space-efficient algorithm for local similarities. *Comput. Appl. Biosci.* 6: 373–381.

Myers EW, Miller W (1988). Optimal alignment in linear space. *Comput. Appl. Biosci.* 4: 11–17.

Needleman SB, Wunsch CD (1970). A general method applicable to the search for similarities in the amino acid sequence of two proteins. *J. Mol. Biol.* 48: 443–453.

Pearson WR, Lipman DJ (1988). Improved tools for biological sequence comparison. *Proc. Natl. Acad. Sci. USA* 85: 2444–2448.

Smith TF, Waterman MS (1981). Identification of common molecular subsequences. *J. Mol. Biol.* 147: 195–197.

Waterman MS, Eggert M (1987). A new algorithm for best subsequences alignment with applications to the tRNA-rRNA comparisons. *J. Mol. Biol.* 197: 723–728.

See also Gap Penalty, Alignment—Multiple, Alignment—Pairwise.

E

EBI
ECEPP
EcoCyc
Electron Density Map
Electronic PCR
Electrostatic Energy
Electrostatic Potential
Elston–Stewart Algorithm
E-M Algorithm
EMBL-EBI
EMBL Nucleotide Sequence Database
EMBnet
EMBOSS
eMOTIF
Empirical Pair Potentials
Empirical Potential Energy Function
ENCprime/SeqCount
End Gap
Energy Minimization
Enhancer
Ensembl
Entrez
Entropy
Enzyme

e-PCR
Epistasis
Error
Error Measures
E–S Algorithm
EST
European Bioinformatics Institute
European Molecular Biology Open
    Software Suite
EVA
E Value
Evolution
Evolutionary Clock
Evolutionary Distance
Evolution of Biological Information
Exclusion Mapping
Exon
Exon Shuffling
Expectation Maximization Algorithm
Expressed Sequence Tag
Expression Profiler
eXtensible Markup Language
Extrinsic Gene Prediction

Dictionary of Bioinformatics and Computational Biology. Edited by Hancock and Zvelebil
ISBN 0-471-43622-4   © 2004 John Wiley & Sons, Inc.

EBI (EMBL-EBI, EUROPEAN BIOINFORMATICS INSTITUTE)
Guenter Stoesser

The European Bioinformatics Institute is an international nonprofit academic organization that forms part of the European Molecular Biology Laboratory (EMBL).

The EBI covers a wide range of activities in research and services in bioinformatics, including computational and structural genomics, microarray informatics as well as building and maintaining comprehensive public biological databases of DNA sequences (EMBL Nucleotide Sequence Database), protein sequences (TrEMBL/SWISS-PROT), the InterPro resource for protein families, macromolecular structures (MSD), genome annotation (Ensembl), gene expression data (ArrayExpress), and scientific literature.

Data resources are complemented by query and computational analysis servers, including SRS for querying of more than 150 linked biological data sets, BLAST/FASTA/Smith–Waterman sequence searching, multiple-sequence alignment using CLUSTALW, and structure prediction methods. Most projects at the EBI involve collaborations with major institutes within Europe and worldwide.

The EBI is located on the Wellcome Trust Genome Campus near Cambridge, United Kingdom, next to the Sanger Institute and the Human Genome Mapping Project Resource Centre (HGMP). Together these institutes constitute one of the world's largest concentrations of expertise in genomics and bioinformatics.

Related Websites

European Bioinformatics Institute	http://www.ebi.ac.uk
EMBL Nucleotide Sequence Database	http://www.ebi.ac.uk/embl/
SWISS-PROT Protein Knowledgebase	http://www.ebi.ac.uk/swissprot/
InterProResource of Protein Domains	http://www.ebi.ac.uk/interpro/
Macromolecular Structure Database	http://msd.ebi.ac.uk/
MicroArray Informatics	http://www.ebi.ac.uk/microarray/
Ensembl Genome Browser	http://www.ensembl.org/
GenomesWeb server	http://www.ebi.ac.uk/genomes/
Database searching, browsing, and analysis tools	http://www.ebi.ac.uk/Tools/index.html

Further Reading

Apweiler R, et al. (2001). The InterPro database, an integrated documentation resource for protein families, domains and functional sites. *Nucleic Acids Res.* 29: 37–40.

Brazma A, Vilo J (2000). Gene expression data analysis. *FEBS Lett.* 480: 17–24.

Dietmann S, et al. (2001). A fully automatic evolutionary classification of protein folds: Dali Domain Dictionary version 3. *Nucleic Acids Res.* 29: 55–57.

Emmert DB, et al. (1994). The European Bioinformatics Institute (EBI) databases. *Nucl. Acids Res.* 26: 3445–3449.

Dictionary of Bioinformatics and Computational Biology. Edited by Hancock and Zvelebil
ISBN 0-471-43622-4   © 2004 John Wiley & Sons, Inc.

Hermjakob H, Apweiler R (2002). TEMBLOR—Perspectives of EBI database services. *Comp. Funct. Genom.* 3: 47–50.

ECEPP
Roland L. Dunbrack

A molecular mechanics force field and energy minimization and simulation program developed by Harold Scheraga and colleagues at Cornell University. The ECEPP force field uses rigid internal geometries which allow a more efficient exploration of conformational space, although this may result in large energies due to unrelaxed atomic overlaps.

Related Website

Eceppak	http://www.tc.cornell.edu/reports/NIH/resource/CompBiologyTools/eceppak/index.asp

Further Reading
Roterman IK, et al. (1989). A comparison of the CHARMM, AMBER and ECEPP potentials for peptides. I. Conformational predictions for the tandemly repeated peptide (Asn-Ala-Asn-Pro). *J Biomol. Struct. Dyn.* 7: 391–419.

See also Energy Minimization.

ECOCYC
Robert Stevens

EcoCyc is an encyclopedia of *Escherichia coli* metabolism, regulation, and signal transduction.

It is a knowledge base whose schema is an ontology of prokaryotic genome, proteins, pathways etc. The ontology is encoded in Ocelot, an OKBC-compliant frame-based knowledge representation language. The individuals conforming to the ontology have been painstakingly curated by hand from the *E. coli* literature. The knowledge base can be used to drive a Web interface for browsing the collected knowledge. Several other prokaryotic organisms have been treated in the same manner and all pathway information has been collected within Metacyc (http://biocyc.org/metacyc/). Given the existence of a GenBank entry for a prokaryotic genome, the annotated genes can be processed with pathway tools, which compares the new genome with information in Metacyc to automatically generate an organism-specific pathway genome database. (*Note:* EcoCyc is not related to OpenCyc or CycCorp.)

Further Reading
Karp P, Paley S (1996). Integrated access to metabolic and genomic data. *J. Comput. Biol.* 3: 191–212.

ELECTRON DENSITY MAP
Liz P. Carpenter

X-ray crystallography is a method for solving the structures of macromolecules from a diffraction pattern of X-rays from crystals of the macromolecule. X-rays are scattered by electrons. The diffraction pattern consists of a series of diffraction spots for which an intensity can be measured and a variety of methods can be used to derive the relative phases of the diffraction spots (see phase problem). From this information a map of the distribution of electrons within the crystal unit cell can be constructed with the following equation:

$$\rho(x, y, z) = \frac{1}{V} \sum_{h} \sum_{k} \sum_{l} |F(h, k, l)| \cos 2\pi [hx + ky + lz - \alpha(h, k, l)]$$

where $\rho(x, y, z)$ is the density of electrons per unit volume at point x, y, z
V is the volume of the unit cell.
h, k, l are the indices of the diffraction spots.
$|F(h, k, l)|$ is the amplitude of the wave giving rise to the h, k, l diffraction spot.
$\alpha(h, k, l)$ is the phase of the wave giving rise to the diffraction spot h, k, l.

This equation can be used to produce a series of equipotential contours, the atoms of the macromolecule being found where there is high electron density. Once an initial electron density map has been obtained, the structure of the macromolecule can be built into the map, defining the relative coordinates of most of the atoms in the structure. This structure is then subjected to refinement with programs designed to adjust the positions of the atoms to maximize the agreement between the observed structure factors and those calculated from the model.

Once a model is available, two types of electron density maps can be calculated: the $2F_{obs} - F_{calc}$ map, which shows where there is highest density throughout the asymmetric unit, and the difference map $mF_{obs} - DF_{calc}$, which shows differences between the observed and calculated electron density. Both maps should be studied on the graphics to identify regions of the structure that could be improved. Several rounds of refinement and model building are usually necessary before the structure converges and no further improvement is possible.

Further Reading
Glusker JP, Lewis M, Rossi M (1994). *Crystal Structure Analysis for Chemists and Biologists.* VCH Publishers, New York.

See also Diffraction of X Rays, Refinement, X-Ray Crystallography for Structure Determination.

ELECTRONIC PCR (E-PCR)
Martin J. Bishop and John M. Hancock

A program to test a DNA sequence for the presence of sequence-tagged sites (STSs).

Electronic Polymerase Chain Reaction (PCR) tests a DNA sequence for the presence of STSs by searching for subsequences that closely match the PCR primers and have the correct order, orientation, and spacing that they could plausibly prime the amplification ssof a PCR product of the correct molecular weight.

Related Website

| e-PCR | http://www.ncbi.nih.gov/genome/sts/epcr.cgi/ |

Further Reading
Schuler GD (1998). Electronic PCR: Bridging the gap between genome mapping and genome sequencing. *Trends Biotechnol.* 16: 456–459.

See also Sequence-Tagged Site.

ELECTROSTATIC ENERGY
Roland L. Dunbrack

The energy of interaction (attraction or repulsion) between charged atoms or molecules. In molecular mechanics functions, this is represented with the Coulomb equation, $E(r_{ij}) = \sum_i \sum_{j>i} kq_iq_j/\varepsilon r_{ij}$, where r_{ij} is the interatomic distance of two charged or partially charged atoms i and j with charges q_i and q_j, respectively, ε is the dielectric constant of the medium, and k is the appropriate constant in units of energy, distance, and charge.

See also Dielectric Constant

ELECTROSTATIC POTENTIAL
Jeremy Baum

When a charge is present on a molecule, chemical group, or atom, it will have an effect on the surroundings. Two components of this effect are typically discussed—the electrostatic potential and the electric field. The electric field is the gradient of the electrostatic potential, which is a scalar quantity. A charge q_A produces an electrostatic potential $\varphi = q_A/\varepsilon r$ a distance r away in a medium whose dielectric constant is ε. A second charge q_B that experiences this potential φ will have energy $\varphi q_B = q_A q_B/\varepsilon r$. In enzymes, the electrostatic potential of the surrounding protein structure can substantially alter the protonation equilibrium of a group and through this control reactivity.

See also Dielectric Constant

ELSTON–STEWART ALGORITHM (E–S ALGORITHM)
Mark McCarthy and Steven Wiltshire

A recursive algorithm used in the context of linkage analysis for calculating the likelihood of pedigree data.

The E–S algorithm is an arrangement of the likelihood function of a pedigree that expressly treats a simple pedigree as a multigenerational series of sibships, with one parent of each sibship itself being a sib in the preceding sibship in the pedigree as a whole. The likelihood is written in terms of penetrance probabilities $P(x1|gi)$ and genotype probabilities $P(gi|\cdot)$, which are the population frequencies of ordered genotypes G_i of individual i, if a founder, or the transmission probabilities of genotypes G_i [as functions of θ, the recombination fractions between the putative disease locus and marker(s)] of individual i, if a nonfounder. Writing the pedigree so that parents precede their offspring, the likelihood is given by

$$L(\theta) = \sum_{g1 \in G1} P(x1|g1)P(g1|\cdot) \sum_{g2 \in G2} P(x2|g2)P(g2|\cdot) \cdots \sum_{gn \in Gn} P(xn|gn)P(gn|\cdot)$$

Beginning with the most recent generation in the pedigree (the right-hand side of the equation), the likelihood, over all genotypes, of each sibship is evaluated for all possible genotypes of the linking parent and is then "peeled" from the pedigree, the conditional likelihoods of the sibships being carried over to the summations for the associated genotypes of the linking parent. This procedure is repeated until the top of the pedigree is reached and the overall likelihood is achieved. The E–S algorithm makes rapid likelihood evaluation for large pedigrees possible by reducing the number of calculations that would otherwise be necessary if the likelihood were expressed as the multiple sum over all genotypes of the products of penetrance and genotype probabilities. As such, it is central to many computer programs for performing linkage analyses. The basic algorithm has been developed recently to accommodate complex pedigrees with and without loops and to allow peeling in all directions. The computational burden of the E–S algorithm increases linearly with pedigree size but increases exponentially with number of markers due to the summation terms over all possible genotypes at each step. Despite recent modifications to facilitate small multipoint calculations, the E–S algorithm still remains limited for large multipoint analysis. In this regard, it is complemented by the Lander–Green algorithm, which can evaluate the likelihood of large numbers of markers, but is restricted by pedigree size. Algorithms that combine elements of both the E–S and Lander–Green algorithms have been published that allow rapid multipoint likelihood calculation for large pedigrees.

Examples: Craig et al. (1996) used two programs that implement the E–S algorithm—REGRESS and LINKAGE—to perform segregation analysis and genomewide linkage analysis of a large kindred to identify loci controlling fetal hemoglobin production. Bektas et al. (2001) used VITESSE, which implements a recent version of the E–S algorithm that has been improved for speed and multipoint capability, to further map a locus on chromosome 12 influencing type 2 diabetes susceptibility.

Related Websites
Programs implementing the E–S algorithm in linkage analyses:

LINKAGE	ftp://linkage.rockefeller.edu/software/linkage/
VITESSE	http://watson.hgen.pitt.edu/

Further Reading

Bektas A, et al. (2001). Type 2 diabetes locus on 12q15. Further mapping and mutation screening of two candidate genes. *Diabetes* 50: 204–208.

Craig JE, et al. (1996). Dissecting the loci controlling fetal haemoglobin production on chromosomes 11p and 6q by the regressive approach. *Nat. Genet.* 12: 58–64.

Elston RC, Stewart J (1971). A general model for the analysis of pedigree data. *Hum. Hered.* 21: 523–542.

O'Connell JR (2001). Rapid multipoint linkage analysis via inheritance vectors in the Elston-Stewart algorithm. *Hum. Hered.* 51: 226–240.

O'Connell JR, Weeks DE (1995). The VITESSE algorithm for rapid exact multi-locus linkage analysis via genotype set-recoding and fuzzy inheritance. *Nat. Genet.* 11: 402–408.

Ott J (1999). *Analysis of Human Genetic Linkage.* The Johns Hopkins University Press, Baltimore, MD, pp. 182–185.

Stewart J (1992). Genetics and biology: A comment on the significance of the Elston-Stewart algorithm. *Hum. Hered.* 42: 9–15.

See also Lander–Green Algorithm, Linkage Analysis.

E-M ALGORITHM. *SEE* EXPECTATION MAXIMIZATION ALGORITHM.

EMBL-EBI. *SEE* EBI.

EMBL NUCLEOTIDE SEQUENCE DATABASE (EMBL-BANK, EMBL DATABASE)

Guenter Stoesser

The EMBL Nucleotide Sequence Database constitutes Europe's major nucleotide sequence resource

It is the European member of the tripartide International Nucleotide Sequence Database Collaboration DDBJ/EMBL/GenBank together with DDBJ (Japan) and GenBank (United States). Collaboratively, the three organizations collect all publicly available nucleotide sequences and exchange sequence data and biological annotations on a daily basis. EMBL-Bank's major data sources are individual research groups, large-scale genome sequencing centers, and the European Patent Office (EPO). Sophisticated submission systems are available for individual scientists (Webin) and large-scale genome sequencing groups. EBI's network services allow free access to the most up-to-date data collection via file transfer protocol (ftp), email, and World Wide Web interfaces. EBI's Sequence Retrieval System (SRS) is an integration system for both data retrieval and applications for data analysis. SRS provides capabilities to search the main nucleotide and protein databases plus many specialized databases by shared attributes and to query across databases. For sequence similarity searching a variety of tools (e.g., FASTA, BLAST) are available which allow external users to compare their own sequences against the latest data in the EMBL Nucleotide Sequence Database and SWISS-PROT. Specialized sequence analysis programs include CLUSTALW for multiple-sequence alignment and inference of phylogenies and GeneMark for gene prediction. The database is located and maintained at the European Bioinformatics Institute (EBI) in Cambridge (United Kingdom).

Related Websites

EMBL Nucleotide Sequence Database	http://www.ebi.ac.uk/embl/
EMBL Submission Systems and related information	http://www.ebi.ac.uk/embl/Submission/
GenomesWeb server	http://www.ebi.ac.uk/genomes/
Database searching, browsing, and analysis tools	http://www.ebi.ac.uk/Tools/
European Bioinformatics Institute	http://www.ebi.ac.uk

Further Reading

Abola EE, et al. (2000). Quality control in databanks for molecular biology. *BioEssays* 22: 1024–1034.

Hingamp P, et al. (1999). The EMBL nucleotide sequence database: Contributing and accessing data. *Mol. Biotechnol.* 12: 255–268.

Stoesser G, et al. (2002). The EMBL nucleotide sequence database. *Nucleic Acids Res.* 30: 21–26.

Zdobnov EM, et al. (2002). The EBI SRS server—recent developments. *Bioinformatics* 18: 368–373.

EMBNET (EUROPEAN MOLECULAR BIOLOGY NETWORK)
John M. Hancock and Martin J. Bishop

EMBnet is a network of bioinformatics sites across Europe and beyond.

It aims to provide access to specialized knowledge in bioinformatics beyond the capability of any individual node. Nodes provide training in bioinformatics and collaborate to develop new tools. EMBnet was formed in 1988 and is registered as a foundation in The Netherlands, the EMBnet Stifting. Although originally envisioned as a European collaboration, it now has nodes in 30 countries spanning every continent except Antarctica (see Table 1). EMBnet nodes are of two types. National nodes are mandated by their national governments to provide bioinformatics services, whereas specialist nodes have specialized knowledge in a particular area of bioinformatics. Some of the specialist nodes are industrial rather than academic, adding another dimension to the collaboration.

Related Website

EMBnet	http://www.embnet.org/

Further Reading

Luo J (2002). Bioinformatics service, education and research: The EMBnet and CBI. European Molecular Biology Network Centre of Bioinformatics. *In Silico Biol.* 2: 173–177.

See also Asia Pacific Bioinformatics Network.

Table 1 Web addresses of EMBnet Nodes

National Nodes			
Country	Node Name	URL	Institute
Argentina	IBBM	sol.biol.unlp.edu.ar	Instituto de Bioquimica y Biologia Molecular
Austria	VBC	www.at.embnet.org	Vienna University Computer Center
Australia	ANGIS	www.au.embnet.org	Australia National Genomic Information System
Canada	CBR-RBC	www.cbr.nrc.ca	National Research Council of Canada, Institute for Marine Biosciences
Belgium	BEN	www.be.embnet.org	Belgian EMBnet Node
China	CBI	www.cn.embnet.org	Centre of Bioinformatics, Peking University
Cuba	CIGB	www.cu.embnet.org	Centro de Ingenieria Genetica y Biotechnologia
Chile	DCC	www.dcc.uchile.cl	University of Chile

Table 1 Web addresses of EMBnet Nodes

National Nodes			
Country	Node Name	URL	Institute
Denmark	BioBase	www.dk.embnet.org	Danish Biotechnological Database
France	InfoBioGen	www.fr.embnet.org	Resource Centre Infobiogen
Finland	CSC	www.fi.embnet.org	National Center for Scientific Computing
Germany	GeniusNet	www.de.embnet.org	German Cancer Research Centre
Greece	IMBB	www.imbb.forth.gr	Institute of Molecular Biology and Biotechnology
Hungary	HEN	www.hu.embnet.org	Hungarian EMBnet Node
India	CDFD	www.in.embnet.org	Centre for DNA Fingerprinting and Diagnostics
Ireland	INCBI	www.ie.embnet.org	Irish National Centre for Bioinformatics
Israel	INN	www.il.embnet.org	Weizmann Institute of Science
Italy	CNR	www.it.embnet.org	Area di Ricerca—Bari
Mexico	CIFN	www.embnet.cifn.unam.mx	Nitrogen Fixation Research Center
Norway	BiO	www.no.embnet.org	Biotechnology Centre of Oslo
Poland	IBB	www.pl.embnet.org	Institute of Biochemistry and Biophysics
Portugal	PEN	www.pt.embnet.org	Portuguese EMBnet Node
Russia	GeneBee	www.ru.embnet.org	Belozersky Institute of Physico-Chemical Biology, Moscow State University
Slovakia	IMB-SAS	www.sk.embnet.org	Institute of Molecular Biology, Slovak Academy of Science
South Africa	SANBI	www.za.embnet.org	South African National Bioinformatics Institute, University of Western Cape
Spain	CNB	www.es.embnet.org	Centro Nationale de Biotecnologia, Campus Universidad Autonoma

Table 1 Web addresses of EMBnet Nodes

National Nodes			
Country	Node Name	URL	Institute
Sweden	LCB	www.se.embnet.org	Linnaeus Centre for Bioinformatics, Uppsala Biomedical Centre
Switzerland	SIB	www.ch.embnet.org	Swiss Institute of Bioinformatics
The Netherlands	CMBI	www.nl.embnet.org	Centre for Molecular and Biomolecular Informatics, University of Nijmegen
United Kingdom	HGMP	www.uk.embnet.org	UK MRC Human Genome Mapping Project Resource Centre

Specialist Nodes		
Name	URL	Institute/Company
EBI	www.ebi.ac.uk	European Bioinformatics Institute
ETI	www.eti.uva.nl	Expert Centre for Taxonomic Identification
ICGEB	www.icgeb.trieste.it	International Centre for Genetic Engineering and Biotechnology
UMBER	www.bioinf.man.ac.uk	University of Manchester
MIPS	www.mips.biochem.mpg.de	Max-Planck-Institute fur Biochemie
Pharmacia	www.pnu.com	Pharmacia & Upjohn
Roche	www.roche.com	F. Hoffmann-La Roche
Sanger Centre	www.sanger.ac.uk	The Sanger Centre

EMBOSS (THE EUROPEAN MOLECULAR BIOLOGY OPEN SOFTWARE SUITE)

John M. Hancock and Martin J. Bishop

A suite of open-source programs intended to provide an open-source alternative to the Wisconsin Package.

The EMBOSS project developed out of the earlier EGCG project, which aimed to provide extended functionality to the Wisconsin Package making use of its freely available libraries. When the Wisconsin Package moved to commercial distribution, access to its source code was no longer available and the EMBOSS project emerged to provide an open-source alternative. EMBOSS provides approximately 100 programs classified under the following general headings: sequence alignment, rapid database searching with sequence patterns, protein motif identification, including domain analysis, nucleotide sequence pattern analysis, codon usage analysis, rapid identification of sequence patterns in large-scale sequence sets, and presentation tools for publication. The package is available for a number of Unix platforms.

Related Website

| EMBOSS | http://www.hgmp.mrc.ac.uk/Software/EMBOSS/ |

Further Reading

Rice P, et al. (2000). EMBOSS: The European Molecular Biology Open Software Suite. *Trends Genet.* 16: 276–277.

See also Wisconsin Package.

EMOTIF
Terri Attwood

A collection of regular expressions representing conserved regions (motifs) derived from sequence families in the Blocks+ and PRINTS databases. Rather than encode the exact information observed at each position in a motif, eMOTIF adopts a "permissive" approach in which alternative residues are tolerated according to a set of prescribed groupings. These groups correspond to various biochemical properties (e.g., charge, size), theoretically ensuring that the resulting motifs have meaningful biological interpretations:

Residue property	Residue groups
Small	Ala, Gly
Small hydroxyl	Ser, Thr
Acid/amide	Asp, Glu, Asn, Gln
Basic	His, Lys, Arg
Aromatic	Phe, Tyr, Trp
Medium hydrophobic	Val, Leu, Ile, Met

Although designed to be more flexible than exact regular expression matching, its inherent permissiveness brings with it a signal-to-noise trade-off; i.e., resulting patterns not only have the potential to make more true-positive matches but will consequently also match more false positives. Therefore, when searching the resource, different levels of stringency are offered from which to infer the significance of matches.

eMOTIF is available from the Department of Biochemistry at the University of Stanford, Stanford, California. Because the database is generated automatically, family annotation is not provided directly but is linked to the relevant entries in PRINTS.

Related Websites

| eMOTIF | http://motif.stanford.edu/emotif/ |
| eMOTIF search | http://motif.stanford.edu/emotif/emotif-search.html |

Further Reading

Huang JY, Brutlag DL (2001). The eMOTIF database. *Nucleic Acids Res.* 29(1): 202–204.

Nevill-Manning CG, Sethi KS, Wu TD, Brutlag DL (1997). Enumerating and ranking discrete motifs. *Proc. Intl. Conf. Intell. Syst. Mol. Biol.* 5: 202–209.

Nevill-Manning CG, Wu TD, Brutlag DL (1998). Highly specific protein sequence motifs for genome analysis. *Proc. Natl. Acad. Sci. USA* 95: 5865–5871.

See also Block, Motif, PRINTS, Regular Expression.

EMPIRICAL PAIR POTENTIALS
Patrick Aloy

Empirical pair potentials (EPPs) are mean-force knowledge-based potentials obtained by applying Boltzmann's equation to raw residue-pair statistics derived from sampling databases of known structures.

During the last years, EPPs have been largely applied to many kinds of biological problems (e.g., fold recognition, macromolecular docking, interaction discovery) and have proven to be a very successful scoring system.

In principle, being derived empirically, EPPs incorporate the dominant thermodynamic effects without explicitly having to model each of them. The "physical" validity of this approach has been largely criticized, as EPPs do not reflect the "true potentials" when applied to test systems with known potential functions. However, from a pragmatic perspective, as long as the potentials favor the native conformation over others, they are good enough.

Further Reading
Reva BA, Finkelstein AV, Skolnik J (2000). Derivation and testing residue-residue mean-force potentials for use in protein structure recognition. *Methods Mol. Biol.* 143: 155–174.

EMPIRICAL POTENTIAL ENERGY FUNCTION
Roland L. Dunbrack

A function that represents conformational energies of molecules and macromolecules, derived from experimental data such as crystal structures and thermodynamic measurements. This is in contrast to *ab initio* potential energy functions, derived purely from quantum mechanical calculations, although some empirical functions may also use *ab initio* calculations to fit parameters in the energy function. The general form of the empirical potential energy function is:

$$U = \sum_{\substack{\text{bonds}}} \frac{1}{2} K_b (b - b)^2 + \sum_{\substack{\text{bond} \\ \text{angles}}} \frac{1}{2} K_\theta (\theta - \theta_o)^2 + \sum_{\substack{\text{dihedral} \\ \text{angles}}} K_\phi [1 + \cos(n\phi - \delta)]$$

$$+ \sum_{\substack{\text{nonbonded} \\ \text{atom pairs} \\ j > i}} 4\varepsilon_{ij} \left[\left(\frac{\sigma_{ij}}{r} \right)^{12} - \left(\frac{\sigma_{ij}}{r} \right)^6 \right] + \sum_{\substack{\text{nonbonded} \\ \text{atom pairs} \\ j > i}} \frac{q_i q_j}{\varepsilon r}$$

The first two terms are harmonic energy terms for all covalent bonds and bond angles, where for each pair and triple of atom types there is a separate value for the force constants K_b and K_θ, respectively, and the equilibrium values b_o and θ_o. For all sets of four connected atoms (the chain 1–2–3–4 with covalent bonds 1–2, 2–3, and 3–4), dihedral angle energy terms are expressed as the sum of periodic functions, each with its own K_ϕ, periodicity $2\pi/n$, and phase displacement δ. The last two terms are van der Waals energy and Coulombic electrostatic energy. These two terms usually skip all 1–3 and 1–4 terms (those in the same bond angle or dihedral angle). Additional terms may be included in some terms, including 1–3 pseudobond harmonic forces, improper dihedral energy terms, and solvation terms.

In molecular mechanics potential energy functions, some hydrogen atoms may not be represented explicitly. Nonpolar carbon groups such as methyl and methylene groups are represented as single atoms in the united-atom approximation with van der Waals radius large enough to represent the entire group. In united-atom potentials, polar hydrogens such as those on NH and OH groups are represented explicitly, so that polar interactions such as hydrogen bonds can be represented accurately. More modern potentials include all hydrogen atoms explicitly.

Related Websites

Empirical Force Field Development Resources (CHARMM parameter development)	http://www.pharmacy.umaryland.edu/faculty/amackere/param/force_field_dev.htm
AMBER	http://sigyn.compchem.ucsf.edu/amber/

Further Reading

MacKerell AD, Jr, et al. (1998). All-atom empirical potential for molecular modeling and dynamics studies of proteins. *J. Phys. Chem.* B102: 3586–3616.

Cornell WD, et al. (1995). A second generation force field for the simulation of proteins, nucleic acids, and organic molecules. *J. Am. Chem. Soc.* 117: 5179–5197.

See also Ab Initio, Improper Dihedral Angle, Solvation Free Energy.

ENCPRIME/SEQCOUNT
Michael P. Cummings

A pair of programs to calculate codon bias summary statistics.

SeqCount (sequence count) produces tables of nucleotide and codon counts and frequencies from protein coding sequence data. ENCprime (effective number of codons prime) calculates the codon usage bias statistics effective number of codons $\hat{N}c$, effective number of codons prime $\hat{N}'c$, scaled χ^2, and $B^*(a)$. The number $\hat{N}'c$ takes into account background nucleotide composition, is independent of gene length, and has a low coefficient of variation.

The input file format is FASTA (SeqCount) or output from SeqCount (ENCprime). The programs are available as American National Institute for Standardization (ANSI) C source code.

Related Website

John Novembre's software page	http://ib.berkeley.edu/labs/slatkin/novembre/software/software.html

Further Reading

Novembre JA (2002). Accounting for background nucleotide composition when measuring codon usage bias. *Mol. Biol. Evol.* 19: 1390–1394.

END GAP
Jaap Heringa

End gaps denote gap regions in sequence alignments at either end of a sequence, as opposed to gaps that are inserted in between residues in an alignment.

An important issue is whether to penalize end gaps preceding or following a sequence when a global alignment containing the structure is constructed. The first implementation of global alignment using the dynamic programming algorithm (Needleman and Wunsch, 1970) and other early versions all applied end-gap penalties, including the CLUSTAL method for multiple alignment. However, this is generally not a desirable situation (Abagyan and Batalov, 1997), as many proteins comprise recurring structural domains, often in combination with different complementary domains. For example, if a sequence corresponding to a two-domain protein is globally aligned with a protein sequence associated with one of the two domains only, applying end gaps would be likely to lead to an incorrect alignment in which the single-domain sequence spans the two-domain sequence. Therefore, most modern alignment routines do not penalize end gaps anymore. Global alignment without penalize end gaps is commonly referred to as semi-global alignment. A situation where applying end gaps might be desirable is when aligning tandem repeats, because the various repeat sequences are in fact consecutive parts of a single sequence, so that internal gaps and end gaps cannot be distinguished. End gaps are not an issue in local alignment (Smith and Waterman, 1981) as this technique aligns the most similar regions in a target sequence set only, so that end gaps do not occur.

Further Reading

Abagyan RA, Batalov S (1997). Do aligned sequences share the same fold? *J. Mol. Biol.* 273: 355–368.

Needleman SB, Wunsch CD (1970). A general method applicable to the search for similarities in the amino acid sequence of two proteins. *J. Mol. Biol.* 48: 443–453.

Smith TF, Waterman MS (1981). Identification of common molecular subsequences. *J. Mol. Biol.* 147: 195–197.

ENERGY MINIMIZATION
Roland L. Dunbrack

A procedure for lowering the energy of a molecular system by changing the atomic coordinates. Energy minimization requires a defined potential energy function expressed as a function of internal and/or Cartesian coordinates. Because the energy function is complicated and nonlinear, energy minimization proceeds in an iterative fashion. The methods differ in how the step direction is chosen and how large the steps are. Most such procedures are guaranteed to find not the global minimum energy of the system but rather a local energy minimum.

Energy minimization methods can be distinguished by whether they use the first and/or second derivatives of the energy function with respect to the coordinates. First-order methods include the steepest-descent method and the conjugate gradient method. The steepest-descent method changes the atomic coordinates of a molecular system in steps in the opposite direction of the first derivative of the potential energy function. This derivative is the gradient of the energy function with respect to the Cartesian coordinates of all atoms. The step size is varied such that it is increased if the new conformation has a lower energy and decreased if it is higher. Steepest-descent minimization will usually not converge but rather sample around a local minimum. The method is fast and is often used to remove close steric contacts or other locally high-energy configurations in protein structures, since it finds the nearest local minimum to the starting position.

The conjugate gradient method is an energy minimization procedure that changes atomic coordinates in steps based on the first derivative of the potential energy and the direction of the previous step. It can be written symbolically as

$$\delta_i = -\mathbf{g}_i + \delta_{i-1} \frac{|\mathbf{g}_i|^2}{|\mathbf{g}_{i-1}|^2}$$

$$\mathbf{r}_{i+1} = \mathbf{r}_i + \alpha \delta_i$$

where $\mathbf{g}_i = \nabla U(\mathbf{r}_i)$ is the gradient of the potential energy function U; α determines the step size and is often chosen by a line search, i.e., trying several values, evaluating the energy, and choosing the best value. The purpose of using the previous step is to account for the curvature of the potential energy function without actually calculating the second derivative of this function. Conjugate gradient minimization generally converges faster than steepest-descent minimization.

Second-order methods that use both the first and second derivatives of the energy function include the Newton–Raphson method and a variation called the adopted-basis Newton–Raphson method. For a one-dimensional system with an approximately quadratic (harmonic) potential, it is possible to move to the minimum in one step:

$$x_{\min} = x_o - \left[\frac{dU}{dx} \right]_{x_o} \Big/ \left[\frac{d^2U}{dx^2} \right]_{x_o}$$

For a molecular system, the change at each step can be expressed as

$$\delta_i = -\mathbf{H}_i^{-1} \mathbf{g}_i$$

where $\mathbf{H}$ is the Hessian matrix of all second derivatives of the potential energy. The (k,l)th entry in the Hessian matrix is

$$\mathbf{H}_{i,kl} = \left[\frac{\partial^2 U}{\partial x_k \partial x_l} \right]_{\mathbf{r}_i}$$

where x_k is the x, y, or z coordinate of any atom in the system. The Newton–Raphson method therefore requires $3N$ first-derivative terms and $(3N)^2$ second-derivative terms as well as a large matrix inversion. As such, it is a very expensive calculation and may behave erratically far from a local minimum.

One solution to the size problem of the Newton–Raphson method is to perform the calculation on a set of linear combinations of the coordinates, where the number of vectors in the set is very small. These vectors are chosen to be in the direction of the largest changes in recent steps:

$$\Delta \mathbf{r}_i^{(p)} = \mathbf{r}_{i-1} - \mathbf{r}_{i-1-p}$$

where p ranges from 1 to N, the number of vectors in the set, usually between 4 and 10.

The Newton–Raphson equations are then solved for this basis set. This is the adopted-basis Newton–Raphson (ABNR) method.

Related Websites

Minimization course	http://www.tau.ac.il/~becker/course/mini.html
Newton–Raphson method	http://www.shodor.org/unchem/math/newton/

Further Reading
Brooks CL III, et al. (1988). *Advances in Chemical Physics*, Vol. 71. Wiley, New York.

ENHANCER
Niall Dillon

Regulatory sequence of a gene responsible for increasing transcription from it.

Operational definition describing a sequence that increases transcription from a promoter that is linked to the enhancer *in cis*. A defining feature of enhancers is that they can exert their effects independently of orientation relative to the promoter and with some flexibility with respect to distance from the promoter. The definition does not specify the nature of the assay used in the enhancer effect or the size of the increase in transcription. Analysis of specific enhancers has shown that they contain multiple binding sites for different transcription factors, which in turn recruit auxiliary factors. Together, these factors form a complex called the enhanceosome.

Further Reading
Merika M, Thanos D (2001). Enhanceosomes. *Curr. Opin. Genet. Dev.* 11: 205–208.
Carey M (1998). The enhanceosome and transcriptional synergy. *Cell* 92: 5–8.

ENSEMBL
Guenter Stoesser

Ensembl is a genome sequence and annotation database including graphical views and Web-searchable data sets.

The project creates and applies software systems which produce and maintain automatic annotation on the genomes of humans and other vertebrates. As new genomes become available (e.g., mouse, fly, rat, and zebrafish), comparative genome views are generated. Ensembl annotates known genes and predicts novel genes, with functional annotation from EMBL-Bank and the InterPro protein family databases and with additional annotation by OMIM disease, SAGE expression, and gene family. Additionally, ab initio gene predictions are generated via the program genescan to create a set of genescan peptides.

Ensembl automatic annotation is available as an interactive web service. Ensembl contigview web pages feature the ability to scroll along entire chromosomes, while viewing all integrated features within a selected region in detail. Users can control which features are displayed and dynamically integrate external data sources such as the Distributed Annotation System (DAS) to easily view and compare annotations from different sources that are distributed across the Internet.

Ensembl has been one of the leading sources of human genome sequence annotation and the Ensembl system is being installed around the world in both industrial and academic environments. Ensembl is a joint project between EMBL-EBI and the Sanger Centre.

Related Websites

Ensembl	http://www.ensembl.org/
Human Genome Central	http://www.ensembl.org/genome/central/
Distributed Annotation System	http://www.biodas.org/

Further Reading

Apweiler R, et al. (2001). The InterPro database, an integrated documentation resource for protein families, domains and functional sites. *Nucleic Acids Res.* 29: 37–40.

Dowell RD, et al. (2001). The Distributed Annotation System. *BMC Bioinform.* 2: 7.

Burge C, Karlin S (1997). Prediction of complete gene structures in human genomic DNA. *J. Mol. Biol.* 268: 78–94.

Hubbard TJP, Birney E (2000). Open annotation offers a democratic solution to genome sequencing. *Nature* 403: 825.

Hubbard TJP, et al. (2002). The Ensembl genome database project. *Nucleic Acids Res.* 30: 38–41.

ENTREZ

Dov S. Greenbaum

A large-scale data retrieval system linked to many of the National Center for Biotechnology Information (NCBI) databases.

Entrez allows for interlinking between all of the databases in its system. Specifically it provides researchers access to the following databases:

1. Online Mendelian Inheritance in Man, OMIM, an on-line version of the catalog of human genes and genetic disorders.
2. PubMed a search tool for accessing literature citations and linking to full-text journal articles.
3. A nucleotide sequence database which collects primary sequences from several sources, including GenBank, RefSeq, and Protein Data Bank (PDB).
4. A protein sequence database compiled from SWISS-PROT, PIR, PRF, PDB, and translations from annotated coding regions in GenBank and RefSeq.
5. A structure database containing over 10,000 three-dimensional macromolecular structures of proteins and polynucleotides.
6. A genome database containing the whole and partially sequenced genomes for over 800 organisms from archaea, eukaryota, bacteria, organelles, and viruses.
7. The PopSet database, which contains alignments of both nucleotide and protein sequences resulting from population, phylogenetic, or mutation studies.
8. A taxonomy database containing the names and some information regarding all the organisms listed at least once in the NCBI databases.
9. A database of online biomedical books.
10. A conserved domain database that can be searched for and used to identify conserved motifs in proteins.
11. UniSTS, a unified nonredundant database of sequence-tagged sites collected from dbSTS, RHdb, GDB, and various human and mouse maps.

12. ProbeSet, a database for the Gene Expression Omnibus (GEO), a gene expression and hybridization array repository.

Related Website

| Entrez | http://www.ncbi.nlm.nih.gov/Entrez/ |

Further Reading

Schuler GD, et al. (1996). Entrez: Molecular biology database and retrieval system. *Methods Enzymol.* 266: 141–162.

ENTROPY
Thomas D. Schneider

Entropy is a measure of the state of a system that can roughly be interpreted as its randomness. Since the entropy concept in thermodynamics and chemistry has units of energy per temperature (joules per kelvin), while the uncertainty measure from Claude Shannon has units of bits per symbol, it is best to keep these concepts distinct.

Related Websites

| Entropy | http://www.math.psu.edu/gunesch/entropy.html |
| Entropy 2 | http://www.entropysite.com |

See also Negentropy, Second Law of Thermodynamics, Shannon Entropy, Uncertainty.

ENZYME
Jeremy Baum

Proteins whose function is to catalyze reactions are referred to as enzymes. Enzymes are classified according to the chemistry of the reaction(s) they catalyze. Many enzymes are not pure peptide but also contain cofactors. Some enzymes may also have a structural function.

See also Cofactor.

E-PCR. *SEE* ELECTRONIC PCR.

EPISTASIS
Mark McCarthy and Steven Wiltshire

Pattern of interaction between two loci contributing to a phenotype whereby the probability of developing the phenotype is contingent on the genotypes at both loci.

Alternatively, describes an interaction in which the effects of one gene override or mask the phenotype of a second gene.

In contrast to dominance (where one allele masks the expression of another allele at the same gene), epistasis refers to the interaction of two different genes. Such interactions may arise, e.g., when two genes overlap in their biological function(s) such that either can compensate for dysfunction of the other, with phenotypic consequences only arising if both are disturbed (i.e., a normal genotype at one locus can compensate for a deficient genotype at the other). In the context of linkage analysis, epistatic interaction models are often contrasted with heterogeneity models (in which disease can result from mutations in either locus). It is possible to include such multilocus models (and the interactions between them) in linkage analyses, particularly for multifactorial traits where multiple susceptibility loci are likely to be involved, on the basis that this may represent a more realistic model of genotype–phenotype relationships. However, such analyses are associated with significant costs in terms of computation and the potential for type 1 error inflation due to multiple testing.

Examples: One classical example is of fruit color in squash, which is controlled by two loci. At the first locus white color is dominant to (any) color; at the second, yellow is dominant to green. Phenotypic expression of this second locus is, however, only possible (i.e., the squash is only green or yellow) in the presence of the homozygous recessive genotype at the first. In their search for type 2 diabetes loci in man, Cox and colleagues (1999) sought evidence of interaction between the two most promising regions emerging from a genome scan for linkage performed in Mexican American sibships. They reported evidence of an epistatic interaction between loci on chromosomes 2 and 15, and this information contributed to efforts to identify calpain-10 as the disease susceptibility locus on 2q.

Further Reading

Cox NJ, et al. (1999). Loci on chromosomes 2 (NIDDM1) and 15 interact to increase susceptibility to diabetes in Mexican Americans. *Nat. Genet.* 21: 213–215.

Ott J (1999). *Analysis of Human Genetic Linkage*, 3rd ed. Johns Hopkins University Press, Baltimore, MD, pp. 297–305.

Rapp JP, et al. (1998). Construction of a double congenic strain to prove an epistatic interaction on blood pressure between rat chromosomes 2 and 10. *J. Clin. Invest.* 101: 1591–1595.

See also Linkage Analysis.

ERROR
Thomas D. Schneider

In communications, an error is the substitution of one symbol for another in a received message caused by noise. Shannon's channel capacity theorem showed that it is possible to build systems with as low an error as desired, but one cannot avoid errors entirely.

Further Reading

Shannon CE (1948). A mathematical theory of communication. *Bell Syst. Tech. J.*, 27: 379–423, 623–656. http://cm.bell-labs.com/cm/ms/what/shannonday/paper.html

Shannon CE (1949). Communication in the presence of noise. *Proc. IRE*, 37: 10–21.

See also Channel Capacity, Message.

ERROR MEASURES (ACCURACY MEASURES, GENERALIZATION, PERFORMANCE CRITERIA, PREDICTIVE POWER)

Nello Cristianini

In a classifier, we call accuracy (error rate) the expected rate of correct (incorrect) predictions made by the model over a data set drawn from the same distribution that generated the training data. The true accuracy can only be estimated from a finite sample, and different methods can be used. Accuracy is usually estimated by using an independent test set of data that was not used at any time during the learning process. However, in the literature more complex accuracy estimation techniques, such as cross-validation and bootstrap, are commonly used, especially with data sets containing a small number of instances.

More specific measures of performance can be defined in special cases, e.g., in the important special case of measuring performance of a two-class classifier. In this case, there are only four possible outcomes of a prediction:

True positives—the prediction is positive and the actual class is positive.
True negatives—the prediction is negative and the actual class is negative.
False positives—the prediction is positive and the actual class is negative.
False negatives—the prediction is negative and the actual class is positive.

Let the number of points falling within each of these categories be represented by the variables TP, TN, FP, and FN. They can be organized in a so-called confusion matrix as follows:

	Predicted negative	Predicted positive
Actual negative	TN	FP
Actual positive	FN	TP

Then we can define the following measures of performance of a binary classifier:

Accuracy	$A = (TN + TP)/(TN + FP + FN + TP)$.
True-positive rate (recall, sensitivity)	$TPR = TP/(FN + TP)$.
True-negative rate (specificity)	$TNR = TN/(TN + FP)$.
Precision	$P = TP/(TP + FP)$.
False-positive rate	$FPR = FP/(FN + TP)$.
False-negative rate	$FNR = FN/(TN + FP)$.

The 2×2 confusion matrix contains all the information necessary to compute the above measures of performance of a two-class classifier. More generally it is possible to define an $L \times L$ confusion matrix, where L is the number of different label values.

It is also interesting to compare the ranking induced on the data by different functions. Often, classifiers split the data by sorting them and choosing a cutoff point.

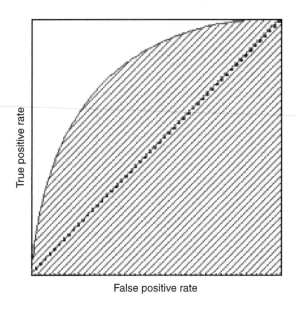

Figure   An example of ROC curve.

The choice of the cutoff (threshold) is the result of a trade-off between sensitivity and specificity in the resulting classifier.

By moving the threshold, one can obtain different levels of sensitivity and specificity. If one plots sensitivity vs. 1−specificity (or equivalently true-positive rate vs, false-positive rate) for all possible choices of cutoff, then the resulting plot is called a receiver operating characteristic curve (ROC curve), a term derived from applications in radar signal processing. Such a curve shows the trade-off between sensitivity and specificity (see the accompanying figure): The closer it lies to the left and top borders, the better is the underlying classifier; the closer it lies to the diagonal, the worse (random guessing gives the diagonal ROC curve).

Finally, one can consider the area delimited by the ROC curve as a measure of performance called the ROC ratio (an area of 1 represents a perfect classifier, of 1/2 a random one). This quantity is sometimes called "discrimination" and is related to the probability that two random points, one taken from the positive and one from the negative class, are both correctly classified by the model.

Further Reading

Metz CE (1978). Basic principles of ROC analysis. *Semin. Nucl. Med.* 8: 283–298.

Mitchell T (1997). *Machine Learning*. McGraw-Hill, New York, 1997.

Provost F, et al. (1998). The case against accuracy estimation for comparing induction algorithms. *Proc. Intl. Conf. Machine Learning.*

See also Classification, Cross-Validation.

E-S ALGORITHM. *SEE* ELSTON–STEWART ALGORITHM.

EST. *SEE* EXPRESSED SEQUENCE TAG.

EUROPEAN BIOINFORMATICS INSTITUTE. *SEE* EBI.

EUROPEAN MOLECULAR BIOLOGY OPEN SOFTWARE SUITE. *SEE* EMBOSS.

EVA
Patrick Aloy

EVA is a Web server for automatically assessing protein structure prediction methods in a large-scale continuous way. Currently, EVA evaluates the performance of a variety of structure prediction methods available on the Internet. The server assesses the accuracy of comparative modeling, threading, secondary-structure, and residue contact prediction methods.

Every week, EVA automatically downloads the new three-dimensional structure entries from the Protein Data Bank (PDB). The sequences of these proteins are sent to the prediction servers and the results are collected and analyzed. An updated summary of the evaluation is published on the Web.

Related Website

EVA	http://cubic.bioc.columbia.edu/eva/

Further Reading
Eyrich VA, Marti-Renom MA, Przybylski D, Madhusudhan MS, Fiser A, Pazos F, Valencia A, Sali A, Rost B (2001). EVA: Continuous automatic evaluation of protein structure prediction servers. *Bioinformatics* 17: 1242–1243.

See also Modeling, Macromolecular; Secondary-Structure Prediction of Protein; Threading.

E VALUE
Jaap Heringa

The expectancy value (*E* value) is a statistical value for estimating the significance of alignments performed in a homology search of a query sequence against a sequence database. It basically estimates for each alignment the number of sequences randomly expected in the sequence database to provide an alignment score, after alignment with the query sequence, at least that of the alignment considered. Such significance estimates are important because sequence alignment methods are essentially pattern-search techniques, leading to an alignment with a similarity score even in case of the absence of any biological relationship. Although similarity scores of unrelated sequences are essentially random, they can behave like "real" scores and, for instance, like real scores are correlated with the length of the sequences compared. Particularly in the context of database searching, it is important to know what scores can be expected by chance and how scores that deviate from random expectation should be assessed. If within a rigid statistical framework a sequence similarity is deemed statistically significant, this provides confidence in deducing that the sequences involved are in fact biologically related. As a result of the complexities of protein sequence evolution and distant relationships observed in nature, any statistical scheme will invariably lead to situations where a sequence is assessed as unrelated while it is in fact homologous (false negative), or the inverse, where a sequence is deemed homologous while it is in fact biologically unrelated (false positive). The derivation of a general statistical framework for evaluating the significance of sequence similarity scores has been a major task. However, a rigid framework has not been established for global alignment and has only partly been completed for local alignment.

Sequence similarity values resulting from global alignments are known to grow linearly with the sequence length, although the growth rate has not been determined. Also, the exact distribution of global similarity scores is yet unknown, and only numerical approximations exist, providing some rough bound on the expected random scores. Since the variance of the global similarity score has not been determined either, most applications derive a sense of the score by using shuffled sequences. Shuffled sequences retain the composition of a given real sequence but have a permuted order of nucleotides or amino acids. The distribution of similarity scores over a large number of such shuffled sequences often approximates the shape of the Gaussian distribution, which is therefore taken to represent the underlying random distribution. Using the mean (m) and standard deviation (σ) calculated from such shuffled similarity scores, each real score S can be converted to the z-score using z-score $= (S - m)/\sigma$. The z-score measures how many standard deviations the score is separated from the mean of the random distribution. In many studies, a z-score >6 is taken to indicate a significant similarity.

A rigid statistical framework for local alignments without gaps has been derived for protein sequences following the work by Karlin and Altschul (1990), who showed that the optimal local ungapped alignment score grows linearly with the logarithm of the product of sequence lengths of two considered random sequences: $S \sim \ln(n \cdot m)/\lambda$, where n and m are the lengths of two random sequences and λ is a scaling parameter that depends on the scoring matrix used and the overall distribution of amino acids in the database. Specifically, λ is the unique solution for x in the equation

$$\sum_{i,j} p_i p_j e^{S_{i,j}x} = 1$$

where the summation is over all amino acid pairs, p_i represents the background probability (frequency) of residue type i, and $s_{i,j}$ is the scoring matrix.

An important contribution for fast sequence database searching has been the realization (Dembo and Karlin, 1991; Dembo et al., 1994; Karlin and Altschul, 1990) that local similarity scores of ungapped alignments follow the extreme-value distribution (EVD) (Gumbel, 1958). This distribution is unimodal but not symmetric like the normal distribution, because the right-hand tail at high scoring values falls off more gradually than the lower tail, reflecting the fact that a best local alignment is associated with a score that is the maximum out of a great number of independent alignments.

Following the EVD, the probability of a score S to be larger than a given value x can be calculated as

$$P(S \geq x) = 1 - \exp(-e^{-\lambda(x-\mu)})$$

where $\mu = (\ln Kmn)/\lambda$ and K is a constant that can be estimated from the background amino acid distribution and scoring matrix [see Altschul and Gish (1996) for a collection of values for λ and *K* over a set of widely used scoring matrices]. Using the equation for μ, the probability for *S* becomes

$$P(S \geq x) = 1 - \exp(-Kmne^{-\lambda x})$$

In practice, the probability $P(S \geq x)$ is estimated using the approximation $1 - \exp(-e^{-x}) \approx e^{-x}$, which is valid for large values of *x*. This leads to a simplification of the equation for $P(S \geq x)$:

$$P(S \geq x) \approx e^{-\lambda(x-\mu)} = Kmne^{-\lambda x}$$

The lower the probability for a given threshold value *x*, the more significant the score *S*.

Although similarities between sequences can be detected reasonably well using methods that do not allow insertions/deletions in aligned sequences, it is clear that insertion/deletion events play a major role in divergent sequences. This means that accommodating gaps within alignments of distantly related sequences is important for obtaining an accurate measure of similarity. Unfortunately, a rigorous statistical framework as obtained for gapless local alignments has not been conceived for local alignments with gaps. However, although it has not been proven analytically that the distribution of *S* for gapped alignments can be approximated with the EVD, there is accumulated evidence that this is the case: For example, for various scoring matrices, gapped alignment similarities have been observed to grow logarithmically with the sequence lengths (Arratia and Waterman, 1994). Other empirical studies have shown it to be likely that the distribution of local gapped similarities follows the EVD (Smith et al., 1985; Waterman and Vingron, 1994), although an appropriate downward correction for the effective sequence length has been recommended (Altschul and Gish, 1996). The distribution of empirical similarity values can be obtained from unrelated biological sequences (Pearson, 1998). Fitting of the EVD parameters λ and *K* (see above) can be performed using a linear regression technique (Pearson, 1998), although the technique is not robust against outliers, which can have a marked influence. Maximum-likelihood estimation (Lawless, 1982; Mott, 1992) has been shown to be superior for EDV parameter fitting and, e.g., is the method used to parameterize the gapped BLAST method (Altschul et al., 1997). However, when low gap penalties are used to generate the alignments, the similarity scores can lose their local character and assume more global behavior such that the EVD-based probability estimates are not valid anymore (Arratia and Waterman, 1994).

In order to be useful in sequence database searches, it is important to establish the probability for a given query sequence to have a significant similarity with at least one of the database sequences. A *p* value is defined as the probability of seeing at least one unrelated score *S* greater than or equal to a given score *x* in a database search over *n* sequences. This probability has been demonstrated to follow the Poisson distribution (Waterman and Vingron, 1994):

$$P(x, n) = 1 - e^{-n \cdot P(S \geq x)}$$

where *n* is the number of sequences in the database. Some database search methods employ the expectation value (or *E* value) of the Poisson distribution, which is defined as the expected number of nonhomologous sequences with score greater than or equal to a score *x* in a database of *n* sequences:

$$E(x, n) = n \cdot P(S \geq x)$$

For example, if the E value of a matched database sequence segment is 0.01, then the expected number of random hits with score $S \geq x$ is 0.01, which means that this E value is expected by chance only once in 100 independent searches over the database. However, if the E value of a hit is 5, then five fortuitous hits with $S \geq x$ are expected within a single database search, which renders the hit not significant. Database searching is commonly performed using an E value between 0.1 and 0.001. Low E values decrease the number of false positives in a database search but increase the number of false negatives, thereby lowering the sensitivity of the search.

Further Reading

Altschul SF, Gish W (1996). Local alignment statistics. *Methods Enzymol.* 266: 460–480.

Altschul SF, et al. (1997). Gapped BLAST and PSI-BLAST: A new generation of protein database search programs. *Nucleic Acids Res.* 25: 3389–3402.

Arratia R, Waterman MS (1994). A phase transition for the score in matching random sequences allowing depletions. *Ann. Appl. Prob.* 4: 200–225.

Dembo A, Karlin S (1991). Strong limit theorems of empirical functionals for large exceedances of partial sums of i.i.d. variables. *Ann. Prob.* 19: 1737.

Dembo A, et al. (1994). Limit distributions of maximal non-aligned two-sequence segmental score. *Ann. Prob.* 22: 2022.

Gumbel EJ (1958). *Statistics of Extremes.* Columbia University Press, New York.

Karlin S, Altschul SF (1990). Methods for assessing the statistical significance of molecular sequence features by using general scoring schemes. *Proc. Natl. Acad. Sci. USA* 87: 2264–2268.

Lawless JF (1982). *Statistical Models and Methods for Lifetime Data.* Wiley, New York.

Mott R (1992). Maximum-likelihood estimation of the statistical distribution of Smith-Waterman local sequence similarity scores. *Bull. Math. Biol.* 54: 59–75.

Pearson WR (1998). Empirical statistical estimates for sequence similarity searches. *J. Mol. Biol.* 276: 71–84.

Smith TF, et al. (1985). The statistical distribution of nucleic acid similarities. *Nucleic Acids Res.* 13: 645–656.

Waterman MS, Vingron M (1994). Rapid and accurate estimates of statistical significance for sequence data base searches. *Proc. Natl. Acad. USA* 91: 4625–4628.

EVOLUTION

A. Rus Hoelzel

Evolution is descent with modification leading to the accumulation of change in the characteristics of organisms or populations.

The importance and influence of ideas about evolution changed forever when Charles Darwin proposed the mechanism of natural selection to explain this process.

Evolution is said to be divergent when related species change such that they no longer resemble each other and convergent when unrelated species change such that they resemble each other more.

Microevolution is thought of as the process of evolutionary change within populations, while macroevolution reflects the pattern of evolution at and above the species level. However, there is a continuum from the processes that affect gene frequencies at the population level through to the pattern of lineages in phylogenetic reconstructions. In each case the direction and rate of evolution is influenced by random processes such as genetic drift and the differential survival of phenotypes through natural selection.

Further Reading

Darwin C (1859). *The Origin of Species by Means of Natural Selection, or the Preservation of Favoured Races in the Struggle for Life*, Penguin Books, London.

Ridley M (1996). *Evolution*. Blackwell Science, Oxford, UK.

Maynard Smith J (1998). *Evolutionary Genetics*, 2nd ed, Oxford University Press, Oxford, UK.

See also Allopatric Evolution, Convergence, Sympatric Evolution.

EVOLUTIONARY CLOCK. *SEE* MOLECULAR CLOCK.

EVOLUTIONARY DISTANCE
Sudhir Kumar and Alan Filipski

An evolutionary distance measure is a numerical representation of the dissimilarity between two species, sequences, or populations.

Evolutionary distances are often computed by comparing molecular sequences or by using allele frequency data. A set of evolutionary distances may be used to construct a phylogenetic tree, and conversely, a phylogenetic tree with branch lengths induces a set of distances on the taxa.

Commonly used distance measures obey the following conditions for all taxa (denoted by i, j, and k): (1) nonnegativity—all $d_{ij} \geq 0$; (2) distinctness—$d_{ij} = 0$ if and only if $i = j$; (3) symmetry—$d_{ij} = d_{ji}$; and (4) triangle inequality—$d_{ij} \leq d_{ik} + d_{jk}$. The set of all pairwise distances among n taxa is commonly represented as a symmetric matrix in which all the diagonal entries are zero and hence is determined by specifying $n(n - 1)/2$ independent values.

The four-point condition is said to hold for a distance measure if and only if the quantity $d_{a,b} + d_{c,d}$ is less than or equal to the maximum of the quantities $d_{a,d} + d_{b,c}$ and $d_{a,c} + d_{b,d}$ for all sets of four taxa a, b, c, and d. This condition requires equality of the largest two of the three quantities $d_{a,b} + d_{c,d}$, $d_{a,d} + d_{b,c}$, and $d_{a,c} + d_{b,d}$. The four-point condition is equivalent to additivity for the distance measure. In this case, there is some phylogenetic tree on the set of taxa such that the given distance between any two taxa is equal to the patristic distance (sum of branch lengths) from one to the other.

The ultrametric inequality is said to be satisfied for a distance measure if and only if $d_{a,c}$ is less than or equal to the maximum of $d_{a,b}$ and $d_{b,c}$ for all triplets of species. This actually implies that, of the three distances, the longer two are equal. If a molecular clock is assumed, then UPGMA may be used to create a rooted tree with all leaf nodes equidistant from the root.

Further Reading

Kumar S, et al. (2001). *MEGA2*: Molecular evolutionary genetics analysis software. *Bioinformatics* 17: 1244–1245.

Nei M, Kumar S (2000). *Molecular Evolution and Phylogenetics*. Oxford University Press, Oxford, UK.

Sokal RR, Sneath PHA (1963). *Principles of Numerical Taxonomy*. W. H. Freeman, San Francisco.

See also Sequence Distance Measures.

EVOLUTION OF BIOLOGICAL INFORMATION
Thomas D. Schneider

The information of patterns in nucleic acid binding sites can be measured as *Rsequence* (the area under a sequence logo). The amount of information needed to find the binding sites, *Rfrequency*, can be predicted from the size of the genome and number of binding sites. *Rfrequency* is fixed by the current physiology of an organism but *Rsequence* can vary. A computer simulation shows that the information in the binding sites (*Rsequence*) does indeed evolve toward the information needed to locate the binding sites (*Rfrequency*) (Schneider, 2000).

Related Website

EV	http://www.lecb.ncifcrf.gov/~toms/paper/ev

Further Reading
Adami C, et al. (2000). Evolution of biological complexity. *Proc. Natl. Acad. Sci. USA* 97: 4463–4468.

Schneider TD (2000). Evolution of biological information. *Nucleic Acids Res.* 28: 2794–2799.

Schneider TD, et al. (1986). Information content of binding sites on nucleotide sequences. *J. Mol. Biol.* 188: 415–431. http://www.lecb.ncifcrf.gov/~toms/paper/schneider1986/.

See also Binding Site, Pattern, Rsequence, Rfrequency, Sequence Logo.

EXCLUSION MAPPING
Mark McCarthy and Steven Wiltshire

Use of linkage analysis to designate genomic regions for which the evidence against linkage is sufficiently strong as to "exclude" that region from involvement in disease susceptibility, at least under a specific disease model.

Just as it is possible to use linkage analysis to determine the likely location of a disease susceptibility gene, it is possible to define genomic locations where such genes are unlikely to be found. Such approaches were originally developed for the analysis of Mendelian diseases in the context of parametric linkage analyses. In this situation, disease locus parameters are specified in advance, and it is trivial to exclude regions, on the basis that some values of the recombination fraction (and hence some genomic locations) are incompatible with the experimental data, under the specified disease locus model. Typically, a LOD score (logarithm of the odds for linkage) of -2 or less is taken as indicating exclusion. In the case of multifactorial traits, where analysis is predominantly based on nonparametric methods and effect size and disease parameters are not specified in advance, exclusion mapping requires designation of a locus effect size (typically parametrized in terms of the locus-specific sibling relative risk, or λ_s). It is then possible to designate those genomic locations which are excluded from containing a susceptibility locus of this magnitude, although such exclusions do not necessarily rule out a locus of more modest effect from that location.

Examples: Hanis et al. (1996) reported a genome scan for type 2 diabetes conducted in Mexican American families. The strongest evidence for linkage was on chromosome 2q. They were able to exclude 99% of the genome from harboring a gene with a locus-specific sibling relative risk of 2.8. Equivalent figures for loci of lesser effect were 71% (for $\lambda_s = 1.6$) and 5% (for $\lambda_s = 1.2$). Chung et al. (1995) used homozygosity mapping in consanguineous families to exclude the mineralocorticoid receptor from a role in the etiology of a form of pseudohypoaldosteronism.

Related Websites
Programs for linkage and exclusion mapping:

ASPEX	ftp://lahmed.stanford.edu/pub/aspex/
GENEHUNTER	http://www-genome.wi.mit.edu/ftp/distribution/software/genehunter

Further Reading
Chung E, et al. (1995). Exclusion of a locus for autosomal recessive pseudohypoaldosteronism type 1 from the mineralocorticoid receptor gene region on human chromosome 4q by linkage analysis. *J. Clin. Endocrinol. Metab.* 80: 3341–3345.

Clerget-Darpoux F, Bonaïti-Pellié C (1993). An exclusion map covering the whole genome: A new challenge for genetic epidemiologists? *Am. J. Hum. Genet.* 52: 442–443.

Hanis CL, et al. (1996). A genome-wide search for human non-insulin-dependent (type 2) diabetes genes reveals a major susceptibility locus on chromosome 2. *Nat. Genet.* 13: 161–171.

Hauser E, et al. (1996). Affected-sib-pair interval mapping and exclusion for complex genetic traits: Sampling considerations. *Genet. Epidemiol.* 13: 117–137.

See also Genome Scans, Linkage Analysis, Multipoint Linkage Analysis.

EXON
Niall Dillon and Laszlo Patthy

Segment of a eukaryotic gene which, when transcribed, is retained in the final spliced mRNA.

Exons are separated from one another by introns which are removed from the primary transcript by RNA splicing.

Eukaryotic genes are split into exons and intervening sequences (introns). The primary RNA transcript contains both exons and introns, and the introns are then removed in a complex RNA splicing reaction which is closely coupled to transcription. The coding region of a gene is contained within the exons, but exons can also include the 5′ and 3′ untranslated regions of the mRNA. Exons or parts of exons of protein coding genes that are translated are collectively referred to as protein coding regions.

Further Reading
Lewin B (2000). *Genes VII.* Oxford University Press, Oxford, UK.

See also Coding Region, Intron.

EXON SHUFFLING
Austin L. Hughes and Laszlo Patthy

The process whereby exons of genes are duplicated or deleted or exons of different genes are joined through recombination in introns.

It has been hypothesized that exon shuffling has played an important role in the evolution of new gene functions, particularly when exons are transferred and encode a domain that constitutes a structural or functional unit. For such cases, the term "domain shuffling" is frequently used. Shuffling of exons or exon sets encoding complete protein modules leads to module shuffling and to the formation of multidomain proteins with an altered domain organization.

Further Reading
Doolittle RF (1995). The multiplicity of domains in proteins. *Annu. Rev. Biochem.* 64: 287–314.

Gilbert W (1978). Why genes in pieces? *Nature* 271: 501.

Patthy, L (1995). Protein Evolution by Exon Shuffling. Molecular Biology Intelligence Unit, R.G. Landes Company, Springer-Verlag. New York.

Patthy, L (1996). Exon shuffling and other ways of module exchange. *Matrix Biol.* 15: 301–310.

Patthy L (1999). Genome evolution and the evolution of exon-shuffling—a review. *Gene* 238: 103–114.

See also Exon, Intron, Intron Phase, Module Shuffling, Multidomain Protein, Protein Module.

EXPECTATION MAXIMIZATION ALGORITHM (E-M ALGORITHM)
Mark McCarthy and Steven Wiltshire

An algorithm, much used in various forms of genetic analysis, for the maximum-likelihood estimation of parameters in situations where all the information required for the estimation is not available (because it either is missing or cannot be observed).

In this iterative two-step algorithm, initial values for the parameters whose values are to be estimated are chosen and the function of the parameters, or some part therefore, evaluated—the E step—to obtain expected values for the missing data. These values are then used in the M step to reestimate the values for the parameters. In turn, these new estimates are used in a second cycle of the algorithm to obtain new expectations of the missing data and, thence, more accurate estimates of the parameters. The cycle is repeated until the algorithm reaches convergence—i.e., the change in the values of the parameters to be estimated is sufficiently small as to no longer exceed some predetermined threshold—providing maximum-likelihood estimates of the parameters of interest.

Examples: The E-M algorithm is used by the EH and ARLEQUIN programs to estimate haplotype frequencies (the parameters of interest) from unphased population data. In this case, the unknown or missing information is the haplotypes in individuals who are doubly heterozygous (in whom phase is ambiguous). Initial estimates of the haplotype frequencies are used to obtain expected values for the haplotype distributions in these individuals (the E step). These expectations are then used to reestimate the haplotype frequencies (the M step). Dawson et al. (2002), in their construction of a first-generation haplotype map of chromosome 19, used the E-M algorithm to calculate haplotype frequencies from the population and combined population/pedigree data. From these, they were able to determine the pattern of pairwise linkage disequilibrium across the chromosome.

Related Websites
Programs that implement the E-M algorithm for linkage disequilibrium mapping:

EH-PLUS	http://www.iop.kcl.ac.uk/IoP/Departments/PsychMed/GEpiBSt/software.stm
ARLEQUIN	http://anthropologie.unige.ch/arlequin/
GOLD	http://www.sph.umich.edu/csg/abecasis/GOLD/

Further Reading
Dawson E, et al. (2002). A first-generation linkage disequilibrium map of human chromosome 22. *Nature* 418: 544–548.

Sham P (1998). *Statistics in Human Genetics*. Arnold, London, pp. 151–157.

Weir BS (1996) *Genetic Data Analysis II*. Sinauer Associates, Sunderland, MA, pp. 67–73.

See also Haplotype, Linkage Disequilibrium, Phase.

EXPRESSED SEQUENCE TAG (EST)
Jean-Michel Claverie

A small (one-read) sequence fragment of a cDNA clone insert, selected at random from a complementary DNA (cDNA) library.

ESTs are used according to two different modes: the *gene discovery* mode and the *expression profiling* mode. In the gene discovery mode several thousands of inserts of random picked clones from a cDNA library are sequenced to establish a "quick-and-dirty" repertoire of the genes expressed in the organism or tissue under study. The value of this approach is rapidly limited by the redundancy originating from the most abundant transcripts. This problem is partially corrected by the use of "normalized" cDNA libraries. Prior to the launch of the human genome sequencing projects, large-scale EST sequencing programs from normalized libraries were performed that allowed a tag to be generated for most human genes. These ESTs played an important role in the annotation of the human genome. In the expression profiling mode the redundancy of ESTs corresponding to the same gene is used as a measurement of expression intensity. ESTs generated from the exact 3′ end (3′ EST; see accompanying figure) of a given transcript give the most accurate estimate of the number of transcripts corresponding to a given gene. 5′ ESTs (randomly covering other parts of the gene transcripts) are usually preferred for gene discovery, as they usually exhibit a higher fraction of coding regions. It is worth recalling that, in contrast with 3′ EST sequences that do correspond to the 3′ extremity transcripts (mostly 3′ UTR), 5′ ESTs very rarely correspond to the 5′ extremity of transcripts.

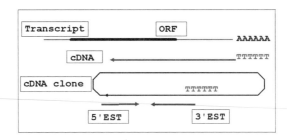

Figure   The EST concept.

Related Websites

Bodymap gene expression database	http://bodymap.ims.u-tokyo.ac.jp/
dbEST EST database	http://www.ncbi.nlm.nih.gov/dbEST/
Cancer Genome Anatomy Project	http://www.ncbi.nlm.nih.gov/ncicgap/

Further Reading

Adams MD, et al. (1991). Complementary DNA sequencing: Expressed sequence tags and human genome project. *Science* 252: 1651–1656.

Adams MD, et al. (1992). Sequence identification of 2,375 human brain genes. *Nature* 355: 632–634.

Bailey LC, et al. (1998). Analysis of EST-driven gene annotation in human genomic sequence. *Genome Res.* 8: 362–376.

Okubo K, et al. (1992). Large scale cDNA sequencing for analysis of quantitative and qualitative aspects of gene expression. *Nat. Genet.* 2: 173–179.

EXPRESSION PROFILER

John M. Hancock and Martin J. Bishop

A suite of programs for the analysis of microarray data.

The suite contains programs for clustering expression data (EPCLUST), relating expression profiles to protein–protein interaction data (EP:PPI), discovering shared sequence patterns (SPEXS), visualizing patterns in biological sequences (PATMATCH), relating expression data to genomic sequences (GENOMES), drawing sequence logos (SEQLOGO), browsing the Gene Ontology and extracting associated genes (EP:GO), and linking applications (URLMAP). Web interfaces are provided for all programs.

Related Website

Expression Profiler	http://ep.ebi.ac.uk/EP/

Further Reading

Brazma A, Vilo J (2000). Gene expression data analysis. *FEBS Lett.* 480: 17–24.

Brazma A, et al. (1998). Predicting gene regulatory elements in silico on a genomic scale. *Genome Res.* 8: 1202–1215.

Kemmeren P, et al. (2002). Protein interaction verification and functional annotation by integrated analysis of genome-scale data. *Mol. Cell* 9: 1133–1143

Möller S, et al. (2001). Prediction of the coupling specificity of G protein coupled receptors to their G proteins. *Bioinformatics* 17: S174–S181.

Vilo J, Kivinen K (2001). Regulatory sequence analysis: Application to interpretation of gene expression. *Eur. Neuropsychopharmacol.* 11: 399–411.

Vilo J, et al. (2000). Mining for putative regulatory elements in the yeast genome using gene expression Data. In *Intelligent Systems for Molecular Biology, 2000.* AAAI Press, Menlo Park, CA, pp. 384–394.

See also Clustering, Gene Ontology, Microarray, Sequence Logo.

EXTENSIBLE MARKUP LANGUAGE. *SEE* XML.

EXTRINSIC GENE PREDICTION. *SEE* GENE PREDICTION, HOMOLOGY BASED.

F

Family-Based Association Analysis
FASTA
FASTP
Features
Fingerprint
FingerPrinted Contigs
Finishing
Finishing Criteria
Fisher Discriminant Analysis
FLUCTUATE
FlyBase
Fold

Fold Library
Fold Recognition
Force Field
Founder Effect
FPC
Frame-Based Language
Free *R*-Factor
Functional Databases
Functional Genomics
Functional Signature
Functome

Dictionary of Bioinformatics and Computational Biology. Edited by Hancock and Zvelebil
ISBN 0-471-43622-4 © 2004 John Wiley & Sons, Inc.

FAMILY-BASED ASSOCIATION ANALYSIS
Mark McCarthy and Steven Wiltshire

Variant of linkage disequilibrium (association) analysis in which family-based controls are used in place of population-based controls.

The archetypal design for linkage disequilibrium (or association) analysis is the case–control study, in which the aim is to determine whether a particular allele (or haplotype) occurs more frequently in chromosomes from cases (i.e., individuals enriched for disease susceptibility alleles) than in chromosomes from a sample of control individuals. These controls are typically unrelated individuals, ostensibly from the same population as the cases, selected either at random or on the basis that they do not have the disease phenotype. In principle, however, such a study design is prone to generate spurious associations if the case and control populations are not tightly matched for genetic background, since latent population substructure may lead to case–control frequency differences which reflect differences in ancestry rather than differences in disease status. In other words, associations may be seen between loci which are not in linkage disequilibrium (in the stricter definition of that term). Family-based association methods were devised to counter this possibility: By using appropriately selected family-based control individuals, it is possible to ensure that positive results are obtained only when the variant genotype and the susceptibility allele are both linked and associated and thereby protect against population substructure. The most frequently used family-based association method, the transmission disequilibrium test (or TDT), is based on the analysis of trios comprising an affected individual and both parents. Analysis proceeds by comparing, in parents who are heterozygous for the typed variant, those alleles transmitted to affected offspring with those not so transmitted. Deviation from expectation under the null hypothesis is, under appropriate circumstances, a simultaneous test of both linkage and association. Other family-based association methods have been devised for use in discordant sibling pairs (e.g., the sibling transmission/disequilibrium test) and in general pedigrees (the pedigree disequilibrium test, or PDT) and for quantitative traits (QTDT). Disadvantages of family-based association methods over the case–control design may include (a) increased collection costs, especially for late-onset conditions where parents may be rare; (b) increased genotyping costs, since, in the TDT study design, three individuals have to be typed to provide the equivalent of one case–control pair; and (c) lower power, particularly in the case of discordant sibling pair (sibpair) methods.

Examples: Huxtable et al(2000) used parent offspring trios to demonstrate both linkage and allelic association between a regulatory minisatellite upstream of the insulin gene and type 2 diabetes and thereby to consolidate previous findings from case–control studies. An added advantage of the use of trios, apparent in this study, was the ability to examine parent-of-origin effects: Susceptibility could be shown to be restricted to alleles transmitted from fathers. Altshuler and colleagues (2000) combined association analysis conducted using a range of case–control and family-based association resources to confirm that the *Pro12Ala* variant of the *PPARG* gene is also implicated in type 2 diabetes susceptibility.

Dictionary of Bioinformatics and Computational Biology. Edited by Hancock and Zvelebil
ISBN 0-471-43622-4 © 2004 John Wiley & Sons, Inc.

Related Websites
Programs for conducting family-based association tests:

TRANSMIT	www-gene.cimr.cam.ac.uk/clayton/software
PDT	http://www.chg.duke.edu/software/index.html
QTDT	http://www.sph.umich.edu/csg/abecasis/QTDT/
ETDT	http://www.mds.qmw.ac.uk/statgen/dcurtis/software.html

Further Reading

Abecasis GR, et al. (2000). A general test of association for quantitative traits in nuclear families. *Am. J. Hum. Genet.* 66: 279–292.

Altshuler D, et al. (2000). The common PPARgamma Pro12Ala polymorphism is associated with decreased risk of type 2 diabetes. *Nat. Genet.* 26: 76–80.

Frayling TM, et al. (1999). Parent-offspring trios: A resource to facilitate the identification of type 2 diabetes genes. *Diabetes* 48: 2475–2479.

Huxtable SJ, et al. (2000). Analysis of parent-offspring trios provides evidence for linkage and association between the insulin gene and type 2 diabetes mediated exclusively through paternally transmitted class III variable number tandem repeat alleles. *Diabetes* 49: 126–130.

Lander ES, Schork NJ (1994). Genetic dissection of complex traits. *Science* 265: 2037–2048.

Martin ER, et al. (2000). A test for linkage and association in general pedigrees: The pedigree disequilibrium test. *Am. J. Hum. Genet.* 67: 146–154.

Spielman RS, Ewens WJ (1998). A sibship test for linkage in the presence of association: The sib transmission/disequilibrium test. *Am. J. Hum. Genet.* 62: 450–458.

Spielman RS, et al. (1993). Transmission test for linkage disequilibrium: The insulin gene region and insulin-dependent diabetes mellitus (IDDM). *Am. J. Hum. Genet.* 52: 506–516.

See also Allelic Association, Association Analysis, Linkage Disequilibrium.

FASTA (FASTP)

Jaap Heringa

Until recently, the most widely used routine for sequence database searching.

FASTA (Pearson and Lipman, 1988) is a more modern implementation of the older FASTP technique. The FASTA program compares a given query sequence with a library of sequences and calculates for each pair the highest scoring local alignment. The speed of the algorithm is obtained by delaying application of the dynamic programming technique to the moment where the most similar segments are already identified by faster and less sensitive techniques. To accomplish this, the FASTA routine operates in four steps. The first step searches for identical words of a user-specified length occurring in the query sequence and the target sequence(s). The technique is based on that of Wilbur and Lipman (1983, 1984) and involves searching for identical words (*k*-tuples) of a certain size within a specified bandwidth along search matrix diagonals. For not-too-distant sequences (>35% residue identity), little sensitivity is lost while speed is greatly increased. The search is performed by a technique known as hash coding or hashing, where a lookup table is constructed for all words in the query sequence, which is then used to compare all encountered words in the target sequence(s). The relative positions of each word in the two sequences are then calculated by subtracting the position in the first sequence from that in the second. Words that have the same offset position reveal a region of alignment between the two sequences. The number of comparisons increases linearly in proportion to average sequence length. In contrast, the time taken in dot-matrix and dynamic programming methods increases as the square of the average sequence length. The k-tuple length is user defined and is usually 1 or 2 for protein sequences (i.e., either the positions of each of the individual 20 amino acids or the positions of each of the 400 possible dipeptides are located). For nucleic acid sequences, the k-tuple is 5–20 and should be longer because short k-tuples are much more common due to the four-letter alphabet of nucleic acids. The larger the k-tuple chosen, the more rapid but less thorough a database search.

Generally, for proteins, a word length of two residues is sufficient ($ktup=2$). Searching with higher *ktup* values increases the speed but also the risk that similar regions are missed. For each target sequence the 10 regions with the highest density of ungapped common words are determined. In the second step, these 10 regions are rescored using the Dayhoff PAM 250 residue exchange matrix (Dayhoff et al., 1983) and the best scoring region of the 10 is reported under *init1* in the FASTA output. In the third step, regions scoring higher than a threshold value and being sufficiently near each other in the sequence are joined, now allowing gaps. The highest score of these new fragments can be found under *initn* in the FASTA output. The fourth and final step performs a full dynamic programming alignment (Chao et al., 1992) over the final region, which is widened by 32 residues at either side, of which the score is written under *opt* in the FASTA output.

Further Reading

Chao K-M, et al. (1992). Aligning two sequences within a specified diagonal band. *Comput. Appl. Biosci.* 8: 481–487.

Dayhoff MO, et al. (1983). Establishing homologies in protein sequences. *Methods Enzymol.* 91: 524–545.

Pearson WR, Lipman DJ (1988). Improved tools for biological sequence comparison. *Proc. Natl. Acad. Sci. USA* 85: 2444–2448.

Wilbur WJ, Lipman DJ (1983). Rapid similarity searches of nucleic acid and protein data banks. *Proc. Natl. Acad. Sci. USA* 80: 726–730.

Wilbur WJ, Lipman DJ (1984). The context dependent comparison of biological sequences. *SIAM J. Appl. Math.* 44: 557–567.

FASTP. *SEE* FASTA.

FEATURES (ATTRIBUTE, DESCRIPTOR, INDEPENDENT VARIABLE, OBSERVATION, PREDICTOR VARIABLES)
Nello Cristianini

Attributes describing a data item.

In machine learning, each data item is typically described by a set of attributes, called features (and also descriptors, predictors, or independent variables). Often they are organized as vectors of fixed dimension, and the features assume numeric values, although it is possible to have symbolic features (e.g., {*small, medium, large*} or {*green, yellow*}) or Boolean features (e.g., {*present, absent*}).

When present, the desired response for each input is instead denoted as a "label" (or response or dependent variable).

The problem of automatically identifying the subset of features that is most relevant for the task at hand is called *feature selection* and is of crucial importance in situations where the initial number of attributes is very large and many of them are irrelevant or redundant, since this may lead to overfitting. This is, e.g., often the case for data generated by DNA microarrays.

Further Reading
Duda RO, Hart PE, Stork DG (2001). *Pattern Classification*. Wiley, New York.
Mitchell T (1997). *Machine Learning*. McGraw-Hill, New York.

See also Machine Learning.

FINGERPRINT (SIGNATURE)
Terri Attwood

A group of ungapped motifs excised from a multiple alignment and used to build a characteristic signature of protein family membership by means of iterative database searching. The process of building fingerprints is largely manual, commencing with the creation of a representative seed alignment and identification of its most conserved regions. The motifs are extracted, converted to residue frequency matrices, and used to trawl a SWISS-PROT/TrEMBL composite database. Results are then examined to determine which sequences have matched all the motifs—if more sequences have matched the fingerprint than were in the seed alignment, the additional sequence information is added to the motifs, and the database is searched again. This process is repeated until the scans converge, i.e., until no more new matches can be found that contain all the constituent motifs.

Fingerprints offer improved diagnostic reliability over single-motif-based sequence analysis approaches (e.g., such as the regular expression searching embodied in PROSITE) because of the mutual context provided by motif neighbors: The more motifs matched, the greater the confidence that a diagnosis is correct provided all the motifs match in the correct order with appropriate distances between them. In contrast to domain-based approaches (such as profiles and hidden Markov models), which perform best in the diagnosis of superfamily membership, fingerprints perform best in the diagnosis of protein families and subfamilies.

Fingerprints underpin the PRINTS database. They may be searched using the FingerPRINTScan suite, which allows queries of individual sequences against the whole database or of a single sequence against individual fingerprints.

Related Websites

About fingerprints	http://www.bioinf.man.ac.uk/dbbrowser/PRINTS/printsman.html
FingerPRINTScan	http://www.bioinf.man.ac.uk/dbbrowser/fingerPRINTScan/
PRINTS	http://www.bioinf.man.ac.uk/dbbrowser/PRINTS/

Further Reading

Attwood TK (2001). A compendium of specific motifs for diagnosing GPCR subtypes. *Trends Pharmacol. Sci.* 22(4): 162–165.

Attwood TK, Blythe M, Flower DR, Gaulton A, Mabey JE, Maudling N, McGregor L, Mitchell A, Moulton G, Paine K, Scordis P (2002). PRINTS and PRINTS-S shed light on protein ancestry. *Nucleic Acids Res.* 30(1): 239–241.

Attwood TK, Findlay JBC (1994). Fingerprinting G-protein-coupled receptors. *Protein Eng.* 7(2): 195–203.

FINGERPRINTED CONTIGS. *SEE* FPC.

FINISHING

Rodger Staden

The process of completing and checking the correctness of a sequencing project, often using a set of finishing criteria.

See also Finishing Criteria.

FINISHING CRITERIA

Rodger Staden

Those engaged in large sequencing projects often have a set of defined criteria which state the quality of the finished sequence they attempt to produce and set out the steps that they have taken to resolve problems. For example: "All regions were double stranded, sequenced with an alternate chemistry, or covered by high-quality data (i.e., phred quality ≥ 30); an attempt was made to resolve all sequencing problems, such as compressions and repeats; all regions were covered by at least one plasmid subclone or more than one M13 subclone; and the assembly was confirmed by restriction digest."

Related Website

Finger Rules	http://www.genome.wustl.edu/gsc/Overview/hgfinrules.php

FISHER DISCRIMINANT ANALYSIS (LINEAR DISCRIMINANT ANALYSIS)

Nello Cristianini

The statistical and computational task of assigning data points to one of two (or more) predefined classes is called *classification* or *discrimination*, and the function that maps data into classes is called a *classifier* or a *discriminant function*.

Typically the data items are denoted by vectors $\mathbf{x} \in \Re^n$ and their classes by "labels" $y \in \{-1, +1\}$. The classifier is then a function $f : \Re^n \to \{-1, +1\}$. When the classifier is based upon a linear function $g_{w,b}(\mathbf{x}) = \langle \mathbf{w}, \mathbf{x} \rangle + b$ parametrized by a vector $\boldsymbol{w} \in \Re^n$ (and possibly by a real number b called "bias"), this statistical task goes under the name of linear discriminant analysis (LDA) and reduces to the problem of selecting a suitable value of (w,b). Many algorithms exist for learning the parameters (w,b) from a set of labeled data, called "training data," among them the Perceptron and the Fisher discriminant (this giving rise to Fisher discriminant analysis, or FDA).

In FDA the parameters are chosen so as to optimize a criterion function that depends on the distance between the means of the two populations being separated and on their variances. More precisely, consider projecting all the multivariate (training) data onto a generic direction w and then separately observing the mean and the variance of the projections of the two classes. If we denote μ_+ and μ_- the means of the projections of the two classes on the direction w and s_+ and s_- their variances, the cost function associated to the direction w is then defined as

$$C(w) = \frac{(\mu_+ - \mu_-)^2}{s_+^2 + s_-^2}$$

FDA selects the direction w that maximizes this measure of separation.

This mathematical problem can be rewritten in the following way. Define the scatter matrix of the ith class as

$$S_i = \sum_{x \in \text{class}(i)} (\mathbf{x} - \overline{\mathbf{x}}_i)(\mathbf{x} - \overline{\mathbf{x}}_i)^T$$

(where by $\overline{\mathbf{x}}$ we denote the mean of a set of vectors) and define the total within-class scatter matrix as $S_W = S_1 + S_{-1}$. In this way one can write the denominator of the criterion as $s_+^2 + s_-^2 = w^T S_W w$ and, similarly, the numerator can be written as $(\mu_+ - \mu_-)^2 = w^T (\overline{\mathbf{x}}_1 - \overline{\mathbf{x}}_{-1})(\overline{\mathbf{x}}_1 - \overline{\mathbf{x}}_{-1})^T w = w^T \overset{\circ}{S}_B w$, where S_B is the between-class scatter matrix.

In this way the criterion function becomes

$$C(w) = \frac{w' S_B w}{w' S_W w}$$

(an expression known as a Reyleigh quotient) and its maximization amounts to solving the generalized eigenproblem $S_B w = \lambda S_w w$.

This method was introduced by R. A. Fisher in 1936 and can also be used in combination with kernel functions, in this way transforming it into an effective nonlinear classifier. It can be proven to be optimal when the two populations are Gaussian and have the same covariance.

Further Reading
Duda RO, et al. (2001). *Pattern Classification*. Wiley, New York.

See also Classification, Neural Network, Support Vector Machines.

FLUCTUATE. *SEE* LAMARC.

FLYBASE
Guenter Stoesser

FlyBase is an integrated database for genomic and genetic data on the major genetic model organism *Drosophila melanogaster* and related species.

Following the publication of the *Drosophila* genome sequence, FlyBase has primary responsibility for the continual reannotation of the *D. melanogaster* genome. FlyBase organizes genetic and genomic data on chromosomal sequences and map locations, on the structure and expression patterns of encoded gene products, and on mutational and transgenic variants and their phenotypes. The publicly funded stock centers are included, as are numerous cross-links to sequence databases and to homologs in other model organism databases. Images and graphical interfaces within FlyBase include interactive and static regional maps as well as anatomical drawings and photomicrographs. FlyBase is one of the founding participants in the Gene Ontology Consortium, which provides ontologies for the description of the molecular functions, biological processes, and cellular components of gene products. The FlyBase Consortium includes *Drosophila* researchers and computer scientists at Harvard University, University of California, Indiana University, University of Cambridge (United Kingdom), and the European Bioinformatics Institute (United Kingdom).

Related Websites

Flybase	http://flybase.bio.indiana.edu/
Berkeley *Drosophila* Genome Project	http://awww.fruitfly.org/
European *Drosophila* Genome Project	http://edgp.ebi.ac.uk/
Gene Ontology Consortium	http://www.geneontology.org/

Further Reading

Adams MD, et al. (2000). The genome sequence of *Drosophila melanogaster*. *Science* 287: 2185–2195.

FlyBase Consortium (2002). The FlyBase database of the *Drosophila* genome projects and community literature. *Nucleic Acids Res*. 30: 106–108.

Matthews K, Cook K (2001). Genetic information online: A brief introduction to FlyBase and other organismal databases. In *Encyclopedia of Genetics*, ECR Reeve, Ed. Fitzroy Dearborn Publishers, Chicago, IL.

Rubin GM, et al. (2000). Comparative genomics of the eukaryotes. *Science* 287: 2204–2215.

FOLD

Andrew Harrison and Christine A. Orengo

The fold of a protein describes the topology of the three-dimensional domain structure that the polypeptide chain adopts as it folds. The topology refers to both the connectivity of the positions (e.g., residues or secondary structures) in the domain structure and the arrangements of those positions in three-dimensional coordinate space. Analysis of protein families has revealed that the structures of homologous proteins are well conserved during evolution, presumably because it is important for the stability and the function of the protein. However, since insertions and deletions of residues can occur between related sequences during evolution, this can sometimes result in extensive structural embellishments. In this sense the folds of the distant structural relatives may not correspond completely over the entire domain. However, it is conventional to describe a group of related domains as having a common fold if their structures are similar over a significant portion. Furthermore, most structural classifications tend to group proteins having similar domain structures over at least 50% to 60% of their chains into the same fold group. Some schools of opinion suggest that it may be appropriate to consider not discrete fold groups but rather a continuum of structures. In 2002, there are approximately 700 different fold groups known, and some researchers have speculated that there may be fewer than 1000 folds adopted in nature due to constraints on secondary-structure packing.

Related Websites

DALI server	http://www2.ebi.ac.uk/dali/
SCOP server	http://scop.mrc-lmb.cam.ac.uk/scop/
3Dee database	http://www.compbio.dundee.ac.uk/3Dee/
CATH server	http://www.biochem.ucl.ac.uk/bsm/cath_new

Further Reading

Orengo CA (1994). Classification of protein folds. *Curr. Opin. Struct. Biol.* 4: 429–440.
Orengo CA (1999). Protein folds, functions and evolution. *J. Mol. Biol.* 293: 333–342.

See also Homology.

FOLD LIBRARY

Andrew Harrison and Christine A. Orengo

Fold libraries comprise representative three-dimensional structures of each known domain fold. Currently fewer than 1000 unique folds are known. However, there are no commonly agreed-upon quantitative descriptions of protein folds, so that determining whether two proteins share a common fold is a somewhat subjective process and will depend on the methods used for aligning the structures and scoring their similarity. For this reason, different numbers of fold groups are reported in the public structure classification resources. Fold libraries are often used with threading methods which perform a one- to three-dimensional alignment and attempt to "thread" a query sequence through a three-dimensional target structure to determine whether it is similar to the structure adopted by the query. It is important to use nonredundant data sets with these methods, both to improve the reliability of the statistics and to reduce the time required to thread the query against all structures in the library, since threading is a very computationally expensive method.

Related Websites

DALI server	http://www2.ebi.ac.uk/dali/
SCOP server	http://scop.mrc-lmb.cam.ac.uk/scop/
3Dee database	http://www.compbio.dundee.ac.uk/3Dee/
CATHWheels	http://www.biochem.ucl.ac.uk/bsm/cathwheels/
CATH server	http://www.biochem.ucl.ac.uk/bsm/cath_new

Further Reading
Thornton JM, Orengo CA, Todd AE, Pearl FM (1999). Protein folds, functions and evolution. *J. Mol. Biol.* 293: 333–342.

See also Alignment Entries, Fold, Threading.

FOLD RECOGNITION
David T. Jones

A general term for any method which attempts to predict the tertiary structure of a protein from its amino acid sequence by selecting the best matching fold to its native structure from a set of alternatives (fold library).

Comparative modeling can be thought of as a special case of fold recognition, in which case the fold is recognized purely from sequence similarity. More generally, however, fold recognition methods are aimed at predicting the structure for proteins with little or no detectable sequence similarity to any proteins of known three-dimensional structure. Fold recognition methods can make use of different sources of information in identifying the most likely matching folds, including sequence profile information, predicted secondary structure, and various statistical scoring functions such as pair potentials and solvation potentials.

Related Websites

GenThreader	http://bioinf.cs.ucl.ac.uk/psipred/index.html
TOPITS	http://www.embl-heidelberg.de/predictprotein/predictprotein.html
UCLA-DOE	http://fold.doe-mbi.ucla.edu/
3D-PSSM	http://www.sbg.bio.ic.ac.uk/~3dpssm/

Further Reading
Fischer D, Eisenberg D (1996). Protein fold recognition using sequence-derived predictions. *Protein Sci* 5(5): 947–955.

Jones DT (1999). GenTHREADER: An efficient and reliable protein fold recognition method for genomic sequences. *J. Mol. Biol.* 287: 797–815.

Kelley LA, MacCallum RM, Sternberg MJE (2000). Enhanced genome annotation using structural profiles in the program 3D-PSSM.
J. Mol. Biol. 299(2): 499–520.

See also Fold, Fold Library, Homology Modeling, Threading.

FORCE FIELD
Roland L. Dunbrack

A potential energy function that expresses energy as a function of atomic positions for use in molecular dynamics simulations or energy minimization.

See also Empirical Potential Energy Function.

FOUNDER EFFECT
A. Rus Hoelzel

The founder effect refers to the distortion in allele frequency that can occur when a population is founded by a small sample from a larger source population.

The high frequency of rare heritable diseases found in some captive breeding populations [e.g., chondrodystrophy (lethal dwarfism) in the breeding population of California condors] and in some human populations (e.g., Finland) has been attributed to small founder populations.

The founder effect has also been proposed as a possible mechanism for promoting speciation, especially in the context of founding new populations on islands. For example, this may have been important in the radiation of some of the approximately 800 *Drosophila* species found in the Hawaiian Islands.

Further Reading
Graur D, Li W-H (1999). *Fundamentals of Molecular Evolution*. Sinauer Associates, Sunderland, MA.
Hartl DL (2000). *A Primer of Population Genetics*. Sinauer Associates, Sunderland, MA.

See also Evolution, Genetic Drift, Population Bottleneck.

FPC (FINGERPRINTED CONTIGS)
John M. Hancock and Martin J. Bishop

A program to combine restriction endonuclease fingerprint patterns of genomic clones to produce a contig.

In one approach to genome sequencing, a preliminary stage is to produce a large number of clones and to characterize them by restriction endonuclease digestion, producing a pattern of bands. The purpose of FPC is to use these patterns to identify overlapping clones and place them in their correct physical order on the chromosome. Clones are clustered into contigs based on a score known as probability of coincidence. For each contig a consensus band (CB) map is produced. Individual clones are then aligned to the CB map, allowing visualization of their chromosomal order. FPC allows editing of contigs, including merging, splitting, and deleting, and markers can be associated with particular clones. Sequence ready clones can be selected for further processing.

Related Website

FPC	http://www.genome.arizona.edu/fpc/

Further Reading
Soderlund C, et al. (1997). FPC: A system for building contigs from restriction fingerprinted clones. *Comput. Appl. Biosci.* 13: 523–535.
Soderlund C, et al. (2000). Contigs built with fingerprints, markers, and FPC V4.7. *Genome Res.* 10: 1772–1787.

See also Contig, Fingerprint.

FRAME-BASED LANGUAGE
Robert Stevens

Type of language for representing ontologies.

In a frame-based language, a "frame" represents a concept—category or individual—in the ontology. Properties or relations of a concept are represented by slots in its frame. Slots can be filled by other classes, symbols, strings, numbers, etc. For example, a frame Enzyme might have a slot Catalyses and a slot filler Chemical reaction. There are usually special slots for the frame name and for the taxonomic "is-a" relation. Slots are inherited along the taxonomic "is-a" relation and default slot fillers can be either inherited or changed in the inheriting frame. Systems such as RiboWeb and EcoCyc are represented in frame-based languages. Protégé-2000 is a well-known editor and toolset for building ontologies and applications using a frame-based language. OKBC is a standard for exchanging information among frame-based languages.

FREE *R*-FACTOR
Liz P. Carpenter

The free *R*-factor is a version of the *R*-factor used in X-ray crystallography to follow the progress of the refinement of a crystallographic structure. The free *R*-factor is calculated in the same way as an *R*-factor, using a set of randomly selected reflections, usually 5% or 10% of the total possible number of reflections. These reflections are excluded from calculations of electron density maps and all refinement steps. The free *R*-factor is therefore independent of the refinement process and is less biased by the starting model. A drop in the free *R*-factor is thought to indicate a significant improvement in the agreement between model and data, whereas a decrease in the *R*-factor could be due to overrefinement.

Generally free *R*-factors below 30% are considered to indicate that a model is correct and values above 40% after refinement suggest that the model has substantial errors. High-resolution macromolecular structures can have free *R*-factors below 20%.

Further Reading
Brünger AT (1992). The free *R* value: A novel statistical quantity for assessing the accuracy of crystal structures. *Nature* 355: 472–474.

Brünger AT (1997). Free *R* value: Cross-validation in crystallography. *Methods Enz.* 277: 366–396.

See also Refinement, *R*-factor, X-Ray Crystallography for Structure Determination.

FUNCTIONAL DATABASES
Dov S. Greenbaum

Databases that serve to provide a functional annotation of genes and their protein products.

An example of a functional database is the MIPS (Munich Information Center for Protein Sequences) database, which focuses on a few genomes (i.e., watercress, yeast, human, and neurospora). SWISS-PROT is also a functional database in that it aims to be a "curated protein sequence database which strives to provide a high level of annotations (such as the description of the function of a protein, its domains structure, post-translational modifications, variants, etc.), a minimal level of redundancy and high level of integration with other databases."

Functional databases require, most importantly, a clear methodology to describe the exact function of an individual gene. In most cases a gene's function will fit into a hierarchical structure of broad terms, subsequently narrowing to very well defined nomenclatures. Many databases now use automated processes to define a protein's function, although these databases are still manually curated.

Related Websites

SWISS-PROT	http://www.expasy.ch/sprot/
EXProt	http://www.cmbi.kun.nl/EXProt/

Further Reading
Kumar A, et al. (2000). TRIPLES: A database of gene function in *Saccharomyces cerevisiae*. *Nucleic Acids Res.* 28: 81–84.
Ursing BM, van Enckevort FH, Leunissen JA, Siezen RJ (2002). EXProt: A database for proteins with an experimentally verified function. *Nucleic Acids Res.* 30: 50–51.

FUNCTIONAL GENOMICS
Jean-Michel Claverie

A research area and the set of approaches dealing with the determination of gene function at a large scale and/or in a high-throughput fashion.

The concept of functional genomics emerged after the completion of the first whole-genome sequences (*Haemophilus influenzae* and *Saccharomyces cerevisiae*) and the unexpected discovery that a large fraction (25% to 50%) of the genes were previously unknown (i.e., were missed by traditional genetics) and had to be functionally characterized. While the goal of classical genetics is to find a sequence for each function, functional genomics tries to identify a function for each sequence. Functional genomics is thus reverse genetics at the genome scale. Functional genomics is taking advantage of a large (and growing) range of techniques, including:

1. Computational techniques (bioinformatics):
   Sequence similarity searches
   Functional signature and motif detection
   Phylogenomics
   Chemical library virtual screening (computational docking of small
   molecule models)
2. Protein interaction measurements:
   Large-scale chemical library screening
   Two-hybrid system assays
3. Gene expression profiling using:
   Expressed sequence tags (ESTs)
   DNA arrays
4. Systematic gene disruption
5. Structural genomics

Each of these approaches only gives a partial view about the functions of the genes, such as "being a kinase," "binding calcium," "being coexpressed with

NF kappa-B," "being involved in liver development," etc. The art of functional genomics is thus the integration of multiple functional hints gathered from many different techniques. These various techniques usually address a different level of gene "function," such as biochemical (e.g., kinase), cellular (e.g., cell signaling), or pertaining to the whole organism (e.g., liver development).

Related Websites

European Science Foundation Functional Genomics	http://www.functionalgenomics.org.uk/
Science magazine functional genomics	http://www.sciencemag.org/feature/plus/sfg/
Functional and comparative genomics fact sheet	http://www.ornl.gov/hgmis/faq/compgen.html

Further Reading

Claverie JM (1999). Computational methods for the identification of differential and coordinated gene expression. *Hum. Mol. Genet.* 8: 1821–1832.

Eisenberg D, et al. (2000). Protein function in the post-genomic era. *Nature* 405: 823–826.

Legrain P, et al. (2001). Protein-protein interaction maps: A lead towards cellular functions. *Trends Genet.* 17: 346–352.

FUNCTIONAL SIGNATURE

Jean-Michel Claverie

A sequence pattern that appears to be uniquely (or almost uniquely) associated with a given function in a protein.

The simplest and most ideal form of functional signature is a set of strictly conserved positions within a family of sequences. Allowing a little more variation leads to "regular expression" signatures (such as PROSITE) and to position weight matrices (PWMs). Modern functional signatures now use more sophisticated implementations such as hidden Markov models.

Related Websites

NCBI BLAST	http://www.ncbi.nlm.nih.gov/BLAST/
PROSITE	http://www.expasy.ch/prosite/
InterPro	http://www.ebi.ac.uk/interpro/
Blocks	http://www.blocks.fhcrc.org/
PFAM	http://www.sanger.ac.uk/Pfam/
PRINTS	http://bioinf.man.ac.uk/dbbrowser/PRINTS/
ProDom	http://prodes.toulouse.inra.fr/prodom/doc/prodom.html
SMART	http://smart.embl-heidelberg.de/

Further Reading

Bateman A, et al. (2002). The Pfam protein families database. *Nucleic Acids Res.* 30: 276–280.

Falquet L, et al. (2002). The PROSITE database, its status in 2002. *Nucleic Acids Res.* 30: 235–238.

Gribskov M, Veretnik S (1996). Identification of sequence pattern with profile analysis. *Methods Enzymol.* 266: 198–212.

See also Gene Family, Motif.

FUNCTOME
Dov S. Greenbaum

The functome is described as the list of potential functions encoded by the genome.

The overall purpose of many genomics efforts, including comparative genomics, is to annotate gene sequences, specifically, to determine the function of unknown genes. The functome is the sum of many other "-omes", as function is described by the conglomeration of other facets of gene annotation, including, e.g., subcellular localization, structure, and interaction partners, and can be inferred partially by expression analysis through clustering genes with similar expression patterns. Additionally, it incorporates other -omes such as the secretome or the metabolome.

Genomics has many tools to determine the function of previously unknown proteins, including annotation transfer from known homologs and describing proteins via their function-specific motifs either in sequence or structure (e.g., transcription factors by DNA-binding motifs). Extreme caution must be applied when using techniques such as those listed above to describe a protein, specifically not to use weak evidence to functionally annotate a gene and thus possibly corrupt the databases.

There are presently many databases collecting functional data on genes.

Further Reading

Greenbaum D, et al. (2001). Interrelating different types of genomic data, from proteome to secretome: 'oming in on function. *Genome Res.* 11: 1463–1468.

See also Functional Databases, Metabolome, Secretome.

G

Gametic Phase Disequilibrium
Gap
Gap4
Gap Penalty
Garnier–Osguthorpe–Robson Method
Gaussian
GC Composition
GCG
GC Richness
GDE
Gel Reading
GenBank
Gene
GeneChip
Gene Cluster
Gene Distribution
Gene Diversity
Gene Duplication
Gene Elongation
Gene Expression Domain
Gene Expression Profile
Gene Family
Gene Flow
Gene Fusion
Gene Fusion Method
Gene Identification
Gene Inactivation
Gene Loss and Inactivation
Gene Neighborhood
Gene Ontology
Gene Ontology Consortium
Gene Prediction, Ab Initio
Gene Prediction Accuracy
Gene Prediction, Comparative
Gene Prediction, Homology Based
Gene Prediction Systems, Integrated
Generalization
Gene Sharing
Gene Size
Genetic Code

Genetic Data Environment
Genetic Drift
Genetic Linkage
Genetic Linkage User Environment
Genetic Network
Genetic Redundancy
GENEWISE
Genome Annotation
GenomeInspector
Genome Scans
Genome Sequence
Genome Sequencing
Genome Size
Genomewide Scans
Genomewide Surveys
Genomics
GENSCAN
Global Alignment
Global Open Biology Ontologies
Globular
GLUE
Glutamic Acid
Glutamine
Glycine
GO
GOBASE
GOBO
GOR Secondary-Structure
    Prediction Method
Gradient Descent
GRAIL Coding Region Recognition Suite
GRAIL Description Logic
Graph Representation of Genetic,
    Molecular, and Metabolic Networks
Grid and the GRIN and GRID Methods
GROMOS
Group I Intron
Group II Intron
Gumball Machine

Dictionary of Bioinformatics and Computational Biology. Edited by Hancock and Zvelebil
ISBN 0-471-43622-4 © 2004 John Wiley & Sons, Inc.

GAMETIC PHASE DISEQUILIBRIUM. *SEE* LINKAGE DISEQUILIBRIUM.

GAP (INDEL)
Jaap Heringa

A meaningful alignment of two or more nucleotide or protein sequences typically contains gaps. The gaps represent sites where the affected sequences are thought to have accumulated an insertion or deletion (indel) during divergent evolution.

In most algorithms for automatic alignment, placing the gaps is penalized through gap penalty parameters, which are set to create a balance between being too strict and being overly permissive in creating gaps. High gap penalties lead to alignments where corresponding motifs might not be in the register, as this would require too many gaps in the alignment, while low penalties typically yield alignments with many spurious gaps created to align almost any identical residue in the target sequences. The latter alignments often show so-called widow amino acids, which are residues isolated from neighboring residues by gaps at either side. Although empirical recommendations exist for gap penalty values associated with many amino acid exchange matrices used in protein sequence alignment routines, there is no formal way yet to deduce gap penalty settings given a particular amino acid exchange matrix.

GAP4
Rodger Staden

A widely used sequencing project management program.

Related Website

Gap4	http://www.mrc-lmb.cam.ac.uk/pubseq/manual/gap4_unix_toc.html

Further Reading

Bonfield JK, et al. (1995). A new DNA sequence assembly program. *Nucleic Acids Res.* 23: 4992–4999.

See also Pregap4.

GAP PENALTY
Jaap Heringa and Laszlo Patthy

A penalty used in calculating the score of a sequence alignment.

Since most alignments are scored assuming a Markov process, where the amino acid matches are considered independent events, the product of the probabilities for each match within an alignment is typically taken. Since many of the scoring matrices contain exchange propensities converted to logarithmic values (log odds), the

Dictionary of Bioinformatics and Computational Biology. Edited by Hancock and Zvelebil
ISBN 0-471-43622-4   © 2004 John Wiley & Sons, Inc.

alignment score can be calculated by summing the log-odd values corresponding to matched residues minus appropriate gap penalties:

$$S_{a,b} = \sum_l s(a_i, b_f) - \sum_k N_k \, \mathrm{gp}(k)$$

where the first summation is over the exchange values associated with l matched residues and the second over each group of gaps of length k, with N_k the number of gaps of length k and $\mathrm{gp}(k)$ the associated gap penalty.

Needleman and Wunsch (1970) used a fixed penalty value for the inclusion of a gap of any length, while Sellers (1974) added a penalty value for each inserted gap position. Most current alignment routines take an intermediate approach by using the formula $\mathrm{gp}(k) = p_i + k \cdot p_e$, where p_i and p_e are the penalties for gap initialization and extension, respectively. Gap penalties of this latter form are known as affine gap penalties. Many researchers use a value of p_i 10 to 30 times larger than that of p_e. The choice of proper gap penalties is also closely connected to the residue exchange matrix used in the analysis. However, no formal model exists to estimate gap penalty values associated with a particular residue exchange matrix. Terminal gaps at the ends of alignments are usually treated differently to internal gaps, reflecting the fact that the N-terminal and C-terminal regions of proteins are quite tolerant to extensions/deletions.

Further Reading
Needleman SB, Wunsch CD (1970). A general method applicable to the search for similarities in the amino acid sequence of two proteins. *J. Mol. Biol.* 48: 443–453.

Sellers PH (1974). On the theory and computation of evolutionary distances. *SIAM J. Appl. Math.* 26: 787–793.

GARNIER–OSGUTHORPE–ROBSON METHOD. *SEE* GOR SECONDARY-STRUCTURE PREDICTION METHOD.

GAUSSIAN
Roland L. Dunbrack

A program for ab initio quantum mechanical calculations on molecules developed by the Nobel Prize winner John Pople and colleagues.

Related Website

The Official Gaussian	http://www.gaussian.com/

Further Reading
Gill PMW, Head-Gordon M, Pople JA (1989). An Efficient Algorithm for the Generation of Two Electron Repulsion Integrals over Gaussian Basis Functions. *Int. J. Quant. Chem.* S33, 269.

See also Ab Initio.

GC COMPOSITION. *SEE* BASE COMPOSITION.

GCG. *SEE* WISCONSIN PACKAGE.

GC RICHNESS. *SEE* BASE COMPOSITION.

GDE (GENETIC DATA ENVIRONMENT)
John M. Hancock

A program providing a graphical user interface to various sequence analysis tools.

GDE is based around a sophisticated sequence alignment editor. Selecting a region of the alignment allows analysis to be applied to that region of the alignment. The aim of the program is to integrate access to freely available sequence analysis software. The basic package provides access to the FASTA and BLAST database-searching tool, the CLUSTAL alignment tool, the PHYLIP package for phylogenetic analysis, and the MFOLD secondary-structure prediction program.

Further Reading

Eisen J (1999). The Genetic Data Environment: A User Modifiable and Expandable Multiple Sequence Analysis package. http://www.tigr.org/~jeisen/Resume/Eisen.GDE.pdf.

Smith SW, et al. (1994). The genetic data environment an expandable GUI for multiple sequence analysis. *Comput. Appl. Biosci.* 10: 671–675.

See also BLAST, CLUSTAL, FASTA, PHYLIP, RNA Folding.

GEL READING. *SEE* READING.

GENBANK
Guenter Stoesser

GenBank is the American member of the International Nucleotide Sequence Database Collaboration DDBJ/EMBL/GenBank together with DDBJ (Japan) and the EMBL nucleotide sequence database (United Kingdom). Collaboratively, the three organizations collect all publicly available nucleotide sequences and exchange sequence data and biological annotations on a daily basis. Major data sources are individual research groups, large-scale genome sequencing centers, and the U.S. Patent Office (USPTO). Stand-alone (Sequin) and Web-based (BankIt) sequence submission systems are available. Entrez is NCBI's search and retrieval system that provides users with integrated access to sequence, mapping, taxonomy, and structural data as well as to biomedical literature via PubMed. Sophisticated software tools allow searching and data analysis. The BLAST program of the National Center for Biotechnology Information (NCBI) is used for sequence similarity searching and is instrumental in identifying genes and genetic features. GenBank is the National Institutes of Health (NIH) genetic sequence database and is maintained at the NCBI, Bethesda, Maryland.

Related Websites

National Center for Biotechnology Information	http://www.ncbi.nlm.nih.gov
GenBank	http://www.ncbi.nlm.nih.gov/Genbank/
GenBank submission systems	http://www.ncbi.nlm.nih.gov/Sequin/ http://www.ncbi.nlm.nih.gov/BankIt/
Genomes	http://www.ncbi.nlm.nih.gov/Genomes/
Tools for data mining	http://www.ncbi.nlm.nih.gov/Tools/
Entrez	http://www.ncbi.nlm.nih.gov/Entrez/

Further Reading

Benson DA, et al. (2002). GenBank. *Nucleic Acids Res.* 30: 17–20.

Ouellette BF, Boguski MS (1997). Database divisions and homology search files: A guide for the perplexed. *Genome Res* 7: 952–955.

Wheeler DL, et al. (2001). Database resources of the National Center for Biotechnology Information. *Nucleic Acids Res.* 29: 11–16.

GENE

John M. Hancock

A segment of genome (typically DNA but also RNA in some viruses) encoding one or more molecular species.

A simple definition of the term hides various levels of ambiguity. The single-sentence definition above includes introns and 5′ and 3′ UTRs in eukaryotic genes, as these are included in immature transcripts, which are themselves molecular species. It also includes sequences encoding RNA molecules. This minimal definition does not include regulatory sequences, although these might reasonably also be included. This is, however, problematic when elements such as enhancers and LCRs may lie kilobases away from their promoters and when other coding sequences may lie between these regulatory sequences and the coding region(s) whose transcription they regulate. Regulatory regions may also control the transcription of more than one coding region, in which case they would have to be included in more than one "gene."

This definition does not exclude the possibility that genes might overlap or that a single gene might encode a variety of products (e.g., via alternative splicing).

Further Reading

Lewin B (2000). *Genes VII*. Oxford University Press, Oxford, UK.

GENECHIP. *SEE* AFFYMETRIX GENECHIP OLIGONUCLEOTIDE MICROARRAY.

GENE CLUSTER
Katheleen Gardiner

A group of sequence-related genes found in close physical proximity in genomic DNA.

For example, ribosomal RNA genes are clustered on the short arms of the five human acrocentric chromosomes. Other well-known gene clusters include the genes for histones and immune system genes such as the major histocompatibility complex.

Further Reading

Cooper DN (1999). *Human Gene Evolution*. Bios Scientific Publishers, Oxford, UK, pp. 265–285.

Ridley M (1996). *Evolution*. Blackwell Science, Cambridge, MA, pp. 265–276.

GENE DISTRIBUTION
Katheleen Gardiner

The variable gene density within chromosomes and chromosome bands.

In the human genome, gene distribution is strongly nonuniform, with G bands relatively gene poor and the most gene-rich regions in T bands. Some regions of R bands on human chromosomes 21 and 22 have been annotated with one gene per 20–40 kb, while in a segment of a large G band on chromosome 21, only two genes were identified within ~7 Mb. The gene distribution varies among chromosomes; >550 genes were annotated within chromosome 22, while in the comparably sized chromosome 21 only 225 genes were found. For their relative sizes, human chromosome 19 appears gene rich and human chromosome 13 gene poor. This correlates with the relative density of R bands.

Further Reading

Caron H, et al. (2001). The human transcriptome map: Clusters of highly expressed genes in chromosome domains. *Science* 291: 1289–1292.

Dunham I (1999). The DNA sequence of human chromosome 22. *Nature* 402: 489–495.

Federico C, Andreozzi L, Saccone S, Bernardi G (2000). Gene density in the Giemsa bands of human chromosomes. *Chromosome Res.* 8: 727–746.

Hattori M, et al. (2000). The DNA sequence of human chromosome 21. *Nature* 405: 311–320.

International Human Genome Sequencing Consortium (2001). Initial sequencing and analysis of the human genome. *Nature* 409: 866–921.

Venter C, et al. (2001). The sequence of the human genome. *Science* 291: 1304–1351.

GENE DIVERSITY
Mark McCarthy and Steven Wiltshire

In population genetics terms, a measure of the extent of genetic variation within populations.

Whereas heterozygosity is often used as a measure of genetic variation at marker loci, gene diversity (sometimes termed *average* heterozygosity) represents a more general measure of variation. Heterozygosity simply describes the proportion of heterozygous individuals in a sample. Inbreeding, or recent admixture, will, however, result in a deficiency of heterozygotes, and this measure will therefore underestimate the true variation within the population. Gene diversity is formed from the sum of squares of allele frequencies: Specifically, where p_i is the frequency of allele i at a locus of interest, the gene diversity at that locus is $1 - \Sigma p^2$. Comparison of these different measures of genetic variation can provide insights into population structure (e.g., evidence for inbreeding).

Related Website

ARLEQUIN	http://lgb.unige.ch/arlequin/software/

Further Reading

Dutta R, et al, (2002). Patterns of genetic diversity at the nine forensically approved STR loci in the Indian populations. *Hum. Biol.* 74: 33–49.

Jorgensen TH, et al. (2002). Linkage disequilibrium and demographic history of the isolated population of the Faroe Islands. *Eur. J. Hum. Genet.* 10: 381–387.

Weir BS (1996). *Genetic Data Analysis 2: Methods for Discrete Population Genetic Data*. Sinauer Associates, Sunderland, MA, pp. 141–159.

See also Polymorphism.

GENE DUPLICATION
Katheleen Gardiner

As a description of a feature on a chromosome: two closely related genes adjacent in genomic DNA.

Successive duplications caused by unequal recombination or crossing over will lead to gene clusters and gene families. After duplication, random mutations can lead to sequence divergence and possibly to functional or expression divergence.

Alternatively, the process by which duplicated genes (which are not necessarily clustered in the genome) arise.

Further Reading

Cooper DN (1999). *Human Gene Evolution*. Bios Scientific Publishers, Oxford, UK, pp. 265–285.

Ridley M (1996). *Evolution*. Blackwell Science, Cambridge, MA, pp. 265–276.

GENE ELONGATION
Austin L. Hughes

An increase in the length of a gene and thus an increase in the length of the polypeptide it encodes.

Gene elongation occurs through internal duplication of a portion or portions of the gene's coding region. Internal duplication within a gene is much more likely if a portion of the gene has a repeating structure. Hypothetical mechanisms of internal duplication include slipped-strand mispairing and unequal crossing over.

Further Reading

Hakani-Covo E, et al. (2002). The evolutionary history of prosaposin: Two successive tandem-duplication events gave rise to the four saposin domains in vertebrates. *J. Mol. Evol.* 54: 30–34.

Hughes AL (1999). Concerted evolution of exons and introns in the MHC-linked tenascin-X gene of mammals. *Mol. Biol. Evol.* 16: 1558–1567.

Saenz de Mierra LE, Perez del la Vega M (2001). Evidence that the N-terminal extension of the Vicieae convicilin genes evolved by intragenic duplications and trinucleotide expansions. *Genome* 44: 1022–1030.

GENE EXPRESSION DOMAIN
Niall Dillon

The region containing a gene or a set of coregulated genes and their associated regulatory sequences.

The organization of eukaryotic gene expression domains is controversial. One type of model proposes a highly organized domain with specialized functional boundaries that separate it from neighboring domains. The discovery that there is frequent overlap of regulatory regions for genes with divergent expression patterns creates a problem for this type of model. An alternative model proposes that the structure of the domain is determined by multiple binding sites for transcription factors scattered throughout the domain. According to this model, functional isolation of genes is mainly due to specificity of interactions between regulatory sequences.

Further Reading

Dillon N, Sabbattini P (2000). Functional gene expression domains: Defining the functional unit of eukaryotic gene regulation. *Bioessays* 22: 657–665.

West AG, Felsenfeld G (2002). Insulators: Many functions, many mechanisms. *Genes Dev.* 16: 271–278.

GENE EXPRESSION PROFILE
Jean-Michel Claverie

The measurement of the abundance of transcripts corresponding to the same (usually large) set of genes in a number of different cell types, organs, tissue, and physiological or disease conditions.

Gene expression profiling provides information on the variation of gene activity across conditions. At a first level, expression profiling allows the identification of genes that are differentially expressed and the activity of which appear linked to a given condition (e.g., organ or disease). At a second level, the comparison of the activity profiles from different genes allows the discovery of coordinated expression patterns (e.g., coexpression) used to establish functional relationships between genes (e.g., coexpression of genes involved in the same biochemical pathway). Beyond these common principles, the mathematical methods used to analyze expression profiles depend on the technology used to measure expression intensities. Two main classes are distinguished: sequence tag counting methods (e.g., ESTs and SAGE) and hybridization methods (e.g., DNA chips).

Related Websites

HGMP Genome	http://www.hgmp.mrc.ac.uk/GenomeWeb/nuc-genexp.html
NCBI SAGEmap	http://www.ncbi.nlm.nih.gov/SAGE/
Bodymap gene expression	http://bodymap.ims.u-tokyo.ac.jp/
GEO	http://www.ncbi.nlm.nih.gov/geo/

Further Reading

Brown PO, Botstein D (1999). Exploring the new world of the genome with DNA microarrays. *Nat. Genet.* 21(1, Suppl.): 33–37.

Claverie JM (1999). Computational methods for the identification of differential and coordinated gene expression. *Hum. Mol. Genet.* 8: 1821–1832.

Okubo K, et al. (1992). Large scale cDNA sequencing for analysis of quantitative and qualitative aspects of gene expression. *Nat. Genet.* 2: 173–179.

GENE FAMILY

Terri Attwood and Katheleen Gardiner

Groups of closely related genes that encode similar products (usually protein but also RNA) and have descended from the same ancestral gene.

Examples of extensive gene families include those that encode G protein–coupled receptors, a ubiquitous group of membrane-bound receptors involved in signal transduction; tubulins, structural proteins involved in microtubule formation; and potassium channels, a diverse group of ion channels important in shaping action potentials and in neuronal excitability and plasticity. By contrast with domain families, members of gene or protein families share significant similarity along the entire length of their sequences and consequently exhibit a high degree of structural and functional correspondence.

In order to characterize new members of gene families and their protein products, various bioinformatics tools have become available in recent years. Standard search tools, such as FASTA and BLAST, are commonly used to trawl the sequence databases for distant relatives. Other, diagnostically more potent approaches involve searches of the protein family databases, notably, PROSITE, PRINTS, and InterPro. PRINTS, in particular, offers the ability to make fine-grained hierarchical diagnoses, from superfamily, through family, down to individual subfamily levels (consequently, this hierarchy also forms the basis of InterPro).

Members of a gene family may be clustered at one or several chromosome locations or may be widely dispersed. Families may be modest in size and distribution. For example, there are 13 crystallin genes in humans, 4 each clustered on chromosome 2 and chromosome 22, and the rest in individual locations; the HOX genes are located on four chromosomes in groups of 9–10 genes. Larger families include several hundred zinc finger genes, with many clustered on chromosome 19, and more than 1000 olfactory receptor genes. Sequence divergence among members of a gene family may result in functional divergence, expression pattern divergence, or pseudogenes.

Related Websites

PROSITE	http://www.expasy.org/prosite/
PRINTS	http://www.bioinf.man.ac.uk/dbbrowser/PRINTS/
InterPro	http://www.ebi.ac.uk/interpro/

Further Reading

Bockaert J, Pin JP (1999). Molecular tinkering of G protein-coupled receptors: An evolutionary success. *EMBO J.* 18(7): 1723–1729.

Cooper DN (1999). *Human Gene Evolution*. Bios Scientific Publishers, Oxford, UK, pp. 265–285.

Gogarten JP, Olendzenski L (1999). Orthologs, paralogs and genome comparisons. *Curr. Opin. Genet. Dev.* 9(6): 630–636.

Greene EA, Pietrokovski S, Henikoff S, Bork P, Attwood TK, Hood L, Bairoch A (1997). Building gene families. *Science* 278: 615–626.

Henikoff S, Greene EA, Pietrokovski S, Attwood TK, Bork P, Hood L (1997). Gene families: The taxonomy of protein paralogs and chimeras. *Science* 278: 609–614.

Ridley M (1996). *Evolution*. Blackwell Science, Cambridge, MA, pp. 265–276.

See also BLAST, Domain Family, FASTA, InterPro, PROSITE, PRINTS.

GENE FLOW
A. Rus Hoelzel

The exchange of genetic factors between populations by migration and interbreeding.

In sexual species, dispersal alone does not necessarily imply gene flow, since successful reproduction is required to transfer alleles. This is also known as *genetic dispersal*. Gene flow among populations reduces differentiation due to the effects of genetic drift and natural selection. Under the specific conditions set out by Wright's Island model (Wright, 1940), the relationship between gene flow and population structure can be defined as follows:

$$Nm = (1 - F_{ST})/(4F_{ST})$$

where N is the effective population size, m is the migration rate (and so Nm is gene flow), and F_{ST} is the interpopulation fixation index (the proportion of the total genetic variance accounting for differences among populations). When there is a high degree of structure among populations (F_{ST} is large), gene flow is low and vice versa.

Further Reading

Hartl DL (2000). *A Primer of Population Genetics*. Sinauer Associates, Sunderland, MA.

Wright S (1940). Breeding structure of populations in relation to speciation. *Am. Nat.* 74: 232–248.

See also Evolution, Genetic Drift.

GENE FUSION
Austin L. Hughes

Gene fusion occurs when two genes are combined to form a new gene. For example, the human *Kua-UEV* gene was created by fusion of two distinct genes which were adjacent to one another.

Further Reading

Long M (2000). A new function evolved from gene fusion. *Genome Res.* 10: 1655–1657.

GENE FUSION METHOD
Patrick Aloy

The gene fusion method relies on the principle that there is selective pressure for certain genes to be fused over the course of evolution. It predicts functionally related proteins, or even interacting pairs, by analyzing patterns of domain fusion.

If two proteins are found separately in one organism (components) and fused into a single polypeptide chain in another (composite), this implies that they might physically interact or, more likely, have a related function. This approach can be automated and applied large scale to prokaryotes but not to higher organisms, where the detection of orthologs becomes a difficult problem.

Related Websites

| ALLFuse | http://maine.ebi.ac.uk/services/allfuse/ |
| EMBL | http://dag.embl- heidelberg.de |

Further Reading

Enright AJ, Iliopoulos I, Kyrpides N, Ouzounis CA (1999). Protein interaction maps for complete genomes based on gene fusion events. *Nature* 402: 86–90.

Marcotte EM, Pellegrini M, Ng HL, Rice DW, Yeates TO, Eisenberg D (1999). Detecting protein function and protein-protein interactions from genome sequences. *Science* 285: 751–753.

See also Ortholog.

GENE IDENTIFICATION. *SEE* GENE PREDICTION, AB INITIO; GENE PREDICTION, COMPARATIVE; GENE PREDICTION, HOMOLOGY BASED; GENE PREDICTION SYSTEMS, INTEGRATED.

GENE INACTIVATION. *SEE* GENE LOSS AND INACTIVATION.

GENE LOSS AND INACTIVATION (GENE INACTIVATION)

Austin L. Hughes

Processes by which, over evolutionary times, genes may be lost from genomes or rendered nonfunctional.

Gene loss and inactivation may occur by deletion of a gene, by mutations that prevent transcription, or by mutations that prevent translation of a functional gene product. Pseudogenes, which are inactivated genes, are abundant in eukaryotic genomes.

See also Pseudogene.

GENE NEIGHBORHOOD

Patrick Aloy

In bacteria and archaea, genes are transcribed in polycistronic mRNAs encoding coregulated and functionally related genes. Operon structure can be inferred from the conserved physical proximity of genes in phylogenetically distant genomes (i.e., conservation of pairs of genes as neighbors or colocalization of genes in potential operons).

The conservation of gene neighbors in several bacterial genomes can be used to predict functional relationships, or even physical interactions, between the encoded proteins.

As for phylogenetic profiles, this is a *computational genomics* approach that can only be applied to entire genomes (i.e., not individual pairs of proteins). Another limitation of this approach is that it cannot be used to predict interactions in eukaryotes.

Related Website

EMBL	http://dag.embl- heidelberg.de

Further Reading

Dandekar T, Snel B, Huynen M, Bork P (1998). Conservation of gene order: A fingerprint of proteins that physically interact. *Trends Biochem. Sci.* 23: 324–328.

Snel B, Lehmann G, Bork P, Huynen M (2000). STRING: A web-server to retrieve and display the repeatedly occurring neighborhood of a gene. *Nucleic Acids Res.* 28: 3442–3444.

See also Operon, Phylogeny.

GENE ONTOLOGY (GO)
Robert Stevens

Gene Ontology aims to provide a structured, controlled vocabulary, with definitions, in the form of a DAG.

The terms in this ontology are used to annotate gene products or their proxies in the form of a gene for the attributes of *molecular function, biological process,* and *cellular component.* Originally developed to provide a shared understanding for the model organisms fly, yeast, and mouse, GO is now used by over 15 model organism databases and non-species-specific databases such as SWISS-PROT and InterPro.

Related Website

geneontology	http://www.geneontology.org

See also Gene Ontology Consortium.

GENE ONTOLOGY CONSORTIUM
Robert Stevens

The Gene Ontology Consortium includes representatives of the database groups who participate in the development of Gene Ontology (GO).

Originally consisting of the model organism databases for fly, yeast, and mouse, it has now expanded to more than a dozen groups covering more than 15 organisms. The consortium manages the content and representation of its ontologies, provides sets of gene product annotation data, and develops software for use with GO. The consortium also supports the development of additional biological ontologies, such as a chemical ontology and a sequence-feature ontology, through its GOBO effort.

Related Website

| Gene Ontology | http://www.geneontology.org |

See also Gene Ontology, Global Open Biology Ontologies.

GENE PREDICTION, AB INITIO (INTRINSIC GENE PREDICTION, TEMPLATE GENE PREDICTION)
Roderic Guigó

Prediction of genes in genomic sequences: inferring the amino acid sequence of the proteins encoded in a query genomic sequence.

Genes are predicted in genomic sequences by a combination of one or more of the following approaches:

1. Analysis of sequence signals potentially involved in gene specification (*search by signal*)
2. Analysis of regions showing sequence compositional bias correlated with coding function (*search by content*)
3. Comparison with known coding sequences (*homology-based gene prediction*)
4. Comparison with anonymous genomic sequences (*comparative gene prediction*)

Ab initio gene prediction includes approaches 1 and 2, in which genes are predicted without direct reference to sequences external to the query. For that reason, it is also often referred to as intrinsic or template gene prediction, as opposed to extrinsic or look-up gene prediction. This includes approaches 3 and 4, in which genes are predicted based on sequences external to the query.

The complexity of gene prediction differs substantially in prokaryotic and eukaryotic genomes. While prokaryotic genes are single continuous ORFs, usually adjacent to each other, eukaryotic genes are separated by long stretches of intergenic DNA and their coding sequences, the exons, are interrupted by large noncoding introns.

Prokaryotic Gene Finding
Gene prediction in prokaryotic organisms (and, in general, in organisms without splicing) involves the determination of ORFs initiated by suitable translation start sites. The translation start site signal, however, does not carry enough information to discriminate ORFs occurring by chance from those corresponding to protein coding genes. Thus, gene identification requires, in addition, the scoring of putative ORFs according to the species-specific protein coding model. For instance, GLIMMER (Salzberg et al., 1998a) and GENEMARK (Borodovsky and McIninch, 1993), two popular programs for prokaryotic gene finding, use nonhomogeneous Markov models as the underlying coding model. In GENEMARK the models are of fixed order five, while GLIMMER uses interpolated Markov models, that is, combinations of Markov models from first through eighth order, weighting each model according to its predictive power. (Other prokaryotic gene predictors appear in the table below.)

One serious problem facing prokaryotic gene prediction is that species-specific coding models (such as bias in codon usage) may be unavailable prior to the sequencing of a complete genome. To overcome this limitation, gene prediction in complete prokaryotic genomes often proceeds in two steps. First, ORFs corresponding to known genes are identified through sequence similarity database searches. These ORFs are then used to infer the species-specific coding model. In the second step, the inferred coding model is used to score the remaining ORFs. This is the approach, for instance, in the program ORPHEUS (Frishman et al., 1998).

Eukaryotic Gene Finding

Gene prediction in eukaryotic genomes involves the identification of start and stop translation signals and splice sites, which are responsible for the exonic structure of genes. These signals, however, do not appear to carry the information required for the recognition of coding exons in genomic sequences (Burge, 1998). Thus, gene prediction in eukaryotes, as in prokaryotes, requires the utilization of some ad hoc protein coding model to score potential coding exons. Indeed, typical computational ab initio eukaryotic gene prediction involves the following tasks:

- Identification and scoring of suitable splicing sites and start and stop codons along the query sequence
- Prediction of candidate exons defined by these signals
- Scoring of these exons as a function, at least, of the signals defining the exon and of one or more coding statistics computed on the exon sequence.
- Assembly of a subset of these exon candidates in a predicted gene structure. The assembly is produced under a number of constraints, including the maximization of a scoring function which depends, at least, on the scores of the candidate exons.

The particular implementation of these tasks (which often are only implicit) varies considerably between programs.

Combining prediction of sequence signals and regions with sequence compositional bias had already been suggested in the mid-1980s as an effective strategy to predict the exonic structure of eukaryotic genes (Nakata et al., 1985). In 1990 Gelfand proved this strategy successfully in nine mammalian genes. The same year, Fields and Soderlund published GENEMODELER, a computer program for gene prediction in *Caenorhabdites elegans*. This was the first computer program publicly available for the prediction of the exonic structure of the eukaryotic genes. In 1992 Guigó et al. published GENEID, the first generic gene prediction program for vertebrate organisms. A number of accuracy measures were introduced to benchmark GENEID in a large set of genomic sequences. In 1993, Snyder and Stormo published GENEPARSER, a predecessor of the successful hidden Markov models gene predictors. In 1994, Xu and others published GRAIL2/Gap, built around the popular GRAIL program for coding region identification. In 1997, Burge and Karlin published GENSCAN, the first program able to deal with large genomic sequences encoding multiple genes in both strands and one that was substantially more accurate that its predecessors.

The current generation of ab initio gene prediction programs are only slightly more accurate than GENSCAN. Their accuracy is remarkable when analyzing single-gene sequences but drops significantly when analyzing large multigenic sequences. When these are from vertebrate organisms, only about 50% of the coding exons are correctly predicted. The computational efficiency of the programs, however, has increased substantially during the past decade. While programs in the early 1990s were limited to the analysis of sequences only a few kilobases long, programs currently available can analyze chromosome-size sequences in a few minutes running on standard workstations.

Table Ab Initio Gene Prediction Programs

Eukaryotic Gene Prediction		
GENEMODELER	Fields and Soderlund, 1990	
GENEID	Guigó et al., 1992	http://www1.imim.es/geneid.html
SORFIND	Hutchinson and Hayden, 1992	http://www.rabbithutch.com/produc ts.html
GENEPARSER	Snyder and Stormo, 1993	http://beagle.colorado.edu/~eesnyder/GeneParser.html
GENEMARK	Borodovski and McIninch, 1993	http://opal.biology.gatech.edu/GeneMark/
GENVIEW 2	Milanesi et al., 1993	http://www.itba.mi.cnr.it/webgene
GREAT	Gelfand and Roytberg, 1993	
GRAIL2/GAP	Xu et al., 1994	http://grail.lsd.ornl.gov/public/tools/
FGENEH	Solovyev et al., 1994	http://www.softberry.com/
GENELANG	Dong and Searls, 1994	http://zeus.pcbi.upenn.edu/genlang/genlang_home.html
XPOUND	Thomas and Skolnick, 1994	
GENIE	Kulp et al., 1996	http://www.fruitfly.org/seq_tools/genie.html
MZEF	Zhang, 1997	http://argon.cshl.org/genefinder/
GENSCAN	Burge and Karlin, 1997	http://genes.mit.edu/GENSCAN.html
MORGAN	Salzberg et al., 1998b	http://www.tigr.org/~salzberg/morgan.html
VEIL	Henderson et al., 1997	http://www.tigr.org/~salzberg/veil.html
HMMGENE	Krogh, 1997	http://www.cbs.dtu.dk/services/HMMgene
GENEFINDER	Wilson et al, unpublished	http://ftp.genome.washington.edu/cgi-bin/Genefinder
Prokaryotic Gene Prediction		
GENEMARK	Borodovsky and McIninch, 1993	http://opal.biology.gatech.edu/GeneMark/

Table Ab Initio Gene Prediction Programs

Eukaryotic Gene Prediction		
ECOPARSE	Krogh et al., 1994	http://www.cbs.dtu.dk/~krogh/EcoParse.info
GLIMMER	Salzberg et al, 1998a	http://www.tigr.org/~salzberg/glimmer/
SELFID	Audic and Claverie, 1998	http://igs-server.cnrs-mrs.fr/~audic/selfid.html
CRITICA	Badger and Olsen, 1999	http://rdpwww.life.uiuc.edu
ORPHEUS	Frishman et al., 1999	http:http://pedant.gsf.de/orpheus/
FRAMED	Schiex et al., 2000	http://www.toulouse.inra.fr/FrameD/cgi-bin/FD

Despite the progress made, computational eukaryotic gene prediction is still an open problem, and the accurate identification of every gene in the higher eukaryotic genomes by computational methods remains a distant goal. Only the determination of the mRNA sequence of a gene guarantees the correctness of its predicted exonic structure. Even in this case, a number of important issues remain to be solved: the completeness of the 3' and 5' ends of the gene, in particular if only a partial cDNA sequence has been obtained, the possibility of alternative primary transcripts, the determination for each of these of the complete catalog of splice isoforms (of which the sequenced mRNA may correspond to only one among multiple instances), the usage of noncanonical splice sites, the precise site (or the alternative sites) at which translation starts, the possibility of the gene being encoded within an intron or nested within a gene, or the possibility of the gene not fully conforming to the standard genetic code, such as in the case of selenoproteins. Currently available gene prediction programs do not deal well with any of these issues.

In the gene prediction literature, the term *gene* is often used to denote only its coding fraction. Similarly, the term *exon* refers only to the coding fraction of coding exons. See Zhang (2002) for a discussion of the types of exons considered in the gene prediction literature

Related Website

A bibliography on computational gene recognition by Wentian Li	http://linkage.rockefeller.edu/wli/gene/

Further Reading

Audic S, Claverie J-M (1998). Self-identification of protein-coding regions in microbial genomes. *Proc. Natl. Acad. Sci. USA* 95: 10,026–10,031.

Badger JH, Olsen GJ (1999). Coding region identification tool invoking comparative analysis. *Mol. Biol. Evol.* 16: 512–524.

Borodovsky M, McIninch J (1993). GenMark: Parallel gene recognition for both DNA strands. *Comput. Chem.* 17: 123–134.

Burge CB (1998). Modeling dependencies in pre-mrna splicing signals. In *Computational Methods in Molecular Biology*, Salzberg S, Searls D, Kasif S, Eds. Amsterdam. Elsevier Science, Amsterdam, pp. 127–163.

Burge CB, Karlin S (1997). Prediction of complete gene structures in human genomic DNA. *J. Mol. Biol.* 268: 78–94.

Burge CB, Karlin S (1998). Finding the genes in genomic DNA. *Curr. Opin. Struct. Biol* 8: 346–354.

Claverie J-M (1997). Computational methods for the identification of genes in vertebrate genomic sequences. *Hum. Mol. Genet.* 6: 1735–1744.

Dong S, Searls DB (1994). Gene structure prediction by linguistic methods. *Genomics* 23: 540–551.

Fickett J (1996). Finding genes by computer: The state of the art. *Trends Genet.* 12: 316–320.

Fields CA, Soderlund CA (1990). gm: A practical tool for automating DNA sequence analysis. *Comput. Appl. Biosci.* 6: 263–270.

Frishman D, et al. (1998). Combining diverse evidence for gene recognition in completely sequenced bacterial genomes. *Nucleic Acids Res.* 26: 2941–2947.

Frishman D, et al. (1999). Starts of bacterial genes: Estimating the reliability of computer predictions. *Gene* 234: 257–265.

Gelfand MS (1990). Computer prediction of exon-intron structure of mammalian pre-mrnas. *Nucleic Acids Res.* 18: 5865–5869.

Gelfand MS, Roytberg MA (1993). Prediction of the exon-intron structure by a dynamic programming approach. *Biosystems* 30: 173–182.

Guigó R (1997). Computational gene identification: An open problem. *Comput. Chem.* 21: 215–222.

Guigó R, et al. (1992). Prediction of gene structure. *J. Mol. Biol.* 226: 141–157.

Henderson J, et al. (1997). Finding genes in DNA with a hidden Markov model. *J. Comput. Biol.* 4: 127–141.

Hutchinson GB, Hayden MR (1992). The prediction of exons through an analysis of spliceable open reading frames. *Nucleic Acids Res.* 20: 3453–3462.

Krogh A (1997). Two methods for improving performance of an HMM and their application for gene finding. *ISMB* 5: 179–186.

Krogh A, et al. (1994). A hidden Markov model that finds genes in *E. coli* DNA. *Nucleic Acids Res.* 22: 4768–4778.

Kulp D, et al. (1996). A generalized hidden Markov model for the recognition of human genes in DNA. In *Intelligent Systems for Molecular Biology*, DJ States, P Agarwal, T Gaasterland, L Hunter, R Smith, Eds. AAAI Press, Menlo Park, CA, pp. 134–142.

Makarov V (2002). Computer programs for eukaryotic gene prediction. *Briefings Bioinformatics* 3: 195–199.

Milanesi L, et al. (1993). GenView: a computing tool for protein-coding regions prediction in nucleotide sequences. In *Proc. of the 2nd International Conference on Bioinformatics, Supercomputing and Complex Genome Analysis*, HA Lim, JW Fickett, CR Cantor, RJ Robins, Eds. World Scientific Publishing, Singapore, pp. 573–588.

Milanesi L (1999). Genebuilder: Interactive in silico prediction of gene structure. *Bioinformatics* 15: 612–621.

Mural RJ (1999). Current status of computational gene finding: A perspective. *Methods Enzymol.* 303: 77–83.

Nakata K, et al. (1985). Prediction of splice junctions in mrna sequences. *Nucleic Acids Res.* 13: 5327–5340.

Salzberg SL, et al. (1998a). Microbial gene identification using interpolated Markov models. *Nucleic Acids Res.* 26: 544–548.

Salzberg SL, et al. (1998b). A decision tree system for finding genes in DNA. *J. Comput. Biol.* 5: 667–680.

Schiex T, et al. (2000). Recherche des gènes et des erreurs de séquençage dans les génomes bactériens GC-riches (et autres…). In *Proc. of JOBIM'2000*, Montpellier, France, pp. 321–328.

Snyder EE, Stormo GD (1993). Identification of coding regions in genomic DNA sequences: An application of dynamic programming and neural networks. *Nucleic Acids Res.* 21: 607–613.

Solovyev VV, et al. (1994). Predicting internal exons by oligonucleotide composition and discriminant analysis of spliceable open reading frames. *Nucleic Acids Res.* 22: 5156–5163.

Thomas A, Skolnick MH (1994). A probabilistic model for detecting coding regions in DNA sequences. *IMA J. Math. Appl. Med. Biol.* 11: 149–60.

Xu Y, et al. (1994). Constructing gene models from accurately predicted exons: An application of dynamic programming. *Comput. Appl. Biosci.* 11: 117–124.

Zhang MQ (1997). Identification of protein coding regions in the human genome based on quadratic discriminant analysis. *Proc. Natl. Acad. Sci. USA* 94: 565–568.

Zhang MQ (2002). Computational prediction of eukaryotic protein-coding genes. *Nature Rev. Genet.* 3: 698–709.

See also Coding Statistic; Gene Prediction, Homology Based; Gene Prediction Accuracy; GENSCAN; Hidden Markov Model.

GENE PREDICTION ACCURACY
Roderic Guigó

Methods of estimating the accuracy of gene predictions.

To evaluate the accuracy of a gene prediction program on a test sequence, the gene structure predicted by the program is compared with the actual gene structure of the sequence, as, for instance, established with the help of an experimentally validated mRNA. Several metrics have been introduced during the past years to compare the predicted gene structure with the real one.

The accuracy can be evaluated at different levels of resolution. Typically, these are the nucleotide, exon, and gene levels. These three levels offer complementary views of the accuracy of the program. At each level, there are two basic measures—sensitivity and specificity—which essentially measure prediction errors of the first and second kind. Briefly, sensitivity (Sn) is the proportion of real elements (coding nucleotides, exons, or genes) that have been correctly predicted, while specificity (Sp) is the proportion of predicted elements that are correct. If TP (true positives) and TN (true negatives) denote the number of coding and noncoding nucleotides (exons/genes) correctly predicted, FN (false negatives) the number of actual coding nucleotides (exons/genes) predicted as noncoding, and FP (false positives) the number of nucleotides (exons/genes) predicted coding that are noncoding, then

$$Sn = \frac{TP}{TP + FN} \quad \text{and} \quad Sp = \frac{TP}{TP + FP}$$

Both sensitivity and specificity take values from 0 to 1, with perfect prediction when both measures are equal to 1. Neither of them alone constitutes a good measure of global accuracy, since one can have high sensitivity with little specificity and vice versa. It is desirable to use a single scalar value to summarize both of them. In the gene-finding literature, the preferred such measure is the correlation coefficient CC at the nucleotide level, which is computed as

$$CC = \frac{(TP \times TN) - (FN \times FP)}{\sqrt{(TP + FN) \times (TN + FP) \times (TP + FP) \times (TN + FN)}}$$

where CC ranges from -1 to 1, with 1 corresponding to a perfect prediction and -1 to a prediction in which each coding nucleotide is predicted as noncoding and vice versa. To uniformly test and compare different programs, a number of standard sequence data sets have been compiled during the past years to benchmark gene prediction software (see Related Websites). Human chromosome 22 has been heavily studied and currently constitutes one of the best large DNA sequence benchmarks. The most accurate computational gene prediction tools have an average accuracy (sensitivity and specificity) at the exon level of about 0.60 in this sequence.

Related Websites

Benchmark data sets	http://www1.imim.es/datasets/genomics96/
AcE gene-finding accuracy evaluation tool	http://bioinformatics.org/project/?group_id=39

Further Reading

Bajic V (2000). Comparing the success of different prediction software in sequence analysis: A review. *Briefings Bioinf.* 1: 214–228.

Baldi P, et al. (2000). Assessing the accuracy of prediction algorithms for classification: An overview. *Bioinformatics* 16: 412–424.

Burset M, Guigó R (1996). Evaluation of gene structure prediction programs. *Genomics* 34: 353–357.

Guigó R, Wiehe T (2003). Gene prediction accuracy in large DNA sequences. In *Frontiers in Computational Genomic*, MY Galperin, EV Koonin, Eds. Caister Academic Press, Norfolk, VA, pp. 1–33.

Guigó R, et al. (2000). Gene prediction accuracy in large DNA sequences. *Genome Res.* 10: 1631–1642.

Reese MG, et al. (2000). Genome annotation assessment in *Drosophila melanogaster. Genome Res.* 10: 483–501.

Rogic S, et al. (2001). Evaluation of gene-finding programs on mammalian sequences. *Genome Res.* 11: 817–832.

GENE PREDICTION, COMPARATIVE
Roderic Guigó

Prediction of genes based on the comparison of genomic sequences.

The rationale behind comparative gene prediction methods is that functional regions, protein coding among them, are more conserved than noncoding ones between genomic sequences from different organisms. This characteristic conservation can be used to identify protein coding exons in the sequences.

The different methods use quite different approaches:

In a first approach, given two homologous genomic sequences, the problem is to infer the exonic structure in each sequence, maximizing the score of the alignment of the resulting amino acid sequences. This problem is usually solved through a complex extension of the classical dynamic programming algorithm for sequence alignment. The program PRO-GEN by Novichkov et al. (2001) and the algorithms by Blayo et al. (2003) and Pedersen and Scharl (2002) are examples of this approach, in which to some extent gene prediction is the result of the sequence alignment.

In a second approach, pair hidden Markov models for sequence alignment and generalized HMMs (GHMMs) for gene prediction are combined into the so-called generalized pair HMMs, as in the programs SLAM (Alexandersson et al., 2003) and DOUBLESCAN (Meyer and Durbin, 2002). In this approach, given two homologous genomic sequences, both gene prediction and sequence alignment are obtained simultaneously.

In a third approach, gene prediction is separated from sequence alignment. First, the alignment is obtained between two homologous genomic sequences using some generic sequence alignment program, such as TBLASTX, SIM96, or GLASS. Then, gene structures are predicted that are compatible with this alignment, meaning that predicted exons fall in the aligned regions. The programs ROSETTA (Batzoglou et al., 2000), CEM (Bafna and Huson, 2000), and SGP1 (Wiehe et al., 2001) are examples of this approach.

The fourth approach does not require the comparison of two homologous genomic sequences. Rather, a query sequence from a target genome is compared against a collection of sequences from a second (informant, reference) genome (which can be a single homologous sequence to the query sequence, a whole assembled genome, or a collection of shotgun reads), and the results of the comparison are used to modify the scores of the exons produced by underlying ab initio gene prediction algorithms. The programs TWINSCAN (Korf et al., 2001) and SGP2 (Parra et al., 2003) are examples of this approach. Also related to this approach is the program CRITICA (Badger and Olsen, 1999), for gene prediction in prokaryotic genomes.

Comparative gene prediction methods have been used extensively for the first time in the analysis of the human and mouse genomes (Mouse Genome Sequence Consortium, 2002). This analysis has underscored the difficulties of using sequence conservation to predict coding regions: Depending on the phylogenetic distance separating the genomes, sequence conservation may extend well beyond coding regions.

Related Websites

CRITICA	http://rdpwww.life.uiuc.edu
PRO-GEN	http://www.anchorgen.com/pro_gen/pro_gen.html
ROSETTA	http://crossspecies.lcs.mit.edu/
SLAM	http://baboon.math.berkeley.edu/~syntenic/slam.html
SGP1	http://195.37.47.237/sgp-1/
SGP2	http://www1.imim.es/software/sgp2/
TWINSCAN	http://genes.cs.wustl.edu

Further Reading

Alexandersson M, et al. (2003). SLAM: Cross-species gene finding and alignment with a generalized pair hidden Markov model. *Genome Res.* 13: 496–502.

Badger JH, Olsen GJ (1999). Coding region identification tool invoking comparative analysis. *Mol. Biol. Evol.* 16: 512–524.

Bafna V, Huson DH (2000). The conserved exon method. In *Proceedings of the Eighth International Conference on Intelligent Systems in Molecular Biology*, ISMB, pp. 3–12.

Batzoglou S, et al. (2000). Human and mouse gene structure: Comparative analysis and application to exon prediction. *Genome Res.* 10: 950–958.

Blayo P, et al. (2003). Orphan gene finding—an exon assembly approach. *Theor. Comp. Sci.* 290: 1407–1431.

Korf I, et al. (2001). Integrating genomic homology into gene structure prediction. *Bioinformatics* 17: S140–S148.

Meyer I, Durbin R (2002). Comparative *ab initio* prediction of gene structure using pair HMMs. *Bioinformatics* 18: 1309–1318.

Mouse Genome Sequence Consortium (2002). Initial sequencing and comparative analysis of the mouse genome. *Nature* 420: 520–562.

Novichkov P, et al. (2001). Gene recognition in eukaryotic dna by comparison of genomic sequences. *Bioinformatics* 17: 1011–1018.

Parra G, et al. (2003). Comparative gene prediction in human and mouse *Genome Res.* 13: 108–117.

Pedersen C, Scharl T (2002). Comparative methods for gene structure prediction in homologous sequences. In *Algorithms in Bioinformatics*, R Guigó, D Gusfield, Eds. Springer-Verlag, Berlin.

Wiehe T, et al. (2000). Genome sequence comparisons: Hurdles in the fast lane to functional genomics. *Briefings Bioinf.* 1: 381–388.

Wiehe T, et al. (2001). SGP-1: Prediction and validation of homologous genes based on sequence alignments. *Genome Res.* 11: 1574–1583.

See also BLAST; Gene Prediction, Ab Initio; Gene Prediction, Homology Based; Hidden Markov Model.

GENE PREDICTION, HOMOLOGY BASED (EXTRINSIC GENE PREDICTION, LOOK-UP GENE PREDICTION, SEQUENCE SIMILARITY–BASED GENE PREDICTION)

Roderic Guigó

Prediction of genes in a DNA sequence based on the comparison of the sequence to known coding sequences.

The rationale behind this approach is that those segments in the genomic query similar to known protein or complementary DNA (cDNA) (EST) sequences are likely to correspond to coding exons. Because the incidence of unspecific EST matches is relatively common, similarity to ESTs is often only considered indicative of coding function when the alignment of the EST on the query sequence occurs across a splice junction.

Translated nucleotide searches, such as those performed by BLASTX, constitute one of the simplest homology-based gene prediction approaches. These searches are particularly relevant when comparing predicted ORFs in prokaryotic genomes. When dealing with the split nature of the eukaryotic genes, however, BLASTX-like searches do not resolve exon splice boundaries well. Thus, a number of programs use the results of translated nucleotide searches against protein sequences (or direct nucletotide searches against ESTs) to modify the exon scoring schema of some underlying ab initio gene prediction programs. GENIE (Kulp et al., 1996), GRAILEXP (Xu and Uberbacher, 1997), AAT (Huang et al., 1997), EBEST (Jiang and Jacob, 1998), GIN (Cai and Bork, 1998), GENEBUILDER (Milanesi et al., 1999), GRPL (Hooper et al., 2000), HMMGENE (Krogh, 2000), EUGENE (Schiex et al., 2001), and GENOMESCAN (Yeh et al., 2001) are some of the published methods.

A more sophisticated approach involves aligning the genomic query against a protein (or cDNA) target (presumably homologous to the protein encoded in the genomic sequence). In these alignments, often referred to as spliced alignments, large gaps corresponding to introns in the query sequence are only allowed at legal splice junctions. SIM4 (Florea et al., 1998), EST_GENOME (Mott, 1997), PROCRUSTES (Gelfand et al., 1996), and GENEWISE (Birney and Durbin, 1997) are examples of this approach.

Related Websites

AAT	http://genome.cs.mtu.edu/aat/aat.html
EBEST	http://rgd.mcw.edu/EBEST/
EUGENE	http://www.inra.fr/bia/T/EuGene/
GENEBUILDER	http://l25.itba.mi.cnr.it/~webgene/genebuilder.html
GENIE	http://www.cse.ucsc.edu/~dkulp/cgi-bin/genie
GENOMESCAN	http://genes.mit.edu/genomescan.html
GRAILEXP	http://compbio.ornl.gov/grailexp/
HMMGENE	http://www.cbs.dtu.dk/services/HMMgene/
GRPL	ftp://snipe.pharmacy.ualberta.ca/pub

Further Reading

Birney E, Durbin R (1997). Dynamite: A flexible code generating language for dynamic programming methods used in sequence comparison. *ISMB* 5: 56–64.

Cai Y, Bork P (1998). Homology-based gene prediction using neural nets. *Anal. Biochem.* 265: 269–274.

Florea L, et al. (1998). A computer program for aligning a cDNA sequence with a genomic DNA sequence. *Genome Res.* 8: 967–974.

Gelfand MS, et al. (1996). Gene recognition via spliced alignment. *Proc. Natl. Acad. Sci. USA* 93: 9061–9066.

Guigó R, et al. (2000). Sequence similarity based gene prediction. In *Genomics and Proteomics: Functional and Computational Aspects*, S Suhai, Ed. Kluwer Academic/Plenum Publishing, New York, pp. 95–105.

Hooper PM, et al. (2000). Prediction of genetic structure in eukaryotic dna using reference point logistic regression and sequence alignment. *Bioinformatics* 16: 425–438.

Huang X, et al. (1997). A tool for analyzing and annotating genomic sequences. *Genomics* 46: 37–45.

Jiang J, Jacob HJ (1998). Ebest: An automatic tool using expressed sequence tags to delineate gene structure. *Genome Res.* 5: 681–702.

Krogh A (2000). Using database matches with hmmgene for automated gene detection in drosophila. *Genome Res.* 10: 523–528.

Kulp D, et al. (1996). A generalized hidden Markov model for the recognition of human genes in DNA. In *Intelligent Systems for Molecular Biology*, DJ States, P Agarwal, T Gaasterland, L Hunter, R Smith, Eds. AAAI Press, Menlo Park, CA, pp. 134–142.

Milanesi L, et al. (1999). Genebuilder: Interactive in silico prediction of gene structure. *Bioinformatics* 15: 612–621.

Mott R (1997). EST_GENOME: A program to align spliced DNA sequences to unspliced genomic DNA. *Comput. Appl. Biosci.* 13: 477–478.

Schiex T, et al. (2001). Eugene: An eukaryotic gene finder that combines several sources of evidence. In *Lecture Notes in Computer Science*, Vol. 2066 (*Computational Biology*), O Gascuel, MF Sagot, Eds. Springer, Vienna, pp. 111–125.

Xu Y, Uberbacher EC (1997). Automated gene identification in large-scale genomic sequences. *J. Comput. Biol.* 4: 325–338.

Yeh R, et al. (2001). Computational inference of homologous gene structures in the human genome. *Genome Res.* 11: 803–816.

See also BLASTX, Comparative Gene Prediction, Dynamic Programming, GENEWISE, Spliced Alignment.

GENE PREDICTION SYSTEMS, INTEGRATED
Roderic Guigó

A number of systems produce and maintain automatic annotation of sequenced genomes.

At the core of these systems there is a gene prediction pipeline which integrates information from different gene prediction programs. In the European Bioinformatics Institute (EBI) Ensembl system, exon candidates are predicted by GENSCAN and confirmed through cDNA and protein sequence searches. GENEWISE is used to delineate the exonic structure of the predicted genes. Celera used a similar pipeline, OTTO, to annotate the human and mouse genomes. Ensembl and OTTO are rather conservative systems, predicting about 25,000–30,000 human genes. The National Center for Biotechnology Information (NCBI) runs a less conservative pipeline based on GENOMESCAN that predicts about 40,000 human genes. The UCSC Genome Browser displays gene prediction by different programs on the human and mouse genomes but does not have a specific gene prediction pipeline.

Similarly, a few stand-alone systems and public servers run a number of gene prediction tools on a sequence query on demand and display the results through a single interface: GENEMACHINE, GENOTATOR, NIX, RUMMAGE, and WEBGENE are a few such tools. Given the current status of computational gene prediction and their rather large ratio of false positives, it is indeed advisable to run a number of different programs on a query sequence and look after consistent predictions. One step further, METAGENE qualifies the predictions by different programs and suggests a consensus prediction.

Related Websites

Ensembl	http://www.ensembl.org
NCBI annotation pipeline	http://www.ncbi.nlm.nih.gov/genome/guide/build.html
UCSC Genome Browser	http://genome.ucsc.edu/
GENEMACHINE	http://genemachine.nhgri.nih.gov/
GENOTATOR	http://www.fruitfly.org/~nomi/genotator/
NIX	http://www.hgmp.mrc.ac.uk/NIX/
RUMMAGE	http://gen100.imb-jena.de/rummage/
WEBGENE	http://www.itba.mi.cnr.it/webgene/
METAGENE	http://rgd.mcw.edu/METAGENE/

Further Reading

Jones J, et al. (2002). A comparative guide to gene prediction tools for the bioinformatics amateur. *Int. J. Oncol.* 20: 697–705.

See also Gene Prediction, Ab Initio.

GENERALIZATION. *SEE* ERROR MEASURES.

GENE SHARING
Austin L. Hughes

Gene sharing occurs when a single gene encodes a protein that has more than one function.

These distinct functions are then said to be "shared" by a single gene. This phenomenon was first named by Piatigorsky and colleagues as a result of studies on the eye lens crystallins of animals. These researchers found that, over the course of animal evolution, soluble proteins having a wide variety of original functions have been recruited to serve the role of crystallins. After an initial period of gene sharing, there is typically a gene duplication, after which one daughter gene is free to specialize as a crystallin while the other retains the original function.

Further Reading

Piatigorsky J, Horwitz J (1996). Characterization and enzyme activity of argininosuccinate lyase/δ-crystallin of the embryonic duck lens. *Biochim. Biophys. Acta* 1295: 158–164.

Piatigorsky J, Wistow GJ (1991). The recruitment of crystallins: New functions precede gene duplication. *Science* 252: 1078–1079.

GENE SIZE
Katheleen Gardiner

The amount of genomic DNA in base pairs containing all exons and introns of a gene.

This includes exons and introns spanned by the $5'$ untranslated region, the coding region, and the $3'$ untranslated region. Gene size in genomic DNA is to be distinguished from that of the mature mRNA or cDNA, which lacks all introns. Gene size is highly variable and is governed by the size and number of introns. Intronless genes can be ~1–5 kb in size (histones), while the largest annotated gene is the 79-intron Duchenne muscular dystrophy (DMD) gene that spans 2.4 Mb. Early estimates of gene sizes have probably been low but can now be determined with greater accuracy from the complete genomic sequence. Estimates of average gene size tend to increase as sequence quality of a genome sequence improves and exons initially thought to be from separate genes are merged.

Further Reading

Caron H, et al. (2001). The human transcriptome map: Clusters of highly expressed genes in chromosome domains. *Science* 291: 1289–1292.

Cooper DN (1999). *Human Gene Evolution.* Bios Scientific Publishers, Oxford, UK.

International Human Genome Sequencing Consortium (2001). Initial sequencing and analysis of the human genome. *Nature* 409: 866–921.

Dunham I (1999). The DNA sequence of human chromosome 22. *Nature* 402: 489–495.

Hattori M, et al. (2000). The DNA sequence of human chromosome 21. *Nature* 405: 311–320.

Venter C, et al. (2001). The sequence of the human genome. *Science* 291: 1304–1351.

GENETIC CODE (STANDARD GENETIC CODE, UNIVERSAL GENETIC CODE)

John M. Hancock

The code by which the molecular machinery of the cell translates protein coding DNA sequences into amino acid sequences.

Bases in protein coding regions of genomes are arranged into codons, which consist of a group of three consecutive bases. When processed by the ribosome,

Table Standard Genetic Code. Codons, which are conventionally represented in their RNA form, are arranged alphabetically for ease of reference. Amino acids are represented by their three-letter codes. Shaded boxes indicate codons that have been shown to be reassigned in the mitochondrial or nuclear genome of at least one species

Codon	AA	Codon	AA	Codon	AA	Codon	AA
AAA	Lys	CAA	Gln	GAA	Glu	UAA	STOP
AAC	Asn	CAC	His	GAC	Asp	UAC	Tyr
AAG	Lys	CAG	Gln	GAG	Glu	UAG	STOP
AAU	Asn	CAU	His	GAU	Asp	UAU	Tyr
ACA	Thr	CCA	Pro	GCA	Ala	UCA	Ser
ACC	Thr	CCC	Pro	GCC	Ala	UCC	Ser
ACG	Thr	CCG	Pro	GCG	Ala	UCG	Ser
ACU	Thr	CCU	Pro	GCU	Ala	UCU	Ser
AGA	Arg	CGA	Arg	GGA	Gly	UGA	STOP
AGC	Ser	CGC	Arg	GGC	Gly	UGC	Cys
AGG	Arg	CGG	Arg	GGG	Gly	UGG	Trp
AGU	Ser	CGU	Arg	GGU	Gly	UGU	Cys
AUA	Ile	CUA	Leu	GUA	Val	UUA	Leu
AUC	Ile	CUC	Leu	GUC	Val	UUC	Phe
AUG	Met	CUG	Leu	GUG	Val	UUG	Leu
AUU	Ile	CUU	Leu	GUU	Val	UUU	Phe

each codon can either encode an amino acid or signal the termination of a protein sequence. The code by which codons represent amino acids is almost universal in nuclear genomes (see the table on page 208) but mitochondria use a different code in which UGA codes for tryptophan, AUA codes for methionine, and AGA and AGG are terminators. In addition, a number of variant codes have been discovered in both mitochondrial and nuclear genomes. These are characterized by the reassignment of individual codons, including termination codons, to new amino acids rather than by wholesale reorganization.

Related Website

Freeland laboratory	http://www.evolvingcode.net/uni_code.php3

Further Reading
Barrell BG, et al. (1979). A different genetic code in human mitochondria. *Nature* 282: 189–194.

Knight RD, et al. (2001) Rewiring the keyboard: Evolvability of the genetic code. *Nat. Rev. Genet.* 2: 49–58.

Lewin B (2000). *Genes VII.* Oxford University Press, Chapters 5–7.

See also Codon.

GENETIC DATA ENVIRONMENT. *SEE* GDE.

GENETIC DRIFT (EVOLUTION, FOUNDER EFFECT, NEUTRAL THEORY)
A. Rus Hoelzel

Genetic drift is the random change in allele frequencies that results from variation in the number of offspring among individuals, and for sexually reproducing species, from the independent segregation of alleles. It is a dispersive force, leading to the loss of variation over time, and this loss is inversely proportional to population size. The following formula describes that relationship for finite populations:

$$H_t = H_o(1 - (1/2N))^t$$

where H_t is heterozygosity at generation t, H_o is the initial probability of being heterozygous, and N is the population size.

Due to the stochastic fluctuations in allele frequency, a given allele may go to either fixation or extinction, resulting in the loss of diversity. A further implication of genetic drift is that the direction of evolutionary change by this force is unpredictable and on balance, neutral.

Further Reading
Hartl DL (2000). *A Primer of Population Genetics.* Sinauer Associates, Sunderland, MA.

GENETIC LINKAGE. *SEE* LINKAGE.

GENETIC LINKAGE USER ENVIRONMENT. *SEE* GLUE.

GENETIC NETWORK. *SEE* NETWORK.

GENETIC REDUNDANCY
Katheleen Gardiner

Members of gene family performing the same function, such that deletion of one gene has no effect on organism viability.

For example, the human genome contains hundreds of rRNA genes in five clusters. The number of rRNA genes in each cluster varies among individuals, and entire clusters can be deleted with no obvious negative consequences on the individual. Redundant genes are nonessential. However, for protein coding genes, being nonessential in a controlled laboratory setting may not translate to being redundant in a normal environment.

Further Reading
Tautz D (1992). Redundancies, development and the flow of information. *Bioessays* 14: 263–266.
Tautz D (2000). A genetic uncertainty problem. *Trends Genet.* 16: 475–477.

GENEWISE
Roderic Guigó

GENEWISE compares a genomic sequence to a protein sequence or to a hidden Markov model (HMM) representing a protein domain.

It performs the comparison at the protein translation level, while simultaneously maintaining a reading frame regardless of intervening introns and sequence errors that may cause frame shifts. Thus, GENEWISE does both the gene prediction and the homology comparison together.

GENEWISE is based around the probabilistic formalism of HMMs. The underlying dynamic programming algorithms are written using the code-generating language DYNAMITE (Birney and Durbin, 1997). GENEWISE is part of the WISE2 package, which includes also ESTWISE, a program to compare EST or complementary DNA (cDNA) sequences to a protein sequence or a protein domain HMM. GENEWISEDB and ESTWISEDB are the database searching versions of GENEWISE and ESTWISE, respectively. They compare a database of DNA sequences (genomic or cDNA) to a database protein or protein domain HMMs.

The database search programs in WISE2 are computationally very expensive, and their use may become prohibitive for large collections of sequences. The WISE2 package provides two scripts, BLASTWISE and HALFWISE, that allow users with average computer resources to run database searches with GENEWISE more sensibly. BLASTWISE compares a DNA sequence to a protein database using BLASTX and then calls GENEWISE on a carefully selected set of proteins. Similarly, HALFWISE compares a DNA sequence to a database of protein domain HMMs, such as PFAM.

Alternatively, specialized commercial hardware exists from Compugen, Paracel, and Time Logic that can approximate a GENEWISE search very closely or even run a full GENEWISE-type algorithm. GENEWISE is at the core of the Ensembl system and it has been extensively used to annotate the human genome sequence.

Related Websites

GENEWISE	http://www.sanger.ac.uk/Software/Wise2/
DYNAMITE	http://www.sanger.ac.uk/Software/Dynamite/

Further Reading

Birney E, Durbin R (1997). Dynamite: A flexible code generating language for dynamic programming methods used in sequence comparison. *ISMB* 5: 56–64.

See also BLAST; Gene Prediction, Homology Based; Gene Prediction Systems, Integrated; Hidden Markov Model; Spliced Alignment.

GENOME ANNOTATION
Jean-Michel Claverie

The process of interpreting a newly determined genome sequence.

Genome annotation is performed in two distinct phases using different software tools. In the first phase, the "raw" genomic sequence is parsed into elementary components such as protein coding regions, tRNA genes, putative regulatory regions, repeats, etc. The software tools used to parse higher eukaryotic sequences (animal and plants) are more sophisticated (yet still more error prone) than the programs required for lower eukaryotes, prokaryotes, and archaea due to the presence of introns and more frequent repeats and pseudogenes. Bacterial "ORFing," for instance, is quite trivial, reaching more than 95% accuracy with the use of low-order Markov models. Higher eukaryotic genome parsing is done by locally fitting a probabilistic multifeature gene model supplemented by the detection of local similarity with known protein or RNA sequences as well as ESTs. High-quality gene parsing still requires visual validation and manual curation for higher eukaryotes. In the second phase of the genome annotation process, protein coding genes are translated into putative protein sequences that are exhaustively compared to existing domain and functional motif databases. "Functional" attributes are mostly inherited from homologous proteins in a few model species (e.g., *Escherichia coli,* yeast, *Drosophila, Caenorhabdits elegans*) that have been previously characterized experimentally. Non–protein coding genes (e.g., tRNA, rRNA, etc.) are also annotated by homology. The detailed functional annotation of genomes is still a research activity, requiring specialized expertise in cellular physiology, biochemistry, and metabolic pathways. Except for the smallest parasitic bacteria, the current genome annotation protocols succeed in predicting a clear function for only half of the genes. Phylogenomic approaches ("guilt-by-association methods") are then used to extend direct functional information through the identification of putative relationships between genes (of unknown or known function).

Related Websites

High-Quality Automated and Manual Annotation of Microbial Proteomes (HAMAP)	http://www.expasy.org/sprot/hamap/

Clusters of Orthologous Groups	http://www.ncbi.nlm.nih.gov/COG/
Oak Ridge National Laboratory genome analysis	http://genome.ornl.gov/
NCBI Genomic Sequence Assembly and Annotation Process	http://www.ncbi.nlm.nih.gov/genome/guide/build.html
Ensembl	http://www.ensembl.org/

Further Reading

Audic S, Claverie JM (1998). Self-identification of protein-coding regions in microbial genomes. *Proc. Natl. Acad. Sci. USA* 95: 10,026–10,031.

Burge C, Karlin S (1997). Prediction of complete gene structures in human genomic DNA. *J. Mol. Biol.* 268: 78–94.

Lukashin AV, Borodovsky M (1998). GeneMark.hmm: New solutions for gene finding. *Nucleic Acids Res.* 26: 1107–1115.

Rogic S, et al. (2001). Evaluation of gene-finding programs on mammalian sequences. *Genome Res.* 11: 817–832.

Apweiler R, et al. (2000). InterPro—an integrated documentation resource for protein families, domains and functional sites. *Bioinformatics* 16: 1145–1150.

GENOMEINSPECTOR
John M. Hancock and Martin J. Bishop

Program to scan DNA sequences for consensus spacings between sequence elements.

The program reads in data from a variety of sources (examples given are data bank entry annotations, GCG FindPatterns output, ConsInspector and MatInspector output) and searches for spacings between elements in megabase-scale DNA sequences. Results are presented as a histogram of spacings. A number of spacings can be displayed at the same time to aid a search for patterns.

Related Website

| GenomeInspector | http://www.gsf.de/biodv/genomeinspector.html |

Further Reading

Quandt K, et al. (1996a). GenomeInspector: Basic software tools for analysis of spatial correlations between genomic structures within megabase sequences. *Genomics* 33: 301–304.

Quandt K, et al. (1996b) GenomeInspector: A new approach to detect correlation patterns of elements on genomic sequences. *Comp. Appl. Biosci.* 12: 405–413.

See also ConsInspector, MatInspector, ModelInspector, PromoterInspector, Wisconsin Package.

GENOME SCANS (GENOMEWIDE SCANS)
Mark McCarthy and Steven Wiltshire

Used, in the context of disease–gene mapping, to describe strategies for susceptibility gene identification which aim to survey the complete genome and to identify chromosomal regions with a high probability of containing susceptibility variants.

With current technology, almost all genome scans focus on identifying regions of linkage by typing a collection of families segregating the disease (or phenotype) of interest using a set of polymorphic markers arranged at regular intervals across the genome. In a typical study, this will entail, for the initial screen at least, around 400 highly polymorphic microsatellite markers, with an average separation of around 10 cM. Since linkage signals typically extend for tens of centimorgans, this should allow, provided sufficient families are typed, some or all of the regions containing major susceptibility genes to be localized. However, since, for multifactorial traits in particular, linkage tends to be relatively underpowered within the data sets generally available, there is increasing interest in conducting genomewide scans that seek evidence of association (or linkage disequilibrium) rather than linkage. Scans for linkage disequilibrium should be capable, in principle, of better power and localization, but very large numbers of markers (hundreds of thousands) will need to be typed in most populations.

Examples: The first example of a genomewide scan for a multifactorial trait was conducted for type 1 diabetes (Davies et al., 1994). This scan revealed evidence for linkage in several chromosomal regions, though the largest signal by far (as expected) fell in the histocompatibility locus antigen (HLA) region of chromosome 6. Since that time around 300 genome scans have been reported for a wide variety of multifactorial (complex) traits.

Further Reading

Altmuller J, et al. (2001). Genomewide scans of complex human disease: True linkage is hard to find. *Am. J. Hum. Genet.* 69: 936–950.

Concannon P, et al. (1998). A second-generation screen of the human genome for susceptibility to insulin-dependent diabetes mellitus. *Nat. Genet.* 19: 292–296.

Davies JL, et al. (1994). A genome-wide search for human type 1 diabetes susceptibility genes. *Nature* 371: 130–136.

Risch N, Merikangas K (1996). The future of genetic studies of complex human diseases. *Science* 273: 1516–1517.

Todd JA, Farrall M (1996). Panning for gold: Genome-wide scanning for linkage in type 1 diabetes. *Hum. Mol. Genet.* 5: 1443–1448.

Wiltshire S, et al. (2001). A genome-wide scan for loci predisposing to type 2 diabetes in a UK population (The Diabetes (UK) Warren 2 Repository): Analysis of 573 pedigrees provides independent replication of a susceptibility locus on chromosome 1q. *Am. J. Hum. Genet.* 69: 553–569.

See also Linkage, Linkage Analysis, Multipoint Linkage Analysis, Positional Candidate Approach.

GENOME SEQUENCE
John M. Hancock and Rodger Staden

The complete sequence of all the DNA in a single cell, unicellular organism, or virus.

The genome sequence of an organism encompasses, for diploid organisms, a complete haploid set, including both sex chromosomes, where relevant, and a single copy of any extrachromosomal genomes, such as mitochondria and plastids. In bacteria it includes a copy of the main chromosome or chromosomes plus those of common plasmids. Commonly, the term may also be applied to the nuclear genome of eukaryotes or the chromosomal genome of bacteria.

Genome sequences may be finished to a high standard but are commonly released in a preliminary form as draft sequences when they are approximately 90%

complete. Some centers release sequence reads as they come out of the sequencing pipeline to provide the quickest possible access to other researchers. This class of sequences are known as high-throughput genome sequences (HTGSs).

The first genome to be completely sequenced was that of the bacteriophage φX174 (Sanger et al., 1977). The first bacterial sequence completed was that of *Haemophilus influenzae* and the first eukaryote the yeast *Saccharomyces cerevisiae*. On July 4, 2003, 121 eubacterial, 36 archaeal, and 21 eukaryotic genomes were publicly available to some degree from the NCBI server.

Related Website

NCBI genomes	http://www.ncbi.nlm.nih.gov:80/entrez/query.fcgi?db=Genome

Further Reading

Fleischmann RD, et al. (1995). Whole-genome random sequencing and assembly of *Haemophilus influenzae* Rd. *Science* 269: 496–512.

Goffeau A, et al. (1996). Life with 6000 genes, *Science*. 274: 546, 563–567.

Sanger F, et al. (1977). Nucleotide sequence of bacteriophage phi X174 DNA. *Nature* 265: 687–695.

See also Finishing Criteria.

GENOME SEQUENCING
Rodger Staden

At present the most common strategy employed by large-scale sequencing projects is to isolate large overlapping segments of the target genome as artificial chromosomes such as BACs and P1-derived artificial chromosome (PACs), to find the order of these clones along the genome (contig mapping) and determine their sequences. However, current sequencing methods can only produce the sequence of around 500–800 bases per experiment and require a short (around 20-base) segment of DNA (a primer) complementary to the sequence immediately 5′ to the target DNA in order to initiate the process. This suggests the need to know segments of the sequence beforehand, but it is circumvented by randomly breaking the artificial chromosome into overlapping pieces of around 2000 bases and cloning them into a sequencing vector. The vector sequences on either side of the cloning site are adjacent to all segments of target DNA, and so the same primer sequences can be used for all sequencing experiments. The sequence obtained from one side of the cloning site will be from the forward strand of the DNA, and that from the other side will be from the reverse strand. When forward and reverse readings are derived from the same clone, they are said to form a read pair. In an assembly, not only do they give useful data from opposite strands of the DNA, but also knowledge that they form a read pair from a clone with a known size range provides a means of checking the ordering and orientation of the readings.

The strategy of randomly breaking the target DNA and cloning the resulting over-lapping segments prior to determining their sequences is known as shotgun sequencing. An alternative way of accessing random start points in the sequence while retaining the facility of using the same primers for every experiment can be obtained by use of transposons.

Further Reading

Primrose SB (1998). *Principles of Genome Analysis: A Guide to Mapping and Sequencing DNA from Different Organisms.* Blackwell, Oxford, UK.

See also Contig Mapping, PRIMER, Transposable Element.

GENOME SIZE (*C* VALUE)
Katheleen Gardiner

The number of base pairs in a haploid genome (one copy of each pair of all chromosomes) in a given species.

Genome size has also been expressed in picograms (1 kb = 10^{-6} pg), but this is less informative now that many genomes have been completely sequenced. Mammalian genomes contain $\sim$3 billion bp, birds 1–2 billion.

Further Reading

Ridley M (1996). *Evolution.* Blackwell Science, Cambridge, MA, pp. 265–276.

See also C-Value Paradox.

GENOMEWIDE SCANS. *SEE* GENOME SCANS.

GENOMEWIDE SURVEYS
Dov S. Greenbaum

Analyses surveying the overall content of genomes, proteomes, etc.

Comparative genomic analysis is becoming increasingly more complex given the overwhelming amount of data now being produced for every organism. Methods have been borrowed from other disciplines to help cope with this data deluge.

Similar to censuses taken by governments, genomewide surveys serve to assess the number and nature of distinct parts—finite parts lists—within individual genomes and between genomes. The power of these surveys lies in the fact that they can condense much of the information unique to a particular genome into key values, thus facilitating genome–genome comparisons.

Surveys assess, in particular, such genomic information as protein folds, domains, metabolic pathways, and the like; these units generally have a limited repertoire available in nature. Through assessing standardized attributes of each unit, e.g., function, usage, and size, these units can be compared with other units both inter- and intragenomically. This allows science, through investigating shared parts, to globally and comprehensively compare and examine organisms and their genomes.

Furthermore, multiple methodologies can be applied to the surveyed information to analyze individual parts, including, e.g., phylogenetic trees. Approximately 500 fold types are presently known and the maximum number is thought to be in the few thousand. By calculating the number of these folds within genomes and subsequently ranking them by their degree of usage by the organism, a researcher can compare diverse genomes, through their specific fold usage, thus determining evolutionary relatedness or environmental conditions that favor distinct fold types.

Related Websites

Partslist	www.partslist.org
GeneCensus	www.genecensus.org

Further Reading

Gerstein M, Hegyi H (1998). Comparing genomes in terms of protein structure: Surveys of a finite parts list. *FEMS Microbiol. Rev.* 22: 277–304.

Hegyi H, et al. (2002). Structural genomics analysis: Characteristics of atypical, common, and horizontally transferred folds. *Proteins* 47: 126–141.

Qian J, et al. (2001). PartsList: A web-based system for dynamically ranking protein folds based on disparate attributes, including whole-genome expression and interaction information. *Nucleic Acids Res-.* 29: 1750–1764.

GENOMICS

Stuart M. Brown

The application of high-throughput automated technologies to molecular biology and the resulting global approach to the analysis of all genes, all transcripts, or all proteins in an organism.

This is a very broad research area that encompasses many different technologies and types of data. The common theme is the rapid collection of large amounts of data from massively parallel experiments using laboratory robotics and bioinformatics. Genomics technologies include:

- Transcriptomics—the measurement of expression of all genes in an organism by microarray analysis of complementary DNA (cDNA)
- Proteomics—the measurement of all of the proteins in an organism and the investigation of all protein–protein interactions
- Toxicogenomics—the analysis of the effects of toxic substances on gene expression patterns and the prediction of toxicity based on gene expression studies in model systems and genetic predispositions to toxin sensitivity
- Pharmacogenomics—the study of the interaction between genotype (measured by genomewide SNP profiling) and the response of individuals to drugs
- Functional genomics—a high-throughput approach to determining the function of every gene and every protein in an organism and to understand the complex networks of metabolic pathways and regulatory control mechanisms
- Comparative genomics—comparisons of whole genomes across species to identify similarities in gene sequences, gene structure, and gene expression

Related Websites

NCBI genomics primer	http://www.ncbi.nlm.nih.gov/about/primer
National Center for Toxicogenomics	http://www.niehs.nih.gov/nct/home.htm
Genomics news and products	http://genomics.biocompare.com/
European genomics efforts	http://www.functionalgenomics.org.uk/

Further Reading
Twyman RM, Primrose SB (2003). *Principles of Genome Analysis*. Blackwell, Oxford, UK.

GENSCAN
Roderic Guigó

GENSCAN is a general-purpose eukaryotic gene prediction program.

For each query sequence, the program determines the most likely gene structure under an underlying generalized hidden Markov model (GHMM). The states in the GENSCAN GHMM correspond to the different functional units of a gene like exons, introns, and splice sites. The transitions between the states ensure that the order in which the model visits various states is biologically consistent. In particular, exons are modeled according to a nonhomogeneous fifth-order Markov model. Introns are modeled using a homogeneous fifth-order Markov model. To model donor and acceptor splice sites, maximal dependence decomposition is used to obtain a series of weight matrices which capture dependencies between positions in these sites.

In addition, the GENSCAN GHMM contains parameters that account for many higher order gene structural properties of genomic sequences, e.g., typical gene density, the typical number of exons per gene, and the distribution of exon sizes for different types of exon. Separate sets of model parameters can be used to account for the many substantial differences in gene density and gene characteristics observed in distinct $C + G\%$ compositional regions of the human genome and the genomes of other vertebrates. Models exist also for maize and *Arabidopsis* sequences.

The publication of GENSCAN in 1996 was a breakthrough in the field of computational gene prediction. GENSCAN was not only the first program capable of analyzing large genomic sequence encoding multiple genes in both strands, but it was also substantially more accurate than the existing programs at that time. GENSCAN is still one of most accurate ab initio gene prediction programs, and it has been used to annotate most of the vertebrate and plant genomes sequenced so far. However, although its accuracy when analyzing single gene sequences is quite remarkable, when analyzing large genomic sequences in vertebrate genomes, the specificity of the program suffers, apparently predicting a large fraction of false-positive exons.

Recently the program GENOMESCAN has been developed that integrates similarity to known protein sequences into GENSCAN.

The GHMM approach has its roots in the program GENEPARSER (Snyder and Stormo, 1993). Other eukaryotic gene prediction programs based on hidden Markov models are GENIE (Kulp et al., 1996), HMMGENE (Krogh, 1997), VEIL (Henderson et al., 1997), FGENESH (Salamov and Solovyev, 2000), and GENEMARK.HMM (Lukashin and Borodovsky, 1998).

Related Website

GENSCAN	http://genes.mit.edu/GENSCAN.html

Further Reading
Burge CB, Karlin S (1997). Prediction of complete gene structures in human genomic DNA. *J. Mol. Biol.* 268: 78–94.

Henderson J, et al. (1997). Finding genes in DNA with a hidden Markov model. *J. Comput. Biol.* 4: 127–141.

Krogh A (1997). Two methods for improving performance of an HMM and their application for gene finding. *ISMB* 5: 179–186.

Kulp D, et al. (1996). A generalized hidden Markov model for the recognition of human genes in DNA. In *Intelligent Systems for Molecular Biology*, DJ States, P Agarwal, T Gaasterland, L Hunter, R Smith, Eds. AAAI Press, Menlo Park, CA, pp. 134–142.

Lukashin A, Borodovsky M (1998). Genemark.hmm: New solutions for gene finding. *Nucleic Acids Res.* 26: 1107–1115.

Salamov A, Solovyev V (2000). Ab initio gene finding in drosophila genomic dna. *Nucleic Acid Res.* 10: 516–522.

Snyder EE, Stormo GD (1993). Identification of coding regions in genomic DNA sequences: An application of dynamic programming and neural networks. *Nucleic Acids Res.* 21: 607–613.

See also Coding Statistic; Gene Prediction, Ab Initio; Gene Prediction Accuracy; Gene Prediction, Homology Based; Hidden Markov Model.

GLOBAL ALIGNMENT
Jaap Heringa

An alignment of two or more sequences over their full lengths, as opposed to local alignment (Smith and Waterman, 1981), which includes a consecutive segment of each sequence.

The first automatic method to generate a global alignment for two protein sequences was based upon the dynamic programming technique developed by Needleman and Wunsch (1970). Global alignment should be carried out when a sequence set is suspected to be linked through homologous relationships resulting from divergent evolution. If the sequences have associated tertiary structures with multiple domains, it is important that the domain organization is colinear through all sequences and that all domains are present in all sequences. If zero end-gap penalty values are used to align the sequences, missing N- or C-terminal domains can mostly be tolerated by the automatic alignment methods.

Related Websites

GAP	http://genome.cs.mtu.edu/align/align.html
NAP	http://genome.cs.mtu.edu/align/align.html

Further Reading
Needleman SB, Wunsch CD (1970). A general method applicable to the search for similarities in the amino acid sequence of two proteins. *J. Mol. Biol.* 48: 443–453.

Smith TF, Waterman MS (1981). Identification of common molecular subsequences. *J. Mol. Biol.* 147: 195–197.

See also Alignment—Multiple, Alignment—Pairwise; Dynamic Programming, Needleman–Wunsch Algorithm.

GLOBAL OPEN BIOLOGY ONTOLOGIES (GOBO)
Robert Stevens

GOBO is an umbrella organization for the collection and dissemination of biological ontologies.

GOBO contains ontologies and points to some other efforts within the community. GOBO seeks a range of ontologies being designed for the general genomics and proteomics domains. Some of these will be generic, other more restricted in scope, e.g., to specific taxonomic groups. The criteria for inclusion in GOBO are:

1. The ontologies are "open" and can be used by all without any constraint other than that their origin must be acknowledged.
2. The ontologies are or can be instantiated in the GO syntax, extensions of this syntax, or DAML + OIL.
3. The ontologies are orthogonal to other ontologies already lodged with GOBO.
4. The ontologies share a unique identifier space.
5. The ontologies include natural-language terms.

Related Website

GOBO	http://www.geneontology.org/doc/gobo.html

GLOBULAR
Roman A. Laskowski

Proteins characterized by a compact three-dimensional conformation of their polypeptide chain are called globular. The compactness is usually achieved by a tight packing of regions of regular secondary structures that places the protein's hydrophobic residues toward the inside of the protein, away from the surrounding solvent, forming a "hydrophobic core," and the polar residues on or close to the surface.

Globular proteins differ markedly from polypeptides with random or simply repetitive conformation. Their overall structure tends to be stable under changes in temperature, pH, or pressure until a point is reached at which they "denature," losing their globular conformation and consequently their biological function.

They form the majority of naturally occurring proteins in solution and are distinct from structural proteins such as the fibrous proteins.

See also Amino Acid, Conformation, Fibrous, Polar, Secondary Structure of Protein.

GLUE (GENETIC LINKAGE USER ENVIRONMENT, GLUE3)
John M. Hancock and Martin J. Bishop

Provides a convenient environment from which to run a variety of popular linkage analysis packages.

The program accepts pedigree data in *pre-makeped* format and other data such as information on loci and locus order, description of loci, information on recombination, and program-specific information in *linkage* format. The program allows editing of data files and as of March 2003 supports the following programs: MLINK, LINKMAP, GENEHUNTER, MERLIN, UNPHASED, TRANSMIT, and SLINK.

Related Website

GLUE	http://www.hgmp.mrc.ac.uk/Registered/Webapp/glue/

See also Linkage Analysis.

GLUTAMIC ACID
Jeremy Baum

Glutamic acid is a polar (often negatively charged) amino acid with side chain $(CH_2)_2CO_2H$ found in proteins. In sequences, written as Glu or E.

Further Reading
Branden C, Tooze J (1991). *Introduction to Protein Structure*. Garland Science, New York.

See also Amino Acid.

GLUTAMINE
Jeremy Baum

Glutamine is a polar amino acid with side chain $—(CH_2)_2CONH_2$ found in proteins. In sequences, written as Gln or Q.

See also Amino Acid.

GLYCINE
Jeremy Baum

Glycine is a small nonpolar amino acid with side chain —H found in proteins. In sequences, written as Gly or G.

See also Amino Acid.

GO. *SEE* GENE ONTOLOGY.

GOBASE (ORGANELLE GENOME DATABASE)
Guenter Stoesser

The Organelle Genome Database (GOBASE) is organizing and integrating data related to organelles.

GOBASE contains all published nucleotide and protein sequences encoded by mitochondrial and chloroplast genomes, selected RNA secondary structures of mitochondria-encoded molecules, genetic maps, taxonomic information for all species whose sequences are present in the database, and organismal descriptions of key protistan eukaryotes. Data are integrated and organized in a formal database structure allowing sophisticated biological queries using terms that are inherent in biological concepts.

Related Websites

GOBASE	http://megasun.bch.umontreal.ca/gobase/
Organelle Genome Megasequencing Program (OGMP)	http://megasun.bch.umontreal.ca/ogmpproj.html

Further Reading

Shimko N, et al. (2001). GOBASE: The organelle genome database. *Nucleic Acids Res.* 29: 128–132.

GOBO. *SEE* GLOBAL OPEN BIOLOGY ONTOLOGIES.

GOR SECONDARY-STRUCTURE PREDICTION METHOD (GARNIER–OSGUTHORPE–ROBSON METHOD)
Patrick Aloy

The Garnier–Osguthorpe–Robson (GOR) is a method for protein secondary-structure prediction from sequence information. It was one of the first automated implementations, a few years after the Chou–Fasman method, and also falls into the categories of "first-generation" structure prediction methods.

The original method (GOR-I) was based on single-residue preferences and also took into account the protein type (i.e., relative abundance of helices, beta strands, turns, and coils) obtained either by circular dichroism or from a preliminary prediction. All these considerations yielded a three-state-per-residue accuracy of 63%. More recent implementations of the GOR method (GOR-III) include propensities calculated for pairs of residues, increasing the accuracy up to 69.7%. However, more realistic calculations of the performance accuracy have placed the GOR-I and GOR-III at levels of 55% and 58%, respectively.

Related Website

GOR method	http://molbiol.soton.ac.uk/compute/GOR.html

Further Reading

Garnier J, Osguthorpe DJ, Robson B (1974). Analysis of the accuracy and implications of simple methods for predicting the secondary structure of globular proteins. *J. Mol. Biol.* 120: 97–120.

Gibrat JF, Garnier J, Robson B (1987). Further developments of protein secondary structure prediction using information theory. New parameters and consideration of residue pairs. *J. Mol. Biol.* 198: 425–443.

See also Chou—Fasman Prediction Method, Secondary-Structure Prediction of Protein.

GRADIENT DESCENT (STEEPEST-DESCENT METHOD)
Nello Cristianini

Simple optimization method for minimizing a function over a large set of parameters whenever such a function is differentiable.

Gradient descent consists of iteratively updating the current estimate by moving it in the direction of steepest descent, provided by the gradient calculations.

If the function $f(x)$ is to be minimized with respect to x and the current estimate of the minimum is x_0, a step in the direction of the steepest descent is given by $x \leftarrow x_0 - \alpha \nabla f(x_0)$ where $\nabla f(x_0)$ is the gradient of f at x_0 and α is a parameter that controls the size of such a step.

For a suitable choice of the parameter α, this procedure is guaranteed to converge to a local minimum. Its major drawback is that if the function is nonconvex, it will not necessarily find the global minimum, and furthermore the solution will be dependent on the initial conditions. Despite this problem, it has been used with success, e.g., forming the basis of neural network training procedures such as backpropagation, which provides a way to compute the gradient in a feedforward neural network.

Further Reading
Bishop C (1996). *Neural Networks for Pattern Recognition*. Oxford University Press, Oxford, UK.
Duda RO, Hart PE, Stork DG (2001). *Pattern Classification*. Wiley, New York.

See also Neural Network.

GRAIL CODING REGION RECOGNITION SUITE
Roderic Guigó

GRAIL is a suite of tools designed to provide analysis and putative annotation of DNA sequences.

The original GRAIL program (GRAIL1) (Uberbacher and Mural, 1991) is a coding region recognition program. It uses a backpropagation neural network with two hidden layers to combine the output of seven coding statistics. These statistics (sensors) are computed on a fixed-length sliding sequence window. GRAIL can be accessed through a World Wide Web interface or, more conveniently, through a dedicated X client (XGRAIL) which allows for a high degree of interactivity.

GRAIL2 (Xu et al., 1994a) is a more sophisticated version of GRAIL. It uses variable-length windows tailored to each potential exon candidate on the query sequence. Thus, while GRAIL1 simply predicts coding regions, GRAIL2 is already a coding exon prediction program.

GAP (Xu et al., 1994b) is the GRAIL gene assembly program. It takes exons predicted by GRAIL2 and assembles them into predicted genes.

The GRAIL programs above have been extended recently to the GRAILEXP system (Xu and Uberbacher, 1997). This features a GRAIL-like exon finder with improved splice site recognition. The gene assembly module has been improved by searching a database of known gene messages and building gene models based on the corresponding alignments. While its primary use is to locate protein coding genes within DNA sequences, GRAILEXP can also locate EST/mRNA alignments, certain types of promoters, polyadenylation sites, CpG islands, and repetitive elements.

Related Websites

GRAIL1	http://compbio.ornl.gov/Grail-1.3/
GRAILEXP	http://grail.lsd.ornl.gov/grailexp/
XGRAILclient	ftp.lsd.ornl.gov/pub/xgrail

Further Reading
Uberbacher EC, Mural RJ (1991). Locating protein-coding regions in human DNA sequences by a multiple sensor-neural network approach. *Proc. Natl. Acad. Sci. USA* 88: 11,261–11,265.
Xu Y, Uberbacher EC (1997). Automated gene identification in large-scale genomic sequences. *J. Comput. Biol.* 4: 325–338.

Xu Y, et al. (1994a). Recognizing exons in genomic sequence using grail ii. In *Genetic Engineering: Principles and Methods*, J Setlow, Ed. Plenum, New York.

Xu Y, et al. (1994b). Constructing gene models from accurately predicted exons: An application of dynamic programming. *Comput. Appl. Biosci.* 11: 117–124.

See also Coding Region Prediction; Coding Statistic; Gene Prediction, Ab Initio; Gene Prediction, Homology Based.

GRAIL DESCRIPTION LOGIC
Robert Stevens

GRAIL is a description logic (DL) with hierarchies for both categories and relations, transitive relations, and restricted forms of concept, including axioms.

The GALEN terminology architecture also includes functionality to support lexical operations and multilingual support. The language has primarily been used to produce models of medical terminology [the GALEN Common Reference Model (CRM)]. A suite of modeling tools known as the OpenKnoME are available for GRAIL modeling. The GRAIL language was also used to encode the TAMBIS ontology. In both GALEN and TAMBIS the reasoning facilities of the DL played a vital role inferring the ontology's taxonomy from the given descriptions.

Related Website

GALEN Common Reference Model	http://www.openGALEN.org

GRAPH REPRESENTATION OF GENETIC, MOLECULAR, AND METABOLIC NETWORKS
Denis Thieffry

Formally, a metabolic or a genetic network can be represented by a connected graph. Vertices (or nodes) then correspond to the different components (genes or metabolic species) of the network, and the edges represent cross-interactions. Interactions can be oriented (one then uses the term *oriented edge*), signed (plus or minus), or labeled in other ways according to functional characteristics (type of molecular mechanism, molecular family, etc.). One then refers to directed, signed, or colored connected graphs, respectively.

In graph theory, one distinguishes between different class of interactive structures or types of graph topology, each consisting of a particular type of connected graph: trees, acyclic graphs, cyclic graphs, etc. A graph can also be characterized on a local level, by the number of edges involving the same vertex, called the degree of a vertex, or, in the case of oriented graphs, the number of incoming edges (fanin) or of out-coming edges (fanout).

When a graph describes regulatory or other types of molecular, metabolic, or genetic interactions, one speaks of a graph of interactions. Vertices then represent the different interacting species (genes, metabolic species, etc.) and the (unoriented) edges represent the interactions between these genes or molecular species.

Graph representation can also be used to represent transitions between different genetic or metabolic states. Vertices then represent the state of activity of a metabolic network (e.g., in terms of the different species concentration or activities) or of the expression of a genetic network (e.g., in terms of the levels of expression of the different genes), and edges represent mandatory or potential transitions between these different states across time. A specific time course of the system is thus represented by a pathway in this dynamical graph. Whenever several edges come out of a single vertex (state), one faces alternative dynamical pathways.

In the context of graph theory, a connected component is defined as a maximal connected state transition subgraph. In the case of a directed graph, one can further define strongly connected components as sets of vertices, such that, each vertex can lead to any other vertex of the set by following the oriented edges. In metabolic or genetic interaction graphs, such connected components may correspond to biologically meaningful functional structures such as protein complexes (graph of protein–protein interactions) and genetic regulatory modules (genetic regulatory graph).

One further defines a circuit of a graph as a subset of the edges of this graph forming a closed (circular) chain. From each vertex, a path then leads back to itself through all other vertices of the circuit. When dealing with directed graphs, one also uses the term *cycle*. In biology, one distinguishes between two types of regulatory circuits, according to the kind of effect each element has on itself through the other elements of the circuit. On the one hand, each element can have an (indirect) inhibitory effect on itself, amounting to a negative circuit. On the other hand, when this effect is activatory, one has a positive circuit.

The effect of an element on itself is direct in the case of an autoregulation (corresponding to a loop in graph theory) and indirect otherwise. To these two classes of regulatory circuits correspond strikingly different dynamic and biological properties. Positive circuits are necessary to generate multistable behavior and are involved in differentiation decisions. Negative circuits are needed to generate sustained oscillatory behavior and are involved in homeostatic properties of molecular and physiological systems. Negative circuits can also contribute to stabilize a steady state.

A Bayesian graph (or Bayesian network) is a directed acyclic graph in which vertices represent assertions and edges conditional probabilities. Such Bayesian networks are used to represent correlations in the expression of genes, which can be computed on the basis of extensive sets of large-scale expression measurements (e.g., transcriptomics). Vertices then represent genes and edges represent potential interactions. This approach has been recently used to reverse engineer gene cross-regulations on the basis of DNA chip expression data.

Further Reading

Diestel R (2000). *Graph Theory*, 2nd ed. Graduate Texts in Mathematics. Springer, Verlag, New York.

Gross J, Yelen J (1999). *Graph Theory and Its Applications*. Discrete Mathematics and Applications Series, KH Rosen, Ed. CRC Press, Boca Raton, FL.

Kanehisa M (2000). *Post-Genome Informatics*. Oxford University Press, New York.

Thomas R, d'Ari R (1990). *Biological Feedback*. CRC Press, Boca Raton, FL.

GRID AND THE GRIN AND GRID METHODS
Marketa J. Zvelebil

A more accurate docking could be carried out if it were possible to calculate the interaction energy while moving interacting molecules in real time. However, the calculation of the interaction energy is very computer intensive. Therefore programs have been written where an energy grid approximating for the larger of the two molecules under investigation can be precalculated. Then the interaction energy can be approximated by calculating the energy between atoms of the moving molecule and the appropriate grid points enabling docking to be performed.

The GRIN and GRID program computes regions within a binding site which has a high affinity for certain types of "'robes." The favorable regions are defined by super-imposing a regular grid upon the binding site. The probe is then placed at the peaks of the grid, and the empirical interaction energy of the probe with the protein is cal-culated. This results in a three-dimensional grid with an energy value at each peak. The favorable regions are then displayed as contours superimposed onto the protein structure showing the regions that are most favorable for the different probes. The three-dimensional grid can then be analyzed to find the best location(s) for the par-ticular probe. These probes can either be single atoms or functional groups, such as methyl groups.

Related Website

Affinity	http://www.msg.ucsf.edu/local/programs/insightII/doc/life/insight2000.1/affinity/affinityTOC.html

Further Reading
McConkey BJ, Sobolev V, Edelman M (2002). The performance of current methods in lig-and–protein docking. *Current Sci.* 83(7): 845–856.

See also DOCK, Docking.

GROMOS
Roland L. Dunbrack

A molecular dynamics simulation and modeling computer program developed by Wilfred van Gunsteren and colleagues. GROMOS (GROningen MOlecular Simulation program library) uses an empirical molecular mechanics potential energy function for studying the structures and dynamics of proteins, DNA, and other molecules.

Related Website

GROMOS	http://www.igc.ethz.ch/gromos/

Further Reading
Gunsteren WF, et al. (1998). *Encyclopedia of Computational Chemistry*, Vol. 2. Wiley, Athens, GA, pp. 1211–1216.

See also Empirical Potential Energy Function.

GROUP I INTRON. *SEE* INTRON.

GROUP II INTRON. *SEE* INTRON.

GUMBALL MACHINE
Thomas D. Schneider

A gumball machine is a model for the packing of Shannon spheres. Each gumball represents one possible message or molecular state (an after sphere). The radius of the gumball represents the thermal noise. The balls are all enclosed inside a larger sphere (the before sphere) whose radius is determined from both the thermal noise and the power dissipated at the receiver (or by the molecule) while it selects that state. The way the spheres are packed relative to each other is the coding.

Related Website

Molecular machines	http://www.lecb.ncifcrf.gov/toms/molecularmachines.html

Further Reading

Schneider TD (1991). Theory of molecular machines. I. Channel capacity of molecular machines. *J. Theor. Biol.* 148: 83–123. http://www.lecb.ncifcrf.gov/~toms/paper/ccmm/.

Schneider TD (1994). Sequence logos, machine/channel capacity, Maxwell's demon, and molecular computers: A review of the theory of molecular machines. *Nanotechnology* 5: 1–18.
http://www.lecb.ncifcrf.gov/~toms/paper/nano2/.

See also Channel Capacity, Molecular Machine Capacity.

H

h2
Haplotype
Hardy–Weinberg Equilibrium
Haseman–Elston Regression
Heavy-Atom Derivative
HE-COM, HE-CP, HE-SD, HE-SS
Helical Wheel
Helix–Coil Transition
Heritability
Heterogeneous Nuclear RNA
HGI
HGVBASE
Hidden Markov Model
Hierarchy
High-Scoring Segment Pair
Histidine
HIV RT and Protease Sequence Database
HIV Sequence Database
HMM
HMMer
HMMSTR

hnRNA
Holophyletic
Homologous Genes
Homologous Superfamily
Homology
Homology-Based Gene Prediction
Homology Modeling
Homology Search
Homozygosity Mapping
Horizontal Gene Transfer
HSP
Human Gene Index
Human Genome Variation Database
Hydrogen Bond
Hydropathy
Hydropathy Profile
Hydrophilicity
Hydrophobicity
Hydrophobicity Plot
Hydrophobic Moment
Hydrophobic Scale

Dictionary of Bioinformatics and Computational Biology. Edited by Hancock and Zvelebil
ISBN 0-471-43622-4   © 2004 John Wiley & Sons, Inc.

H2. *SEE* HERITABILITY.

HAPLOTYPE
Mark McCarthy and Steven Wiltshire

The set of alleles carried on a single chromosome: equivalently, the set of alleles inherited by an individual from a parent.

There is growing interest in the use of haplotype information for association analysis. One reason is that analysis at the haplotype level may, on occasion, be more powerful than analyses that examine loci individually. This is particularly so when a gene contains several interacting susceptibility variants, since the combined phenotypic effect may be best understood in terms of haplotypes. Where there is linkage disequilibrium between neighboring loci, certain haplotypes will occur within a population more frequently than expected (on the basis of the individual allele frequencies). Recent evidence suggesting that linkage disequilibrium in human populations is arranged into haplotype "blocks" (regions of marked linkage disequilibrium and limited haplotype diversity separated by intervals of limited disequilibrium and extensive diversity) has sparked interest in developing a "haplotype map" of the genome, charting such blocks and their haplotypic composition. Such a map should prove a valuable guide for efforts to develop large-scale association analysis. Since current linkage disequilibrium patterns and haplotype structures reflect the cumulative history and migration of the human species, haplotype analyses also play a role in helping to reconstruct this history.

Examples: Horikawa et al. (2000) demonstrated association between variants in the calpain-10 gene and type 2 diabetes in Mexican American and European subjects, but the strongest relative risk was observed with particular haplotype combinations: These appeared to explain risk much better than any single marker allele. Rioux et al. (2001) were able to localize a Crohn's disease susceptibility locus to a single haplotype block on chromosome 5q and to identify the most strongly associated haplotype; however, almost complete linkage disequilibrium between multiple variants on this haplotype frustrated efforts to determine which variant site was functionally responsible.

Related Websites
Programs for haplotype determination:

PHASE	http://www.stats.ox.ac.uk/mathgen/software.html
HAPLOTYPER	www.people.fas.harvard.edu/~junliu/
SNPHAP	www-gene.cimr.cam.ac.uk/clayton/software/

Repository for planned haplotype map data:

HapMap	http://www.ncbi.nlm.nih.gov/SNP/HapMap/index.html

Dictionary of Bioinformatics and Computational Biology. Edited by Hancock and Zvelebil
ISBN 0-471-43622-4   © 2004 John Wiley & Sons, Inc.

Further Reading

Gabriel SB, et al. (2002). The structure of haplotype blocks in the human genome. *Science* 296: 2225–2229.

Horikawa Y, et al. (2000). Genetic variation in the gene encoding calpain-10 is associated with type 2 diabetes mellitus. *Nat. Genet.* 26: 163–175.

Johnson GCL, et al. (2001). Haplotype tagging for the identification of common disease genes. *Nat. Genet.* 29: 233–237.

Patil N, et al. (2001). Blocks of limited haplotype diversity revealed by high-resolution scanning of human chromosome 21. *Science* 294: 1719–1723.

Rioux JD, et al. (2001). Genetic variation in the 5q31 cytokine gene cluster confers susceptibility to Crohn disease. *Nat. Genet.* 29: 223–228.

See also Association Analysis, Linkage Disequilibrium, Phase.

HARDY–WEINBERG EQUILIBRIUM

A. Rus Hoelzel

Description of the equilibrium frequencies in a population of a gene having two or more alleles.

Formulated by an English mathematician, G. H. Hardy, and a German physician, W. Weinberg, this equilibrium condition is based on the observation that given certain conditions (random mating and the lack of factors that change allele frequency such as gene flow, mutation, natural selection, and genetic drift), genotype frequencies for diploid organisms can be represented by binomial (two alleles) or multinomial (more than two alleles) functions of allele frequencies. Given two alleles A_1 and A_2 with frequencies p and q, respectively, the following relationship holds:

Genotypes:	A_1A_1		$A_1 A_2$		$A_2 A_2$
Genotype frequencies: (expressed in terms of allele frequencies)	p^2	$+$	$2pq$	$+$	$q^2 = 1$

This relationship is stable over time and will revert to the same genotype frequencies after one generation of random mating or to new genotype frequencies if allele frequencies are changed. Important implications for the study of population genetics include the fact that genotype frequencies can be determined from allele frequencies alone and that inference about evolutionary processes can be made on the basis of deviation from Hardy–Weinberg equilibrium genotype frequencies.

Further Reading

Hartl DL (2000). *A Primer of Population Genetics.* Sinauer Associates, Sunderland, MA.

See also Evolution, Natural Selection.

HASEMAN–ELSTON REGRESSION (HE-SD, HE-SS, HE-CP, HE-COM)

Mark McCarthy and Steven Wiltshire

An allele-sharing regression-based method for detecting linkage between a marker and a quantitative trait locus.

The original Haseman–Elston regression method (HE-SD) regressed the squared difference of the sibs' quantitative trait values (Y_{ij}) on the estimated proportion of alleles shared identical by descent (IBD) by the sibling pair (sibpair) ($\hat{\pi}_{ij_{ij}}$):

$$Y_{ij} = \alpha + \beta \hat{\pi}_{ij}$$

The gradient of the regression line, β, equals $-2\sigma_g^2(1 - 2\theta)^2$, where θ is the recombination fraction between the quantitative trait locus (QTL) and the marker, $(1 - 2\theta)^2$ is the correlation between the proportion of alleles shared IBD at the QTL and the marker locus, and $2\sigma_g^2$ is the proportion of the trait variance attributable to the QTL. The hypothesis $\beta = 0$ (no linkage) can be tested with the ratio of β to its standard error, distributed as Student's t statistic. A significantly negative value of β is evidence for linkage.

Several recent modifications of the original squared differences Haseman–Elston method exist, differing in the nature of the dependent variable: HE-CP, in which Y_{ij} is the cross product of the mean-centered sibs' trait values; HE-SS, in which Y_{ij} is the square of the sum of the sibs' mean-centered trait values; and HE-COM, in which Y_{ij} is a weighted combination of the squared differences and squared sums. Methods combining squared sums and squared differences have the greatest power. All these methods have been implemented for sibpairs or small sibships (total number of children born to a set of parents). However, a combined SD–SS implementation that is applicable to general pedigrees has been developed recently.

Examples: Daniels et al (1996) used the original HE regression to examine quantitative phenotypes related to asthma and atopy and found evidence for QTL on chromosome 13 influencing serum levels of immunoglobulin E (IgE) and on chromosome 7 influencing bronchial responsiveness to metacholine challenge. Using HE-COM (the combined squared sums/squared differences implementation), Wiltshire et al. (2002) observed evidence for a QTL on chromosome 3 influencing adult stature in a sample of United Kingdom pedigrees.

Related Websites
Programs for performing original or combined HE regression analyses:

SAGE	http://darwin.cwru.edu/pub/sage.html
GENEHUNTER2	www-genome.wi.mit.edu/ftp/distribution/software/genehunter/
QMS2	http://qms2.sourceforge.net/
MERLIN	http://www.sph.umich.edu/csg/abecasis/Merlin/

Further Reading
Daniels SE, et al. (1996). A genome-wide search for quantitative trait loci underlying asthma. *Nature* 383: 247–250.

Elston RC, et al. (2000). Haseman and Elston revisited. *Genet. Epidemiol.* 19: 1–17.

Drigalenko E (1998). How sib pairs reveal linkage. *Am. J. Hum. Genet.* 63: 1242–1245.

Haseman JK, Elston RC (1972). The investigation of linkage between a quantitative trait and a marker locus. *Behav. Genet.* 2: 3–9.

Sham PC, Purcell S (2001). Equivalence between Haseman-Elston and variance components linkage analyses for sib pairs. *Am. J. Hum. Genet.* 68: 1527–1532.

Sham PC, et al. (2002). Powerful regression-based quantitative trait linkage analysis of general pedigrees. *Am. J. Hum. Genet.* 71: 238–253.

Visscher PM, Hopper JL (2001). Power of regression and maximum likelihood methods to map QTL from sib-pair and DZ twin data. *Ann. Hum. Genet.* 65: 583–601.

Wiltshire S, et al. (2002). Evidence for linkage of stature to chromosome 3p26 in a large UK family data set ascertained for type 2 diabetes. *Am. J. Hum. Genet.* 70: 543–546.

See also Allele-Sharing Methods, Genome Scan, Identity by Descent, Quantitative Trait, Variance Components.

HEAVY-ATOM DERIVATIVE
Liz P. Carpenter

In X-ray crystallography it is often necessary to obtain multiple data sets on the same type of crystal of a macromolecule to allow the structure to be solved. One technique for modifying crystals is to soak them in solutions of heavy-atom compounds containing, e.g., mercury, platinum, or gold ions. Often a few sites on the macromolecule crystal will be occupied by well-ordered heavy atoms. If this substitution occurs without a significant change in the crystal or macromolecule structure, it is possible to obtain data sets of derivatized and native crystals which are similar except for the presence of extra heavy atoms. The differences in diffraction will be dominated by the effects of the heavy atoms, since scattering is proportional to the number of electrons. Comparison of the data sets will allow the phase problem to be overcome so that the structure of the macromolecule can be determined

Related Websites

Top 25 heavy-atom derivatives	http://www.x12c.nsls.bnl.gov/x12c/mr_table.html
Heavy-atom database	http://www.bmm.icnet.uk/had/heavyatom.html
"Heavy-atom derivatives" by Bart Hazes	http://eagle.mmid.med.ualberta.ca/tutorials/HA/

Further Reading
Blundell TL, Johnson LN (1976). *Protein Crystallography.* Academic Press, New York.

See also Isomorphous Replacement (MIR, SIR, SIRAS).

HE-COM, HE-CP, HE-SD, HE-SS. *SEE* HASEMAN–ELSTON REGRESSION.

HELICAL WHEEL
Terri Attwood

Typically, a circular graph containing five turns of α-helix (~20 residues) around which successive residues in a protein sequence are plotted 100° apart (3.6 residues per 360° turn), as shown in the accompanying figure; different parameters can be used to depict helices with different pitch (e.g., 120° for a 3_{10} helix, 160° for β-strands). The principal use of such graphs is in the depiction of amphipathic helices

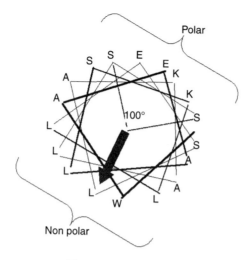

Figure Helical wheel

(i.e., those with both hydrophobic and hydrophilic character). Helical potential is recognized by the clustering of hydrophilic and hydrophobic residues in distinct polar and nonpolar arcs.

While helical wheels are good tools for illustrating amphiphilic character, they are of limited general use because not all helices are amphipathic. Moreover, that a sequence plotted on such a graph appears to have a "sidedness" does not mean that it is helical—in other words, these tools are illustrative, not predictive.

Related Website

| Helical wheel plotting tool | http://marqusee9.berkeley.edu/kael/helical.htm |

Further Reading

Attwood TK, Parry-Smith DJ (1999). *Introduction to Bioinformatics.* Addison-Wesley Longman, Harlow, Essex, United Kingdom.

Schiffer M, Edmundson AB (1967). Use of helical wheels to represent the structures of proteins and to identify segments with helical potential. *Biophys. J.* 7: 121–135.

Shultz GE, Schirmer RH (1979). *Principles of Protein Structure.* Springer-Verlag, New York.

See also Alpha Helix, Amphipathic, Beta Strand.

HELIX–COIL TRANSITION
Patrick Aloy

It is known that protein folding and stability largely depend on the stability of secondary-structure elements. Experiments have suggested that a substantial fraction of secondary-structure elements are formed at a very early stage in folding, much faster than tertiary structure. Thus, fluctuating embryos of secondary structure may be formed in unfolded (or at least not fully collapsed) proteins.

Considering these observations, to understand the protein-folding process, we have to study the factors that determine the secondary-structure elements' formation. To date, attempts to describe the energetics of systems formed by short polypeptide chains have concentrated on α-helix formation.

The *helix–coil transition* model, the most commonly used, is based on statistical mechanics. In its simplest version, only two parameters are considered, a nucleation factor and an elongation factor responsible for the helical tendency of a particular sequence. Recently, information about side-chain interactions has been implemented in the model, closing the gap between predictions and experimental results.

Further Reading
Muñoz V, Serrano L (1994). Elucidating the folding problem of helical peptides using empirical parameters. *Nature Struct. Biol.* 1: 399–409.

HERITABILITY (DEGREE OF GENETIC DETERMINATION, H2)
Mark McCarthy and Steven Wiltshire

The heritability of a quantitative trait is that proportion of the total variation in a quantitative trait attributable to genetic factors.

If the total variance of the trait is given by σ_p^2 and the total genetic variance (including additive, dominance, and epistatic components) is denoted σ_g^2, then broad heritability (or degree of genetic determination—the proportion of the trait variance attributable to all genetic components) is given by σ_g^2/σ_p^2. Narrow heritability (often just called heritability) is the proportion of the trait attributable to the additive genetic variance σ_a^2 and is given by σ_a^2/σ_p^2. Heritability estimates vary between 0 (no genetic component to the trait) and 1 (no environmental component to the trait). Heritability can be estimated with standard linear regression (of offspring on parent or midparent), correlation (between half sibs or full sibs), DeFries–Fulker regression (from twin data), variance components analysis (of pedigree data), and structural equation modeling (of twin or other relative data).

Example: Hirschhorn et al. (2001) calculated heritability estimates for adult stature from samples of Finnish, Swedish, and Quebec pedigrees using variance components analysis to be >95%, 80%, and 70%, respectively.

Related Websites
Programs that can be used for calculating heritabilities:

SOLAR	http://www.sfbr.org/sfbr/public/software/solar/index.html
SAGE	http://darwin.cwru.edu/pub/sage.html
MX	http://griffin.vcu.edu/mx/

Further Reading
DeFries JC, Fulker DW (1985). Multiple regression analysis of twin data. *Behav. Genet.* 15: 467–473.

DeFries JC, Fulker DW (1988). Multiple regression analysis of twin data: Etiology of deviant scores versus individual differences. *Acta Genet. Med. Gemellol.* 37: 205–216.

Falconer DS, MacKay TFC (1996). *Introduction to Quantitative Genetics*. Prentice-Hall, Harlow, UK, pp. 160–183.

Hirschhorn JN, et al. (2001). Genomewide linkage analysis of stature in multiple populations reveals several regions with evidence of linkage to adult height. *Am. J. Hum. Genet.* 69: 106–116.

Plomin R, et al. (2001). *Behavioral Genetics*. Worth, New York.

See also Polygenic Inheritance, Quantitative Trait, Variance Components.

HETEROGENEOUS NUCLEAR RNA. *SEE* HNRNA.

HGI. *SEE* HUMAN GENE INDEX.

HGVBASE (HUMAN GENOME VARIATION DATABASE)
Guenter Stoesser

The Human Genome Variation Database (HGVBASE) includes genomic variation data and summarizes known variations in the human genome as a nonredundant set of records.

The primary purpose of HGVBASE is to facilitate genotype–phenotype association analyses that explore how single-nucleotide polymorphisms (SNPs) and other common sequence variations may influence phenotypes such as common disease risk and drug response differences. Variations represented in HGVBASE encompass sequence changes known or suspected to exist in the human genome, including but not limited to SNPs, indels, and simple tandem repeats and regardless of whether they are known or not known to be functionally neutral or pathogenic. Online search tools facilitate data interrogation by sequence similarity, keyword queries, and genome coordinates. All variants have been uniquely mapped to the human draft genome sequence and are referenced to positions in DDBJ/EMBL/GenBank database entries. Additionally, HGVBASE serves as a central depository for mutation collection efforts undertaken with the Human Genome Variation Society (HGVS). HGVBASE is collaborating to develop data exchange protocols with other public variation and mutation databases, such as dbSNP.

Related Websites

HGVBASE	http://hgvbase.cgb.ki.se
Human Genome Variation Society	http://www.hgvs.org
Sequence Variation Database project	http://www.ebi.ac.uk/mutations/

Further Reading
Fredman D, et al. (2002). HGVbase: A human sequence variation database emphasizing data quality and a broad spectrum of data sources. *Nucleic Acids Res.* 30: 387–391.

HIDDEN MARKOV MODEL (HMM)
Irmtraud M. Meyer

A hidden Markov model (HMM) is a probabilistic method for linearly analyzing sequences.

A hidden Markov model M consists of a fixed number of states $Q = \{q_1, \ldots, q_n\}$ which are connected by directed transitions (see accompanying figure). Each state i reads a fixed number of letters Δ_i from an input string S of length L over an alphabet A. States that read no letters are called silent states. Each run of M can be

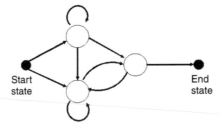

Figure Example of a simple HMM states.

described by a state sequence $\Pi = (\pi_1, \ldots, \pi_m)$ (the state path) which starts in a special silent state (the start state) and proceeds from state to state via transitions between the states until all letters of the input sequence S have been read. The state path ends in another special silent state (the end state). Each state path can be translated into a labeling of the input sequence by assigning to each letter the label of the state that has read the letter. With each transition from a state s to a state t there is associated a transition probability $p_s(t)$ and with every reading of a substring S' by a state s there is associated an emission probability $p_s(S')$. The probabilities of all transitions emerging from each state add up to 1 and the emission probabilities for reading all possible substrings add up to 1 for every state. The overall probability of a sequence given a state path is the product of the individual transition and emission probabilities encountered along the state path.

For a given hidden Markov model M and an input sequence S, a variety of dynamic programming algorithms can be used to infer different probabilities: The Viterbi algorithm can be used to determine the most probable state path Π^* (the Viterbi path). This algorithm has a memory and time requirement of the order of the sequence length. Using the forward-and-backward algorithm, we can calculate the conditional probability that sequence position i is labeled by state k (the posterior probability of state k at position i in the sequence). From these probabilities, the most probable label of each letter in the input sequence can be derived.

When setting up an HMM for a given classification problem, one first defines the states and the transitions between them. The states reflect the different classes which one aims to distinguish, and the transitions between them determine all possible linear successions in which the labels may occur in the input sequences. Once the structure of the HMM has been fixed, the transition and emission probabilities have to be specified. This can be done in a variety of ways depending on which information is available on a sufficiently large and representative set of sequences (the training set). If state paths are know which reproduce the correct classification of the training sequences, we can use maximum-likelihood estimators to derive the emission and transition probabilities of the HMM. If these state paths are not known, the emission and transition probabilities can be estimated in an iterative way using the Baum–Welch algorithm. This algorithm iteratively optimizes the parameters such that the training sequences have the maximum likelihood under the model. Unfortunately, the Baum–Welch algorithm can only be shown to converge to a local maximum which may therefore depend on the initial values of the parameters. The Baum–Welch algorithm is a special case of the expectation maximization (EM) algorithm. Instead of optimizing the likelihood of the sequences given the model, as is done by the Baum–Welch algorithm, we can alternatively choose to maximize the likelihood of the sequences under the model considering only the Viterbi paths (Viterbi training).

HMMs can be expressed as regular grammars. These grammars form the class of transformational grammars within the Chomsky hierarchy of transformational grammars which have the most restrictions. A regular grammar is a set of production rules which analyze a sequence from left to right. The next and more general class of grammars are context-free grammars. They comprise regular grammars as special cases but are also capable of modeling long-range correlations within a sequence (this feature is, e.g., needed to model the long-range correlations imposed by the secondary structure of an RNA sequence). The computational parsing automatons corresponding to regular grammars are finite-state automatons.

Variants of HMMs

The above concept of HMMs can be easily extended to analyze several (k) input sequences instead of just one (for $k = 2$ sequences they are called pair hidden Markov models). Each state of a k HMM then reads substrings of fixed but not necessarily the same length from up to k sequences. The time and memory requirements of the Viterbi algorithm are of the order of the product of the sequence lengths. The memory requirement can be reduced by a factor of one sequence length by using the Hirschberg algorithm.

Another variant of HMMs are hidden semi-Markov models (HSMMs), which explicitly model the duration of time spent in a state by a probability distribution. The time requirement of the Viterbi algorithm for an HSMM is of the order of the cube of the sequence length in the most general case. The memory requirement is still of the order of the sequence length.

Profile HMMs are another type of HMM. They model a known alignment of several related sequences and are used to test whether or not another sequence belongs to the set of sequences.

Use of HMMs in Bioinformatics

HMMs are used in bioinformatics for almost any task that can be described as a process which analyzes sequences from left to right. Applications of HMMs to biological data include sequence alignment, gene finding, protein secondary-structure prediction, and the detection of sequence signals such as, e.g., translation initiation sites. HMMs are used for aligning sequences in order to detect and test potential evolutionary, functional, or other relationships between several sequences (e.g., DNA-to-DNA alignment, protein-to-DNA alignment, protein-to-protein alignment, or the alignment of a family of already aligned related proteins to a novel protein sequence). In ab initio gene prediction, one input DNA sequence is searched for genes, i.e., classified into protein coding intergenic and intergenic subsequences which form valid splicing patterns. In homology-based gene prediction, a known protein sequence is used to predict the location and splicing pattern of the corresponding gene in a DNA sequence. Recently, comparative gene-finding methods which search two related DNA sequences simultaneously for pairs of evolutionarily related genes have been developed. HMMs are particularly well suited for complex tasks such as gene finding as they can incorporate probabilistic information derived by dedicated external programs (programs which, e.g., predict potential splice sites or translation start signals) into the HMM in order to improve its predictive power.

Further Reading

The two books listed introduce HMMs in the context of applications to biological data and contain many references which provide good entry points into the original literature. Durbin et al. (1998) is equally well suited for readers with a biological and mathematical background. Waterman (1995) assumes that readers have a mathematical background.

Durbin R, et al. (1998). *Biological Sequence Analysis*. Cambridge University Press, Cambridge, MA.

Rabiner L (1989). A tutorial on hidden Markov models and selected applications in speech recognition. *Proc. IEEE* 77(2): 257–285.

Waterman MS (1995). *Introduction to Computational Biology*. Chapman & Hall.

HIERARCHY
Robert Stevens

A grouping of concepts in treelike structures.

A tree structure is a hierarchy in which there is only one parent for each item (also known as a strict hierarchy, simple hierarchy, or pure hierarchy. A hierarchy has an explicit notion of top and bottom and so is often referred to as "directed." A tree may be seen in the accompanying figure.

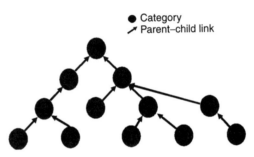

Figure   Organization of concepts in a tree.

HIGH-SCORING SEGMENT PAIR (HSP)
Jaap Heringa

Gapless local pairwise alignments found by the BLAST method to score at least as high as a given threshold score.

This term was introduced by the authors of the homology searching method BLAST. An HSP is the fundamental unit of BLAST algorithm output and consists of two sequence fragments of arbitrary but equal length whose alignment is locally maximal and for which the alignment score is beyond a given cutoff score. The alignment score is dependent on a scoring system which is based on a residue exchange matrix and expected frequencies of residue type. In the context of the BLAST algorithm, each HSP consists of a segment from the query sequence and one from a database sequence. The sensitivity and speed of the programs can be adjusted using the standard BLAST parameters W, T, and X (Altschul et al., 1990). Selectivity of the programs can be adjusted via the cutoff score.

The approach to similarity searching taken by the BLAST program is first to look for similar segments (HSPs) between the query sequence and a database sequence, then to evaluate the statistical significance of any matches found, and finally to report only those matches that satisfy a user-selectable threshold of significance. Multiple HSPs involving the query sequence and a single database sequence can be treated statistically in a variety of ways. The BLAST program uses "sum" statistics (Karlin and Altschul, 1993), where it is possible that the statistical significance attributed to a set of HSPs is higher than that of any individual member of the set. This corresponds to the fact that occurrence of similarity over a wider region in the alignment of a query and database sequence raises the chance that the latter sequence is a true homolog to the former. A match will only be reported to the user if the significance score is below the user-selectable threshold (E value).

Further Reading

Altschul SF, et al. (1990). Basic local alignment search tool. *J. Mol. Biol.* 215: 403–410.

Karlin S, Altschul SF (1993). Applications and statistics for multiple high-scoring segments in molecular sequences. *Proc. Natl. Acad. Sci. USA* 90: 5873–5877.

HISTIDINE
Jeremy Baum

Histidine is a polar (sometimes positively charged) amino acid with side chain—CH_2CC found in proteins. In sequences, written as His or H.

See also Amino Acid.

HIV RT AND PROTEASE SEQUENCE DATABASE. *SEE* STANFORD HIV RT AND PROTEASE SEQUENCE DATABASE.

HIV SEQUENCE DATABASE
Guenter Stoesser

The human immunodeficiency virus (HIV) databases contain data on HIV genetic sequences, immunological epitopes, drug resistance–associated mutations, and vaccine trials.

HIV's website gives access to a large number of tools that can be used to analyze these data. The HIV Sequence Database's primary activities are the collection of HIV and simian immunodeficiency virus (SIV) sequence data (since 1987), their curation, annotation and computer analysis, and the production of sequence analysis software. Data and analyses are published in electronic form and also in a yearly printed publication, the HIV Sequence Compendium. A companion database, the HIV Molecular Immunology Database provides a comprehensive annotated listing of defined HIV epitopes. The HIV Sequence Database is maintained at the Los Alamos National Laboratory, Los Alamos, NM.

Related Website

HIV Database	http://hiv-web.lanl.gov

Further Reading

Kuiken CL, et al. (1999). *Human Retroviruses and AIDS 1999: A Compilation and Analysis of Nucleic Acid and Amino Acid Sequences.* Theoretical Biology and Biophysics Group, Los Alamos National Laboratory, Los Alamos, NM.

Kuiken C, et al. (2000). *HIV Sequence Compendium 2000.* Theoretical Biology and Biophysics Group, Los Alamos National Laboratory, Los Alamos, NM.

HMM. *SEE* HIDDEN MARKOV MODEL.

HMMER

John M. Hancock and Martin J. Bishop

Package using profile hidden Markov models (HMM) to carry out database searching.

HMMer accepts a multiple alignment as input and uses this to construct a HMM profile of the sequence represented by the alignment. This HMM can then be used to search sequence databases. Version 2.2 of the package (August 2001) consists of nine main programs: hmmalign, which aligns sequences to a preexisting model; hmmbuild, which constructs an HMM from a sequence alignment; hmmcalibrate, which calculates E scores for the database search; hmmconvert, which converts a HMM into a variety of formats; hmmemit, which emits sequences from a given HMM; hmmfetch, which gets an HMM from a database; hmmindex, which indexes an HMM database; hmmpfam, which searches an HMM database for matches to a sequence; and hmmsearch, which searches a sequence database for matches to an HMM.

Related Website

| HMMer | http://hmmer.wustl.edu/ |

Further Reading
Eddy SR (1998). Profile hidden Markov models. *Bioinformatics* 14: 755–763.

See also Hidden Markov Model.

HMMSTR

Patrick Aloy

HMMSTR is a method for predicting protein structure ab initio. It relies on a database of sequence-structure motifs (I sites) and hidden Markov models. The main difference with the linear hidden Markov models used in sequence analysis is that HMM-STR has a highly branched topology able to extract structural features. The creation of this database has been possible only in the last few years and is mainly due to the dramatic increase in the number of proteins with known three-dimensional structure.

HMMSTR has represented a breakthrough in the field because, for the first time, ab initio predictions can provide us with biologically useful results.

Although HMMSTR was designed for predictions of higher complexity (i.e., tertiary structure), it is very accurate at identifying secondary-structure elements, reaching a three-state-per-residue accuracy of 74.3%.

Related Website

| HMMSTR | http://honduras.bio.rpi.edu/isites/hmmstr/ |

Further Reading
C. Bystroff, Y Shao (2002). *Bioinformatics* 18: S54–S61.

See also Ab Initio, Hidden Markov Model.

hnRNA (HETEROGENEOUS NUCLEAR RNA)
John M. Hancock

In eukaryotes, immature transcripts before processing to form the mRNA is complete.

Eukaryotic transcripts may undergo splicing to remove introns, polyadenylation to add a poly(A) tail, and capping. All intermediates in this process before the production of the mature mRNA are classed as hnRNA.

Further Reading
Lewin B (2000). *Genes VII.* Oxford University Press, Oxford, UK.

See also mRNA, Transcription.

HOLOPHYLETIC. *SEE* MONOPHYLETIC.

HOMOLOGOUS GENES
Austin L. Hughes and Laszlo Patthy

Two genes (or genomic regions) are said to be homologous if they are descended from a common ancestral gene (or genomic region).

In the case of protein coding genes, the proteins encoded by homologous genes can be said to be homologous. The best evidence of homology is sequence similarity. Strictly speaking, homology is an either–or condition. Thus, it is incorrect to speak of "percent homology" between two sequences except in cases of exon shuffling where genes share a certain percentage of homologous sequences.

Homologous genes or proteins of different species that evolved from a common ancestral gene by speciation are said to be orthologs, homologous genes acquired through horizontal transfer of genetic material between different species are called xenologs, and homologous genes that arose by duplication within a genome are called paralogs.

See also Homology.

HOMOLOGOUS SUPERFAMILY
Andrew Harrison and Christine A. Orengo

A homologous superfamily consists of a set of proteins related by divergent evolution to a common ancestral protein. However, in contrast to a protein family, the superfamily contains more distant relatives, and these may have no detectable sequence similarity. In some remote homologs, the function may also have changed, particularly in paralogous proteins, i.e., proteins which have arisen by a gene duplication event during evolution. In order to detect these very remote homologs, sensitive sequence profiles must be used or the structures of proteins must be compared, where they are known. Since structure is much more highly conserved than sequence during evolution, distant homologs can usually be detected from their structural similarity. However, to distinguish homologs from analogs, it is necessary to search for additional evidence to support homology. This may be some unusual sequence pattern associated with ligand binding or some rare structural motif unlikely to have arisen twice by chance within the same analogous structure.

Related Website

| HOMSTRAD | http://www- cryst.bioc.cam.ac.uk/data/align/ |

Further Reading
Thornton JM, Todd AE, Milburn D, Borkakoti N, Orengo CA (2000). From structure to function: Approaches and limitations. *Nature Struct. Biol.* 7(11): 991–994.

See also Paralog.

HOMOLOGY
Jaap Heringa and Laszlo Patthy

Biological entities (systems, organs, genes, proteins) are said to be homologous if they share a common evolutionary ancestor.

The term *homology* in the evolution of molecular sequences implies the presence of a common ancestor between the sequences and hence assumes divergent evolution. Whereas sequence similarity is a quantification of an empirical relationship of sequences expressed using a gradual scale, homology is a binary state: A pair of sequences is homologous or not (except in cases of exon shuffling, where genes share a certain proportion of homologous sequences). Homologous proteins constitute a protein family. Since protein tertiary structures are more conserved during evolution than their coding sequences, homologous sequences are assumed to share the same protein fold and the same or similar functions. Although it is possible in theory that two proteins evolve different structures and functions from a common ancestor, this situation cannot be traced so that such proteins are seen as unrelated. However, numerous cases exist of homologous protein families where subfamilies with the same fold have evolved distinct molecular functions.

Multidomain proteins that arose by intragenic duplication of segments encoding a protein domain display internal homology. Mosaic proteins that arose through transfer of genetic material encoding a protein module from one gene to another gene may display a complex network of homologies, with different regions having distinct evolutionary origins. Homologous protein domains derived through gene duplications, intragenic duplications, or intergenic transfer of an ancestral domain constitute a domain family.

The term homology is often used in practice when two sequences have the same structure or function, although in the case of two sequences sharing a common function this ignores the possibility that the sequences are analogs resulting from convergent evolution. Unfortunately, it is not straightforward to infer homology from similarity because enormous differences exist between sequence similarities within homologous families. Many protein families of common descent comprise members sharing pairwise sequence similarities that are only slightly higher than those observed between unrelated proteins. This region of uncertainty has been characterized to lie in the range of 15% to 25% sequence identity (Doolittle, 1981) and is commonly referred to as the "twilight zone" (Doolittle, 1987). There are even some known examples of homologous proteins with sequence similarities below the randomly expected level given their amino acid composition (Pascarella and Argos,

1992). As a consequence, it is impossible to prove using sequence similarity that two sequences are not homologous.

Further Reading

Doolittle RF (1981). Similar amino acid sequences: Chance or common ancestry. *Science* 214: 149–159.

Doolittle RF (1987). *Of URFS and ORFS. A Primer on How to Analyze Derived Amino Acid Sequences*. University Science Books, Mill Valley, CA.

Fitch WM (2000). Homology: A personal view on some of the problems. *Trends Genet.* 16(5): 227–231.

Pascarella S, Argos P (1992). A data bank merging related protein structures and sequences. *Protein Eng.* 5: 121–137.

Reeck GR, de Haen C, Teller DC, Doolittle RF, Fitch, WM, Dickerson RE, Chambon P, McLachlan AD, Margoliash E, Jukes TH, Zuckerkandl E (1987). "Homology" in proteins and nucleic acids: A terminology muddle and a way out of it. *Cell* 50(5): 667.

See also Domain Family, Mosaic Protein, Multidomain Protein, Ortholog, Paralog, Protein Domain, Protein Family, Protein Module, Sequence Alignment, Xenolog.

HOMOLOGY-BASED GENE PREDICTION. *SEE* GENE PREDICTION, HOMOLOGY BASED.

HOMOLOGY MODELING. *SEE* COMPARATIVE MODELING.

HOMOLOGY SEARCH
Jaap Heringa

A homology search refers to scouring a sequence database to find sequences that are likely to be homologous to a give query sequence. Comparative sequence analysis is a common first step in the analysis of sequence–structure–function relationships in protein and nucleotide sequences. In the quest for knowledge about the role of a certain unknown protein in the cellular molecular network, comparing the query sequence with the many sequences in annotated protein sequence databases often leads to useful suggestions regarding the protein's three-dimensional (3D) structure or molecular function. This extrapolation of the properties of sequences in public databases that are identified as "neighbors" by sequence analysis techniques has led to the putative characterization (annotation) of very many sequences. Although progress has been made, the direct prediction of a protein's structure and function is still a major unsolved problem in molecular biology. Since the advent of the genome sequencing projects, the method of indirect inference by comparative sequence techniques and homology searching has only gained in significance.

A typical application to infer knowledge for a given query sequence is to compare it with all sequences in an annotated sequence database. A database search can be performed for a nucleotide or amino acid sequence against an annotated database of nucleotides (e.g., EMBL, GenBank, DDBJ) or protein sequences (e.g., SWISS-PROT, PIR (Protein Information Resource), TrEMBL, GenPept, NR-NCBI, NR-ExPasy). Although the actual pairwise comparison can take place at the nucleotide or peptide level, the most effective way to compare sequences is at the peptide level (Pearson,

1996). This requires that nucleotide sequences must first be translated in all six reading frames followed by comparison with each of these conceptual protein sequences. Although mutation, insertion, and deletion events take place at the DNA level, reasons why comparing protein sequences can reveal more distant relationships include the following: (i) Many mutations within DNA are synonymous, which means that these do not lead to a change of the corresponding amino acids. As a result of the fact that most evolutionary selection pressure is exerted on protein sequences, synonymous mutations can lead to an overestimation of the sequence divergence if compared at the DNA level. (ii) The evolutionary relationships can be more finely expressed using a 20×20 amino acid exchange table than using exchange values among four nucleotides. (iii) DNA sequences contain noncoding regions, which should be avoided in homology searches. Note that the latter is still an issue when using DNA translated into protein sequences through a codon table. However, a complication arises when using translated DNA sequences to search at the protein level because frame shifts can occur, leading to stretches of incorrect amino acids and possibly elongation of sequences due to missed stop codons.

The widely used dynamic programming (DP) technique for sequence alignment is too slow for repeated homology searches over large databases and may take multiple CPU hours for a single query sequence on a standard workstation. Although some special hardware has been designed to accelerate the DP algorithm, this problem has triggered the development of several heuristic algorithms that represent shortcuts to speed up the basic alignment procedure, including the currently most widely used methods to scour sequence databases for homologies: PSI-BLAST (Altschul et al., 1997), an extension of the BLAST technology (Altschul et al., 1990), and FASTA (Pearson and Lipman, 1988). A particular significant feature of these rapid-search routines is that they incorporate statistical estimates of the significance of each pairwise alignment score between a query and database sequence relative to random-sequence scores (p and E values). Owing to recent advances in computational performance, procedures for sequence database homology searching have been developed based on the more computationally intensive formalism of hidden Markov modeling (HMM), including SAM-T98 (Karplus et al., 1998) and HMMer2 (Eddy, 1998).

Further Reading
Altschul SF, et al. (1990). Basic local alignment search tool. *J. Mol. Biol.* 215: 403–410.
Altschul SF, et al. (1997). Gapped BLAST and PSI-BLAST: A new generation of protein database search programs. *Nucleic Acids Res.* 25: 3389–3402.
Eddy SR (1998). Profile hidden Markov models. *Bioinformatics* 14: 755–763.
Karplus K, et al. (1998). Hidden Markov models for detecting remote protein homologies. *Bioinformatics* 14: 846–856.
Pearson WR, Lipman DJ (1988). Improved tools for biological sequence comparison. *Proc. Natl. Acad. Sci. USA* 85: 2444–2448.
Pearson WR (1996). Effective protein sequence comparison. *Methods Enzymol.* 266: 227–258.

HOMOZYGOSITY MAPPING
Mark McCarthy and Steven Wiltshire

A variant of linkage analysis which provides a powerful tool for mapping rare recessive traits within inbred families or populations

A high proportion of the presentations of rare autosomal recessive traits will occur in consanguineous families (e.g., the offspring of first-cousin marriages), since such pedigree relationships enhance the chance that two copies of the same rare allele will segregate into the same individual. Since disease in this situation results from the relevant part of the same ancestral chromosome being inherited from both parents, the susceptibility gene will lie within a localized region of homozygosity (or, as it is sometimes termed, autozygosity). This region can be detectable by undertaking a genome scan for linkage using polymorphic markers. As with any linkage analysis, the strength of the evidence for the localization of the susceptibility gene is determined using appropriate linkage programs. The HOMOZ program is an example of such a linkage program designed for use in the specialized situation. The particular advantage of homozygosity mapping is that substantial evidence for linkage can be obtained from relatively few affected individuals.

Examples: Novelli et al (2002) describe their success in mapping mandibuloacral dysplasia, a rare recessive disorder to chromosome 1q21, by homozygosity mapping in five consanguineous Italian families and the subsequent identification of etiological mutations in *LMNA*.

Related Website

HOMOZ (linkage program optimized for homozygosity mapping)	http://www-genome.wi.mit.edu/ftp/distribution/software/

Further Reading

Abney M, et al. (2002). Quantitative-trait homozygosity and association mapping and empirical genomewide significance in large, complex pedigrees: Fasting serum-insulin level in the Hutterites. *Am. J. Hum. Genet.* 70: 920–934.

Faivre L, et al. (2002). Homozygosity mapping of a Weill-Marchesani syndrome locus to chromosome 19p13.3-p13.2. *Hum. Genet.* 110: 366–370.

Kruglyak L, et al. (1995). Rapid multipoint linkage analysis of recessive traits in nuclear families, using homozygosity mapping. *Am. J. Hum. Genet.* 56: 519–527.

Lander ES, Botstein D (1986). Mapping complex genetic traits in humans: New methods using a complete RFLP linkage map. *Cold Spring Harbor Symp. Quant. Biol.* 51: 49–62.

Novelli G, et al. (2002). Mandibuloacral dysplasia is caused by a mutation in LMNA-encoding Lamin A/C. *Am. J. Hum. Genet.* 71: 426–431.

See also Genome Scan, Linkage Analysis.

HORIZONTAL GENE TRANSFER
Austin L. Hughes and Laszlo Patthy

Horizontal (or lateral) gene transfer occurs when a gene is transferred from the genome of one species to that of another species.

In bacteria, a well-understood mechanism of horizontal gene transfer is transformation, by which certain bacterial cells are able to acquire DNA from the environment and incorporate it into their genomes. Horizontal transfer of genes may have played an important role in bacterial evolution. There is also evidence that DNA viruses parasitic on vertebrates have acquired from their hosts genes encoding proteins that function to modulate the host's immune response. These horizontally transferred genes are believed to benefit the virus by disrupting the host's immune response.

A gene acquired through horizontal gene transfer that is homologous with a gene of the recipient species is called a xenolog.

Further Reading
Eisen JA (2000). Horizontal gene transfer among microbial genomes: New insights from complete genome analysis. *Curr. Opin. Genet. Dev.* 10: 606–611.
Lalani AS, McFadden G (1999). Evasion and exploitation of chemokines by viruses. *Cytokine Growth Factor Rev.* 10: 219–233.
Ochman H, et al. (2000). Lateral gene transfer and the nature of bacterial innovation. *Nature* 405: 299–304.

See also Homology, Xenolog.

HSP. *SEE* HIGH-SCORING SEGMENT PAIR.

HUMAN GENE INDEX (HGI)
John M. Hancock and Rodger Staden

A database of gene data available at The Institute for Genomic Research.

This database aims to integrate data on expression patterns, cellular roles, functions, and evolutionary relationships, taking data from public sources and combining them with data generated at TIGR (The Institute for Genomic Research).

Related Website

| HGI | http://www.tigr.org/tdb/tgi/hgi/ |

HUMAN GENOME VARIATION DATABASE. *SEE* HGVBASE.

HYDROGEN BOND
Jeremy Baum

An interaction between two nonhydrogen atoms, one of which (the proton donor) has a covalently bonded hydrogen atom. The other nonhydrogen atom (the proton acceptor) has electrons not involved directly in covalent bonds that can interact favorably with the hydrogen atom. The energy of this interaction is typically up to 10 kJ/mol for uncharged groups but can be significantly more if charges are present.

This interaction is extremely common in biological systems, as it is a fundamental interaction in water, and within proteins. The oxygen atoms of C=O groups and the nitrogen atoms of N—H groups in protein backbones act as proton acceptors and proton donors, respectively. The secondary structures of proteins occur as a result of hydrogen bonding between such backbone groups.

See also Backbone, Secondary Structure.

HYDROPATHY (HYDROPHOBICITY)
Terri Attwood

Broadly, having the property of hydrophobicity, a low affinity for water. Depending on the context in which the property is being considered, hydropathy may be defined in different ways. For example, it may be estimated from chemical behavior or amino acid physicochemical properties (e.g., solubility in water, chromatographic migration) or from environmental characteristics of residues in three-dimensional structures, where it carries more complex information than that conferred by a single physical or chemical property (e.g., solvent accessibility, propensity to occupy the protein interior).

There are many different amino acid hydropathic rankings (some examples are given below). At the simplest level, "internal, external, ambivalent" (where internal = [FILMV], external = [DEHKNQR], and ambivalent = [ACGPSTWY]) provides a useful hydrophobicity alphabet. More sensitive scales assign quantitative values to individual amino acids.

Scale	Residue ranking
Von Heinje	FILVWAMGTSYQCNPHKEDR
Eisenberg	IFVLWMAGCYPTSHENQDKR
Kyte	IVLFCMAGTSWYPHDNEQKR
Rose	CFIVMLWHYAGTSRPNDEQK
Sweet	FYILMVWCTAPSRHGKQNED

Although the distribution of residues within the available scales is broadly similar, there is no general consensus regarding which residues appear at the most hydrophobic and most hydrophilic extremes. Nevertheless, such scales can be used to plot hydropathy profiles to provide an overview of the hydropathic character of a given protein sequence. There is no "right" scale but, used together, they can help to yield a consensus of the most significant hydrophobic features within a sequence.

Related Website

Hydropathy scales	http://molvis.chem.indiana.edu/C687_S99/hydrophob_scale.html

Further Reading

Attwood TK, Parry-Smith DJ (1999). *Introduction to Bioinformatics.* Addison-Wesley Longman, Harlow, Essex, United Kingdom.

Kyte J, Doolittle RF (1982). A simple method for displaying the hydropathic character of a protein. *J. Mol. Biol.* 157(1), 105–132.

See also Hydrophilic, Hydrophobic.

HYDROPATHY PROFILE (HYDROPHOBICITY PLOT, HYDROPHOBIC PLOT)
Terri Attwood

A graph that provides an overview of the hydropathic character of a protein sequence, wherein the x axis represents the sequence and the y axis the hydrophobicity score. In creating such a profile, a sliding window (of user-defined width) is scanned across the query sequence and, for each window length, a score is calculated from the hydrophobicity values of each constituent residue according to a particular hydropathic ranking.

Typical profiles show characteristic peaks and troughs, corresponding to the most hydrophobic and most hydrophilic parts of the protein. They are therefore especially useful for identifying putative transmembrane domains, as denoted by runs of 20–25 hydrophobic residues. Because of the variation in hydrophobicity scales, to achieve the most reliable results, it is advisable to use several different rankings and to seek a consensus between the resulting range of different profiles—this provides different perspectives on the same sequence, potentially allowing characteristics that are not apparent with one scale to be visualized with one or more of the others [e.g., a relatively hydrophilic transmembrane (TM) domain]. A combined view is important, as many erroneous TM assignments have been associated, for example, with database sequences based on incautious interpretations of individual profiles.

Related Websites

TM domain prediction tools	http://www.expasy.org/tools/#transmem
Hydropathy profile illustration	http://www.bioinf.man.ac.uk/dbbrowser/bioactivity/hydroprofile.gif

Further Reading

Attwood TK, Parry-Smith DJ (1999). *Introduction to Bioinformatics*. Addison-Wesley Longman, Harlow, Essex, United Kingdom.

Kyte J, Doolittle RF (1982). A simple method for displaying the hydropathic character of a protein. *J. Mol. Biol.* 157(1): 105–132.

See also Hydrophilic, Hydrophobic.

HYDROPHILICITY
Jeremy Baum

The term hydrophilicity is used in contrast to hydrophobicity and is applied to polar chemical groups that have favorable energies of solvation in water and tend to have water/octanol partition coefficients favoring the aqueous phase.

See also Hydrophobicity, Polar.

HYDROPHOBICITY. *SEE* HYDROPATHY.

HYDROPHOBICITY PLOT. *SEE* HYDROPATHY PROFILE.

HYDROPHOBIC MOMENT
Jeremy Baum

In an extension of the concept of the hydrophobic scale, Eisenberg and co-workers defined a vector quantity for each amino acid residue in a protein structure. This was based on the direction defined from the backbone to the end of the side chain, taken together with a hydrophobic scale. These residue vectors could be summed for pieces of secondary structure such as α-helices and in experimentally determined structures were found to have characteristic orientations relative to the overall protein fold. Such hydrophobic moments were proposed to be useful in diagnosing incorrect theoretical folds.

Further Reading
Eisenberg D, Weiss RM, Terwilliger TC (1982). *Nature* 299: 371–374.

See also Backbone, Hydrophobicity, Secondary Structure, Side Chain.

HYDROPHOBIC SCALE
Jeremy Baum

Many workers have deduced numerical scales to define the hydrophobicity of certain chemical groups and of amino acids. Some of these scales have their origins in experimental measurements of, e.g., water/octanol partition coefficients.

See also Hydropathy.

I

I2H
IBD
IBS
Identical by Descent
Identical by State
Identity by Descent
Identity by State
IMGT
Imprinting
Improper Dihedral Angle
Indel
Independent Variable
Individual
Individual Information
Information
Information Theory
Initiator Sequence
Insertion–Deletion Region
Insertion Sequence
In Silico Biology
Instance

Insulator
Integrated Gene Prediction Systems
Intelligent Data Analysis
Interactome
Intergenic Sequence
International Immunogenetics Database
International Society for
  Computational Biology
Interolog
InterPreTS
InterPro
InterProScan
Interspersed Sequence
Intrinsic Gene Prediction
Intron
Intron Phase
ISCB
Isochore
Isoleucine
Isomorphous Replacement
IUPAC-IUB Codes

Dictionary of Bioinformatics and Computational Biology. Edited by Hancock and Zvelebil
ISBN 0-471-43622-4   © 2004 John Wiley & Sons, Inc.

I2H
Patrick Aloy

The in silico two-hybrid system (I2H) predicts protein–protein interactions from sequence information. It relies on the detection of pairs of positions in two different proteins that show a correlated mutational behavior. These positions might correspond to compensatory mutations that stabilize mutations in one protein by changing the other.

Correlated mutations have proved useful for detecting residues involved in interacting surfaces in individual proteins. Recently, this method has been extended to the prediction of protein interactions based on differential accumulation of correlated mutations in interacting partners.

Related Website

| I2H | http://www.pdg.cnb.uam.es/i2h/ |

Further Reading

Göbel U, Sander C, Schneider R, Valencia A (1994). Correlated mutations and residue contacts in proteins. *Proteins* 18: 309–317.

Pazos F, Valencia A (2002). In silico two-hybrid system for the selection of physically interacting protein pairs. *Proteins* 47: 219–227.

IBD. *SEE* IDENTICAL BY DESCENT.

IBS. *SEE* IDENTICAL BY STATE.

IDENTICAL BY DESCENT (IBD, IDENTITY BY DESCENT)
Mark McCarthy and Steven Wiltshire

A state of nature in which alleles from separate individuals are descended from, and therefore are copies of, the same ancestral allele.

The term IBD is most frequently used in describing the relationship between individuals in terms of allele sharing at particular loci. For any given pair of relatives, there are clear biological constraints to the maximum and minimum numbers of alleles that can be shared IBD at any given locus. For instance, monozygotic twins share both alleles IBD at all loci, whereas full sibs and dizygotic twins may share two, one, or zero alleles IBD at any given locus. A parent and its offspring share one allele IBD at all loci. The parameter $\cap\pi_{ij}$—the estimated proportion of alleles shared IBD by relatives i and j—is the basis of allele-sharing methods of linkage analysis: The correlation coefficient between $\cap\pi_{ij}$ at two loci is $2\psi - 1$, where $\Psi = \theta^2 + (1 - \theta)^2$ θ being the recombination fraction between the two loci in question. Estimates of the proportion of alleles shared IBD averaged over many markers may also be used

Dictionary of Bioinformatics and Computational Biology. Edited by Hancock and Zvelebil
ISBN 0-471-43622-4 © 2004 John Wiley & Sons, Inc.

to confirm (or refute) familial relationships between purported siblings. Alleles that are IBD have the same DNA sequence and are therefore identical by state also.

Related Websites
Programs for estimating the proportions of alleles shared IBD by purported siblings:

RELATIVE	ftp://linkage.rockefeller.edu/software/relative/
RELPAIR	http://www.sph.umich.edu/group/statgen/software/

Further Reading
Boehnke M, Cox NJ (1997). Accurate inference of relationships in sib-pair linkage studies. *Am. J. Hum. Genet.* 61: 423–429.

Sham P (1998). *Statistics in Human Genetics.* Arnold, London, pp. 98–106.

See also Allele-Sharing Methods, Homology, Identical by State, Linkage Analysis, Recombination.

IDENTICAL BY STATE (IBS, IDENTITY BY STATE)
Mark McCarthy and Steven Wiltshire

A state of nature in which alleles from separate individuals are identical in DNA sequence.

Alleles that are IBS may or may not be descended from the same ancestral copy: If they are copies of the same ancestral allele, then they are also identical by descent (IBD). Consider a nuclear family of two sibs with paternal genotype AB and maternal genotype BC at a given locus. If the first sib inherits allele A from the father and B from the mother and the second sib inherits allele B from the father and C from the mother, then the pair of sibs share one allele—B—IBS (but zero IBD). If the sibs' genotypes had been AB and AC, they would share one allele IBS and one allele—A—IBD; if both sibs had been AB, they would share both alleles IBS and both IBD also. It is not always possible to determine the IBD status of a relative pair even though the IBS status is unequivocal: If both parents have genotype AB and both offspring have genotype AB, then the two sibs share both alleles IBS but either no alleles IBD (if each inherited alleles A and B from different parents) or both alleles IBD (if each inherited alleles A and B from the same parent). Early allele-sharing methods for linkage analysis of extended pedigrees utilized IBS sharing, although more recent, powerful methods are now based on allele-sharing IBD.

Further Reading
Sham P (1998). *Statistics in Human Genetics.* Arnold, London, pp. 98–106.

See also Allele-Sharing Methods, Identical by Descent, Linkage Analysis.

IDENTITY BY DESCENT. *SEE* IDENTICAL BY DESCENT.

IDENTITY BY STATE. *SEE* IDENTICAL BY STATE.

IMGT (INTERNATIONAL IMMUNOGENETICS DATABASE)
Guenter Stoesser

The International Immunogenetics Database (IMGT) provides specialized information on immunoglobulins (Ig), T cell receptors (TcR), and major histocompatibility complex (MHC) molecules of all vertebrate species.

IMGT includes two databases: LIGM-DB, a comprehensive database of nucleic acid sequences for immunoglobulins and T-cell receptors from human and other vertebrates, with translations for fully annotated sequences, and HLA-DB, a database for sequences of the human MHC, referred to as HLA (human leukocyte antigen). The IMGT server (Montpellier, France) provides common access to all immunogenetics data. IMGT was established in 1989 by the Université Montpellier II and the CNRS (Montpellier, France). IMGT is produced in collaboration with the European Bioinformatics Institute and data are integrated with EMBL nucleotide sequence entries.

Related Websites

| IMGT | http://imgt.cines.fr:8104/ |
| IMGT/HLA Sequence Database | http://www.ebi.ac.uk/imgt/hla/ |

Further Reading

Lefranc MP (2001). IMGT, the international ImMunoGeneTics database. *Nucleic Acids Res.* 29: 207–209.

Robinson J, et al. (2001). IMGT/HLA Database—a sequence database for the human major histocompatibility complex. *Nucleic Acids Res.* 29: 210–213.

IMPRINTING
Mark McCarthy and Steven Wiltshire

Process whereby the expression of a gene is determined by its parental origin.

Imprinted genes represent exceptions to the usual rule that both paternally and maternally inherited copies of a gene are equally expressed in the offspring. The imprints inherited from parents are erased in the germ line and then reset in the gametes by differential methylation of the genes concerned. Since imprinting of a gene or region generally results in reduced function, maternal imprinting generally refers to the situation when only the paternal copy is expressed. Around 50 imprinted loci have been identified in humans and other higher mammals. The evolutionary basis for the development of imprinting remains uncertain, but the observation that many imprinted genes are important regulators of early growth suggests that it may have evolved to mediate the conflict between the differing costs and benefits of excessive fetal growth for father and mother. Breakdown of normal imprinting mechanisms (which may arise via defects in methylation, duplication, translocation, or defects in meiosis that lead to uniparental disomy) such that individuals inherit either zero or two expressed copies of a usually imprinted gene is responsible for a number of diseases.

Examples: Classical examples of imprinting include the chromosome 15q region whereby deletion of the paternally (expressed) copy of the imprinted *SNRPN* gene (and of other contiguous genes) leads to Prader-Willi syndrome. In contrast, deletion of the same region on the maternal chromosome leads (due to loss of maternal

UBE3A brain-specific expression) to Angelman syndrome. Loss of *IGF2* imprinting in somatic cells is associated with development of a renal malignancy of childhood (Wilms' tumor). Finally, expression of two paternal copies of the imprinted *ZAC* gene on chromosome 6q24 (through duplication or uniparental disomy) leads to the condition of transient neonatal diabetes mellitus.

Related Website

Catalogue of imprinted genes	http://cancer.otago.ac.nz/IGC/Web/home.html

Further Reading

Gardner RJ, et al. (2000). An imprinted locus associated with transient neonatal diabetes mellitus. *Hum. Mol. Genet.* 9: 589–596.

Morrison IM, et al. (2001). The imprinted gene and parent-of-origin effect database. *Nucleic Acids Res.* 29: 275–276.

Nicholls RD, Knepper JL (2001). Genome organization, function, and imprinting in Prader-Willi and Angelman syndromes. *Annu. Rev. Genomics Hum. Genet.* 2: 153–175.

IMPROPER DIHEDRAL ANGLE

Roland L. Dunbrack

A dihedral angle defined on four atoms A, B, C, D that do not form a connected chain in the order A, B, C, D. A harmonic potential energy term is used in molecular mechanics energy functions to keep sp^2-hybridized groups planar. For instance, the improper dihedral angle $C\alpha$—C—O—N_{i+1} may be used to keep the backbone carbonyl group planar.

Related Website

Improper dihedral angle potentials	http://www.dl.ac.uk/TCS/Software/DL_POLY/USRMAN/node49.html

See also Dihedral Angle.

INDEL

Terri Attwood

A commonly used term denoting evolutionary insertion and deletion events. DNA sequences evolve not only as a result of point mutations but also as a consequence of sequence expansions and/or contractions. It is frequently the case, therefore, that evolutionarily related sequences have different lengths (see accompanying figure).

Homologs are normally compared by creating sequence alignments. During this process, where indels have occurred, "gap" characters must be inserted in order to bring corresponding regions of sequence into the correct register. The more divergent the sequences, the greater the number of gaps likely to be needed to achieve the optimal result. This presents particular challenges to automatic alignment programs, which assign different values both for an insertion event itself (gap opening) and for its size (gap extension). Indels are thus the principal reason why reliable automatic alignment of divergent sequences is still so difficult.

The consequence of the presence of indels within a sequence alignment is that islands of conservation become visible against a backdrop of mutational change. Thus indels tend to mark the boundaries of the core structural and functional motifs within protein

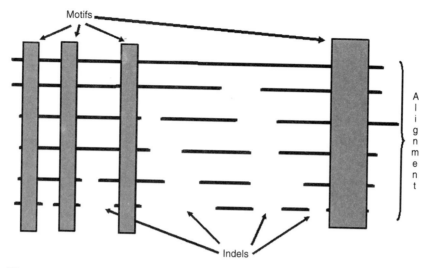

Figure

structures. They may therefore be used to guide the design of appropriate diagnostic signatures for particular aligned families by highlighting the critically conserved regions. In protein sequences, many indel regions correspond to loop sites in the associated tertiary structures, as protein cores do not easily accommodate insertions or deletions.

Further Reading
Doolittle RF, Ed. (1990). Molecular evolution: Aligning protein and nucleic acid sequences section. In *Methods in Enzymology*, Vol. 183. Academic Press, San Diego, CA.
Lawrence CB, Goldman DA (1988). Definition and identification of homology domains. *Comput. Appl. Biosci.* 4(1): 25–33.

INDEPENDENT VARIABLE. *SEE* FEATURE.

INDIVIDUAL (INSTANCE)
Robert Stevens

A particular member of a category because it either possesses the properties of that category or is part of the enumeration of category members.

"England" is an individual of the category Country; "Professor Michael Ashburner" is an individual of the category Person; a single *Escherichia coli* cell is an individual of the category Bacterium. The distinction between category and individual is often difficult to determine. Some systems refer to individuals as "instances." For most purposes the two words are interchangeable.

INDIVIDUAL INFORMATION
Thomas D. Schneider

Individual information is the information that a single binding site contributes to the sequence conservation of a set of binding sites. This can be graphically displayed by a sequence walker. It is computed as the decrease in surprisal between the before state and the after state. The technical name is R_i.

Further Reading
Schneider TD (1997). Information content of individual genetic sequences. *J. Theor. Biol.*, 189: 427–441.http://www.lecb.ncifcrf.gov/toms/paper/ri/.
Schneider TD (1999). Measuring molecular information. *J. Theor. Biol.*, 201: 87–92. http://www.lecb.ncifcrf.gov/toms/paper/ridebate/.

See also Information; Binding Site; Sequence Walker; Before State; After State.

INFORMATION
Thomas D. Schneider

Information is measured as the decrease in uncertainty of a receiver or molecular machine in going from the before state to the after state. It is usually measured in bits per second or bits per molecular machine operation.

Related Website

Information is not entropy, information is not uncertainty!	http://www.lecb.ncifcrf.gov/toms/ information.is.not.uncertainty.html

See also Evolution of Biological Information; Information Theory; Before State, After State; Bit; Molecular Machine; Uncertainty.

INFORMATION THEORY
Thomas D. Schneider

Information theory is a branch of mathematics founded by Claude Shannon in the 1940s. The theory addresses two aspects of communication: "How can we define and measure information?" and "What is the maximum information that can be sent through a communications channel?" (channel capacity). John R. Pierce (1980), an engineer at Bell Labs in the 1940s, wrote an excellent introductory book about information theory.

Related Websites

Information theory references	http://www.lecb.ncifcrf.gov/~toms/ bionet.info-theory.faq.html REFERENCES_Information Theory
Primer on information theory	http://www.lecb.ncifcrf.gov/~toms/paper/primer
Information theory resources	http://www.lecb.ncifcrf.gov/~toms/itresources.html

Further Reading
Cover TM, Thomas JA (1991). *Elements of Information Theory*. Wiley, New York.
Pierce JR (1980). *An Introduction to Information Theory: Symbols, Signals and Noise*, 2nd ed. Dover Publications, New York.

See also Molecular Information Theory, Channel Capacity.

INITIATOR SEQUENCE
Niall Dillon

Short consensus sequence located around the transcription initiation site of RNA polymerase II transcribed genes that lack a TATA box and involved in specifying the site of initiation of transcription.

Two different elements have been identified that are involved in specifying the transcriptional initiation site of polymerase II transcribed genes in eukaryotes. These are the TATA box (located 30–35 bp upstream from the transcription start site) and the initiator. Promoters can contain one or both of these elements and a small number have neither. The initiator and the TATA box bind the multi-protein complex TFIID, which is involved in specifying the initiation site.

Further Reading

Groschedl R, Birnstiel M (1980). Identification of regulatory sequences in the prelude sequences of an H2A histone gene by the study of specific deletion mutants in vivo. *Proc. Natl. Acad. Sci. USA* 77: 1432–1436.

Smale S, Baltimore D (1989). The "initiator" as a transcription control element. *Cell* 57: 103–113.

Smale S, et al. (1998). The initiator element: A paradigm for core promoter heterogeneity within metazoan protein-coding genes. *Cold Spring Harbor Symp. Quant. Biol.* 63: 21–31.

See also TATA Box.

INSERTION–DELETION REGION. *SEE* INDEL, LOOP PREDICTION/MODELING.

INSERTION SEQUENCE
Dov S. Greenbaum

A discrete segment that can transpose between locations within a genome.

There are over 500 known insertion sequences in bacteria alone. Within simple organisms they are known to mutate genes, transmit resistance to antibiotics, and promote gene acquisition.

Insertion sequences have also become integral parts of bacterial chromosomes where they play a role in activities such as chromosome rearrangement and plasmid integration.

Transpositional activity, controlled by weak *Tpase* promoters, of insertion sequences are usually held at a low level as a high level of transposition and the mutations they cause would be detrimental to the host cell.

Transposition is also controlled somewhat by host factors such as chaperones, *LexA*, and replication initiator *DnaA*.

Further Reading

Lawrence JG, Ochman H (1992). The evolution of insertion sequences within enteric bacteria. *Genetics* 131: 9–20.

Mahillon J, Chandler, M (1998). Insertion sequences. *Microbiol. Mol. Biol. Rev.* 62: 725–774.

See also Transposable Element.

IN SILICO BIOLOGY
John M. Hancock

A subdiscipline of bioinformatics/computational biology concerned with the integration of genomic and functional genomic information.

Whereas the terms *bioinformatics* and *computational biology* have become irreducibly intertwined, the newer term *in silico biology* (in silico referring to the silicon basis of the computer chip and setting this form of biology in contrast to the classic modes of biological investigation in vivo and in vitro, although glass is also a form of silicon!) tends to refer to bioinformatic investigations that are integrative in nature and related to high-throughput genomics and functional genomics. It is therefore closer in meaning to *systems biology*.

Further Reading
Palsson B. (2000). The challenges of in silico biology. *Nat. Biotechnol.* 18: 1147–1150.

See also Bioinformatics, Network, System Biology.

INSTANCE. *SEE* INDIVIDUAL.

INSULATOR
John M. Hancock

A regulatory element responsible for insulating a gene from the regulatory effects of elements affecting other genes.

Further Reading
Gdula DA, et al. (1996). Genetic and molecular analysis of the gypsy chromatin insulator of *Drosophila*. *Proc. Natl. Acad. Sci. USA* 93: 9378–9383.
Kuhn EJ, et al. (2003). A test of insulator interactions in *Drosophila*. *EMBO J.* 22: 2463–2271.
Prioleau MN, et al. (1999). An insulator element and condensed chromatin region separate the chicken beta-globin locus from an independently regulated erythroid-specific folate receptor gene. *EMBO J.* 18: 4035–4048.

INTEGRATED GENE PREDICTION SYSTEMS. *SEE* GENE PREDICTION SYSTEMS, INTEGRATED.

INTELLIGENT DATA ANALYSIS. *SEE* PATTERN ANALYSIS.

INTERACTOME
Dov S. Greenbaum

The list of interactions between all macromolecules of the cell.

This includes, but is not limited to, protein–protein, protein–DNA, and protein–complex interactions. As many relatively organisms seem to have genes and protein counts on par with more complex organisms, it is thought that much of the complexity results, not from the number of proteins, but rather from the degree and complexity of their interactions.

Given that much of a protein's function can sometimes be deduced from its interactions with other macromolecules in the cell, the elucidation of the interactome is important for the further annotation of the genome. Additionally, the sum total of interactions in an organism can be used as a tool to compare distinct organisms, defining their specific networks of protein interactions and distinguishing genomes from each other on the basis of differing interactions.

There are presently many high-throughput techniques for discovering protein–protein interactions, e.g., yeast two-hybrid or Tandem-Affinity Purification (TAP) tagging. Due to limitations in the experiments, many high-throughput experiments are prone to high levels of both false positives and false negatives. It is imperative when studying interactomes that protein interactions be verified through multiple methods to minimize the risk of incorporating false and misleading information into the databases.

Presently there are many databases which serve to collect, catalog, and curate the known protein–protein interactions. These include DIP, MIPS, and BIND.

Related Websites

DIP (Database of Interacting Proteins)	http://dip.doe-mbi.ucla.edu/
BIND The Biomolecular Interaction Network Database	http://www.binddb.org/BIND
MIPS interaction tables	http://mips.gsf.de/proj/yeast/CYGD/db/

Further Reading
Gerstein M, et al. (2002). Proteomics. Integrating interactomes. *Science* 295: 284–287.

Ito T, et al. (2000). A comprehensive two-hybrid analysis to explore the yeast protein interactome. *Proc. Natl. Acad. Sci. USA* 98: 4569–4574.

INTERGENIC SEQUENCE
Katheleen Gardiner

Regions in genomic DNA between genes.

The region between the promoter of one gene and the 3′ end of the adjacent gene when two genes are arranged head to tail or the distance between promoters when two genes are arranged head to head. Intergenic distances are highly variable in size and related to base composition. In GC-rich regions, they tend to be short (a few kilobase pairs) or nonexistent (where adjacent genes overlap). In AT-rich regions, intergenic distances can be very large (tens to hundreds of kilobase pairs). Intergenic regions may contain regulatory sequences or may be functionless.

Further Reading
Dunham I (1999). The DNA sequence of human chromosome 22. *Nature* 402: 489–495.

Hattori M, et al. (2000). The DNA sequence of human chromosome 21. *Nature* 405: 311–320.

INTERNATIONAL IMMUNOGENETICS DATABASE. *SEE* IMGT.

INTERNATIONAL SOCIETY FOR COMPUTATIONAL BIOLOGY (ISCB)

John M. Hancock and Martin J. Bishop

ISCB is an academic society dedicated to computational biology, which it defines as "advancing the understanding of living systems through computation."

The society organizes an annual meeting, Intelligent Systems for Molecular Biology; sponsors other conferences in computational biology; and affiliated with the journal *Bioinformatics*. It is also committed to training and education in the subject and raising the field's profile in the scientific and wider community.

Related Website

ISCB	http://www.iscb.org/

INTEROLOG (INTERLOG)

Dov S. Greenbaum

Homolog sets of proteins that are know to interact within a different organism.

It is thought that much of the diversity between different organisms can be traced to their protein–proteins interactions. Although there are many high-throughput techniques designed to study protein interactions, there is also an effort to transfer annotation from one set of interactions to similar protein sets.

Logically, if two proteins interact in one organism, two functionally and structurally similar proteins can be thought to interact in another, thus hypothesizing a novel interaction without strict experimental information, similar to transferring structural annotation based on sequence homology. With the knowledge of interologs, it may be possible to predict the function of a gene.

Related Website

Interlog	http://ymbc.ym.edu.tw/proteome/interact/interlog.htm

Further Reading

Matthews LR, et al. (2001). Identification of potential interaction networks using sequence-based searches for conserved protein-protein interactions or "interologs." *Genome Res.* 11: 2120–2126.

INTERPRETS

Patrick Aloy

InterPreTS (Interaction Prediction through Tertiary Structure) is a Web-based tool to predict protein–protein interactions using three-dimensional (3D) structure information. The method assesses the fit of two potential interacting partners on a complex of known 3D structure and thus can only be applied to those pairs of proteins for which the structure of a homologous complex is known.

In brief, after identifying the residues that make atomic contacts in the complex of known structure, the method checks whether the query protein sequences preserve these interactions by means of empirical potentials. It then estimates a statistical significance for the potential interaction based on a background of random sequences.

Since InterPreTS uses information extracted from 3D complexes, all the interacting pairs predicted are in physical contact. The method also provides molecular details of how the predicted interactions are likely to occur (i.e., which residues are in contact). It can be applied either to individual pairs of proteins or on a large scale (i.e., whole genomes).

Related Website

| InterPreTS | http://www.russell.embl-heidelberg.de/interprets/ |

Further Reading

Aloy P, Russell RB (2002). Interrogating protein-interaction networks through structural biology. *Proc. Natl. Acad. Sci. USA* 99: 5896–5901.

Aloy P, Russell RB (2003). InterPreTS: Protein interaction prediction through tertiary structure. *Bioinformatics* 19: 161–162.

INTERPRO
Terri Attwood

An integrated documentation resource for protein families, domains, and functional sites developed initially as a means of rationalizing the complementary approaches of the PROSITE, PRINTS, PFAM, and ProDom databases. The project was subsequently extended to include SMART and TIGRFAM, and links to Blocks are now also provided. InterPro thus combines databases with different underpinning methodologies and a varying degree of biological information—its annotation is drawn almost exclusively from PRINTS and PROSITE.

InterPro's member databases use different analytical approaches, including regular expressions, fingerprints, blocks, profiles, and hidden Markov models (HMMs). Diagnostically, these methods have different areas of optimum application: Some focus on functional sites (e.g., PROSITE); some focus on divergent domains (e.g., PFAM); and others focus on families, specializing in hierarchical definitions from superfamily down to subfamily levels in order to pinpoint specific functions (e.g., PRINTS). ProDom exploits a different approach, instead using PSI-BLAST automatically to cluster similar sequences in SWISS-PROT and TrEMBL. This allows the resource to be relatively comprehensive, because it does not depend on manual crafting, validation, and annotation of family discriminators.

InterPro is made available from the European Bioinformatics Institute, United Kingdom. Each entry has a unique accession number and includes functional descriptions and literature references, with links back to the relevant member database(s). All matches against SWISS-PROT and TrEMBL are listed with links to graphical views of the results. By uniting the member databases, InterPro capitalizes on their individual strengths, producing a powerful integrated diagnostic tool that is greater than the sum of its parts. The database may be searched with query sequences using the InterProScan tool or via simple keyword or Sequence Retrieval System (SRS) text searches.

Related Websites

InterPro	http://www.ebi.ac.uk/interpro/
InterPro text search	http://www.ebi.ac.uk/interpro/search.html
InterProScan	http://www.ebi.ac.uk/interpro/scan.html

Further Reading

Apweiler R, Attwood TK, Bairoch A, Bateman A, Birney E, Biswas M, Bucher P, Cerutti L, Corpet F, Croning MDR, Durbin R, Falquet L, Fleischmann W, Gouzy J, Hermjakob H, Hulo N, Jonassen I, Kahn D, Kanapin A, Karavidopoulou Y, Lopez R, Marx B, Mulder NJ, Oinn TM, Pagni M, Servant F, Sigrist CJA, Zdobnov EM (2001). The InterPro database, an integrated documentation resource for protein families, domains and functional sites. *Nucleic Acids Res.* 29(1): 37–40.

Tatusov RL (1997). A genomic perspective on protein families *Science* 278: 631–637.

Zdobnov EM, Apweiler R (2001). InterProScan—an integration platform for the signature-recognition methods in InterPro. *Bioinformatics* 17(9): 847–848.

See also Blocks, Hidden Markov Model, PFAM, PRINTS, ProDom, PROSITE, Protein Domain, Protein Family.

INTERPROSCAN
John M. Hancock and Martin J. Bishop

A tool to search sequences using the range of protein signature recognition resources incorporated into InterPro.

InterProScan is a Perl-based program. It can carry out its operations in parallel and in a distributed manner. Output is in raw, XML, text, or HTML (hypertext mark-up language) formats. The program accepts DNA as well as protein sequences. In the case of DNA sequences, translations in all six frames are scanned.

Related Website

InterProScan	http://www.ebi.ac.uk/interpro/

Further Reading

Zdobnov EM, Apweiler R (2001). InterProScan—an integration platform for the signature-recognition methods in InterPro. *Bioinformatics* 17: 847–848.

See also InterPro.

INTERSPERSED SEQUENCE (LOCUS REPEAT, LONG-PERIOD INTERSPERSION, LONG-TERM INTERSPERSION, SHORT-PERIOD INTERSPERSION, SHORT-TERM INTERSPERSION)
Dov S. Greenbaum

Sequences comprising repeated sequences interspersed with unique sequences.

The coding regions of the genomes of many insects and plants have either short- or long-period interspersion patterns. Typically, smaller genomes contain long-period interspersions and larger genomes have short-period interspersions.

A long-period interspersion is defined as repetitious DNA sequences greater than 5.6 kb that alternate with longer stretches of nonrepetitious DNA. Short-period interspersions are composed of unique sequences of DNA that alternate with shorter repetitious sequences.

Further Reading
Britten RJ, Davidson EH (1976). DNA sequence arrangement and preliminary evidence on its evolution. *Fed. Proc.* 35: 2151–2157.

INTRINSIC GENE PREDICTION. *SEE* GENE PREDICTION, AB INITIO.

INTRON
Niall Dillon and John M. Hancock

Sequence within a gene that separates two exons.

Eukaryotic genes are split into exons and intervening sequences (introns). The primary RNA transcript contains both exons and introns and the introns are then spliced out in a complex splicing reaction which is closely coupled to transcription. In the majority of genes, the sequence of the intron begins with GT and ends with AG (the GT–AG rule).

Three classes of introns are recognized: nuclear introns, group I introns, and group II introns.

Nuclear introns are the common type of introns in eukaryotic nuclear genes and require the spliceosome [a macromolecular complex of proteins and small nuclear RNAs (snRNAs)] to be spliced.

Group I introns are self-splicing introns that have a conserved secondary structure. They are found in RNA transcripts of protozoa, fungal mitochondria, bacteriophage T4, and bacteria.

Group II introns are found in fungal mitochondria, higher plant mitochondria, and plastids. They have a conserved secondary structure and may or may not require the participation of proteins in the splicing reaction.

Further Reading
Lewin B (2000). *Genes VII.* Oxford University Press, Oxford, UK.

See also Exon.

INTRON PHASE
Austin L. Hughes

The phase of an intron is a way of describing the position of the intron relative to the reading frame of the exons.

There are three types of intron phases depending on where the intron is located. In phase 0, the intron lies between codons. In phase 1, the intron occurs between the first and second bases of a codon. In phase 2, the intron occurs between the second and third bases of a codon. An exon in which both upstream and downstream introns have the same phase is called a symmetrical exon. If an exon from one gene is inserted into an intron of another gene (as has been hypothesized to occur in cases of exon shuffling), the exon must be symmetrical if the new exon is to be functional in its new location.

Further Reading

Long M, et al. (1995). Intron phase correlations and the evolution of the intron/exon structure of genes. *Proc. Natl. Acad. Sci. USA* 92: 12495–12499.

Patthy L (1991). Modular exchange principles in proteins. *Curr. Opin. Struct. Biol.* 1: 351–362.

See also Exon Shuffling, Intron.

ISCB. *SEE* INTERNATIONAL SOCIETY FOR COMPUTATIONAL BIOLOGY.

ISOCHORE
Katheleen Gardiner

Within a genome, a large DNA segment (>300 kb) that is homogeneous in base composition.

Genomes of warm-blooded vertebrates are mosaics of isochores belonging to four classes: The L isochores are AT rich ($\sim$ 35% to 42% GC) and the H1, H2, and H3 isochores are increasingly GC rich (42% to 46%, 47% to 52%, and >52%, respectively). Isochores may be reflected in chromosome band patterns. G bands appear to be uniformly composed of L isochores, while R bands heterogeneously contain all classes. The brightest R bands, T bands, contain H2 and H3 isochores.

Further Reading

Bernardi G (2000). Isochores and the evolutionary genomics of vertebrates. *Gene* 241: 3–17.

Eyre-Walker A, Hurst LD (2001). The evolution of Isochores. *Nat. Rev. Genet.* 2: 549–555.

Oliver JL, Bernaola-Galvan P, Carpena P, Roman-Roldan R (2001). Isochore chromosome maps of eukaryotic genomes. *Gene* 276: 47–56.

Pavlicek A, Paces J, Clay O, Bernardi G (2002). A compact view of isochores in the draft human genome sequence. *FEBS Lett.* 511: 165–169.

ISOLEUCINE
Jeremy Baum

A nonpolar amino acid with side chain —CH(CH$_3$)C$_2$H$_5$ found in proteins. In sequences, written as Ile or I.

See also Amino Acid.

ISOMORPHOUS REPLACEMENT
Liz P. Carpenter

Isomorphous replacement is a technique used to overcome the phase problem in X-ray crystallography. In order to solve the structure of a macromolecule from the diffraction of X rays by a crystal of the molecule, it is necessary to obtain phases as well as intensity information for each diffraction spot in the data set. This information cannot be measured directly during the data collection. It can, however, be obtained by modifying the crystals with heavy atoms and collecting data with derivatized and native crystals. Often several data sets are collected on different heavy-atom derivatives, and these are combined with a native data set in a technique called multiple isomorphous replacement (MIR). If only one derivative is used, the technique is called single isomorphous replacement (SIR), and if anomalous dispersion information is included, the method is called single isomorphous replacement with anomalous scattering (SIRAS).

If the differences between the data sets are due to the presence of a few well-defined heavy atoms bound to the macromolecule, then the differences can be used to identify the positions of the heavy atoms in the asymmetric unit. The positions are then refined, and from the location of the heavy atoms, approximate phases can be calculated for each reflection in the data set. Once initial phase information has been obtained, an electron density map can be calculated and the structure can be build into the map.

See also Diffraction of X Rays, Electron Density Map, Heavy-Atom Derivative, Phase Problem, X-ray Crystallography for Structure Determination.

Table   Commonly Used IUPAC-IUB Codes for Nucleic Acid Bases

Code	Base
a	Adenine
c	Cytosine
g	Guanine
t	Thymine in DNA; uracil in RNA
m	a or c
r	a or g
w	a or t
s	c or g
y	c or t
k	g or t
v	a or c or g; not t
h	a or c or t; not g
d	a or g or t; not c
b	c or g or t; not a
n	a or c or g or t

Table   Commonly Used IUPAC-IUB Codes for Amino
Acids

Code		Amino Acid
3 Letter	1 Letter	
Ala	A	Alanine
Arg	R	Arginine
Asn	N	Asparagine
Asp	D	Aspartic acid (Aspartate)
Cys	C	Cysteine
Gln	Q	Glutamine
Glu	E	Glutamic acid (Glutamate)
Gly	G	Glycine
His	H	Histidine
Ile	I	Isoleucine
Leu	L	Leucine
Lys	K	Lysine
Met	M	Methionine
Phe	F	Phenylalanine
Pro	P	Proline
Ser	S	Serine
Thr	T	Threonine
Trp	W	Tryptophan
Tyr	Y	Tyrosine
Val	V	Valine
Asx	B	Aspartic acid or asparagine
Glx	Z	Glutamine or glutamic acid
Xaa	X	Any amino acid.
TERM		Termination codon

IUPAC-IUB CODES (AMINO ACID ABBREVIATIONS, NUCLEOTIDE BASE CODES)

John M. Hancock

Codes used to represent nucleotide bases and amino acid residues.

The process of sequencing DNA or protein molecules does not always produce unambiguous results. Similarly, the representation of consensus sequences of either type necessitates a simple way of representing sequence redundancy at a position. To produce a standard nomenclature that could accommodate these requirements, the IUPAC (International Union of Pure and Applied Chemistry) and IUB (International Union of Biochemistry) Commission on Biochemical Nomenclature agreed on standard nomenclatures for nucleic acids and amino acids. For nucleic acids they agreed on a single nomenclature, while for amino acids they agreed on two parallel nomenclatures, a single letter and a three-letter code, the three-letter code being more descriptive.

Related Websites

Amino acid nomenclature	http://www.chem.qmul.ac.uk/iupac/AminoAcid/
Nucleic acid nomenclature	http://www.chem.qmul.ac.uk/iupac/misc/naabb.html

Further Reading

Cornish-Bowden, A (1985). Nomenclature for incompletely specified bases in nucleic acid sequences: Recommendations 1984. *Nucleic Acids Res.* 13: 3021–3030.

IUPAC-IUB Joint Commission on Biochemical Nomenclature (JCBN) (1984). Nomenclature and symbolism for amino acids and peptides. Recommendations 1983. *Eur. J. Biochem.* 138: 9–37.

See also Amino Acid, Base Pair.

J

JACKKNIFE
Dov S. Greenbaum

A method used when attempting to determine a confidence level of a proposed relationship within a phylogeny.

Jackknifing is similar to bootstrapping but differs in that it only removes (iteratively and randomly) one value and uses the new smaller data set to calculate the phylogenetic relationship. After multiple repetitions a consensus tree can be built. The degree of variance of the sample can then be determined by comparing each branch in the original tree to the number of times it was found via the jackknifing algorithm.

See also Bootstrap Analysis, Cross-Validation.

JUMPING GENE. *SEE* TRANSPOSABLE ELEMENT.

Dictionary of Bioinformatics and Computational Biology. Edited by Hancock and Zvelebil
ISBN 0-471-43622-4   © 2004 John Wiley & Sons, Inc.

K

Kappa Virtual Dihedral Angle
Karyotype
KB
Kernel-Based Learning Methods
Kernel Function
Kernel Machine
Kernel Methods
k-Fold Cross-Validation
KIF
Kilobase
Kinetochore

Kingdom
Kin Selection
k-Means Clustering
Knowledge
Knowledge Base
Knowledge-Based Modeling
Knowledge Interchange Format
Knowledge Representation Language
Kozak Sequence
KRL

Dictionary of Bioinformatics and Computational Biology. Edited by Hancock and Zvelebil
ISBN 0-471-43622-4 © 2004 John Wiley & Sons, Inc.

KAPPA VIRTUAL DIHEDRAL ANGLE
Roman A. Laskowski

Defined by four successive Cα atoms along a protein chain: $C\alpha_{i-1}-C\alpha_i-C\alpha_{i+1}-C\alpha_{i+2}$.

KARYOTYPE
Katheleen Gardiner

The number and structure of the complement of chromosomes from a cell, individual, or species.

It is determined by examining metaphase chromosomes under the light microscope after they have been stained to produce chromosome banding patterns. Structural features include the relative sizes of each chromosome and the locations of each centromere plus the identity of the sex chromosomes. Each species has a unique normal karyotype that is invariant among individuals of the same sex. For example, the human karyotype is composed of 23 pairs of chromosomes, 22 autosomes and 2 sex chromosomes. Members of an autosome pair appear identical, contain the same set of genes, and are referred to as homologous chromosomes. Sex chromosomes are usually significantly different in size and gene content.

Further Reading

Miller OJ, Therman E (2000). *Human Chromosomes*. Springer, Vienna.

Scriver CR, Beaudet AL, Sly WS, Valle D, Eds. (1995). *The Metabolic and Molecular Basis of Inherited Disease*, McGraw-Hill, New York.

Standing Committee on Human Cytogenetic Nomenclature (1985). *An International System for Human Cytogenetic Nomenclature*. Karger, Basel, Switzerland.

Wagner RP, Maguire MP, Stallings RL (1993). *Chromosomes—A Synthesis*. Wiley-Liss, New York.

KB. *SEE* BASE PAIR.

KERNEL-BASED LEARNING METHODS. *SEE* KERNEL METHODS.

KERNEL FUNCTION
Nello Cristianini

Function of two arguments that returns the inner product between the images of such arguments in a vector space.

If the map embedding the two arguments is called ϕ, we can write a kernel function as $k(x, z) = \langle \phi(x), \phi(z) \rangle$. The arguments can be vectors, strings, or other data structures. These functions are used in machine learning and pattern analysis as the essential building block of algorithms in the class of kernel methods.

Dictionary of Bioinformatics and Computational Biology. Edited by Hancock and Zvelebil
ISBN 0-471-43622-4 © 2004 John Wiley & Sons, Inc.

A simple example of a kernel function is given by the following map from a two-dimensional space to a three-dimensional one: $\phi(x_1, x_2) = (x_1^2, x_2^2, \sqrt{2}x_1x_2)$. The inner product in such a space can easily be computed by a kernel function without explicitly rewriting the data in the new representation.

Consider two points:

$$x = (x_1, x_2)$$

$$y = (y_1, y_2)$$

Then consider the kernel function obtained by squaring their inner product:

$$\langle x, z \rangle^2 = (x_1z_1 + x_2z_2)^2$$
$$= x_1z_1^2 + x_2z_2^2 + 2x_1z_1x_2z_2$$
$$= \langle (x_1^2, x_2^2, \sqrt{2}x_1x_2), (z_1^2, z_2^2, \sqrt{2}z_1z_2) \rangle$$
$$= \langle \phi(x), \phi(z) \rangle$$

This corresponds to the inner product between two three-dimensional vectors and is computed without explicitly writing their coordinates. By using a higher exponent, it is possible to virtually embed those two vectors in a much higher dimensional space at a very low computational cost.

More generally, one can prove that every symmetric and positive-definite function $k(x, z)$ is a valid kernel, i.e., a map $\phi : X \to R^n$ exists such that $k(x, z) = \langle \phi(x), \phi(z) \rangle$, where $x \in X$ and $z \in X$. Examples of kernels include the Gaussian $k(x, z) = e^{-\frac{\|x-z\|^2}{2\sigma^2}}$, the general polynomial kernel $k(x, z) = (\langle x, z \rangle + 1)^d$, and many others, including kernels defined over sets of sequences that have been used for bioinformatics applications such as remote protein homology detection.

Related Website

Kernel Machines	www.kernel-machines.org

Further Reading

Cristianini N, Shawe-Taylor J. (2000). *An Introduction to Support Vector Machines*. Cambridge University Press, Cambridge.

Shawe-Taylor J, Cristianini N (2004) *Kernel Methods of Pattern Analysis*, Cambridge University Press, Cambridge, UK.

See also Kernel Methods, Support Vector Machines.

KERNEL MACHINE. *SEE* KERNEL METHODS.

KERNEL METHODS (KERNEL-BASED LEARNING METHODS, KERNEL MACHINE)

Nello Cristianini

A class of algorithms for machine learning and pattern analysis based on the notion of the kernel function.

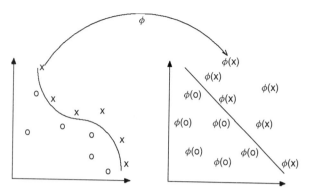

Figure The embedding of data into a feature space may simplify the classification problem.

The main idea underlying the use of kernel methods is to embed the data into a vector space where an inner product is defined. Linear relations among the images of the data in such a space are then sought. The mapping (or embedding) is performed implicitly by defining a kernel function $K(x, z) = \langle \phi(x), \phi(z) \rangle$.

By mapping the data into a suitable space, it is possible to transform nonlinear relations within the data into linear ones. Furthermore, the input domain does not need to be a vector space, so that statistical methods developed for data in the form of inputs can be applied to data structures such as strings by using kernels.

Kernel-based algorithms only require information about the inner products between data points. Such information can often be obtained at a computational cost that is independent of the dimensionality of that space. The kind of relations detected by kernel methods include classifications, regressions, clustering, principal components, canonical correlations, and many others.

The accompanying figure shows the basic idea of a kernel-induced embedding into a feature space: that by using a nonlinear map some problems can be simplified.

In the kernel approach, the linear functions used to describe the relations found in the data are written in the form $f(x) = \sum \alpha_i K(x_i, x)$, called the dual form, where K is the chosen kernel, α_i the dual coordinates, and x_i the training points.

Most kernel-based learning algorithms, such as support vector machines, reduce their training phase to optimizing a convex cost function, hence avoiding one of the main computational pitfalls of neural networks. Since they often make use of very high dimensional spaces, kernel methods run the risk of overfitting. For this reason, their design needs to incorporate principles of statistical learning theory that help identify the crucial parameter that needs to be controlled in order to avoid this risk.

Related Websites

Kernel Machines	www.kernel-machines.org
Kernel Methods	www.kernel-methods.net

Further Reading

Cristianini N, Shawe-Taylor J (2000). *An Introduction to Support Vector Machines.* Cambridge University Press, Cambridge, MA.

Shawe-Taylor J, Cristianini N (2004) *Kernel Methods for Pattern Analysis,* Cambridge University Press, Cambridge, UK.

K-FOLD CROSS-VALIDATION. *SEE* CROSS-VALIDATION.

KIF. *SEE* KNOWLEDGE INTERCHANGE FORMAT.

KILOBASE. *SEE* BASE PAIR.

KINETOCHORE
Katheleen Gardiner

Region of the centromere to which spindle fibers attach during mitosis and meiosis and required for segregation of chromatids to daughter cells.

The kinetochore is composed of chromatin (DNA and proteins) complexed with additional specialized proteins.

Further Reading
Miller OJ, Therman E (2000). *Human Chromosomes*. Springer, Vienna.

Schueler MG, Higgens AW, Rudd MK, Gustashaw K, Willard HF (2001). Genomic and genetic definition of a functional human centromere. *Science* 294: 109–115.

Scriver CR, Beaudet AL, Sly WS, Valle D, Eds. (1995). *The Metabolic and Molecular Basis of Inherited Disease*, McGraw-Hill, New York.

Sullivan BA, Schwartz S, Willard HF (1996). Centromeres of human chromosomes. *Environ. Mol. Mutagen.* 28: 182–191.

Wagner RP, Maguire MP, Stallings RL (1993). *Chromosomes—A Synthesis*. Wiley-Liss, New York.

KINGDOM
Dov S. Greenbaum

Traditionally the highest level of taxonomic classification.

Until the twentieth century biology divided life into two classes, animal and plant. In 1969 Robert Whittaker proposed dividing the kingdoms into five separate groups. An alternative approach, by Carl Woese, was to divide all organisms into three domains or kingdoms. Woese, stating that the "five-kingdom scheme is essentially a mixture of taxonomic apples and oranges," sought to create a theory that allowed useful predictions to be made from its organization—"The incredible diversity of life on this planet, most of which is microbial, can only be understood in an evolutionary framework." Although each of the three kingdoms is composed of distinct and diverse organisms, they have unique and unifying characteristics.

Eukaryotes (including the Whittaker kingdoms Protista, Plantae, Animalia, and Fungi) are defined by those organisms with nuclei, DNA bound by histones and forming chromosomes, cytoskeletons, and internal membranes (i.e., organelles). Prokaryotes and archaea, both lacking the above components, were once considered a single group, until Woese and colleagues found, using the 16S ribosomal RNA, that archaea, which lived at high temperatures or produced methane, could be defined as a separate group from the other bacteria. Archaea are divided into extreme thermophiles, extreme halophiles, and methanogens: Further analysis has shown that archaea differ dramatically from bacteria and are often a mosaic of bacterial and eukaryotic features; thus genes and structures are similar to either prokaryotes or eukaryotes or alternatively, they do not have similarities with either of the kingdoms.

Although subsequent systems have been proposed, none have usurped this present system. The classical phylogenetic tree based on the ribosome (a cellular component common to all kingdoms whose RNA changes slowly—thus the greater variation between the genetic sequences, the greater the evolutionary distance) shows how each of these kingdoms cluster.

Related Websites

2k23	http://www.biology.iupui.edu/biocourses/N100/2k23domain.html
Woese	http://www.life.uiuc.edu/micro/woese.html

Further Reading

Olsen GJ, Woese CR (1993). Ribosomal RNA: A key to phylogeny. *FASEB J.* 7: 113–123.

Woese CR, Kandler O, Wheelis ML (1990). Towards a natural system of organisms: Proposal for the domains Archaea, Bacteria, and Eucarya. *Proc. Natl. Acad. Sci. USA* 87: 4576–4579.

KIN SELECTION
A. Rus Hoelzel

A mechanism for the adaptive evolution of genes through the survival of relatives.

The central idea is said to have been worked out by the British geneticist J. B. S. Haldane in a pub one evening, where he announced that he would lay down his life for two brothers, eight cousins, etc. Through kin selection, it is possible to explain the evolution of altruism, whereby an altruist provides a fitness benefit to a recipient at a fitness cost to the altruist. This can evolve by kin selection if the ratio of the recipient's gains to the altruist's cost is greater than the reciprocal of the coefficient of relatedness between the two, and this is known as Hamilton's rule (Hamilton 1964):

$$B/C > 1/r$$

It is based on the supposition that evolution should make no distinction between genes in direct compared to indirect descendant kin. Provided that the representation of the gene increases in the next generation (through inclusive fitness), traits that reduce individual fitness can evolve.

Further Reading

Hamilton WD (1964). The genetical theory of social behaviour. *J. Theor. Biol.* 7: 1–52.

See also Evolution, Natural Selection.

K-MEANS CLUSTERING
John M. Hancock

A popular algorithm for clustering data.

The k-means clustering algorithm is a simple means to generate a user-specified number of clusters from a given data set. Initially, data vectors are assigned randomly so that each cluster has an equal number of members. The centroid of each of these clusters is then calculated, as are the distances from each data vector to the centroid of all clusters. If a given data vector is closest to the centroid of its cluster, it remains there; otherwise it is moved to the cluster to whose centroid it is closest. This process is repeated until there is no more movement between clusters. The method is dependent on the original clusters (which can be overcome to some extent by replication) and the distance measure employed.

k-means clustering is widely used in the analysis of microarray data because of its simplicity and speed.

Further Reading
Quackenbush J (2001). Computational analysis of microarray data. *Nature Rev. Genet.* 2: 418–427.

KNOWLEDGE
Robert Stevens

What we understand about a domain of interest.

Normally, we look at three levels of interpretation: data, information, and knowledge. Data can be thought of as the stimuli that reach our senses—the patterns of light that reach our eyes or sound waves that reach our ears. Similarly, data are the patterns of bits held in a computer or the raw results gathered from an experiment. The notion of information adds interpretation to these data, often in the form of classification: These patterns of light are a person; these sounds are words (nouns, verbs, etc.); these bits are an integer or a "Person" object. Knowledge is what we understand about a piece of information. The information entity *person* is called "Robert Stevens"; he is a middle-aged man; he has a bachelor's degree in biochemistry and a doctorate in computer science; he is a lecturer in bioinformatics at the University of Manchester. It is this knowledge that we wish to capture in an ontology.

KNOWLEDGE BASE
Robert Stevens

Classically, a knowledge base is an ontology together with facts about the categories in it and their instances.

The ontology provides a model or schema, and the instances collected must conform to that schema. Systems such as Protégé-2000 provide an ontology modeling environment and the ability to create forms for collecting instances conforming to the model. It is the use of reasoning to make inferences over the facts that distinguishes a knowledge base from a database.

KNOWLEDGE-BASED MODELING

KNOWLEDGE INTERCHANGE FORMAT (KIF)
Robert Stevens

Knowledge Interchange Format (KIF) provides a declarative language for describing knowledge and for the interchange of knowledge among disparate programs. KIF has a declarative semantics (i.e., the meaning of expressions in the representation can be understood without appeal to an interpreter for manipulating those expressions). It is logically comprehensive (i.e., it provides for the expression of arbitrary sentences in first-order predicate calculus), and it provides for the representation of knowledge about knowledge. KIF is not intended as a primary language for interaction with human users (though it can be used for this purpose). Different programs can interact with their users in whatever forms are most appropriate to their applications (e.g., frames, graphs, charts, tables, diagrams, natural language).

Related Website

| KIF | http://logic.stanford.edu/kif/kif.html |

Further Reading

Bechhofer S (2002). Ontology Language Standardisation Efforts; OntoWeb Deliverable D4.0. http://ontoweb.aifb.uni-karlsruhe.de/About/Deliverables/d4.0.pdf.

KNOWLEDGE REPRESENTATION LANGUAGE (KRL)
Robert Stevens

An ontology is couched in terms of concepts and relationships. This conceptualization of a domain's knowledge needs to be encoded so that it can be stored, transmitted, and used by humans, computers, or both. In biological databases, the usual form for representing knowledge is English. This form of representation is not very amenable to computational processing, as well as having the disadvantage of ambiguity in interpretation by humans. Therefore, a knowledge representation language (KRL) with defined semantics, amenable to computational processing, should be used to encode knowledge. There are many kinds of KRLs, and they have different capabilities and expressiveness. We can divide KRLs into three broad categories:

An *informal* KRL may be specified by a catalog of terms that are either undefined or defined only by statements in a natural language. The Gene Ontology would fall into this category.

A *formal* ontology is specified by a collection of names for concept and relation types organized in a partial ordering by the type–subtype relation. RiboWeb and Ecocyc are examples of this kind of ontology.

Formal ontologies are further distinguished by the way the relationship between subtype and supertypes is established. In an asserted formal ontology all subtype–supertype relations are asserted by the ontology author. In an axiomatized ontology additional subtype–supertype relationships can be inferred by a reasoner based on axioms and definitions stated in a formal language, such as a description logic (DL). The TAMBIS ontology is an example of this type.

Formal ontology languages have a formal semantics, which lend a precise meaning to statements in the language. These statements can be interpreted and manipulated by a machine without ambiguity.

There is no agreed-upon vocabulary for talking about KRLs. Different languages use different terms for the same notion or the same term for different notions. The table below summarizes some of these variations in usage across KRLs.

Further Reading

Ringland GA, Duce DA (1988). *Approaches to Knowledge Representation: An Introduction.* Wiley, Chichester.

Table   Terminology Used in Knowledge Representation Languages

Meaning	DLs	OWL	GRAIL	Frames	UMLclass Diagrams
Category	Concept	Class	Category	Class	Class
Individual	Individual	Individual	(Individual)	Instance	Instance
Relationship type	Role	Property	Attribute	Slot	Relation
Relationship	Existential restriction	Same	Criterion	Filled slot	Relation
Range constraint	Universal restriction	Same	(Sanction[1])	Various facets depending on value type; typically "allowed classes"	(Implied)
Sanctioned	–	–	Sanctioned	(Implied)	(Implied)
Some	Some	Has class some Values From	Topic Necessarily which	Filled slot[2]	1..* mandatory
Only	All	To class all Values From	(Sensibly) "sanctions"	Range facet	–
At least n	At least n	At least n	–	Minimum cardinality facet = n	$n..*$
At most n	At most n	At most n	–	Maximum cardinality facet = n	$x..n$
Default	–	–	(Extrinsics)	(Implied)	–
Any	–	–	(Extrinsics)	(Ambiguous)	(Ambiguous)
And	And	And	(Implied)	(Implied)	(Implied)
Or	Or	Or	–	–	–
Not	Not	Complement	–	–	–

[1] Sanctions in GRAIL provide a check on domain and range constraints but are not used in classification.
[2] The semantics of filling a slot with a class are ambiguous in most frame systems. Usually the most satisfactory translation into DLs or OWL is as existential restrictions.

KOZAK SEQUENCE
Niall Dillon

Consensus sequence located around the site of initiation of eukaryotic mRNA translation.

The Kozak sequence is involved in recognition of translation start site by the ribosome. Using comparisons of large numbers of genes and site-directed mutagenesis, the optimal Kozak consensus sequence has been identified as CCA/GCC<u>A</u>UGG. The small 40S ribosomal subunit, carrying the methionine tRNA and translation initiation factors, is thought to engage the mRNA at the capped 5′ end and then scan linearly until it encounters the first AUG (initiator) codon. The Kozak sequence modulates the ability of the AUG codon to halt the scanning 40S subunit.

Further Reading

Kozak M (1987). An analysis of 5′-noncoding sequences from 699 vertebrate messenger RNAs. *Nucleic Acids Res.* 15: 8125–8148.

Kozak M (1997). Recognition of AUG and alternative initiator codons is augmented by G in position +4 but is not generally affected by the nucleotides in positions +5 and +6. *EMBO J.* 16: 2482–2492.

KRL. *SEE* KNOWLEDGE REPRESENTATION LANGUAGE.

L

L1 Element
Label
Labeled Data
Laboratory Information Management
   System
LAMARC
Lander–Green Algorithm
Langevin Dynamics
LaserGene
Last Universal Common Ancestor
Lattice
Laue Method
LCR
LD
Leave- One- Out
Leucine
Lexicon
L–G Algorithm
LINE
Linear Discriminant Analysis
Linkage

Linkage Analysis
Linkage Disequilibrium
Linkage Disequilibrium Analysis
Local Alignment
Local Similarity
Locus Control Region
LocusLink
Locus Repeat
LOD Score
Logarithm-of-Odds Score
LogDet
Log-Odds Score
Long-Period Interspersion, Long-Term
   Interspersion
Look-Up Gene Prediction
Loop
Loop Prediction/Modeling
Low-Complexity Region
LUCA
Lysine

Dictionary of Bioinformatics and Computational Biology. Edited by Hancock and Zvelebil
ISBN 0-471-43622-4 © 2004 John Wiley & Sons, Inc.

L1 ELEMENT
Katheleen Gardiner

Human long interspersed repetitive element (LINE), originally (and in some cases possibly still) capable of transposition and present in approximately one hundred thousand copies in the human genome.

Full-length L1 sequences are 6.4 kb long, but the majority of copies are truncated at the 5′ end and may be as short as tens to a few hundred base pairs. L1s display a nonuniform distribution with the highest concentrations in the AT-rich regions of Giemsa bands and a subset of reverse bands.

Further Reading
Cooper DN (1999). *Human Gene Evolution*. Bios Scientific, Oxford, UK, pp. 265–285.
Ridley M (1996). *Evolution*. Blackwell Science, Cambridge, MA, pp. 265–276.
Rowold DJ, Herrera RJ (2000). Alu elements and the human genome. *Genetics* 108: 57–72.

See also LINE, SINE.

LABEL (DEPENDENT VARIABLE, LABELED DATA, RESPONSE)
Nello Cristianini

Term used in machine learning to designate an attribute of data that requires to be predicted.

In machine learning and statistical data analysis, a common task is to learn a function $y = f(x)$ from example pairs (x, y). The second part of this pair is called the response, or the label of the data point x. Data sets of examples of the type (x, y) are called *labeled*. In statistics the labels are often called *dependent variables*.

This form of machine learning goes under the name *supervised learning*, since the desired response for each training point is given as a form of "supervision" to the learning process.

It is opposite to "unsupervised" learning, where only the observations $\{x\}$ are given, and the algorithm is requested to discover generic relations within them, e.g., clusters.

Further Reading
Mitchell T (1997). *Machine Learning*. McGraw-Hill, New York.

See also Feature, Machine Learning, Pattern Analysis.

LABELED DATA. *SEE* LABEL.

LABORATORY INFORMATION MANAGEMENT SYSTEM (LIMS)
John M. Hancock

A type of database designed to capture information about standard processes taking place in a laboratory.

Historically, LIMS was employed in analytical laboratories, process testing laboratories, quality assurance laboratories, etc., where the accurate tracking of

Dictionary of Bioinformatics and Computational Biology. Edited by Hancock and Zvelebil
ISBN 0-471-43622-4 © 2004 John Wiley & Sons, Inc.

experiments was essential. Functions of a LIMS can include direct acquisition of data from instruments, recording the details of experiments carried out, and producing reports. By keeping a permanent record, LIMS ensures that critical information is not lost or overwritten. These sorts of functionality have not traditionally been needed in biological research laboratories, but the advent of high-throughput functional genomics and other facilities has resulted in an increase in LIMS use in these settings. Many LIMS products are commercial, and some are available from suppliers of high-throughput equipment, although these are often not appropriate for the full range of experiments that may be carried out in a functional genomics laboratory. Alternatively LIMS may be developed in-house for large specialized facilities, although this is labor intensive and expensive for fully functional systems.

Related Websites

Limsfinder	http://www.limsfinder.com/
Listing of LIMS	http://www.cato.com/biotech/bio-software.html#Laboratory

Further Reading
Goodman N, et al. (1998). The LabBase system for data management in large scale biology research laboratories. *Bioinformatics* 14: 562–574.

Strivens MA, et al. (2000). Informatics for mutagenesis: The design of mutabase—a distributed data recording system for animal husbandry, mutagenesis, and phenotypic analysis. *Mamm. Genome* 11: 577–583.

Taylor S, et al. (1998). Automated management of gene discovery projects. *Bioinformatics* 14: 217–218.

LAMARC (FLUCTUATE, MIGRATE, RECOMBINE)
Michael P. Cummings

A set of programs that use coalescence theory and Markov chain Monte Carlo methods to estimate population genetic parameters. FLUCTUATE estimates effective population size and growth rate (FLUCTUATE subsumes the functionality of an earlier program, COALESCE); MIGRATE estimates the effective sizes of and migration rates among populations using nonrecombining sequences, microsatellite data, or enzyme electrophoretic data; and RECOMBINE estimates effective population size and per-site recombination rate for DNA sequence data. Estimation of these parameters involves simplifying assumptions (e.g., constant population size, exponential growth rate, nonrecombining sequences). An alpha version of a single comprehensive program, LAMARC, embodies much of the capabilities of the previously mentioned separate programs.

FLUCTUATE, MIGRATE, and RECOMBINE are available as American National Standards Institute (ANSI) C source code and executables for several platforms. LAMARC is available as C++ source code and executables for several platforms. MIGRATE has parallel processing capabilities based on message-passing interface (MPI).

Related Website

LAMARC	http://evolution.genetics.washington.edu/lamarc.html

Further Reading

Beerli P, Felsenstein J (1999). Maximum-likelihood estimation of migration rates and effective population numbers in two populations using a coalescent approach. *Genetics* 152: 763–773.

Beerli P, Felsenstein J (2001). Maximum likelihood estimation of a migration matrix and effective population sizes in *n* subpopulations by using a coalescent approach. *Proc. Natl. Acad. Sci. USA* 98: 4563–4568.

Kuhner MK, et al. (1995). Estimating effective population size and mutation rate from sequence data using Metropolis-Hastings sampling. *Genetics* 140: 1421–1430.

Kuhner MK, et al. (1998). Maximum likelihood estimation of population growth rates based on the coalescent. *Genetics* 149: 429–434.

Kuhner MK, et al. (2000). Maximum likelihood estimation of recombination rates from population data. *Genetics* 156: 1393–1401.

LANDER–GREEN ALGORITHM (L–G ALGORITHM)
Mark McCarthy and Steven Wiltshire

Used in the context of linkage analysis, an algorithm for calculating the likelihood of a pedigree based on a hidden Markov chain of inheritance distributions for a map of ordered marker loci.

For each position in a given marker map, the L–G algorithm defines the distribution of binary inheritance vectors describing, for the entire pedigree at once, all possible inheritance patterns of alleles at that position in terms of their parental origin. In the absence of any genotype data, all inheritance vectors compatible with Mendelian segregation have equal prior probabilities. However, in the presence of genotype data, the posterior distribution of inheritance vectors, conditional on the observed genotypes, will have nonequal vector probabilities. In ideal circumstances—fully informative markers, no missing data, no missing individuals—it may be possible to identify unambiguously the one true inheritance vector for any given position: In real-life data, this is unusual. The L–G algorithm models these inheritance distributions at positions across the marker map, in terms of a hidden Markov model, with the transition probabilities between states (the inheritance distributions) being functions of the recombination fractions between the loci. The likelihood of the inheritance distribution at any locus, conditional on the whole marker set, can be calculated from this model.

This likelihood function forms the basis of several software implementations of parametric and nonparametric linkage analyses. The computational burden of the L–G algorithm increases linearly with marker number but increases exponentially with pedigree size and is therefore ideally suited to multipoint linkage analyses in medium-sized pedigrees. In this regard, it complements the Elston–Stewart algorithm, which can calculate likelihoods rapidly for large pedigrees but can cope with relatively few markers at a time.

Examples: Hanna et al. (2002) used GENEHUNTER PLUS—a widely used program which implements the L–G algorithm—to perform genomewide parametric and nonparametric linkage analyses of seven extended pedigrees for loci influencing susceptibility to obsessive-compulsive disorder. Wiltshire et al. (2001) used ALLEGRO—a related program—to study a large collection of United Kingdom sibpair families for evidence of linkage to type 2 diabetes.

Related Websites
Software that implements the L–G algorithm in multipoint linkage analyses:

GENEHUNTER 2	www-genome.wi.mit.edu/ftp/distribution/software/genehunter/
GENEHUNTER PLUS	ftp://galton.uchicago.edu/pub/kong/
Merlin	http://www.sph.umich.edu/csg/abecasis/Merlin/index.html

Further Reading
Abecasis GR, et al. (2002). Merlin—rapid analysis of dense genetic maps using sparse gene flow trees. *Nat. Genet.* 30: 97–101.

Hanna GL, et al. (2002). Genome-wide linkage analysis of families with obsessive-compulsive disorder ascertained through pediatric probands. *Am. J. Med. Genet.* 114: 541–552.

Kruglyak L, et al. (1995). Rapid multipoint linkage analysis of recessive traits in nuclear families, including homozygosity mapping. *Am. J. Hum. Genet.* 56: 519–527.

Kruglyak L, et al. (1996). Parametric and nonparametric linkage analysis: A unified multipoint approach. *Am. J. Hum. Genet.* 58: 1347–1363.

Lander ES, Green P (1987). Construction of multilocus genetic maps in humans. *Proc. Natl. Acad. Sci. USA* 84: 2363–2367.

Wiltshire S, et al. (2001). A genome-wide scan for loci predisposing to type 2 diabetes in a UK population (The Diabetes (UK) Warren 2 Repository): Analysis of 573 pedigrees provides independent replication of a susceptibility locus on chromosome 1q. *Am. J. Hum. Genet.* 69: 553–569.

See also Allele-Sharing Methods, Elston–Stewart Algorithm, Genome Scans, Linkage Analysis.

LANGEVIN DYNAMICS
Roland L. Dunbrack

A form of molecular dynamics simulation in which terms are added to the force in Newton's equation to represent neglected degrees of freedom such as a solvent. One term added represents frictional drag on atoms due to the solvent and is therefore proportional to the velocity but opposite in sign. The other term represents random kicks of the system by neglected atoms.

Related Website

Introduction to macromolecular simulation	http://cmm.info.nih.gov/intro_simulation/course_for_html.html

See also Molecular Dynamics Simulations.

LASERGENE
John M. Hancock and Martin J. Bishop

A desktop package providing access to basic sequence manipulation and analysis tools.

The LaserGene package is made up of a number of individual programs that provide different sets of functions for sequence manipulation and analysis. Data are stored in a proprietary format but can be imported and exported in some commonly used formats. The different modules are designed to assist with sequence assembly, gene discovery and annotation, protein structure prediction, DNA and protein sequence alignment and phylogeny reconstruction, Polymerase Chain Reaction (PCR) primer design, restriction site mapping, editing and annotation of sequences, and database searching using BLAST.

Related Website

DNASTAR	http://www.dnastar.com/

See also MacVector, VectorNTI, Wisconsin Package.

LAST UNIVERSAL COMMON ANCESTOR. *SEE* ANCESTRAL GENOME.

LATTICE
Robert Stevens

1. In mathematics, a directed acyclic graph which is "closed" at both top and bottom (see the accompanying figure). (Formally, a directed acyclic graph in which every node has a unique least common subsumer and greatest common subsume.)
2. Informally, in knowledge representation, a multiple hierarchy where the hierarchy is based upon one "taxonomic" relationship (usually subtype) and forms a framework for other "nontaxonomic" relationships. (For knowledge representation the distinction between a "lattice" and an "acyclic directed graph" makes little difference, since any directed acyclic graph can be trivially converted to a lattice by providing a "bottom" node linked to each leaf node. However, software often depends on the extensive body of mathematical theory around lattices, which depends on their being closed at the bottom and at the top.

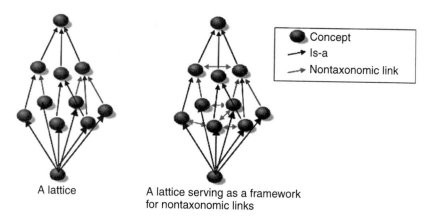

A lattice

A lattice serving as a framework
for nontaxonomic links

Figure A directed acyclic graph.

LAUE METHOD
Liz P. Carpenter

The Laue method is an X-ray crystallography technique for studying structures of molecules in three dimensions using a beam of X rays with more than one wavelength. The usual X-ray experiment uses only one wavelength, and a limited number of diffraction spots are obtained per image. By using a spectrum of wavelengths, more data can be recorded on each image. This allows for more rapid data collection provided that the diffraction spots are not superimposed.

The Laue method has been used for very rapid data collection on crystals with an unstable small molecule, substrate, or intermediate bound in the active site. Data can be collected on a millisecond time scale.

Further Reading

Amoros JL, Buerger MJ, Amoros MC (1975). *The Laue Method* Academic, New York.

Drenth J (1999). *Principles of Protein X-ray Crystallography*, 2nd ed. Springer Verlag, New York.

See also Diffraction of X Rays, X-Ray Crystallography.

LCR. *SEE* LOCUS CONTROL REGION.

LD. *SEE* LINKAGE DISEQUILIBRIUM.

LEAVE- ONE- OUT. *SEE* CROSS-VALIDATION.

LEUCINE
Jeremy Baum

Leucine is a nonpolar amino acid with side chain $—CH_2CH(CH_3)_2$ found in proteins. In sequences, written as Leu or L.

See also Amino Acid.

LEXICON
Robert Stevens

The collection of terms used to refer to the concepts of an ontology. This can also be thought of as the vocabulary delivered by an ontology. It is often important to make concepts independent of any one language; for example the concept Leg would be linked to "leg" in an English lexicon and "jambe" in a French lexicon. The lexicon is also where information about synonyms and other linguistic information is held. (Not all ontologies distinguish clearly between the lexicon and the ontology proper. The practice is, however, to be encouraged.)

L–G ALGORITHM. *SEE* LANDER–GREEN ALGORITHM.

LINE (LONG INTERSPERSED NUCLEAR ELEMENT)
Dov S. Greenbaum

A class of large transposable elements common in the human genome.

Long interspersed elements differ from other retrotransposons in that they do not have long terminal direct repeats (LTRs). In humans, LINEs are one of the major sources of insertional mutagenesis in both somatic and germ cells.

There are two major LINE transposable elements in the human genome, L1 and L2. The approximately half a million repeats take up more than 15% of the human genome.

While other transposons have enhancers and promoters within their LTRs, LINEs differ in that they only have internal promoters.

LINES come in two flavors: full-length copies of these elements and derivatives that have lost some of their sequence. The full-length copies (approximately 6 kb) contain two open reading frames: one that is similar to the viral gag gene and another with sequence similarity to reverse transcriptase. These genes allow LINEs to retrotranspose themselves and, to a much smaller degree, other DNA sequences, resulting in processed pseudogenes.

Further Reading

Cooper DN (1999). *Human Gene Evolution*. Bios Scientific, Oxford, UK, pp. 265–285.

Kazazian HHJ, Moran JV (1998). The impact of L1 retrotransposons on the human genome. *Nat. Genet.* 19: 19–24.

Moran JV, et al. (1996). High frequency retrotransposition in cultured mammalian cells. *Cell* 87: 917–927.

Ridley M (1996). *Evolution*. Blackwell Science, pp. 265–276.

Rowold DJ, Herrera RJ (2000). Alu elements and the human genome. *Genetics* 108: 57–72.

See also L1, SINE.

LINEAR DISCRIMINANT ANALYSIS. *SEE* FISHER DISCRIMINANT ANALYSIS.

LINKAGE (GENETIC LINKAGE)
Mark McCarthy and Steven Wiltshire

Genetic linkage describes the phenomenon whereby loci lying on the same chromosome do not show independent segregation during meiosis.

Assume that an individual inherits, at two loci, A and B, alleles A_p and B_p from its father and A_m and B_m from its mother. If the individual produces gametes which are nonrecombinant (containing either A_p and B_p or A_m and B_m) and recombinant (containing either A_p and B_m or A_m and B_p) in equal proportions, loci A and B are being inherited independently of one another, and there is no linkage between them. In other words, they are unlinked and the probability of recombination between them, designated θ, is 0.5. We may conclude that these two loci are on separate chromosomes (or possibly far apart on the same chromosome). If nonrecombinant gametes are produced in greater proportion to recombinant gametes, the segregation of loci A and B is not independent, and there is genetic linkage between them. The recombination fraction between them, θ, is < 0.5, and we conclude that the two loci are on the same chromosome. When $\theta = 0$, the loci concerned are in complete genetic linkage: There is no recombination between them, and only nonrecombinant gametes are produced. The proportion of recombinant gametes increases with the "genetic" distance between the loci, reflecting the increased probability that a crossover event will have intervened during meiosis. The evidence for linkage between loci can be quantified with a LOD (logarithm-of-odds) score. Detecting and quantifying the evidence for linkage between loci form the basis of linkage analysis but require that it be possible to distinguish between the alleles at each locus.

Example: In a genomewide parametric linkage analysis of a large Australian aboriginal pedigree for loci influencing susceptibility to type 2 diabetes, Busfield et al. (2002) detected linkage to marker *D2S2345* with a maximum two-point LOD score of 2.97 at a recombination fraction of 0.01.

Further Reading
Busfield F, et al. (2002). A genomewide search for type 2 diabetes-susceptibility genes in indigenous Australians. *Am. J. Hum. Genet.* 70: 349–357.

Lander ES, Schork NJ (1994). Genetic dissection of complex traits. *Science* 265: 2037–2048.

Ott J (1999). *Analysis of Human Genetic Linkage.* Johns Hopkins University Press, Baltimore, MD, pp. 1–23.

See also Linkage Analysis, LOD Score, Recombination.

LINKAGE ANALYSIS
Mark McCarthy and Steven Wiltshire

Method for identifying the genomic position of genetic loci which relies on the fact that genes which lie close to each other (i.e., are in linkage) are only rarely separated by meiotic recombination and will therefore cosegregate within pedigrees.

By charting the segregation of variant sites within families, it becomes possible to identify loci which are in linkage and determine their relative genomic location. One use of such methods has been to build genetic maps of polymorphic markers. Given such marker maps, it becomes possible to define the chromosomal location of loci of unknown position, such as disease susceptibility genes, by comparing the pattern of segregation of disease (and by inference of disease susceptibility genes) with that of markers of known position. Broadly, there are two main flavors of linkage analysis. Parametric linkage analysis, typically used in the analysis of Mendelian diseases, requires specification of a credible disease locus model (in which parameters such as dominance, allele frequency, and penetrance are prescribed), and the evidence for linkage (typically expressed as a LOD, or logarithm-of-odds, score) is maximized with respect to the recombination distance between the loci. Software programs such as LINKAGE and GENEHUNTER are available for this. Nonparametric linkage analysis, generally favored for multifactorial traits, requires no prior specification of a disease locus model and seeks to detect genomic regions at which phenotypically similar relatives show more genetic similarity than expected by chance (see allele-sharing methods).

Examples: In their classical paper, Gusella and colleagues (1983) demonstrated linkage between Huntington's disease and a marker on chromosome 4q. A decade later, positional cloning efforts finally identified the etiological gene, huntingtin. Hanis and colleagues (1996) reported their genome scan for type 2 diabetes in Mexican American families that identified a significant linkage to diabetes on chromosome 2q. Subsequent work has indicated that variation in the calpain-10 gene is the strongest candidate for explaining this linkage.

Related Websites
Programs for linkage analysis:

LINKAGE	ftp://linkage.rockefeller.edu/software/linkage/
GENEHUNTER	http://www.fhcrc.org/labs/kruglyak/Downloads/
Merlin	http://www.sph.umich.edu/csg/abecasis/Merlin/index.html
Jurg Ott's Linkage	http://linkage.rockefeller.edu/

Further Reading
Gusella JF, et al. (1983). A polymorphic DNA marker genetically linked to Huntington's disease. *Nature* 306: 234–238.

Hanis CL, et al. (1996). A genome-wide search for human non-insulin-dependent (type 2) diabetes genes reveals a major susceptibility locus on chromosome 2. *Nat. Genet.* 13: 161–171.

Kong A, et al. (2002). A high-resolution recombination map of the human genome. *Nat. Genet.* 31: 241–247.

Kruglyak L, et al. (1996). Parametric and non-parametric linkage analysis: A unified multipoint approach. *Am. J. Hum. Genet.* 58: 1347–1363.

Lander ES, Schork NJ (1994). Genetic dissection of complex traits. *Science* 265: 2037–2048.

Morton NE (1956). The detection and estimation of linkage between the genes for elliptocytosis and the Rh blood type. *Am. J. Hum. Genet.* 8: 80–96.

Ott J (1999). *Analysis of Human Genetic Linkage*, 3rd ed. Johns Hopkins University Press, Baltimore, MD.

Sham P (1998). *Statistics in Human Genetics*. Arnold, London, pp. 51–144.

See also Allele-Sharing Methods, Genome Scans, Linkage, Multipoint Linkage Analysis.

LINKAGE DISEQUILIBRIUM (GAMETIC PHASE DISEQUILIBRIUM, LD)

Mark McCarthy and Steven Wiltshire

The statistical association of alleles at two (or more) polymorphic loci.

Consider two biallelic loci, with alleles A,a (frequencies P_A, P_a) and B,b (frequencies P_B, P_b). If, within a population of individuals (or chromosomes), alleles at these two loci are seen to be independent of one other, then the frequency of the haplotype AB (P_{AB}) will be the product of the respective allele frequencies, $P_A P_B$. However, if the alleles are not independent of one another, the frequency of the haplotype AB will differ from $P_A P_B$ by a nonzero amount, D_{AB}:

$$D_{AB} = P_{AB} - P_A P_B$$

If $D_{AB} \neq 0$, then the alleles at loci A and B are in linkage, or gametic phase, disequilibrium; if $D_{AB} = 0$, then the alleles are said to be in linkage equilibrium. For the example of two biallelic markers, $D_{ab} = D_{AB}$ and $D_{aB} = D_{Ab} = -(D_{AB})$. Additional linkage disequilibrium parameters need to be calculated for markers with more than two alleles and for considerations of more than two markers.

A host of different measures of LD exist with differing properties. Often, the extent of LD is expressed as a proportion, D', of its maximum or minimum possible values given the frequencies of the alleles concerned:

$$D'_{AB} = D_{AB} / \max(-P_A P_B, -P_a P_b) \text{ if } D_{AB} < 0 \text{ or } D'_{AB}$$
$$= D_{AB} / \min(P_a P_B, P_A P_b) \text{ if } D_{AB} > 0.$$

A second useful measure of LD is r^2, calculated from D; thus

$$r^2 = D^2 / P_A P_a P_B P_b$$

where $r^2 N$ is distributed as χ^2, where N is the total number of chromosomes.

LD is a feature of a particular population of individuals (or more strictly, their chromosomes) and arises first through mutation (which creates a new haplotype) and then is sustained and bolstered by events and processes modifying the genetic composition of a population during its history—these include periods of small population size ("bottlenecks"), genetic admixture (due to interbreeding with a distinct population), and stochastic effects ("genetic drift"). At the same time, any LD established is gradually dissipated by the actions of recombination and further mutation. In most current human populations, LD extends over a typically short range (tens of kilobases).

Note that, although the term linkage disequilibrium is often reserved to describe associations between linked loci, it is sometimes used more broadly, in which case it equates to allelic association. Association (or LD) mapping exploits LD (in the stricter sense) to map disease susceptibility genes.

Examples: Abecasis et al. (2001) calculated pairwise between markers in three genomic regions and observed patterns of measurable LD extending for distances exceeding 50 kb. Gabriel et al. (2002) and Dawson et al. (2002) have used pairwise measures of LD to investigate fine-scale LD structure of human chromosomes.

Related Websites
Programs for calculating D from population data:

EH-PLUS	http://www.iop.kcl.ac.uk/IoP/Departments/PsychMed/GEpiBSt/software.stm
ARLEQUIN	http://anthropologie.unige.ch/arlequin/
GOLD	http://www.sph.umich.edu/csg/abecasis/GOLD/

Further Reading

Abecasis GR, et al. (2001). Extent and distribution of linkage disequilibrium in three genomic regions. *Am. J. Hum. Genet.* 68: 191–197.

Dawson E, et al. (2002). A first-generation linkage disequilibrium map of human chromosome 22. *Nature* 418: 544–548.

Devlin B, Risch N (1995). A comparison of linkage disequilibrium measures for fine-scale mapping. *Genomics* 29: 311–322.

Gabriel SB, et al. (2002). The structure of haplotype blocks in the human genome. *Science* 296: 2225–2229.

Ott J (1999). *Analysis of Human Genetic Linkage.* Johns Hopkins University Press, Baltimore, MD, pp. 280–291.

Weir BS (1996). *Genetic Data Analysis*, Vol. 2. Sinauer Associates, Sunderland, MA, pp. 112–133.

See also Allelic Association, Haplotype, Phase, Recombination.

LINKAGE DISEQUILIBRIUM ANALYSIS. *SEE* ASSOCIATION ANALYSIS.

LOCAL ALIGNMENT (LOCAL SIMILARITY)
Jaap Heringa

An alignment of the most similar consecutive segments of two or more sequences, as opposed to global alignment, which is carried out over complete sequences.

A local alignment should be attempted when the target sequences have different lengths, so that different domain organizations or different repeat copy numbers could be present. Local alignment is also the method of choice when permuted domains are suspected. It is further appropriate when sequences are extremely divergent such that evolutionary memory might be retained in some local fragments only. The first automatic method to generate a local alignment between two protein sequences was devised by Smith and Waterman (1981) as an adaptation of the dynamic programming technique developed by Needleman and Wunsch (1970) for global alignment. Fast sequence database search methods, such as BLAST and FASTA, are heuristic approximations of the Smith–Waterman local alignment technique. The Smith–Waterman algorithm has been extended in various techniques to compute a list of top-scoring pairwise local alignments (Huang and Miller, 1991; Huang et al., 1990; Waterman and Eggert, 1987). Alignments produced by the latter techniques are nonintersecting; i.e., they have no matched pair of amino acids in common.

Related Websites

SIM	http://genome.cs.mtu.edu/align/align.html
LAP2	http://genome.cs.mtu.edu/align/align.html

Further Reading

Huang X, Miller W (1991). A time-efficient, linear-space local similarity algorithm. *Adv. Appl. Math.* 12: 337–357.

Huang X, et al. (1990). A space-efficient algorithm for local similarities. *Comput. Appl. Biosci.* 6: 373–381.

Needleman SB, Wunsch CD (1970). A general method applicable to the search for similarities in the amino acid sequence of two proteins. *J. Mol. Biol.* 48: 443–453.

Smith TF, Waterman MS (1981). Identification of common molecular subsequences. *J. Mol. Biol.* 147: 195–197.

Waterman MS, Eggert M (1987). A new algorithm for best subsequences alignment with applications to the tRNA-rRNA comparisons. *J. Mol. Biol.* 197: 723–728.

See also Alignment—Multiple, Alignment of Protein Sequences, Pairwise Alignment.

LOCAL SIMILARITY. *SEE* LOCAL ALIGNMENT.

LOCUS CONTROL REGION (LCR)
Niall Dillon

Operational definition which describes a set of sequences that give transcription from a linked gene in transgenic mice that is (1) at a level per transgene copy that is similar to that observed for the endogenous genes, (2) independent of the position of integration of the gene in the mouse genome, and (3) observed in transgenics that carry a single copy of the transgene. LCRs are generally associated with the presence of multiple DNase I hypersensitive sites.

Further Reading
Fraser P, Grosveld F (1998). Locus control regions, chromatin activation and transcription. *Curr. Opin. Cell. Biol.* 10: 361–365.

Grosveld F, et al. (1987). Position-independent, high-level expression of the human beta-globin gene in transgenic mice. *Cell* 51: 975–985.

LOCUSLINK. *SEE* UNIGENE/LOCUSLINK.

LOCUS REPEAT. *SEE* INTERSPERSED SEQUENCE.

LOD SCORE (LOGARITHM-OF-ODDS SCORE)
Mark McCarthy and Steven Wiltshire

Logarithm of odds for linkage between two loci against no linkage.

LOD score is a general term for the decimal logarithm of a likelihood-ratio-based linkage statistic and is a measure of the evidence for linkage. The precise nature of the LOD score depends on the nature of the analysis, namely parametric or nonparametric linkage analysis.

In parametric linkage analysis, typically used in the study of Mendelian diseases, the classical LOD score, $Z(\theta)$, is given by:

$$Z(\theta) = \log_{10}[L(\theta < 0.5)/L(\theta = 0.5)]$$

where $L(\theta)$ is the likelihood of the data given the value of θ, the recombination fraction between the two loci. The LOD score reaches its maximum value, Z_{max}, at the maximum-likelihood estimate of θ. Several variants of the parametric LOD score exist, including the HLOD (heterogeneity LOD), in which a proportion of the pedigrees analyzed are modeled as having no linkage between the two loci; and ELOD (expected LOD), which yields a measure of the informativeness of the pedigree for linkage. Generally, a parametric LOD score of 3 or more is taken as a significant evidence for linkage.

There are two common LOD score statistics in use in nonparametric, allele-sharing methods of qualitative trait linkage analysis: the *maximum LOD score* (or MLS) and the *allele-sharing LOD score* (or LOD*). The MLS is based on the likelihood ratio of observed allele sharing for zero, one, and two alleles IBD (identical by descent) to that expected under the null hypothesis of no linkage:

$$\text{MLS} = \log_{10}[(z_0^{n0} z_1^{n1} z_2^{n2})/(0.25^{n0} 0.50^{n1} 0.25^{n2})]$$

where z_0, z_1, and z_2 are the allele-sharing proportions for zero, one, and two alleles IBD under the alternative hypothesis and n_0, n_1, and n_2 are the number of sib-pairs falling into each of the three IBD categories. The MLS can be maximized over one (z_0) or two (z_0 and z_1) parameters depending on whether dominance at the trait locus is being modeled. The allele-sharing LOD score is a reparameterization of the nonparametric linkage score calculated by the program GENEHUNTER in terms of an allele-sharing parameter, $\hat{\delta}$. Two versions of this reparameterization exist—linear and exponential.

One of the two variants of linkage analysis of quantitative traits, namely variance components analysis, also uses LOD scores to measure the evidence for linkage of a quantitative trait locus to marker loci.

Examples: Busfield et al. (2002) used parametric linkage analysis in a large aboriginal Australian population to detect evidence for linkage to type 2 diabetes susceptibility loci, quoting classical LOD scores. Pajukanta et al. (2000) used allele-sharing methods to perform a genome scan for loci influencing premature coronary heart disease in Finnish populations, quoting MLS scores. Hanna et al. (2000) used both parametric and nonparametric linkage analyses to detect loci in a genome scan for susceptibility loci influencing obsessive-compulsive disorder, quoting both classical parametric LOD score and nonparametric allele-sharing LOD scores.

Further Reading
Busfield F, et al. (2002). A genomewide search for type 2 diabetes-susceptibility genes in indigenous Australians. *Am. J. Hum. Genet.* 70: 349–357.

Hanna GL, et al. (2002). Genome-wide linkage analysis of families with obsessive-compulsive disorder ascertained through pediatric probands. *Am. J. Med. Genet.* 114: 541–552.

Kong A, Cox NJ (1997). Allele-sharing models: LOD scores and accurate linkage tests. *Am. J. Hum. Genet.* 61: 1179–1188.

Nyholt DR (2000). All LODs are not created equal. *Am. J. Hum. Genet.* 67: 282–288.

Ott J (1999). *Analysis of Human Genetic Linkage*. Johns Hopkins University Press, Baltimore, MD.

Pajukanta P, et al. (2000). Two loci on chromosomes 2 and X for premature coronary heart disease identified in early- and late-settlement populations of Finland. *Am. J. Hum. Genet.* 67: 1481–1493.

See also Linkage, Linkage Analysis, Recombination.

LOGARITHM-OF-ODDS SCORE. *SEE* LOD SCORE.

LOGDET. *SEE* PARALINEAR DISTANCE.

LOG-ODDS SCORE. *SEE* AMINO ACID EXCHANGE MATRIX.

LONG-PERIOD INTERSPERSION, LONG-TERM INTERSPERSION. *SEE* INTERSPERSED SEQUENCE.

LOOK-UP GENE PREDICTION. *SEE* GENE PREDICTION, HOMOLOGY BASED.

LOOP
Marketa J. Zvelebil

Secondary structures such as α-helices and β-strands or sheets are connected to each other by structural segments called loops. Although sometimes referred to as coils, loops often have at least some definite structure while coil regions are unstructured. Insertions and deletions (indels) often occur in loop (or coil) regions. Loops are often found on the surface of the protein (but not always) and are therefore rich in polar or charged residues. Loops often are involved in ligand binding or form part of an active site of the protein.

Related Websites

Loop Database	http://mdl.ipc.pku.edu.cn/moldes/oldmem/liwz/home/loop.html
Protein loop classification	http://ibb.uab.es/loops/

Further Reading
Berezovsky IN, Grosberg AY, Trifonov EN (2000). Closed loops of nearly standard size: Common basic element of protein structure. *FEBS Lett.* 466: 283–286.
Ring CS, Cohen FE (1994). Conformational sampling of loop structures using genetic algorithms. *Israel J. Chem.* 34: 245–252.
Ring CS, Kneller DG, Langridge R, Cohen FE (1992). Taxonomy and conformational analysis of loops in proteins. *J. Mol. Biol.* 224: 685–699.

See also Coil, Indel, Loop Prediction/Modeling.

LOOP PREDICTION/MODELING
Roland L. Dunbrack

A step in comparative modeling of protein structures in which loop or coil segments between secondary structures are built. Alignment of a target sequence to be modeled and a structure to be used as a template or parent will produce insertions and deletions most commonly in coil segments of the template protein. Since the target and template loops are of different lengths, loops of the correct length and sequence must be constructed and fitted onto the template. This is performed first by identifying anchor points, or residues that will remain fixed at or near their Cartesian coordinate positions in the template structure. The loop is then constructed to connect the anchor points with the appropriate sequence.

Most loop prediction methods can be classified as either database methods or construction methods. In database methods, the Protein Data Bank is searched for loops of the correct length that will span the anchor points within some tolerance. The loop is then positioned and adjusted to fit the actual anchors in the template structure. In construction methods, a segment is built, not with reference to a loop in a known structure, but rather as one that will fit the anchor points with reasonable stereochemical quality. This can be performed with Monte Carlo simulations, molecular dynamics, or energy minimization techniques.

Related Websites

SLOOP	http://www-cryst.bioc.cam.ac.uk/~sloop/Info.html
Jackal	http://trantor.bioc.columbia.edu/~xiang/jackal/

Further Reading
Fiser A, et al. (2000). Modeling of loops in protein structures. *Prot. Sci.* 9: 1753–1773.

Moult J, James MNG (1986). An algorithm for determining the conformation of polypeptide segments in proteins by systematic search. *Proteins* 1: 146–163.

Vlijmen HWTv, Karplus M (1997). PDB-based protein loop prediction: Parameters for selection and methods for optimization. *J. Mol. Biol.* 267: 975–1001.

Xiang Z, et al. (2002). Extending the accuracy limits of prediction for side-chain conformations. *Proc. Natl. Acad. Sci. USA* 99: 7432–7437.

See also Anchor Points, Comparative Modeling, Loop, Target.

LOW-COMPLEXITY REGION
Jaap Heringa and Patrick Aloy

A low-complexity region within a protein or nucleotide sequence is a sequence region with a biased amino acid or nucleotide composition that differs significantly from the general composition observed in either the sequence or a database of sequences.

Low-complexity regions are a major source of alignment error, particularly with amino acid sequences (Altschul et al., 1994). If such a region within a query sequence is included during a homology search of that query sequence against a sequence database, biologically unrelated database sequences containing similarly biased regions are likely to score spuriously high, often rendering the search meaningless. For this reason, the PSI-BLAST program filters out biased regions of query sequences by default using the segment sequences (SEG) method (Wootton and Federhen., 1993, 1996). Because the SEG parameters have been set conservatively to avoid masking potentially important regions, some bias may remain so that compositionally biased artifactual hits can still occur, e.g., with protein sequences that have a known bias, such as myosins or collagens. The SEG filtering method can be used with parameters that eliminate nearly all biased regions, and the user can also apply other filtering procedures, such as COILS (Lupas, 1996), which delineates putative coiled-coil regions, before submitting the appropriately masked sequence to PSI-BLAST. Low-complexity sequences found by filtering are substituted by PSI-BLAST using the letter N in nucleotide sequences (e.g., NNNNNNNNNNNNN) and the letter X in protein sequences (e.g., XXXXXXXXX).

A well-known low-complexity region is the poly-glutamine repeat which forms insoluble fibers of beta sheets that can cause neurodegenerative diseases.

Related Website

Low complexity	http://www.ncbi.nlm.nih.gov/BLAST/blast_FAQs.html#LCR
SEG Filter	ftp://ncbi.nlm.nih.gov/pub/seg/seg/

Further Reading
Altschul SF, et al. (1994). Issues in searching molecular sequence databases. *Nat. Genet.* 6: 119–129.
Lupas A (1996). Prediction and analysis of coiled-coil structures. *Methods Enzymol.* 266: 513–525.
Perutz MF (1999). Glutamine repeats and neurodegenerative diseases: Molecular aspects. *Trends Biochem. Sci.* 24: 58–63.
Wootton JC, Federhen S (1993). Statistics of local complexity in amino acid sequences and sequence databases. *Comput. Chem.* 17: 149–163.
Wootton JC, Federhen S (1996). Analysis of compositionally biased regions in sequence databases. *Methods Enzymol.* 266: 554–571.

See also BLAST, Sequence Complexity.

LUCA. *SEE* ANCESTRAL GENOME.

LYSINE
Jeremy Baum

Lysine is a positively charged amino acid with side chain $—(CH_2)_4NH_3$ found in proteins. In sequences, written as Lys or K.

See also Amino Acid.

M

MacClade
Machine Learning
Macroarray
MacroModel
MacVector
MAD Phasing
Main Chain
Majority-Rule Consensus Tree
MALDI-TOF-MS
Map Function
Mapping By Admixture Linkage
   Disequilibrium
MAR
Marker
Markov Chain
Masked Sequence
Mathematical Modeling of
   Molecular/Metabolic/Genetic Networks
MatInspector
Matrix Attachment Region
Maximal Margin Classifiers
Maximum-Likelihood Phylogeny
   Reconstruction
Maximum-Parsimony Principle
MaxSprout
MB
MEGA
Megabase
Mendelian Disease
MEROPS
Message
Messenger RNA
Metabolic Network
Metabolome
Metabonome
Metadata
Meta-predict-protein
Methionine
MGD

MGED Ontology
Microarray
Microarray Image Analysis
Microarray Normalization
Microarray Profiling of Gene Expression
Microsatellite
Midnight Zone
MIGRATE
MIME Types
Minimum-Evolution Principle
Minisatellite
MirrorTree
ModBase
Modeling, Macromolecular
ModelInspector
Modeller
Model Order Selection
Model Selection
Models, Molecular
Modeltest
Modular Protein
Modularity
Module Shuffling
MOE
Molecular Clock
Molecular Coevolution
Molecular Drive
Molecular Dynamics Simulations
Molecular Evolutionary Mechanisms
Molecular Information Theory
Molecular Machine
Molecular Machine Capacity
Molecular Machine Operation
Molecular Mechanics
Molecular Network
Molecular Replacement
MOLPHY
Monophyletic
Monte Carlo Simulations

Dictionary of Bioinformatics and Computational Biology. Edited by Hancock and Zvelebil
ISBN 0-471-43622-4 © 2004 John Wiley & Sons, Inc.

MOPAC   Multidomain Protein
Mosaic Protein   Multifactorial Trait
Motif   Multilayer Perceptron
Motif Searching   Multiple Anomalous Dispersion Phasing
Mouse Genome Database   Multiple Hierarchy
MPsrch   Multipoint Linkage Analysis
MrBayes   Mutation Matrix
mRNA

MACCLADE
Michael P. Cummings

A program for the study of phylogenetic trees and character evolution.

The program has a rich set of features for graphical-based tree manipulation and exploration (e.g., move, clip, collapse branches) and examination of character patterns by mapping characters on trees or through charts and diagrams. The program accommodates multiple-input and multiple-output formats for nucleotide and amino acid sequence data as well as general data types and has broad data-editing capabilities. Many options are available for tree formatting and printing.

The program features an extensive graphical user interface and is available as an executable file for the Apple MacIntosh. A lengthy book by Maddison and Maddison (2001) provides extensive details about program features, examples of their use, and background information.

Related Website

| MacClade | http://macclade.org/macclade.html |

Further Reading

Maddison DR, Maddison WP (2001). *MacClade 4: Analysis of Phylogeny and Character Evolution*. Sinauer Associates, Sunderland, MA.

MACHINE LEARNING
Nello Cristianini

Branch of artificial intelligence concerned with developing computer programs that can learn and generalize from examples.

By this is meant the acquisition of domain-specific knowledge, resulting in increased predictive power. Limited to the setting when the examples are all given together at the start, it is a valuable tool for data analysis. Many algorithms have been proposed, all aimed at detecting relations (patterns) in the training data and exploiting them to make reliable predictions about new, unseen data.

Two stages can be distinguished in the use of learning algorithms for data analysis. First a training set of data is provided to the algorithm and used for selecting a "hypothesis." Then such a hypothesis is used to make predictions on unseen data or tested on a set of known data to measure its predictive power.

A major problem in this setting is that of overfitting or overtraining, when the hypothesis selected reflects specific features of the particular training set due to chance and not to the underlying source generating it. This happens mostly with small training samples and leads to reduced predictive power.

Motivated by the need to understand overfitting and generalization, in the last few years significant advances in the mathematical theory of learning algorithms have brought this field very close to certain parts of statistics, and modern machine learning methods tend to be less motivated by heuristics or analogies with biology (as was the case—at least originally—for neural networks or genetic algorithms) and more by theoretical considerations (as is the case for support vector machines and graphical models).

Dictionary of Bioinformatics and Computational Biology. Edited by Hancock and Zvelebil
ISBN 0-471-43622-4 © 2004 John Wiley & Sons, Inc.

The output of a machine learning algorithm is called a hypothesis or sometimes a model. A common type of hypothesis is a classifier, i.e., a function that assigns inputs to one of a finite number of classes.

Further Reading
Cristianini N, Shawe-Taylor J (2000). *An Introduction to Support Vector Machines*. Cambridge University Press, Cambridge, UK.
Duda RO, Hart PE, Stork DG (2001). *Pattern Classification*. Wiley, New York.
Mitchell T (1997). *Machine Learning*. McGraw-Hill, New York.

See also Classification in Machine Learning, Overfitting, Pattern Recognition, Supervised and Unsupervised Learning.

MACROARRAY
John M. Hancock

A microarray-like system in which DNA is spotted onto nylon membranes of relatively large dimensions and radioactive targets are used.

Macroarrays are relatively easier for small laboratories to make, can be reused, and can be quantitated using PhosphorImage technology, which is widely available.

Related Website

| BioArray software | http://strc.herts.ac.uk/bio/pan/BioArray/ |

Further Reading
Hornberg JJ, et al. (2002). Analysis of multiple gene expression array experiments after repetitive hybridizations on nylon membranes. *Biotechniques* 33:108–113.

See also Microarray.

MACROMODEL
Roland L. Dunbrack

A computer program for molecular mechanics calculations on organic molecules and proteins distributed by Schrödinger. MacroModel provides access to the AMBER and OPLS force fields, among others.

Related Website

| MacroModel | http://www.schrodinger.com/Products/macromodel.html |

Further Reading
Mohamadi F, et al. (1990). MacroModel: An integrated software system for modeling organic and bioorganic molecules using molecular mechanics. *J. Comp. Chem.* 11: 440–467.

MACVECTOR
John M. Hancock

A desktop package for Macintosh computers that provides basic sequence manipulation and analysis facilities.

The package provides many of the functions of the Wisconsin Package with a more user-friendly graphical user interface. Functions include searching remote databases over the Internet, multiple-sequence analysis, evolutionary analysis, primer analysis, mapping, motif searching, protein analysis, and file import and export.

Related Website

Commercial	http://www.accelrys.com/products/macvector/index.html

See also LaserGene, VectorNTI, Wisconsin Package.

MAD PHASING. *SEE* MULTIPLE ANOMALOUS DISPERSION PHASING.

MAIN CHAIN
Jeremy Baum

The set of all backbone atoms in a single peptide chain.

MAJORITY-RULE CONSENSUS TREE. *SEE* CONSENSUS TREE.

MALDI-TOF-MS (MASS SPECTROSCOPY)
Jean-Michel Claverie and Dov S. Greenbaum

Technology used in proteomics for the characterization of unknown proteins.

Matrix-assisted laser desorption/ionization time-of-flight mass spectrometry is used for the ionization of large polypeptides and, as such, has been used to analyze protein fragments extracted from two-dimensional gels.

MALDI, developed in the 1970s and early 1980s, excites a matrix [of laser light-absorbing small organic molecules (e.g., sinapinic acid) which has been dried together with the protein target] with a high-intensity but short laser pulse (nitrogen, ultraviolet, or infrared). The matrix prevents the analyte from degradation, thus allowing for analysis of large molecules. The matrix, absorbing the energy, is vaporized and ionized into a dense plume. These ions then enter a vacuum, the flight tube, and are accelerated by a strong field. An instrument then measures the time of flight (TOF) for the individual ions to arrive at the detector. Given that the TOF is related to the mass-to-charge ratio of the ion (it is proportional to the square root of their m/z), a mass spectrum can be generated. These results are then subjected to a database search of protein data to determine what protein was being analyzed.

The technique is used in conjunction with two-dimensional gel electrophoresis to identify and (quantify) the protein spots in proteomics experiments. The spots of interest are excised and submitted to in situ proteolysis, and the cleavage product is extracted from the gel matrix, purified, and concentrated before being submitted to mass spectrometry analysis. Proteins (spots) are then identified in sequence databases by mass spectrometric peptide mapping. State-of-the-art techniques now combine liquid chromatography with electrospray–ionization tandem mass spectrometry. Tandem mass spectra contain highly specific information on the fragmentation pattern as well as sequence information. This information has been used to search databases of translated protein sequences as well as nucleotide databases such as expressed sequence tags (ESTs). With these techniques, it is now possible to perform high-throughput protein identification at picomolar-to-subpicomolar levels from protein mixtures.

Related Websites

Mass Spectroscopy home page	http://i-mass.com/
PROWL software and database	http://prowl.rockefeller.edu/
C M. Smith's Resources page	http://restools.sdsc.edu/
Tutorial	http://ms.mc.vanderbilt.edu/tutorial.htm
MALDI	http://www.srsmaldi.com/Maldi/Maldi.html

Further Reading

Dongre AR, et al. (1997). Emerging tandem-mass-spectrometry techniques for the rapid identification of proteins. *Trends Biotechnol.* 15: 418–425.

Hamdan M, Galvani M, Righetti PG (2001). Monitoring 2-D gel-induced modifications of proteins by MALDI-TOF mass spectrometry. *Mass Spectrom. Rev.* 20: 121–141.

Jonsson AP (2001). Mass spectrometry for protein and peptide characterisation. *Cell. Mol. Life Sci.* 58: 868–884.

Kowalski P, Stoerker J (2000). Accelerating discoveries in the proteome and genome with MALDI TOF MS. *Pharmacogenomics* 1: 359–366.

Nordhoff E, et al. (2001). Large-gel two-dimensional electrophoresis-matrix assisted laser desorption/ionization-time of flight-mass spectrometry: An analytical challenge for studying complex protein mixtures. *Electrophoresis* 22: 2844–2855.

Pevzner PA, et al. (2001). Efficiency of database search for identification of mutated and modified proteins via mass spectrometry. *Genome Res.* 11: 290–299.

Yates JR III (1998). Database searching using mass spectrometry data. *Electrophoresis* 19: 893–900.

MAP FUNCTION
Mark McCarthy and Steven Wiltshire

A mathematical formula for converting between a recombination fraction and a genetic, or map, distance, measured in Morgans (M) or centimorgans (cM).

A map function describes the mathematical relationship, for two loci under study, between the nonadditive recombination fraction, θ, and the additive map distance, x. The map distance is the mean number of crossovers occurring between the loci on a single chromatid per meiosis. The map function is required because θ has a maximum of 0.5, while, on larger chromosomes at least, several crossovers are possible. The most commonly used map functions differ in the extent to which they allow for the phenomenon of interference (i.e., the tendency for crossover events rarely to occur in close proximity to each other). Haldane's map function assumes no interference (i.e., all crossovers are independent) and is given by

$$x_H = -\frac{1}{2} \ln(1 - 2\theta)$$

for $0 \le x_H < 0.5$, ∞ otherwise, with the inverse being

$$\theta = \frac{1}{2}[1 - \exp(-2|x_H|)]$$

Kosambi's map function assumes a variable level of interference, expressed as 2θ:

$$x_K = \frac{1}{2} \ln\left(\frac{1 + 2\theta}{1 - 2\theta}\right)$$

for $0 \le x_K < 0.5$, ∞ otherwise, with the inverse being

$$\theta = \frac{1}{2} \frac{\exp(4x_K) - 1}{\exp(4x_K) + 1}$$

Other map functions exist, differing in their modeling of interference (Carter–Falconer, Felsenstein) or the assumptions regarding the distribution of chiasmata between loci (Sturt). However, Kosambi's and Haldane's map functions, described above, are the most widely used, with Kosambi's function tending to give more realistic distances, although at small values of θ, there is little practical difference between the two.

Examples: Consider two loci with $\theta = 0.3$. Haldane's map function gives $x_H = 45.8$ cM, whereas Kosambi's function gives $x_K = 34.7$ cM. If the recombination fraction were 0.03, the distances would be $x_H = 3.1$ cM and $x_K = 3.0$ cM.

Kosambi's map function has been used in the construction of Généthon (Dib et al. 1996), Marshfield (Broman et al. 1998), and deCODE (Kong et al. 2002) marker maps. Haldane's map function is used in programs such as GENEHUNTER, ALLEGRO, and MERLIN during multipoint IBD (identity-by-descent) estimation between markers since the Lander–Green algorithm, used by these programs to calculate pedigree likelihoods, implicitly considers recombination events in adjacent intervals to be independent (i.e., no interference).

Further Reading

Broman KW, et al. (1998). Comprehensive human genetic maps: Individual and sex-specific variation in recombination. *Am. J. Hum. Genet.* 63: 861–869.

Dib C, et al. (1996). A comprehensive genetic map of the human genome based on 5,264 microsatellites. *Nature* 380: 152–154.

Kong A, et al. (2002). A high-resolution recombination map of the human genome. *Nat. Genet.* 31: 241–247.

Ott J (1999). *Analysis of Human Genetic Linkage.* Johns Hopkins University Press, Baltimore, MD, pp. 17–21.

Sham P (1998). *Statistics in Human Genetics.* Arnold, London, pp. 54–58.

See also Linkage Analysis, Marker, Recombination.

MAPPING BY ADMIXTURE LINKAGE DISEQUILIBRIUM. *SEE* ADMIXTURE MAPPING.

MAR. *SEE* MATRIX ATTACHMENT REGION.

MARKER
Mark McCarthy and Steven Wiltshire

Any polymorphism of known genomic location such that it can be used for disease–gene mapping.

Polymorphic sites, whether or not they represent sites of functional variation, are essential tools for gene-mapping efforts as they can be used to follow chromosomal segregation and infer the likely position of recombinant events within pedigrees and populations. To be useful, markers must (a) have a known and unique chromosomal location; (b) show at least a moderate degree of polymorphism; (c) have a low mutation rate; and (d) be easily typed (most usually by PCR-based methods). The types of markers most frequently used in genetic analyses are microsatellites (usually di-, tri-, or tetra-nucleotide repeats), which are often highly polymorphic and well suited to genomewide linkage analysis, and single-nucleotide polymorphisms (SNPs), which, though less polymorphic, are more frequent, are thought to represent most functional variation, and are best suited to association analysis.

Examples: Researchers in Iceland (Kong et al., 2002) recently published a high-density genetic map of over 5000 microsatellites, providing a dense framework of robust and polymorphic markers suitable for linkage studies. In addition, well over a million SNPs have been placed onto the human genome assembly (Sachidanandam et al., 2001), and these provide a valuable resource for association analyses.

Related Websites

Genethon microsatellite map	ftp://ftp.genethon.fr/pub/Gmap/Nature-1995/
Marshfield microsatellite map	http://www.marshfieldclinic.org/research/genetics/Map_Markers/maps/IndexMapFrames.html
DECODE microsatellite marker map	http://www.nature.com/cgi-taf/DynaPage.taf?file=/ng/journal/v31/n3/full/ng917.html
DbSNP (database of SNPs)	http://www.ncbi.nlm.nih.gov/SNP/

Further Reading

Dib C, et al. (1996). A comprehensive genetic map of the human genome based on 5,264 microsatellites. *Nature* 380: 152–154.

Kong A, et al. (2002). A high-resolution recombination map of the human genome. *Nat. Genet.* 31: 241–247.

Kruglyak L (1997). The use of a genetic map of biallelic markers in linkage studies. *Nat. Genet.* 17: 21–24.

Sachidanandam R, et al. (2001). A map of human genome sequence variation containing 1.42 million single nucleotide polymorphisms. *Nature* 409: 928–933.

Utah Marker Development Group (1995). A collection of ordered tetranucleotide-repeat markers from the human genome. *Am. J. Hum. Genet.* 57: 619–628.

See also Genome Scans, Polymorphism.

MARKOV CHAIN
John M. Hancock

A Markov chain is a series of observations in which the probability of an observation occurring is a function of the previous observation or observations.

A DNA sequence can be considered to be an example of a Markov chain as the likelihood of observing a base at a particular position depends strongly on the nature of the preceding base. This is the basis of the nonrandom dinucleotide frequency distribution of most DNA sequences.

Strictly, the above definition defines a first-order Markov chain (Markov chain of order 1). Higher order Markov chains can also be defined, in which the probability of an observation depends on a larger number of preceding observations. Again, DNA sequences can be taken as an example as base frequencies are strongly influenced by the preceding five bases, making DNA sequences equivalent to Markov chains of order 5.

The process of generating a Markov chain is known as a Markov process.

Related Website

Markov Chain Primer	http://crypto.mat.sbg.ac.at/~ste/diss/node6.html

See also Dinucleotide Frequency.

MASKED SEQUENCE. *SEE* REPEATMASKER

MATHEMATICAL MODELING OF MOLECULAR/ METABOLIC/GENETIC NETWORKS
Denis Thieffry

To integrate data on molecular, metabolic, or genetic interactions between individual components, one can write mathematical equations modeling the evolution of each molecular species (variable) across time and possibly space. This allows study, simulation, and even predictions about the temporal behavior of the system in response to various types of perturbations. The mathematical formalisms used can be grouped into several classes according to levels of detail (qualitative or quantitative) and basic assumptions (deterministic or stochastic).

In the case of quantitative mathematical modeling, ordinary differential equations (ODEs) are most often used, in particular when dealing with (bio)chemical reactions; these express the evolution (time derivative) of each variable of the system (e.g., the concentration of a molecular species) as a function of other variables of the system. When dealing with regulatory systems, these functions are usually nonlinear, involving products or powers of variables to model cooperative behavior. Nonlinearity complicates the analysis of such sets of equations, and the modeler has to rely on the use of numerical simulation techniques. Parameters (supposedly constant) are included in the equations to modulate the weight of the different terms involved in each equation. On the basis of this formal description, one can select parameter values and an initial state (a set of values for the different variables) to perform a simulation with the help of publicly available software (see Table 1). Simulations can be shown in the form of the evolution of the system as a function of time or in the form of a trajectory in the space defined by the ranges of values for the variables of the system (variable space or phase portrait). Steady states can be located from numerical iteration methods used by physicists.

To represent spatial effects such as diffusion or transport, one can use partial differential equations (PDEs) expressing the variation of the different molecular species in different directions (usually the three axes of the Euclidean space, or referring to polar coordinates). As in the case of ODE systems, it is possible to simulate PDEs, but one then needs to specify boundary conditions in addition to the values of the different parameters of the system. One famous, historical example of PDE application to (bio)chemical problems can be found in the work of Alan Turing (1952) on pattern formation defined by reaction–diffusion systems.

A modeler who wishes to focus on qualitative aspects of the behavior of a regulatory system can use Boolean equations instead of quantitative differential equations. Each component is represented by a Boolean variable which can take only two different values: 0 if the component is absent or inactive, 1 otherwise. The effect of other components on the evolution of a given variable is then described in terms of a logical function built with operators such as AND, OR, etc. When studying the evolution of these logical systems, two opposite treatments of time are used: (1) A synchronous approach considers all equations at once and update all variables

Table 1 Examples of Simulation Software

Name	Type of Simulation/Analysis	Universal Resource Locator
Gepasi	Continuous biochemical simulations	http://gepasi.dbs.aber.ac.uk/softw/gepasi.html
StochSim	Stochastic simulation of molecular networks	http://info.anat.cam.ac.uk/groups/comp-cell/StochSim.html
XPPAUTO	Differential equations and bifurcation analyses	http://www.math.pitt.edu/~bard/xpp/xpp.html
GRIND	Integration of differential equations	http://theory.bio.uu.nl/rdb/grind.html
DDLab	Boolean network simulation	http://www.ddlab.com/
GINsim	Logical simulations and analysis	http://gin.univ-mrs.fr/
GNA	Piecewise linear model simulations	http://bacillus.inrialpes.fr/gna/

simultaneously according to the values calculated on the basis of the functions (left-hand terms in the logical equations). (2) An asynchronous approach consists in selecting only one transition each time the system is in a state in which several variable changes are possible. In the context of these different updating assumptions, the system will present the same stable states, defined as the states where there is equality between the value of the logical variable and the corresponding logical function. To these different updating assumptions, however, correspond different dynamics in terms of transient or cyclic behavior. The logical approach has been generalized to encompass multilevel variables, (i.e., variables taking more than two integer values: 0, 1, 2, . . .) as well as the explicit consideration of threshold values (at the interface between logical values). Finally, logical parameters can be introduced in order to cover families of logical functions.

When studying interactions between components present at very low numbers, it is not possible anymore to assume a continuous range of concentration. It is then necessary to take into account explicitly the probability of encounters between two molecules to describe their interaction. This is done by using stochastic equations. In this context, the system is described as a list of molecular states plus the distributions of probabilities associated to each state transition. As in the case of nonlinear differential equations, stochastic systems can generally not be treated analytically and the modeler has to rely on numerical simulations. Relatively heavy from a computational point of view, such simulations have been performed only for a limited number of experimentally well-defined networks, including the genetic network controlling the lysis–lysogeny decision in bacteriophage lambda and the phosphorylation cascades involved in bacterial chemotaxis.

In the context of dynamical systems, one says that a state is steady when the time derivative of all variables are nil (in the case of ODEs or PDEs) or when the values of all the logical variables equal those of the corresponding functions. In the differential context (ODEs), it is possible to further evaluate the stability of a steady state by analyzing the effects of perturbations around this steady state (usually on the basis of a linear approximation of the original equations). Mathematically, one can delineate these dynamical properties by analyzing the Jacobian matrix of the system, which gives the partial derivative of each equation (row) according to each variable (column) (an approach called stability linear analysis). On the basis of this matrix, one can compute the characteristic equation of the system. The types of roots of this equation at steady state, called the *eigenvalues*, are characteristic of specific types of steady state (see Table 2 in the case of two-dimensional systems).

Once the steady states have been located, other numerical tools enable the modeler to progressively follow their displacement, change of nature, or disappearance as a parameter of the system is modified, something which is represented in the form of a bifurcation diagram.

Related to the notion of steady state is that of an attractor, defined as a set of states (points in the phase space), invariant under the dynamics, toward which neighboring states asymptotically approach in the course of dynamic evolution. An attractor is defined as the smallest unit which cannot be decomposed into two or more attractors. The (inclusive) set of states leading to a given attractor is called its basin of attraction. In the simplest case, a system can have a unique, stable state, which then constitutes the unique attractor of the system, the rest of the phase space defining its basin of attraction.

The list of steady states, their properties, and the extension of the corresponding basins of attractions define the main qualitative dynamical features of a dynamical

Table 2   Main Types of Steady States in Two-Dimensional ODE Systems

Steady States (Two Dimensions)	Local Dynamical Properties	Roots of Characteristic Equations (Eigenvalues)
Stable node	Attractive along all directions	Two real negative roots
Stable focus	Attractive in a periodic way	Two complex roots with negative real parts
Unstable focus	Repulsive in a periodic way	Two complex roots with positive real parts
Saddle point	Attractive along one direction, repulsive along the orthogonal direction	One positive and one negative real root
Peak	Repulsive along all directions	Two positive real roots

system. These features depend on the structure of the equations (presence of feedback circuits, of nonlinearities, etc.) but also on the values selected for the different parameters of the system. When a system has several alternative steady states for given parameter values, one speaks of multistationarity, a property often advanced to explain differentiation and development. Alternatively, cyclic attractors are associated with molecular clocks such as those controlling the cell cycle or circardian rhythms. When the most important dynamical features are only marginally affected by ample changes of parameter values, one says that the system is robust. Robustness appears to be a general property of many biological networks.

More recently, metabolic networks have also been described in terms of Petri nets, which can be defined as digraphs (i.e., graphs involving two different types of vertices, each connecting to the other, exclusively). The first type of vertex, called *places*, corresponds to molecular species, whereas the second type of vertices, called *transitions*, represents reactions. The places contain resources (molecules), which can circulate along the edges connecting places to transitions, according to rules associated to the transition vertices. Starting from an initial state, transition rules are used to perform numerical simulations under deterministic or stochastic assumptions. Petri nets have recently been applied to metabolic graphs comprising up to several thousand nodes.

Applied exclusively to metabolic networks, the metabolic control analysis (MCA) is a phenomenological quantitative sensitivity analysis of fluxes and metabolite concentrations. In MCA, one evaluates the relative control exerted by each step (enzyme) on the system's variables (fluxes and metabolite concentrations). This control is measured by applying a perturbation to the step being studied and measuring the effect on the variable of interest after the system has settled to a new steady state. Instead of assuming the existence of a unique rate-limiting step, it assumes that there is a definite amount of flux control and that this is spread quantitatively among the component enzymes.

Further Reading
Heinrich R, Schuster S (1996). *The Regulation of Cellular Systems*. Chapman & Hall, New York.
Kaplan D, Glass L (1995). *Understanding Nonlinear Dynamics*. Springer-Verlag, New York.
Kauffman SA (1993). *The Origins of Order: Self-Organization and Selection in Evolution*. Oxford University Press, New York.

MATINSPECTOR
John M. Hancock and Martin J. Bishop

Program to identify matches to a matrix description of a protein binding site within a long DNA sequence.

The program takes a matrix generated using the program MatInd. MatInspector scans both strands of a number of sequences in a sequence file for matches to a number of matrices. Matches are assigned a quality rating. More sophisticated searches for constellations of binding sites can be carried out using ModelInspector.

Related Website

MatInspector	http://www.gsf.de/biodv/matinspector.html

Further Reading

Quandt K, et al. (1995). MatInd and MatInspector—New fast and versatile tools for detection of consensus matches in nucleotide sequence data. *Nucleic Acids Res.* 23: 4878–4884.

See also ConsInspector, GenomeInspector, ModelInspector, PromoterInspector.

MATRIX ATTACHMENT REGION (MAR, SAR, SCAFFOLD ATTACHMENT REGION)
Katheleen Gardiner

Sites of chromatin attachment to the nuclear scaffold in the interphase nucleus.

MARs are AT rich and ~200 bp in length. The nuclear scaffold is the site of RNA synthesis and MARs function to anchor genes near the nuclear membrane to facilitate transcription.

Further Reading

Razin SV (2001). The nuclear matrix and chromosomal DNA loops: Is there any correlation between partitioning of the genome into loops and functional domains? *Cell. Mol. Biol. Lett.* 6: 59–69.

Van Drunen CM, et al. (1999). A bipartite sequence associated with matrix/scaffold attachment regions. *Nucleic Acids Res.* 27: 2924–2930.

MAXIMAL MARGIN CLASSIFIERS. *SEE* SUPPORT VECTOR MACHINES.

MAXIMUM-LIKELIHOOD PHYLOGENY RECONSTRUCTION
Sudhir Kumar and Alan Filipski

A method for selecting a phylogenetic tree relating a set of sequence data based on an assumed probabilistic model of evolutionary change and the maximum-likelihood principle of statistical inference.

Maximum likelihood is a standard statistical method for parameter estimation using a model and a set of empirical data. The parameter values are chosen so as to maximize the probability of observing the data given the hypothesis. In phylogenetics, this method has been applied to finding the phylogeny with highest likelihood and estimating branch lengths and other parameters of the evolutionary process.

In maximum-likelihood phylogeny reconstruction, the parameters that are estimated by the maximum-likelihood principle are not the topologies themselves but rather the branch lengths. The topology that gives the highest maximum-likelihood set of branch lengths is chosen as the best topology. Since sequence states are known only at the leaf nodes of the tree, likelihood is computed over all possible internal node states. Because many topologies need to be enumerated and tested, the method can become time consuming.

Further Reading

Hasegawa M, et al. (1991). On the maximum likelihood method in molecular phylogenetics. *J. Mol. Evol.* 32: 443–445.

Nei M, Kumar S (2000). *Molecular Evolution and Phylogenetics*. Oxford University Press, Oxford, UK.

Yang Z (1996). Maximum-likelihood models for combined analyses of multiple sequence data. *J. Mol. Evol.* 42: 587–596.

Yang Z (1997). PAML: A program package for phylogenetic analysis by maximum likelihood. *Comput. Appl. Biosci.* 13: 555–556.

See also Phylogenetic Tree.

MAXIMUM-PARSIMONY PRINCIPLE (OCCAM'S RAZOR, PARSIMONY)
Sudhir Kumar, Alan Filipski, and Dov S. Greenbaum

A method of inferring a phylogenetic tree topology over a set of sequences in which the tree that requires the fewest total number of evolutionary changes to explain the observed sequences is chosen.

Parsimony is the law of Occam's razor, whereby the simplest explanation is preferred to a more complex explanation. Thus, in phylogeny reconstruction a tree that requires the fewest number of evolutionary steps between two sequences is preferred to more complex trees.

An efficient algorithm for estimating the number of nucleotide or amino acid changes required for any tree topology is available. This does not require the estimation of ancestral states to compute the number of changes (although these states may be inferred as well). Finding the best topology under this principle requires enumeration of candidate topologies. Only informative sites need be considered; a site is called *informative* (or *parsimony informative*) in maximum-parsimony analysis if at least two different states occur at least twice each at that site.

Weighted parsimony is a modification of the maximum-parsimony method in which differences in evolutionary rates among sites and among different nucleotides are incorporated. For instance, transitional changes may be given a smaller weight than transversional changes, because transitions are known to occur much more frequently than transversions. The matrix containing relative costs for change from one base to another for use in weighted-parsimony analysis is called the *step matrix*.

Two measures of how well a character (site) supports a given tree topology under the maximum-parsimony criterion are the consistency index and the retention index.

For a given single site, the consistency index is equal to m/s, where m is the minimum number of changes needed under any topology to generate the observed set of states at the site and s is the minimum number of changes needed to generate the observed set of states at the site given the current topology. This value lies between 1 and zero, with a value of 1 indicating complete congruence. For all sites in a sequence alignment, a composite consistency index is computed as M/S, where M is the sum of m over all sites and S is the sum of s over all sites.

The retention index is a measure of the amount of homoplasy at a given site with respect to a given phylogenetic tree of sequences. It is given by $r = (g - s)/(g - m)$, where g is the maximum number of substitutions at the site required for any tree topology under the parsimony principle, s is the minimum number of substitutions needed to generate the observed set of states at the site given the current topology, and m is the minimum number of substitutions needed under any tree topology to generate the observed set of states at the site. When the site is least informative for the current tree, the value is zero. An overall retention index may be computed as $(G - S)/(G - M)$ where M, G, and S are the sums of m, g, and s, respectively, over all informative sites.

Maximum-parsimony methods do not do well in the presence of too much homoplasy, i.e., identity of character state that is not attributable to common ancestry. This can arise in molecular sequences by multiple substitutions or parallel and convergent changes. Another shortcoming of parsimony analysis is the phenomenon known as long-branch attraction. Nonsister taxa may erroneously cluster together in the phylogenetic tree because of similarly longer branches when compared to their true sister taxa. This effect was first demonstrated for the maximum-parsimony method but is known to occur in other statistically rigorous methods as well.

Related Websites

PROTPARS	http://evolution.genetics.washington.edu/phylip/doc/protpars.html
PAUP	http://paup.csit.fsu.edu/

Further Reading

Densmore LD III (2001). Phylogenetic inference and parsimony analysis. *Methods Mol. Biol.* 176: 23–36.

Fitch WM (1971). Toward defining course of evolution—minimum change for a specific tree topology. *Syst. Zool.* 20: 406–416.

Hendy MD, Penny D (1989). A framework for the quantitative study of evolutionary trees. *Syst. Zool.* 38: 297–309.

Kluge AG, Farris JS (1969). Quantitative phyletics and evolution of anurans. *Syst. Zool.* 18: 1–32.

Kumar S, et al. (2001). *MEGA2*: Molecular evolutionary genetics analysis software. *Bioinformatics* 17: 1244–1245.

Maddison WP, Maddison DR (1992). *MacClade: Analysis of phylogeny and character evolution.* Sinauer Associates, Sunderland, MA.

Nei M, Kumar S (2000). *Molecular Evolution and Phylogenetics.* Oxford University Press, Oxford, UK.

Swofford DL (1998). *PAUP*: Phylogenetic Analysis Using Parsimony (and Other Methods).* Sinauer Associates, Sunderland, MA.

Stewart CB (1993). The powers and pitfalls of parsimony. *Nature* 361: 603–607.

See also Phylogenetic Tree.

MAXSPROUT
Roland L. Dunbrack

A program for building a complete protein model from an input file containing only the coordinates for the Cα atoms developed by Lisa Holm. MaxSprout uses a database fitting procedure to build the full set of backbone atom coordinates. The accompanying program Torso builds side chains onto the backbone using a rotamer library and a Monte Carlo procedure.

Related Website

| MaxSprout | http://www.ebi.ac.uk/maxsprout/ |

Further Reading
Holm L, and Sander C (1991). Database algorithm for generating protein backbone and sidechain coordinates from a Cα trace: Application to model building and detection of coordinate errors. *J. Mol. Biol.* 218: 183–194.

See also Backbone, Rotamer Library.

MB. *SEE* BASE PAIR.

MEGA (MOLECULAR EVOLUTIONARY GENETICS ANALYSIS)
Michael P. Cummings

A program for conducting comparative analysis of DNA and protein sequence data.

The program features phylogenetic analysis using parsimony or a choice of a variety of distance models and methods [neighbor joining, minimum evolution, UPGMA (unweighted pair-group method with arithmetic mean)]. There are capabilities for summarizing sequence patterns (e.g., nucleotide substitutions of various types, insertions and deletions, codon usage). The program also allows testing of the molecular clock hypothesis and allows some tree manipulation.

The program has a graphical user interface and is available as Windows executable.

Related Website

| MEGA | http://www.megasoftware.net/ |

Further Reading
Kumar S, et al. (2001). MEGA2: Molecular evolutionary genetics analysis software. *Bioinformatics* 17: 1244–1245.

MEGABASE. *SEE* BASE PAIR.

MENDELIAN DISEASE
Mark McCarthy and Steven Wiltshire

Diseases (or traits) that display classical Mendelian segregation patterns within families, usually because susceptibility in any given pedigree is determined by variation at a single locus.

Mendelian traits are usually the result of mutations causing dramatic changes in the function of genes or their products, with substantial consequences for phenotype. As a result, these conditions tend to be rare, although such deleterious alleles may be maintained through a variety of mechanisms, including high mutation rates or (in the case of some recessive disorders) evolutionary advantages associated with the heterozygous state. Depending on their chromosomal location and whether or not disease is associated with loss of one or two copies of the normal gene, Mendelian traits are characterized as autosomal or sex-linked and dominant or recessive. The genes responsible for over 1300 Mendelian traits have now been identified through a combination of positional cloning and candidate gene methods: These are catalogued in databases such as OMIM. The characteristic feature of Mendelian traits is a close correlation between genotype and phenotype (in contrast to multifactorial traits), although genetic heterogeneity (where a Mendelian trait is caused by defects in more than one gene) and the effects of modifier loci and environment mean that the distinction between Mendelian and multifactorial traits can be somewhat arbitrary.

Examples: The identification of the *CFTR* gene as the basis for the development of cystic fibrosis—one of the most common, severe Mendelian traits—is one of the classical stories in modern molecular biology (Kerem et al., 1989). Retinitis pigmentosa and maturity-onset diabetes of the young (Owen and Hattersley, 2001) are examples of Mendelian traits featuring substantial locus heterogeneity. Finally, thalassemia (Weatherall, 2000) represents a group of conditions which combine features of both Mendelian and multifactorial traits.

Related Websites

Catalogues of human genes and associated phenotypes:

Online Mendelian Inheritance in Man	http://www.ncbi.nlm.nih.gov/omim/
LocusLink	http://www.ncbi.nlm.nih.gov/LocusLink/

Further Reading

Kerem B, et al. (1989). Identification of the cystic fibrosis gene: Genetic analysis. *Science* 245: 1073–1080.

Owen K, Hattersley AT (2001). Maturity-onset diabetes of the young: From clinical description to molecular genetic characterization. *Best Pract. Res. Clin. Endocrinol. Metab.* 15: 309–323.

Peltonen L, McKusick V (2001). Genomics and medicine. Dissecting human disease in the postgenomic era. *Science* 291: 1224–1229.

Weatherall DJ (2000). Single gene disorders or complex traits: Lessons from the thalassaemias and other monogenic diseases. *Brit. Med. J.* 321: 1117–1120.

See also Candidate Gene, Linkage, Linkage Analysis, Multifactorial Trait.

MEROPS

Rolf Apweiler

MEROPS is a specialized protein sequence database covering peptidases.

MEROPS provides a catalogue and structure-based classification of peptidases. An index by name or synonym gives access to a set of PepCards files, each providing information on a single peptidase, including classification and nomenclature, and hypertext links to the relevant entries in other databases. The peptidases are classified into families based on statistically significant similarities between the protein sequences in the "peptidase unit," the part most directly responsible for activity. Families that are thought to have common evolutionary origins and are known or expected to have similar tertiary folds are grouped into clans. MEROPS also provides sets of files called FamCards and ClanCards describing the individual families and clans. Each FamCard document provides links to other databases for sequence motifs and secondary and tertiary structures and shows the distribution of the family across the major taxonomic kingdoms.

Related Website

MEROPS	http://merops.iapc.bbsrc.ac.uk/

Further Reading
Rawlings ND, et al. (2002). MEROPS: The protease database. *Nucl. Acids Res.* 30: 343–346.

See also Fold, Protein Family, Sequence of Proteins.

MESSAGE
Thomas D. Schneider

In communications theory, A message is a series of symbols chosen from a predefined alphabet.

In molecular biology the term *message* usually refers to a messenger RNA. In molecular information theory, a message corresponds to an after state of a molecular machine. In information theory, Shannon proposed to represent a message as a point in a high-dimensional space (see Shannon, 1949). For example, if we send three independent voltage pulses, their heights correspond to a point in three-dimensional space. A message consisting of 100 pulses corresponds to a point in 100-dimensional space. Starting from this concept, Shannon derived the channel capacity.

Further Reading
Shannon CE (1949). Communication in the presence of noise. *Proc. IRE*, 37: 10–21.

See also Molecular Machine Capacity, Noise, Shannon Sphere.

MESSENGER RNA. *SEE* MRNA.

METABOLIC NETWORK. *SEE* NETWORK.

METABOLOME (METABONOME)
Dov S. Greenbaum

The quantitative complement of all small molecules present in a cell in a specific physiological state.

It has been claimed that knowledge of the translatome and transcriptome alone is not enough to fully describe the cell in any given state and that this requires the further analysis of the metabolome as well. While those studying the metabolome have the advantage that the overall size of the population is generally less than either that of the translatome or transcriptome, there is no direct link between the genome and the metabolome. Still, researchers have looked at the effect that deleted genes have on the metabolome.

The metabolome is generally measured experimentally using biochemical analyses, including but not limited to mass spectrometry, nuclear magnetic resonance (NMR), and Fourier transform infrared spectrometry.

Metabolic flux analysis attempts to quantify the intracellular metabolic fluxes by measuring extracellular metabolite concentrations in combination with intracellular reaction stoichiometry (assuming that the system is in a steady state).

The analysis of the metabolome of each individual organism will give insight into the metabolic processes of the cells and allow for comparisons of multiple organisms by way of their common metabolites and small molecules.

Related Website

Metabolome	http://www.mpimp-golm.mpg.de/fiehn/index-e.html

Further Reading

Fiehn O (2001). Combining genomics, metabolome analysis, and biochemical modelling to understand metabolic networks. *Comp. Funct. Genom.* 2: 155–168.

Mendes P (2002). Emerging bioinformatics for the metabolome. *Brief Bioinform.* 3: 134–145.

Tweeddale H (1998). Effect of slow growth on metabolism of *Escherichia coli*, as revealed by global metabolite pool ("metabolome") analysis. *J. Bacteriol.* 180: 5109–5116.

METABONOME. *SEE* METABOLOME.

METADATA
Robert Stevens

Metadata is literally "data about data." An ontology may be thought of as providing metadata about the data about individuals held in a knowledge base. A database schema forms part of the metadata for a database. The ontology or schema itself may have further metadata, such as version number and source. An important class of metadata concerns the authorship and editorial status of works. The Dublin Core is a standard for editorial metadata used to describe library material such as books and journal articles. It includes such information as author, date, publisher, year, subject, and a brief description.

Related Website

Dublin Core	http://dublincore.org

META-PREDICT-PROTEIN
Patrick Aloy

Meta-predict-protein is a server developed and maintained by Volker Eyrich and Burkhard Rost at Columbia University. The user can paste or upload his or her sequence and the server runs a variety of prediction methods, including secondary structure, transmembrane regions, threading, and homology modeling, and returns the results via email after a short period of time.

The existing metaservers are very useful for extracting as much information as possible from a query sequence without having to run many methods independently.

Related Website

| Meta | http://cubic.bioc.columbia.edu/predictprotein/submit_meta.html |

See also Modeling, Macromolecular; Secondary-Structure Prediction of Protein; Threading.

METHIONINE
Jeremy Baum

Methionine is a nonpolar amino acid with side chain —$(CH_2)_2SCH_3$ found in proteins. In sequences, written as Met or M.

See also Amino Acid.

MGD (MOUSE GENOME DATABASE)
Guenter Stoesser

The Mouse Genome Database (MGD) represents the genetics and genomics of the laboratory mouse, a key model organism.

MGD provides integrated information of genotype (nucleotide sequences) to phenotype, including curation about genes and gene products as well as relationships between genes, sequences, and phenotypes. MGD maintains the official mouse gene nomenclature and collaborates with human and rat genome groups to curate relationships between these genomes and to standardize the representation of genes and gene families. MGD includes information on genetic markers, molecular segments [probes, primers and Yeast Artificial Chromosomes (YACs), STSs], description of phenotypic effects of mutations from the Mouse Locus Catalog (MLC), comparative mapping data, graphical displays of linkage, cytogenetic and physical maps, and experimental mapping data. Collaborations include the Gene Expression Database (GXD), the Mouse Genome Sequencing (MGS) project, and the Mouse Tumor Biology (MTB) database as well as SWISS-PROT (EBI) and LocusLink (NCBI) to provide an integrated information resource for the laboratory mouse. MGD is part of the Mouse Genome Informatics (MGI) project based at the Jackson Laboratory in Bar Harbor, Maine.

Related Websites

MGD	http://www.informatics.jax.org
Mouse Gene Expression Database (GXD)	http://www.informatics.jax.org/menus/ expression_menu.shtml
Mouse Genome Sequencing (MGS) project	http://www.informatics.jax.org/mgihome/MGS/ mgs.shtml

Further Reading

Blake JA, et al. (2001). The Mouse Genome Database (MGD): Integration nexus for the laboratory mouse. *Nucleic Acids Res.* 29: 91–94.

Blake JA, et al. (2002). The Mouse Genome Database (MGD): The model organism database for the laboratory mouse. *Nucleic Acids Res.* 30: 113–115.

Bult CJ, et al. (2001). Web-based access to mouse models of human cancers: The mouse tumor biology (MTB) database. *Nucleic Acids Res.* 29: 95–97.

Hill DP, et al. (2001). Biological annotation of mammalian systems: Implementing gene ontologies in mouse genome informatics. *Genomics* 74: 121–128.

Ringwald M, et al. (2001). The mouse gene expression database. *Nucleic Acids Res.* 29: 98–101.

MGED ONTOLOGY

Robert Stevens

The primary purpose of the MGED Ontology is to provide standard terms for the annotation of microarray experiments. The terms taken from the MGED ontology will be compliant with the MAGE standard for the representation of microarray experiments. These terms will enable structured queries of elements of the experiments. Furthermore, the terms will also enable unambiguous descriptions of how the experiment was performed. The terms will be provided in the form of an ontology, where terms are organized into classes with properties and will be defined. A standard ontology format will be used. For descriptions of biological material (biomaterial) and certain treatments used in the experiment, terms may come from external ontologies. Software programs utilizing the ontology are expected to generate forms for annotation, populate databases directly, or generate files in the established MAGE-ML format. Thus, the ontology will be used directly by investigators annotating their microarray experiments as well as by software and database developers and therefore will be developed with these very practical applications in mind.

Related Websites

MGED Ontology	http://mged.sourceforge.net/ontologies/index.php
MAGE standards	http://www.mged.org/Workgroups/MAGE/mage.html)

MICROARRAY (BIOCHIP, DNA CHIP, DNA MICROARRAY, GENE EXPRESSION ARRAY)

Stuart M. Brown

Technology for measuring expression levels of large numbers of genes simultaneously.
A microarray is a collection of tiny DNA spots attached to a solid surface that is used to measure gene expression by hybridization with a labeled complementary DNA (cDNA) mixture. Global measurement of differential gene expression across treatments, developmental stages, disease states, etc., is relevant to a very wide range of biological and clinical disciplines. The technology for microarrays evolved from Southern blots, where fragmented DNA is immobilized to a substrate and then probed with a known gene or fragment.

In general, a DNA *array* is an ordered arrangement of known DNA sequences chemically bound to a solid substrate. A *microarray* has *features* (spots with many copies of a specific DNA sequence) that are typically less than 200 μm in diameter, so that a single array may have tens of thousands of spots on a square centimeter of a glass slide. The most common use of this technology is to measure gene expression, in which case the bound DNA sequences are intended to represent individual genes. The array is then hybridized with labeled cDNA that has been reverse transcribed from a cellular RNA extract. The amount of labeled cDNA bound to each spot on the array is measured, generally by fluorescent scanning, and comparisons are made as to the relative amount of fluorescence for each gene in cDNA samples from various experimental treatments.

The DNA sequences bound on the array are referred to as *probes* because they are of known sequence while the experimental sample of labeled cDNA is called the *target* as it is uncharacterized. The bound probe DNA may be entire cloned cDNA sequences or short oligonucleotides designed to represent a portion of a cDNA. If oligonucleotides are used, they may be synthesized directly on the array or in individual batches and applied to the array by a robotic spotter.

Bioinformatics challenges for the analysis of microarray data include image analysis to quantify the fluorescent intensity of each spot and subtract local background, normalization between two colors on one array or across multiple arrays, calculation of fold change and statistical significance for each gene across experimental treatments, clustering of gene expression patterns across treatments, and functional annotation of coregulated gene clusters. An additional challenge is the development of data annotation standards, file formats, and databases to allow for the comparison of experiments performed by different investigators using different biological samples, different treatments, and different microarray technologies.

Related Websites

Leming Shi's Genome Chip resources	http://www.gene-chips.com/
GRID IT—Introduction to microarrays	http://www.bsi.vt.edu/ralscher/gridit/intro_ma.htm
Chips	http://www.bio.davidson.edu/courses/genomics/chip/chip.html

Further Reading

Bowtell D, Sambrook J. (2003). *DNA Microarrays: A Molecular Cloning Manual*. Cold Spring Harbor Laboratory, Cold Spring Harbor, NY.

Nature Genetics (1999). Vol. 21, Suppl., pp. 1–60.

Schena M, et al. (1995). Quantitative monitoring of gene expression patterns with a complementary DNA microarray. *Science* 270: 467–470.

Schena M (2002). *Microarray Analysis*. Wiley, New York.

See also Affymetrix GeneChip Oligonucleotide Microarray, Gene Expression Profile, Microarray Image Analysis, Microarray Normalization, Spotted cDNA Microarray.

MICROARRAY IMAGE ANALYSIS
Stuart M. Brown

The application of image analysis software to quantitate the signals due to hybridization of a complex mixture of fluorescently tagged target complementary DNA (cDNA) to each of the spots of a microarray.

Once a microarray chip has been hybridized with a fluorescently labeled cDNA and scanned, a digital image file is produced. The initial step in data analysis requires that the signal intensity of each spot be quantitated using some form of densitometry. The image analysis software must achieve several tasks: identify the centers and boundaries of each spot on the array, integrate the signal within the boundaries of each spot, identify and subtract background values, and flag abnormal spots that may contain image artifacts.

Many different algorithms for edge detection and signal processing have been adapted to microarray analysis (e.g., adaptive threshold segmentation, intensity histograms, spot shape selection, and pixel selection), but no rigorous comprehensive analysis of the various methods has been conducted. In general, investigators require a highly automated image-processing solution that applies a common template and a uniform analysis procedure to every hybridization image in a microarray experiment in order to minimize error and variability being added at the image analysis stage of the protocol. Spotted arrays are subject to considerably more variability than in situ synthesized oligonucleotide arrays since there is the potential for each spot to take on an irregular shape and for signals from adjacent spots to overlap. Many of the manufacturers of fluorescent scanners used for microarray analysis have created image analysis software optimized for the characteristics of their scanners. A variety of free software for microarray image analysis has also been produced by academic groups, including Scanalyze by Mike Eisen and Spot by Jean Yang.

Related Website

Comparison of microarray image analysis software	http://ihome.cuhk.edu.hk/~b400559/arraysoft_image.html

Further Reading
Kamberova G, Shah S (2002). *DNA Array Image Analysis: Nuts & Bolts*. DNA Press, San Diego.
Yang YH, et al. (2001). Analysis of cDNA microarray images. *Brief Bioinform.* 2: 341–349.

MICROARRAY NORMALIZATION
Stuart M. Brown

A mathematical transformation of microarray data to allow for accurate comparisons between scanned images of arrays that have different overall brightness.

A microarray experiment can measure the differential abundance of transcripts from many different genes within the total population of messenger RNA (mRNA) across some set of experimental conditions. However, this requires making comparisons of the images created by scanning fluorescent intensities of labeled target mRNA bound by a particular probe in several different array hybridizations. Spotted arrays that are hybridized with a mixture of two complementary DNA (cDNA) extractions labeled with two different fluorescent dyes must be scanned in two colors. Since the two dyes were added to the cDNAs in separate labeling reactions and the two dyes have different fluorescent properties, the relative signal strengths of the two colors on each spot can only be compared after the two images are normalized to the same average brightness.

There are many experimental variables (systematic errors) that can affect the intensity of fluorescent signal detected in a microarray hybridization experiment, including the chemical activity of the fluorescent label, labeling efficiency, hybridization and washing efficiency, and performance of the scanner. Many of these variables have a nonspecific effect on all measurements in a particular array, so that they can be controlled by scaling the intensity values across a set of array images to the same average brightness. However, the variability in fluorescent measurements between arrays often have nonlinear and/or intensity-dependent relationships.

The most common method of microarray normalization is to divide all measurements in each array by the mean or median of that array and then multiply by a common scaling factor. The Affymetrix Microarray Analysis Suite software scales probe set signals from GeneChip arrays to a median value of 2500 (MAS 4) or 500 (MAS 5) by default. A more accurate normalization that corrects for intensity-dependent variation can be calculated by fitting the data for two arrays to a smooth curve on a graph of the intensity of each probe on array A vs. array B using a local regression model (loess). The differences $A - B$ are then normalized by subtracting the fitted curve (quantile normalization). This method can be further refined by fitting the curve to the graph of $A - B$ (fold change) vs. $A + B$ (intensity) for each probe. In the dChip software, Li and Wong normalize a set of arrays to a smooth curve fit to the expression levels of an "invariant set" of probes with small rank differences across the arrays (Figure 1). The method of normalization applied to a set of array data can dramatically affect the fold change and statistical significance observed for genes with low levels of expression.

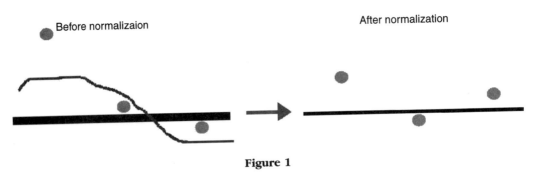

Figure 1

Related Websites

Standardization and normalization of microarray data	http://pevsnerlab.kennedykrieger.org/snomadinput.html
Bioconductor	http://www.bioconductor.org

Further Reading

Bolstad BM, et al. (2003). A comparison of normalization methods for high density oligonucleotide array data based on variance and bias. *Bioinformatics* 19: 185–193.

Colantuoni C, et al. (2002). SNOMAD (Standardization and NOrmalization of MicroArray Data): Web-accessible gene expression data analysis. *Bioinformatics* 18: 1540–1541.

Gautier L, et al. (2002). Textual description of Affy. www.bioconductor.org/repository/devel/vignette/affy.pdf.

Schuchhardt J, et al. (2000). Normalization strategies for cDNA microarrays. *Nucleic Acids Res.* 28: E47.

MICROARRAY PROFILING OF GENE EXPRESSION
Stuart M. Brown

The simultaneous measurement of the level of expression of multiple genes.

The goal of a microarray data analysis is generally to use bioinformatics techniques to identify patterns of common expression among groups of genes across multiple treatments or among different samples across many genes. After primary image analysis and normalization, a microarray experiment can be described by a matrix of gene expression data—rows for each gene and columns for each experimental sample. The rows of the matrix can be clustered to bring together genes with similar patterns of expression across all of the samples. These clusters of coregulated genes may represent particular biochemical or regulatory pathways, and it would be highly useful to capture this functional gene expression information into the public database annotation of each gene (i.e., "induced 3-fold in breast cancer tumor cells"). It is also possible to search the genomic sequences flanking the coding regions of each of the genes in a coexpressed cluster for motifs that are the targets for DNA-binding regulatory proteins. Coexpressed genes may share common promoter elements.

Similarly, it is useful to cluster the columns for the various treatments and/or replicates in the experiment. Clusters would then represent samples with similar patterns of gene expression. These patterns could then be used for the classification of new, unknown samples. One application is to diagnose cancer subtypes by expression profile.

Related Website

Comparative gene expression	http://www.cs.wustl.edu/~jbuhler//research/array/experiment.html

Further Reading

Claverie JM (1999). Computational methods for the identification of differential and coordinated gene expression. *Hum. Mol. Genet.* 8: 1821–1832.

MICROSATELLITE
Terri Attwood and Katheleen Gardiner

DNA sequences that are internally repetitive.

Microsatellites consist of tandem (end-to-end) repeats of short-sequence motifs, such as CA. Because of their hypervariability and their wide distribution, many are useful in genetic linkage studies and as genetic disease markers. The frequencies of the different repeated motifs, mutation rates, and microsatellite length distributions differ between species.

Concentrations and types of repetitive sequences found within genomes vary between species. Concentrations depend to a limited extent on genome size, but differences in the molecular mechanisms generating them, and in some cases selective pressures, may also be important.

It should be noted that repetitive sequences are to be distinguished from repeated sequences, which are sequences that occur large numbers of times within an individual genome, although many repeated sequences are also internally repetitive. Cryptically simple regions are also detected, corresponding to local regions rich in a small number of short motifs that are not arranged in a tandem manner.

Repetitive (and repeated) sequences can be detected by matching to a preexisting database or by searching for tandemly arranged patterns. A further development is to identify cryptically simple regions by identifying regions within sequences that contain statistically significant concentrations of sequence motifs. A more abstract feature of a sequence, its repetitiveness, can also be measured, e.g., by the relative simplicity factor of the SIMPLE program.

Related Websites

RepeatMasker	http://www.genome.washington.edu/UWGC/analysistools/repeatmask.htm
CENSOR	http://www.girinst.org/Censor_Server.html
Tandem Repeats Finder	http://c3.biomath.mssm.edu/trf.html
SIMPLE	http://www.biochem.ucl.ac.uk/bsm/simple/

Further Reading

Charlesworth B, Sniegowski P, Stephan W (1994). The evolutionary dynamics of repetitive DNA in eukaryotes. *Nature* 371: 215–220.

Dib C, et al. (1996). A comprehensive genetic map of the human genome based on 5264 microsatellites. *Nature* 380: 152–154.

Goldstein DB, Schlötterer C (1999). *Microsatellites: Evolution and Applications.* Oxford University Press, New York.

Richard G-F, Paques F (2000). Mini- and microsatellite expansion: The recombination connection. *EMBO Rep.* 1: 122–126.

See also Genome Sequence, Minisatellite, SIMPLE.

MIDNIGHT ZONE

Terri Attwood

The region of protein sequence identity where sequence comparisons fail completely to detect structural similarity. In many cases, protein sequences have diverged to such an extent that their evolutionary relationships are apparent only at the level of shared structural features. Such characteristics cannot be detected, even using the most sensitive sequence comparison methods. Consequently, the midnight zone denotes the theoretical limit to the effectiveness of conventional sequence analysis techniques—beyond this point, threading algorithms tend to be employed to determine whether a given sequence is likely to be compatible with a given fold.

Because the midnight zone is populated by protein structures between which there is no detectable sequence similarity, it is sometimes unclear whether their structural relationships are the result of divergent or convergent evolution. Conservation patterns of residues at aligned positions within a range of structurally similar but sequentially dissimilar proteins have been studied. It turns out that while aligned positions that are only moderately conserved are likely to have arisen by chance, those that are highly conserved reveal distinct features associated with structure and function: A relatively high fraction of structurally aligned positions are buried within the protein core, and glycine, cysteine, histidine, and tryptophan are significantly overrepresented, suggesting residue-specific structural and functional roles.

Further Reading

Friedberg I, Kaplan T, Margalit H (2000). Glimmers in the midnight zone: Characterization of aligned identical residues in sequence-dissimilar proteins sharing a common fold. *Proc. Intl. Conf. Intell. Syst. Mol. Biol.* 8: 162–170.

Friedberg I, Margalit H (2002). Persistently conserved positions in structurally similar, sequence dissimilar proteins: Roles in preserving protein fold and function. *Protein Sci.* 11(2): 350–360.

Rost B (1998). Marrying structure and genomics. *Structure* 6(3): 259–263.

See also Threading.

MIGRATE. *SEE* LAMARC.

MIME TYPES
Eric Martz

Providers of information sent through the Internet label their information with MIME types in order to inform the recipients how to process that information. MIME types are attached as prefixes to transmitted information in order to identify the format in which it is represented and therefore, to some extent, the type of information contained.

MIME stands for Multipart Internet Mail Extensions, which were originally developed to enable content beyond plain text (US-ASCII) to be included in electronic mail. The ambiguities and limitations of MIME types have encouraged development of a better information exchange standard. As the Web evolves, XML (eXtensible Markup Language) is replacing MIME types.

When information is requested through the Internet, the user or receiving program needs to know how to process and interpret the information received. When a server computer sends a data file requested by a client computer or Web browser, the server typically prefixes the data with a tag that specifies the type of information, or MIME type, contained in the file. Typically, the server makes this determination based on name of the data file requested, specifically, the filename extension (last three characters, following a period), using a list maintained on the server. Each server must be configured with a suitable list matching the files it serves.

If the client or browser does not handle files properly, one possible reason is that the server is serving the file with an incorrect MIME type. Typically, the server administrator needs to add some new MIME types to the server's list. The MIME type specified by the server can be seen in Netscape (View, Page Info). Another possible

Table Some Common MIME Types

MIME Type	Customary Filename Extension	Type of Information in File
Text/plain	.txt	Text with no special formatting
Text/html	.htm,. html	Text containing HyperText Markup Language (HTML) tags that specify formatting in a Web browser
Image/gif	.gif	Image in graphics interchange format (GIF)
Chemical/x-mmcif	.mcif	Macromolecular crystallographic interchange format for annotated atomic coordinates
Chemical/x-pdb	.pdb	Atomic coordinates in Protein Data Bank format (most covalent bonds not explicit)
Chemical/x-mdl-molfile	.mol	Atomic coordinates with explicit bond information, typically for small molecules
Application/x-spt	.spt	Script commands for the RasMol family of molecular visualization programs
Application/x-javascript	.js	Javascript (commands in the browser's programming language)

reason is that the browser lacks the ability to process that type of information correctly, e.g., perhaps a plugin such as Chime or a helper application such as RasMol needs to be installed. The client computer and its Web browsers also contain lists of MIME types from which they determine the appropriate function or program to invoke when a particular MIME type of information is received.

Related Websites

Rzepa HS, Chemical MIME	http://www.ch.ic.ac.uk/chemime/
Internet Assigned Numbers Authority, MIME media types	http://www.iana.org/assignments/media-types/

Further Reading
Rzepa HS, Murray-Rust P, Whitaker BJ (1998). The application of Chemical Multipurpose Internet Mail Extensions (Chemical MIME) Internet standards to electronic mail and World-Wide Web information exchange. *J. Chem. Inf. Comp. Sci.* 38: 976–982.

See also Visualization, Molecular; XML.

Eric Martz is grateful for help from Eric Francoeur, Peter Murray-Rust, Byron Rubin, and Henry Rzepa.

MINIMUM-EVOLUTION PRINCIPLE
Sudhir Kumar and Alan Filipski

A phylogenetic reconstruction principle based on minimizing the sum of branch lengths of a tree.

Of all possible tree topologies over a given set of taxa, the minimum-evolution (ME) topology is one that requires the smallest sum of branch lengths, as estimated by least-mean-squared error or similar methods. Mathematically, it can be shown that, when unbiased estimates of evolutionary distances are used, the expected value of the sum of branch lengths is smaller for the true topology than that for a given wrong topology. In general, phylogeny reconstruction by this method requires the enumeration of many candidate trees and so can take considerable time for large problems, unless heuristics are used to reduce the size of the search space. A number of heuristic variants of this approach are available that limit the space of phylogenetic trees searched, e.g., the neighbor-joining method.

Further Reading

Kumar S, et al. (2001). MEGA2: Molecular evolutionary genetics analysis software. *Bioinformatics* 17: 1244–1245.

Nei M, Kumar S (2000). *Molecular Evolution and Phylogenetics*. Oxford University Press, Oxford, UK.

Rzhetsky A, Nei M (1993). *METREE: Program Package for Inferring and Testing Minimum Evolution Trees*, Version 1.2. Pennsylvania State University, University Park, PA.

Swofford DL (1998). *PAUP*: Phylogenetic Analysis Using Parsimony (and Other Methods)*. Sinauer Associates, Sunderland, MA.

See also Branch-Length Estimation, Neighbor-Joining Method, Phylogenetic Tree.

MINISATELLITE
Katheleen Gardiner

Tandem repeats of intermediate-length motifs approximately 10–100 bp in length.

Similar to microsatellites in that they are dispersed throughout the genome (although most common near telomeres) and variable among individuals. For these reasons they have found application in genetic linkage and disease mapping as well as in forensics. Minisatellites may show much higher levels of length polymorphism than microsatellites or may show none at all. Their level of polymorphism appears to relate to the local frequency of recombination, especially gene conversion.

Further Reading

Buard J, Jeffreys A (1997). Big, bad minisatellites. *Nat. Genet.* 15: 327–328.

Charlesworth B, Sniegowski P, Stephan W (1994). The evolutionary dynamics of repetitive DNA in eukaryotes. *Nature* 371: 215–220.

Jeffreys A, Wilson V, Thein SL (1985). Hypervariable minisatellite regions in human DNA. *Nature* 314: 67–73.

Richard G-F, Paques F (2000). Mini- and microsatellite expansion: The recombination connection. *EMBO Rep.* 1: 122–126.

MIRRORTREE
Patrick Aloy

MirrorTree is a method to predict protein–protein interactions based on the assumption that interacting protein pairs tend to coevolve. In such cases, the phylogenetic trees corresponding to the two interacting partners will show a greater degree of similarity that those derived from noninteracting proteins.

The standard method to quantify phylogenetic tree similarities consists of measuring the linear correlation between the distance matrices used to construct the trees.

In order to successfully apply the MirrorTree method, the user will need accurate multiple-sequence alignments for the two proteins considered, including sequences from the same species.

According to the very nature of the method, it should predict physical interactions between proteins; however, this is not always the case.

Related Website

| MirrorTree | http://www.pdg.cnb.uam.es/mirrortree/ |

Further Reading
Goh CS, Bogan AA, Joachimiak M, Walther D, Cohen FE (2000). Co-evolution of proteins with their interaction partners. *J. Mol. Biol.* 299: 283–293.

Pazos F, Valencia A (2001). Similarity of phylogenetic trees as indicator of protein-protein interaction. *Protein Eng.* 14: 609–614.

See also Amino Acid Exchange Matrix.

MODBASE
Roland L. Dunbrack

A database of comparative models constructed with the Modeller program for protein sequences in completed genomes. The database provides models indexed by accession number or accessible by sequence database search and also provides information on predicted model quality.

Related Website

| ModBase | http://guitar.rockefeller.edu/modbase/ |

Further Reading
Pieper U, Eswar N, Stuart AC, Ilyin VA, Sali A (2002). MODBASE, a database of annotated comparative protein structure models. *Nucleic Acids Res.* 30: 255–259.

See also Comparative Modeling, Modeller.

MODELING, MACROMOLECULAR
Eric Martz

Generating a three-dimensional structural model for an amino acid or nucleotide sequence from empirical structure data or from theoretical considerations.

The most reliable and accurate sources for macromolecular models are empirical, most commonly X-ray crystallography or nuclear magnetic resonance (NMR). Even these involve modeling to fit experimental results and vary in reliability (e.g., resolution, disorder) and occasionally err in interpretation. Furthermore, crystallography is often precluded by the difficulty of obtaining high-quality singular crystals, while NMR is generally limited to molecules no greater than 30 kilodaltons that remain soluble without aggregation at high concentrations. Therefore, for the vast majority of proteins, the only way to obtain a model is from theory.

Theoretical modeling can be based upon an empirical structural for a template molecule having sufficient sequence similarity to the unknown target, if one is available. This is called *comparative modeling* (or *homology modeling/knowledge based-modeling*) and is usually reliable at predicting the fold of the backbone, thereby identifying residues that are buried or on the surface or clustered together. It is unreliable at predicting details of side-chain positions.

In the absence of a sequence-similar empirical template (by far the most common case), the only option for obtaining a three-dimensional model is ab initio theoretical modeling. Typically, it is based in part upon generalizations and patterns derived from known empirical structures. Theoretical modeling has fair success at predicting protein secondary structure but rather limited success at predicting tertiary or quaternary structure of proteins. A series of international collaborations performs Critical Assessment of techniques for protein Structure Prediction (CASP)

Related Websites

Nature of 3D structural data	http://www.rcsb.org/pdb/experimental_methods.html
Homology modeling for beginners	http://molvis.sdsc.edu/protexpl/homolmod.htm
Protein Structure Prediction Center (CASP series)	http://predictioncenter.llnl.gov/

Further Reading

Baker D, Sali A. (2001). Protein structure prediction and structural genomics. *Science* 294: 93–96.

Sali, A (1995). *Curr. Opin. Struct. Biol.* 6: 437.

See also Atomic Coordinate File; Comparative Modeling; Docking; Models, Molecular; Secondary Structure Prediction of Protein.

Eric Martz is grateful for help from Eric Francoeur, Peter Murray-Rust, Byron Rubin, and Henry Rzepa.

MODELINSPECTOR
John M. Hancock

Program to screen a DNA sequence for groups of protein binding site sequences with a given spacing.

The program, which belongs to the MatInspector/GenomeInspector family of programs, makes use of models which consist of definitions of the set of binding sites to be screened for, their order, strand orientation, threshold match quality, and spacing. Models can either be learned (using ModelGenerator) or input explicitly (using FastM)

Related Website

| ModelInspector | http://www.gsf.de/biodv/modelinspector.html |

Further Reading

Frech K, et al. (1997). A novel method to develop highly specific models for regulatory units detects a new LTR in GenBank which contains a functional promoter. *J. Mol. Biol.* 270: 674–687.

See also ConsInspector, GenomeInspector, MatInspector, PromoterInspector.

MODELLER
Roland L. Dunbrack

A program for comparative modeling of protein structure developed by Andrej Sali and colleagues. Modeller builds models by satisfaction of spatial restraints, such as atom–atom distances and dihedral angles, obtained from the template structure and statistical analysis of the Protein Data Bank. The model is built by a molecular dynamics protocol with an energy function composed of the restraints and the CHARMM molecular mechanics force field.

Related Website

| Modeller | http://guitar.rockefeller.edu/modeller/modeller.html |

Further Reading

Fiser A, et al. (2000). Modeling of loops in protein structures. *Prot. Sci.* 9: 1753–1773.

Sali A, Blundell TL (1993). Comparative protein modelling by satisfaction of spatial restraints. *J. Mol. Biol.* 234: 779–815.

Sali A, Potterton L, Yuan F, vanVlijmen H, Karplus M (1995). Evaluation of comparative protein structure modeling by MODELLER. *Proteins* 23: 318–326.

See also CHARMM, Comparative Modeling, Dihedral Angle, Template.

MODEL ORDER SELECTION. *SEE* MODEL SELECTION.

MODEL SELECTION (COMPLEXITY REGULARIZATION, MODEL ORDER SELECTION)
Nello Cristianini

In machine learning, the problem of finding a class of hypothesis whose complexity matches the features (size, noise level, complexity) of the data set at hand.

Too rich a model class will result in overfitting, too poor a class will result in underfitting. Often the problem is addressed by using a parametrized class of hypotheses and then tuning the parameter using a validation set.

In neural networks this coincides with the problem of choosing the network's architecture, in kernel methods with the problem of selecting the kernel function. Prior knowledge about the problem can often guide in the selection of an appropriate model.

Further Reading
Duda RO, Hart PE, Stork DG (2001). *Pattern Classification.*, Wiley, New York.
Mitchell T (1997). *Machine Learning.* McGraw-Hill, New York.

MODELS, MOLECULAR
Eric Martz

Three-dimensional depictions of molecular structures.

There are many styles for rendering molecular models, including those with atomic-level detail such as space-filling models, ball-and-stick models, and wire-frame models (also called "stick models" or "skeletal models") as well as surface models. Effectively perceiving the polymer chain folds of macromolecules requires simplified models such as backbone models and schematic ("cartoon") models. Such models can be made of solid physical materials or represented on a computer screen. Physical models allow both visual and tactile perception of molecular properties, while computer models can be rendered in different styles and color schemes at will. Molecular models are usually based upon an atomic coordinate file.

Color schemes are typically employed in molecular models in order to convey additional information. Elements can be identified in models with atomic-level detail. The most common element color scheme originated with space-filling physical models, termed Corey–Pauling–Koltun (CPK) models. In this scheme, carbon is black or gray, nitrogen is blue, oxygen is red, sulfur and disulfide bonds are yellow, phosphorus is orange, and hydrogen is white. This scheme has been adopted in most visualization software. Since most negative charge in proteins is on oxygen atoms and most positive charge on protonated amino radicals, red and blue are common colors for anionic and cationic charges, respectively, used, e.g., on surface models. Since red and blue, when mixed, make magenta, magenta is logical when a single color is desired to distinguish polar or charged regions from apolar regions. Gray (signifying the preponderance of carbon) is the customary color for apolar regions. In RasMol and its derivatives (visualization), alpha helices are colored red and beta strands yellow. These and other color schemes have been unified in a proposed set of open-source, nonproprietary standard color schemes for molecular models, designated the DRuMS color schemes (see Related Websites).

Related Websites

World Index of Molecular Visualization Resources	http://molvisindex.org
DruMS color schemes	http://www.umass.edu/molvis/drums
History of macromolecular visualization	http://www.umass.edu/microbio/rasmol/history.htm
ModView	http://guitar.rockefeller.edu/modview/modview2.shtml
Physical molecular models	http://www.netsci.org/Science/Compchem/feature14b.html

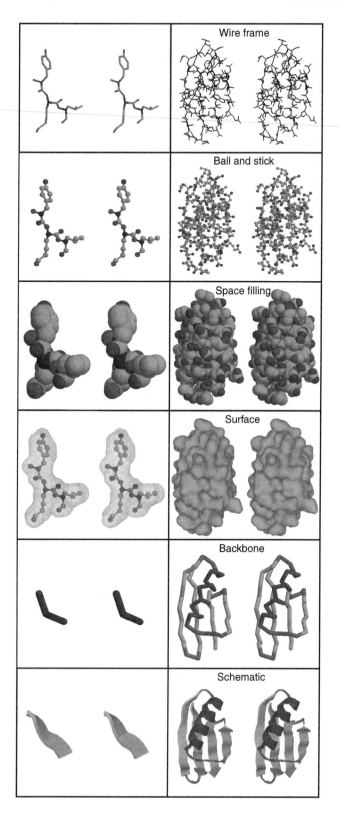

Further Reading

Ilyin VA, Pieper U, Stuart AC, Martí-Renom MA, McMahan L, Sali A (2002). ModView, visualization of multiple protein sequences and structures. *Bioinformatics* 19: 165–166.

Petersen Q. (1970). Some reflections on the use and abuse of molecular models. *J. Chem. Ed.* 47: 24–29.

Smit, D. (1960). *Bibliography on Molecular and Crystal Structure Models*, Vol. 14, *NBS Monograph*. National Bureau of Standards, Washington, DC.

Walton A (1978). *Molecular Crystal Structure Models*. Ellis Horwood, Chichester, England.

See also Alpha Helix; Atomic Coordinate File;, Beta Strand; Modeling, Macromolecular; Space-Filling Models; Visualization, Molecular.

Eric Martz is grateful for help from Eric Francoeur, Peter Murray-Rust, Byron Rubin, and Henry Rzepa.

MODELTEST
Michael P. Cummings

A program that assists in evaluating the fit of a range of nucleotide substitution models to DNA sequence data through a hierarchical series of hypothesis tests.

Two test statistics, likelihood ratio and Akaike information criterion (AIC), are provided to compare model pairs that differ in complexity. The program is used together with PAUP, typically as part of data exploration prior to more extensive phylogenetic analysis.

The program is available as executable files for several platforms.

Further Reading

Akaike H (1974). A new look at the statistical model identification. *IEEE Trans. Autom. Contr.* 19: 716–723.

Posada D, Crandall KA (1998). MODELTEST: Testing the model of DNA substitution. *Bioinformatics* 14: 817–818.

See also PAUP*.

Figure Six types of molecular models (q.v.). On the left is a three-amino-acid peptide (Tyr–Lys–Leu, residues 3–5 of the protein on the right). On the right is a small protein, the immunoglobulin G-binding domain of streptococcal protein G (PDB identification code 1PGB). Notice how the schematic and backbone models enable the main chain of the protein to be discerned, in contrast to the overly detailed atomic-resolution models (wire-frame, ball-and-stick, space-filling) that are more suitable for small molecules such as the tripeptide at left, or details of functional sites on proteins (not shown). Atoms and bonds are colored with the CPK scheme (carbon gray, alpha carbons dark gray, nitrogen blue, oxygen red). In the schematic model (and protein backbone), beta strands are yellow and the alpha helix is red. Images prepared with Protein Explorer (http://proteinexplorer.org).

All models are in divergent stereo. To see in stereo: nearly touch the paper with your nose between the images of a pair, relaxing your eyes while gazing into the distance (so the images are blurry). Slowly move the paper away from your eyes. You will see three images of the model; attend to the center of the three images. As the paper is moved just far enough away to focus clearly, the center image will appear three dimensional. Seeing in this way is very difficult for about one in three people; it is easier for the majority, but takes some practice. Special stereo-viewing lenses, if available, may also be used.

MODULAR PROTEIN
Laszlo Patthy

Multidomain proteins containing multiple copies and/or multiple types of protein modules.

See also Multidomain Protein, Protein Module.

MODULARITY
Marketa J. Zvelebil

Modularity describes functional modules (or units) which can be interconnected to form various parts or whole pathways.

Modularity plays an important part in robustness in biological systems and is used in systems biology.

Further Reading
Kitano H (2002). Systems biology: A brief overview. *Science* 295: 1662–1664.
Lauffenburger DA (2000). Cell signaling pathways as control modules: Complexity for simplicity? *Proc. Natl. Acad. Sci. USA* 97(10): 5031–5033.

See also Robustness, System Biology.

MODULE SHUFFLING
Laszlo Patthy

During evolution protein modules may be shuffled through either exon shuffling or exonic recombination to create multidomain proteins with various domain combinations.

Related Website

SMART	http://smart.embl-heidelberg.de

Further Reading
Patthy L (1991). Modular exchange principles in proteins. *Curr. Opin. Struct. Biol.* 1: 351–361.
Patthy L (1996). Exon shuffling and other ways of module exchange. *Matrix Biol.* 15: 301–310.
Schultz J, et al. (2000). SMART: A Web-based tool for the study of genetically mobile domains *Nucleic Acids Res.* 28: 231–234.

See also Exon Shuffling, Multidomain Protein, Protein Module.

MOE
Roland L. Dunbrack

A program for comparative modeling, quantum chemistry, chemical database access, and drug design developed by Paul Labute and colleagues of the Chemical Computing Group (Montreal).

Related Website

Chemical Computing Group	http://www.chemcomp.com/

Further Reading
Labute P (2000). A widely applicable set of descriptors. *J. Mol. Graph. Model.* 18: 464–477.

See also Comparative Modeling.

MOLECULAR CLOCK (EVOLUTIONARY CLOCK, RATE OF EVOLUTION)

Sudhir Kumar and Alan Filipski

The approximate linear relationship between the evolutionary distance and time. The slope of the regression line is the rate of evolution.

Evolutionary rate is often measured in terms of the number of substitutions per site per unit time. It is obtained by dividing the evolutionary distance by twice the divergence time between taxa (a factor of 2 appears because evolutionary distance captures changes in both evolutionary lineages). In protein coding DNA sequences, rates may be determined for both synonymous and nonsynonymous change by using the appropriate distance measure.

The molecular clock hypothesis is consistent with the neutral theory of evolutionary change in the presence of relatively constant functional constraints. A rooted phylogenetic tree satisfies the molecular clock assumption if all leaf nodes are equidistant from the root.

Tests are available to determine if two lineages are evolving at the same rate. Muse and Weir (1992) developed a test to measure the significance of observed rate differences between two taxa based on likelihood ratios. Tajima (1993) presents a simple nonparametric test for rate equality. Many more complex tests have also been developed.

Further Reading
Kumar S, Subramanian S (2002). Mutation rates in mammalian genomes. *Proc. Natl. Acad. Sci. USA* 99: 803–808.

Li W-H (1997). *Molecular Evolution.* Sinauer Associates, Sunderland, MA.

Muse SV, Weir BS (1992). Testing for equality of evolutionary rates. *Genetics* 132: 269–276.

Nei M, Kumar S (2000). *Molecular Evolution and Phylogenetics.* Oxford University Press, Oxford, UK.

Tajima F (1993). Simple methods for testing the molecular evolutionary clock hypothesis. *Genetics* 135: 599–607.

Takezaki N, et al. (1995). Phylogenetic test of the molecular clock and linearized trees. *Mol. Biol. Evol.* 12: 823–833.

Zuckerkandl E, Pauling L. (1965). Evolutionary divergence and convergence in proteins. In *Evolving Genes and Proteins*, V Bryson, H Vogel, Eds. Academic, New York, pp. 97–166.

See also UPGMA.

MOLECULAR COEVOLUTION. *SEE* COEVOLUTION.

MOLECULAR DRIVE. *SEE* CONCERTED EVOLUTION.

MOLECULAR DYNAMICS SIMULATIONS
Roland L. Dunbrack

Simulation of molecular motion, performed by solving Newton's equations of motion for a system of molecules. The simulation requires a potential energy function that expresses the energy of the system as a function of its coordinates. Simulations usually start with all atoms given a random velocity and direction of motion drawn from a distribution at the temperature of the simulation.

To simulate an infinite system, periodic boundary conditions are used by enclosing a molecular system in a box and virtually reproducing the system infinitely in each Cartesian direction. For instance, molecules on the right face of the box feel the presence (in terms of the energy function) of atoms on the left face of the box from a virtual copy of the box to the right of the system.

Molecular dynamics simulations can be performed at constant pressure rather than constant volume. The volume V of the unit cell is allowed to fluctuate under a piston of mass M, kinetic energy $K = M(dV/dT)^2/2$, and potential energy $E(V) = PV$.

Related Websites

AMBER	http://sigyn.compchem.ucsf.edu/amber/
CHARMM	http://www.scripps.edu/brooks/charmm_docs/charmm.html
GROMOS	http://www.igc.ethz.ch/gromos/

Further Reading

Karplus M, McCammon JA (2002). Molecular dynamics simulations of biomolecules. *Nat. Struct. Biol.* 9: 646–652.

Nose S, Klein ML (1986). Constant-temperature-constant-pressure molecular-dynamics calculations for molecular solids: Application to solid nitrogen at high pressure. *Phys. Rev. B. Condensed Matter* 33: 339–342.

See also AMBER, CHARMM, Empirical Potential Energy Function.

MOLECULAR EVOLUTIONARY MECHANISMS
Dov S. Greenbaum

The processes giving rise to changes in the sequences of DNA, RNA, and protein molecules during evolution.

There are many mechanisms through which evolutionary changes can take place. These mechanisms can result in a change in the gene itself—which may or may not have a phenotypic effect—affect the expression of the gene, or on a more global scale affect the frequency of a gene within a population.

These include:

a. Selective pressure—influential extrinsic factors such as environmental factors. Selective pressure has many results:
    i. Directional
    ii. Stabilizing
    iii. Disruptive

b. Recombination/duplication—the combination of different gene segments or the duplication of a gene or a gene segment.

c. Replication

d. Deletion

e. Mutation:

    i. Neutral—a neutral mutation does not affect the function of a gene.

    ii. Null—the wild type function is lost due to this mutation (also called a loss of function mutation).

    iii. Spontaneous—spontaneous mutations arise randomly, with very minimal frequency (1/106) genomic sequences.

    iv. Point—sometimes natural, but sometimes experimentally induced, a point mutation causes a mutation in a precise area of the sequence and to only one nucleotide. A transition point mutation is when a purine is substituted for another purine or a pyrimidine for another pyrimidine. A transversion occurs when a purine is substituted for a pyrimidine or vice versa.

    v. Frameshift—where there is a shift in the coding sequence (i.e., the insertion or deletion of a nucleotide). Given that each set of three nucleotides codes for an amino acid, the resulting shift causes the sequence to code for totally different amino acids after the insertion or deletion event; this results in a new reading frame.

    vi. Constitutive—the gene in question is turned on or off independent of upstream regulatory factors.

f. Domain shuffling—a new protein is the outcome of bringing different independent domains together.

g. Rearrangement at the chromosome level. This refers to changes that can be seen cytogenetically. They include:

    i. Deletion—loss of sequence. These can be large losses stretching from single bases to genes to millions of bases.

    ii. Duplication—copying of a sequence. When this occurs to the whole gene, a paralog is formed.

    iii. Insertion—inserting a new sequence into the middle of gene sequence. This can have minimal or major effects.

    iv. Inversion—flipping the sequence so that it is read backward.

    v. Translocation—moving the sequence to another area in the genome.

    vi. Transposition—occurs when a segment of a chromosome is broken off and moved to another place within that chromosome.

h. Resistance transfer—the transfer of drug resistance through methods such as horizontal gene transfer can significantly change the lifestyle of microorganisms.

i. Horizontal transfer (lateral transfer)—the process of transferring a gene between two independent organisms. This is in contradistinction to the transfer of genes from a parent to offspring—vertical gene transfer.

j. Silencing—the inactivation of a gene.

Studying these natural events along with an analysis of population genetics allow one to study the evolution of an organism.

Further Reading

Graur D, Li W-H (1999). *Fundamentals of Molecular Evolution.* Sinauer Associates, Sunderland, MA.

MOLECULAR INFORMATION THEORY
Thomas D. Schneider

Molecular information theory is information theory applied to molecular patterns and states. For a review see Schneider (1994).

Related Website

Tom Schneider's site	http://www.lecb.ncifcrf.gov/~toms/

Further Reading
Schneider TD (1994). Sequence logos, machine/channel capacity, Maxwell's demon, and molecular computers: A review of the theory of molecular machines. *Nanotechnology* 5: 1–18.

See also After State, Information.

MOLECULAR MACHINE
Thomas D. Schneider

A molecular machine can be defined precisely by six criteria:

1. A molecular machine is a single macromolecule or macromolecular complex.
2. A molecular machine performs a specific function for a living system.
3. A molecular machine is usually primed by an energy source.
4. A molecular machine dissipates energy as it does something specific.
5. A molecular machine "gains" information by selecting between two or more after states
6. Molecular machines are isothermal engines.

Further Reading
Schneider TD (1991). Theory of molecular machines. I. Channel capacity of molecular machines. *J. Theor. Biol.* 148: 83–123. http://www.lecb.ncifcrf.gov/~toms/paper/ccmm/.

See also Molecular Machine Capacity, Molecular Machine Operation.

MOLECULAR MACHINE CAPACITY
Thomas D. Schneider

The maximum information, in bits per molecular operation, that a molecular machine can handle is the molecular machine capacity. When translated into molecular biology, Shannon's channel capacity theorem states that "by increasing the number of independently moving parts that can interact cooperatively to make decisions, a molecular machine can reduce the error frequency (rate of incorrect choices) to whatever arbitrarily low level is required for survival of the organism, even when the machine operates near its capacity and dissipates small amounts of power" (Schneider, 1991).

This theorem explains the precision found in molecular biology, such as the ability of the restriction enzyme EcoRI to recognize 5'-GAATTC-3' while ignoring all other sites. The derivation is in Schneider (1991).

Further Reading
Schneider TD (1991). Theory of molecular machines. I. Channel capacity of molecular machines. *J. Theor. Biol.* 148: 83–123. http://www.lecb.ncifcrf.gov/~toms/paper/ccmm/.

See also Bit Information, Channel Capacity, Molecular Machine, Molecular Machine Operation.

MOLECULAR MACHINE OPERATION
Thomas D. Schneider

A molecular machine operation is the thermodynamic process in which a molecular machine changes from the high-energy before state to a low-energy after state. There are four standard examples:

- Before DNA hybridization the complementary strands have a high relative potential energy; after hybridization the molecules are noncovalently bound and in a lower energy state.
- The restriction enzyme EcoRI selects 5'-GAATTC-3' from all possible DNA duplex hexamers. The operation is the transition from being anywhere on the DNA to being at a GAATTC site.
- The molecular machine operation for rhodopsin, the light-sensitive pigment in the eye, is the transition from having absorbed a photon to having either changed configuration (in which case one sees a flash of light) or failed to change configuration.
- The molecular machine operation for actomyosin, the actin and myosin components of muscle, is the transition from having hydrolyzed an adenosine triphosphate (ATP) to having either changed configuration (in which the molecules have moved one step relative to each other) or failed to change configuration.

Further Reading
Schneider TD (1991). Theory of molecular machines. I. Channel capacity of molecular machines. *J. Theor. Biol.* 148: 83–123. http://www.lecb.ncifcrf.gov/~toms/paper/ccmm/.

See also After State, Before State, Molecular Machine.

MOLECULAR MECHANICS
Roland L. Dunbrack

Representation of molecules as mechanical devices with potential energy terms such as harmonic functions, torsional potentials, electrostatic potential energy, and van der Waals energy. Molecular mechanics programs for macromolecules usually include routines for energy minimization as a function of atomic coordinates and molecular dynamics simulations.

Related Websites

Guide to molecular mechanics	http://cmm.info.nih.gov/modeling/guide_documents/molecular_mechanics_document.html
AMBER	http://www.amber.ucsf.edu/amber/amber.html

Further Reading
Kollman P, Massova I, Reyes C, Kuhn B, Huo S, Chong L, Lee M, Lee T, Duan Y, Wang W, Donini O, Cieplak P, Srinivasan J, Case DA, Cheatham TE III (2000). Calculating structures and free energies of complex molecules: Combining molecular mechanics and continuum models. *Accts. Chem. Res.* 33: 889–897.

See also Electrostatic Energy, Energy Minimization, Molecular Dynamics Simulations.

MOLECULAR NETWORK. *SEE* NETWORK.

MOLECULAR REPLACEMENT
Liz P. Carpenter

Molecular replacement is a technique used in X-ray crystallography to solve the structures of macromolecules for which a related structure is known. The X-ray diffraction experiment gives a set of indexed diffraction spots each of which has a measured intensity I. The observed structure factor amplitude, $|F_{obs}|$, is proportional to the square root of the intensity.

A diffraction pattern can be calculated from any molecule placed in the unit cell in any position using the equation

$$F(hkl) = \sum_{\substack{j=1 \\ \text{all atoms}}}^{\text{all atoms}} f_j \exp(2\pi i[hk + ky + lz])$$

where $F(hkl)$ is the structure factor for the diffraction spot with indices hkl, f_j is the atomic scattering factor for the jth atom, and x, y, z are the coordinates of the jth atom in the unit cell.

The atomic scattering factor is the scattering from a single atom in a given direction and it can be calculated from the atomic scattering factor equation. In general,

$$f_j\left(\sin\frac{\theta}{\lambda}\right) = \sum_1^4 a_i \exp\left[-b_i\left(\frac{\sin\theta}{\lambda}\right)^2\right] + c$$

where a_i, b_i, and c are known coefficients.

Molecular replacement is a trial-and-error method in which a related structure is moved around inside the unit cell; the diffraction pattern is calculated using the equation above for each trial position and compared to the observed diffraction pattern. If the correlation between the observed and calculated pattern is above the background, this indicates that a solution has been found. The phases calculated from the correctly positioned model can be combined with the observed diffraction pattern to give an initial map of the electron density of the unknown structure.

The search for a molecular replacement solution requires testing both different orientations of the model and different translations of the model within the unit cell. This would be a six-dimensional search, covering three rotation parameters and three translational parameters. Fortunately this problem can be broken down into two consecutive three-dimensional searches. First suitable rotation angles are tested using a rotation function search and then the top solutions from the rotation function search are tested in a translation function search to identify the position of the molecule within the unit cell.

Further Reading
Crowther RA (1972). The fast rotation function. In *The Molecular Replacement Method*, MG Rossman, Ed., Gordon and Breach, New York.
Navaza J (1994). *Acta Cryst.* A50: 157–163.

Read RJ (2001). Pushing the boundaries of molecular replacement with maximum likelihood. *Acta Cryst.* D57: 1373–1382.

MOLPHY
John M. Hancock and Martin J. Bishop

A package for molecular phylogenetics with emphasis on the maximum-likelihood approach.

The package consists of a number of C programs augmented by Perl scripts. The main programs are ProtML, for maximum-likelihood phylogeny inference from protein sequences, and NucML, for maximum-likelihood phylogeny inference from nucleic acid sequences. Additionally, the package contains programs to extract basic statistics of the two types of sequences (ProtST and NucST) and an implementation of the neighbor-joining method of phylogeny reconstruction (NJdist). The scripts carry out a number of routine formatting and editing functions.

Related Website

MOLPHY	http://www.ism.ac.jp/software/ismlib/softother.e.html

See also Maximum-Likelihood Phylogeny Reconstruction, Neighbor-Joining Method.

MONOPHYLETIC (HOLOPHYLETIC)
John M. Hancock

In a phylogenetic tree, a monophyletic group is a group of nodes that share a common ancestor.

The term *holophyletic* is a special case in which no nodes have been knowingly excluded from the group to form a paraphyletic group. A monophyletic group can be either holophyletic (e.g., the vertebrates) or paraphyletic (fish) depending on usage.

See also Phylogenetic Tree, Taxonomic Classification.

MONTE CARLO SIMULATIONS
Roland L. Dunbrack

A method for randomly sampling from a probability distribution used in many fields, including statistics, economics, and chemistry. Monte Carlo simulations must begin with an initial configuration of the system. A new configuration is drawn randomly from a proposal distribution and is either accepted or rejected depending on whether the new configuration is more or less likely than the current configuration. If the new configuration is more probable, the new configuration is always accepted. If the new configuration lowers the probability, the new state is accepted with probability $p_{new}/p_{current}$. For molecular systems, the moves can be changes in Cartesian atom positions or in internal coordinates such as bond lengths, bond angles, and dihedral angles. The energy of the system is used to calculate the probabilities of each move. If the energy decreases, the move is always accepted. If the energy is increased, then the probability of accepting the move is calculated with the change in energy of the system, $\exp(-(E_{new} - E_{current})/k_B T)$, where T is the temperature of the system and k_B is the Boltzmann constant.

Related Website

MCPRO	http://zarbi.chem.yale.edu/

Further Reading
Jorgensen WL (1998). *The Encyclopedia of Computational Chemistry*, Vol. 3. Wiley, Chichester, United Kingdom, pp. 1754–1763.

MOPAC
Roland L. Dunbrack

A computer program package for semiempirical quantum mechanical energy calculations on molecules, from Schrödinger, Inc. MOPAC uses the AM1 ("Austin model 1") force field developed by M. J. S. Dewar and colleagues at the University of Texas in Austin.

Related Website

MOPAC	http://www.schrodinger.com/Products/mopac.html

Further Reading
Dewar MJS, et al. (1985). AM1: A new general purpose quantum mechanical molecular model. *J. Am. Chem. Soc.* 107: 3202–3209.

MOSAIC PROTEIN
Laszlo Patthy

Multidomain proteins containing multiple types of protein domains of distinct evolutionary origin.

Such chimeric proteins that arose by fusion of gene segments of two or more different genes are also referred to as modular proteins, and the constituent domains are usually referred to as protein modules.

See also Modular Protein, Multidomain Protein, Protein Domain, Protein Module.

MOTIF
Terri Attwood and Dov S. Greenbaum

Motifs are characteristic conserved, short regions of peptide or nucleic acid sequence.

Motifs are scientifically valuable when shared by either evolutionarily related proteins or proteins that have the same or similar function. Within a protein motifs can be discerned in the primary, secondary, and tertiary structures.

At the level of protein sequence, a motif can be defined as a consecutive string of residues in a protein sequence whose general character is repeated, or conserved, in all sequences in a multiple alignment at a particular position, as illustrated below:

```
YVTVQH KK LRT PL
AATMKF KK LRH PL
YIFATT KS LRT PA
```

```
VATLRY KK LRQ PL
YIFGGT KS LRT PA
WVFSAA KS LRT PS
WIFSTS KS LRT PS
YLFSKT KS LQT PA
YLFTKT KS LQT PA
```

When building a multiple alignment, as more distantly related sequences are included, it is often necessary to insert gaps to bring equivalent parts of adjacent sequences into the correct register. As a result of this process, islands of conservation (motifs) tend to emerge within a sea of mutational change. The motifs (typically 15–20 residues in length) tend to correspond to the core structural and functional elements of the protein. Their conserved nature allows them to be used to diagnose family membership of uncharacterized sequences through the application of a range of sequence analysis techniques: For example, regular expressions are used to encode single motifs and are the basis of the PROSITE database; fingerprints and blocks encode multiple motifs and are the basis of the PRINTS and Blocks resources. None of the available analytical techniques should be regarded as the best: Each exploits sequence motifs in different way—each therefore offers a different perspective and a different (complementary) diagnostic opportunity. The best strategy is therefore to use them all.

Secondary-structure motifs, minimally comprising at least two or three substructures (i.e., coil, helix, or sheet), are somewhat arbitrary in that they can describe a whole fold or only a small segment of a peptide. A simple motif could be the meander, which comprises three up–down beta helices. Many of these motifs together comprise the larger motif of a beta barrel. A common structural motif in DNA-binding proteins is the zinc finger motif characterized by fingerlike loops of amino acids stabilized by zinc atoms.

Motifs are helpful in comparing widely different genomes by contributing to the biological parts lists for genomes, thus providing a common ground for genome–genome comparisons.

Related Websites

ScanProsite	http://www.expasy.org/tools/scanprosite/
FingerPRINTScan	http://www.bioinf.man.ac.uk/dbbrowser/fingerPRINTScan/
Blocks search	http://blocks.fhcrc.org/blocks/blocks_search.html
TRANSFAC	http://transfac.gbf.de/TRANSFAC/

Further Reading

Attwood TK (2000). The quest to deduce protein function from sequence: The role of pattern databases. *Int. J. Biochem. Cell Biol.* 32(2): 139–155.

Attwood TK, Parry-Smith DJ (1999). *Introduction to Bioinformatics.* Addison Wesley Longman, Harlow, Essex, United Kingdom.

Mulder NJ, Apweiler R (2002). Tools and resources for identifying protein families, domains and motifs. *Genome Biol.* 3: REVIEWS2001.

See also Alignment—Multiple, Gap, Sequence Motifs—Prediction and Modeling.

MOTIF SEARCHING
Jaap Heringa

A collection of related sequences contains more information than a single sequence. Taking advantage of this multiple-sequence information for feature extraction and databank searching is quite natural and has been explored by many researchers. Two closely related problems have to be addressed: (i) how to represent the collective information contained in many sequences and (ii) how to quantify the similarity between the representation of multiple sequences and each individual sequence in the databank to be searched.

With the growth of structural information available, it has become clear that certain regions of a protein molecular structure are less liable to change than others. Knowledge of this kind can be used to elucidate certain characteristics of a protein's architecture such as buried *versus* exposed location of a segment or the presence of specific secondary structural elements. The most salient aspect, however, involves the protein's functionality: The most conserved protein region very often serves as a ligand binding site, a target for posttranslational modification, the enzyme catalytic pocket, and the like. Detecting such sites in newly determined protein sequences could save immense experimental effort and help characterize functional properties. Moreover, using the conserved regions of a protein only rather than its whole sequence for databank searching can reduce background noise and help considerably to establish distant relationships.

In principle, the purpose of any method of sequence comparison, such as the alignment of two sequences, is to find primary-structure conservation and establish reliable regions of local similarity. Sequence motifs are usually considered in the context of multiple-sequence comparison, when certain contiguous sequence spans are shared by a substantial number of proteins. Staden (1988) gave the following classification for protein sequence motifs:

- exact match to a short defined sequence;
- percentage match to a defined short sequence;
- match to a defined sequence, using a residue exchange matrix and a cutoff score;
- match to a weight matrix with cutoff score;
- direct repeat of fixed length; and
- a list of allowed amino acids for each position in the motif.

Sequence patterns are sometimes referred to as a combination of several elementary motifs separated by intervening sequence stretches (Staden, 1988), although this distinction between pattern and motif is not commonly adhered to.

Protein sequence motifs are often derived from a multiple-sequence alignment, so that the quality of the alignment determines the correctness of the patterns. Defining consensus motifs from a multiple alignment is mostly approached using majority and plurality rules, so that a consensus sequence is generally expressed as a set of heterogeneous rules of some kind for each alignment position. Taylor (1986a) used Venn diagrams to make partially overlapping subdivisions of the amino acid residue types into classes, such as CHARGED, SMALL, HYDROPHOBIC, POSITIVE, etc. Any position within a sequence pattern can thus be described by a logical rule of the type TINY.or.POLAR_non-AROMATIC.or.PROLINE. The stringency of the restraint imposed on a particular position should depend on how crucial a given feature is for the protein family under scrutiny. Allowed elements can range from an absolutely required residue type to a gap. Often an initial alignment of sequences is generated on the basis of structural information available (e.g., superposition of Cα atoms) to ensure reliability (Taylor, 1986b). After creating the template from the initial alignment, it can then be extended to include additional related proteins. This process is repeated iteratively until no other protein sequence can be added without giving up on essential features. In the absence of structural information, methods have been developed based on pairwise sequence comparison as a starting point for determining sequence patterns, after which a multiple alignment is built by aligning more and more sequences to the pattern, which is allowed to develop at the same time (e.g., Patthy, 1987; Smith and Smith, 1990). Indirect measures of conservation have also been attempted as criteria for motif delineation, such as the existence of homogeneous regions in a protein's physical property profiles (e.g., Chappey and Hazout, 1992).

Frequently, a sequence motif is proven experimentally to be responsible for a certain function. A collection of functionally related sequences should then yield a discriminating pattern occurring in all the functional sequences and not occurring in other unrelated sequences. Such patterns often consist of several sequentially separated elementary motifs, which, e.g., join in the three-dimensional structure to form a functional pocket. Such discriminating motifs were studied extensively early on for particular cases such as helix–turn–helix motifs (Dodd and Egan, 1990) or G-protein-coupled receptor fingerprints (Attwood and Findlay, 1993).

A consistent semimanual methodology for finding characteristic protein patterns was developed by Bairoch (1993). The aim of the approach was to make the derived patterns as short as possible but still sensitive enough to recognize the maximum number of related sequences and also sufficiently specific to reject, in a perfect case, all unrelated sequences. A large collection of motifs gathered in this way is available in the PROSITE databank (Hofmann et al., 1999). Associated with each motif is an estimate of its discriminative power. For many PROSITE entries, published functional motifs serve as a first approximation of the pattern. In other cases, a careful analysis of a multiple alignment for the protein family under study is performed. The initial motif is searched against the SWISS-PROT sequence databank and search hits are analyzed. The motif is then empirically refined by extending or shrinking it in length to achieve the minimal amount of false positives and maximal amount of true positives. For example, the putative AMP- (adenosine monophosphate) binding pattern ([LIVMFY]-x(2)-[STG]-[STAG]-G-[ST]-[STEI]-[SG]-x-[PASLIVM]-[KR]) given in PROSITE is found in 150 motifs over 112 sequences from the SWISS-PROT databank (Release 38); 137 hits in 99 sequences are true positives and 13 hits in 13 sequences are known false positives. The PROSITE databank and related software is an invaluable and generally available tool for detecting the function of newly sequenced and uncharacterized proteins. The PROSITE database also includes the extended profile formalism of Bucher et al. (1996) for sensitive profile searching.

A description of sequence motifs in a regular expression such as in the PROSITE database, involves enumeration of all residue types allowed at particular alignment positions or by less stringent rules previously described. This necessarily leads to some loss of information, because particular sequences might not be considered. Also, the formalism is not readily applicable to all protein families. Exhaustive information about a conserved sequence span can be stored in the form of an ungapped sequence block. Henikoff and Henikoff (1991) derived a comprehensive collection of such blocks (known as the Blocks databank) from groups of related proteins as specified in the PROSITE databank (Hofmann et al., 1999) using the technique of Smith et al. (1990). Searching protein sequences against the Blocks library of conserved sequence blocks or an individual block against the whole sequence library (Wallace and Henikoff, 1992) provides a sensitive way of establishing distant evolutionary relationships between proteins. Recently, the Blocks database was extended (Henikoff and Henikoff, 1999) by using, in addition to PROSITE, nonredundant information from four more databases: the PRINTS protein motifs database (Attwood and Beck, 1994) and the protein domain sequence databases PFAM-A (Bateman et al., 1999), ProDom (Corpet et al., 1999), and Domo (Gracy and Argos, 1998).

With the availability of fast and reliable sequence comparison tools, methods have been developed to compile automatically as many patterns as possible from the full protein sequence databank. These methods often rely on automatic clustering and multiple alignment of protein sequence families followed by application of motif extraction tools. Recent programs include the methods DOMAINER (Sonnhammer and Kahn, 1994), MKDOM (Gouzy et al., 1997), and DIVCLUS (Park and Teichmann, 1998). Pattern-matching methods based upon machine learning procedures have also been attempted, in particular neural networks (e.g., Frishman and Argos, 1992), as these have the ability to learn an internal nonlinear representation from presented examples. The method GIBBS (Lawrence et al., 1993) detects conserved regions with a residual degree of similarity from unaligned sequences and is based on iterative statistical sampling of individual sequence segments. The motifs found are multiply aligned by the method. Also, the program MEME (Bailey and Elkan, 1994) employs unsupervised motif searching but using an expectation maximization (EM) algorithm. Although MEME motifs are ungapped, the program can find multiple occurrences in individual sequences, which do not need to be encountered within each input sequence.

The evolution of pattern derivation methods and resulting motifs in turn has triggered the development of tools of varying degrees of complexity to search individual sequences and whole-sequence databanks with user-specified motifs. Particularly, many programs are available for searching sequences against the PROSITE databank; a full list of publicly available and commercial programs for this purpose is supplied together with PROSITE (Hofmann et al., 1999).

A further extension of the motif-searching techniques is provided by methods that compare sequence templates with target sequences while allowing gaps for insertions and deletions (e.g., Rhode and Bork, 1993). These methods are similar to profilelike methods. Alternatively, it is possible to search a databank with multiple ungapped motifs independently and then merge the hit lists for the motifs to find consistent occurrences of certain databank sequences (Attwood and Findlay, 1993).

Further Reading

Attwood TK, Beck ME (1994). PRINTS—a protein motif fingerprint database. *Protein Eng.* 7: 841–848.

Attwood TK, Findlay JBC (1993). Design of a discriminating fingerprint for G-protein-coupled receptors. *Protein Eng.* 6: 167–176.

Bailey TL, Elkan C (1994). Fitting a mixture model by expectation maximization to discover motifs in biopolymers. In *Proceedings of the Second International Conference on Intelligent Systems for Molecular Biology*. AAAI Press, Menlo Park, CA, pp. 28–36.

Bairoch A (1993). The PROSITE dictionary of sites and patterns in proteins, its current status. *Nucleic Acids Res.* 21: 3097–3103.

Bateman A, et al. (1999). Pfam 3.1: 1313 multiple alignments and profile HMMs match the majority of proteins. *Nucleic Acids Res.* 27: 260–262.

Bucher P, et al. (1996). A flexible motif search technique based on generalized profiles. *Comput. Chem.* 20: 3–24.

Chappey C, Hazout S (1992). A method for delineating structurally homogeneous regions in protein sequences. *Comput. Appl. Biosci.* 8: 255–260.

Corpet F, et al. (1999). Recent improvements of the ProDom database of protein domain families. *Nucleic Acids Res.* 27: 263–267.

Dodd IB, Egan JB (1990). Improved detection of helix-turn-helix DNA-binding motifs in protein sequences. *Nucleic Acids Res.* 18: 5019–5026.

Frishman DI, Argos P (1992). Recognition of distantly related protein sequences using conserved motifs and neural networks. *J. Mol. Biol.* 228: 951–962.

Gouzy J, et al. (1997). XDOM, a graphical tool to analyse domain arrangements in protein families. *Comput. Appl. Biosci.* 13: 601–608.

Gracy J, Argos P (1998). Automated protein sequence database classification. II. Delineation of domain boundaries from sequence similarities. *Bioinformatics* 14: 174–187.

Henikoff S, Henikoff JG (1991). Automated assembly of protein blocks for database searching. *Nucleic Acids Res.* 19: 6565–6572.

Henikoff S, Henikoff JG (1999). New features of the Blocks database server. *Nucleic Acids Res.*, 27: 226–228.

Hofmann K, et al. (1999). The PROSITE database, its status in 1999. *Nucleic Acids Res.* 27: 215–219.

Lawrence CE, et al. (1993). Detecting subtle sequence signals: A Gibbs sampling strategy for multiple alignment. *Science* 262: 208–214.

Park J, Teichmann SA (1998). DIVCLUS: An automatic method in the GEANFAMMER package that finds homologous domains in single- and multi-domain proteins. *Bioinformatics* 14: 144–150.

Patthy L (1987). Detecting homology of distantly related proteins with consensus sequences. *J. Mol. Biol.* 198: 567–577.

Rhode K, Bork P (1993). A fast sensitive pattern-matching approach for protein sequences. *Comput. Appl. Biosci.* 9: 183–189.

Smith HO, et al. (1990). Finding sequence motifs in groups of functionally related proteins. *Proc. Natl. Acad. Sci. USA* 87: 826–830.

Smith RF, Smith TF (1990). Automatic generation of primary sequence patterns from sets of related sequences. *Proc. Natl. Acad. Sci. USA* 87: 118–122.

Sonnhammer ELL, Kahn D (1994). Modular arrangement of proteins as inferred from analysis of homology. *Protein Sci.* 3: 482–492.

Staden R (1988). Methods to define and locate patterns of motifs in sequences. *Comput. Appl. Biosci.* 4: 53–60.

Taylor WR (1986a). The classification of amino acid conservation. *J. Theor. Biol.* 119: 205–218.

Taylor WR (1986b). Identification of protein sequence homology by consensus template alignment. *J. Mol. Biol.* 188: 233–258.

Wallace JC, Henikoff S (1992). PATMAT: A searching and extraction program for sequence, pattern and block queries and databases. *Comput. Appl. Biosci.* 8: 249–254.

MOUSE GENOME DATABASE. *SEE* MGD.

MPSRCH
John M. Hancock and Martin J. Bishop

A parallel implementation of the Smith–Waterman algorithm for database searching.

By using dynamic programming rather than heuristic methods, MPsrch will find the best local alignments between any pair of sequences. The method is only implemented for protein sequences.

Related Websites

MPsrch	http://www.ebi.ac.uk/MPsrch/
Aneda	http://www.anedabio.com/

Further Reading
Smith TF, Waterman MS (1981). Identification of common molecular subsequences. *J. Mol. Biol.* 147: 195–197.

See also Dynamic Programming, Smith–Waterman Algorithm.

MRBAYES
Michael P. Cummings

A program for phylogenetic analysis of nucleotide or amino acid sequence data using a Bayesian approach.

A metropolis-coupled Markov chain Monte Carlo (MCMCMC) algorithm is used with multiple chains, all but one of which is heated. The chains are used to sample model space through a process of parameter modification proposal and acceptance/rejection steps (also called cycles or generations). The heating raises the posterior probability by a factor, β, which has the effect of increasing the magnitude of change between steps in the Markov chain. After each cycle an exchange between a heated and unheated chain is evaluated similar to the other proposal and acceptance/rejection mechanism. The motivation for MCMCMC is to increase mixing. After the process becomes stationary, the frequency with which parameter values are visited in the process represents an estimate of their underlying posterior probability. A choice of several commonly used likelihood models is available, as are choices for starting tree (user defined and random), data partitions (e.g., by codon position), and Markov chain Monte Carlo parameters. The accessory programs, sumt and sump, are used to process the output files to summarize trees and parameters, respectively.

The program is used through a command line interface, and a user can execute a series of individual commands without interaction by including appropriate commands in the MrBayes block (NEXUS file format). The programs are written in American National Standards Institute (ANSI) C and are available as source code and executables for some platforms.

Related Website

MrBayes	http://morphbank.ebc.uu.se/mrbayes/

Further Reading
Huelsenbeck JP, Ronquist F (2001). MRBAYES: Bayesian inference of phylogenetic trees. *Bioinformatics* 17: 754–755.

See also BAMBE.

MRNA (MESSENGER RNA)
John M. Hancock

The final product of transcription and subsequent processing.

A messenger RNA typically contains a single open reading frame, although in bacteria it may contain more than one (known as a polycistronic message). In eukaryotes mRNA molecules are derived from the immature transcript, typically by the removal of introns and the addition of a poly(A) tail and a methylated cap. In bacteria no maturation takes place and translation can take place on the nascent mRNA. In eukaryotes, mature mRNAs are exported from the nucleus to the cytoplasm where they bind to ribosomes to participate in translation.

Further Reading
Lewin B (2000). *Genes VII*. Oxford University Press, Oxford, UK.

See also hnRNA, Transcription.

MULTIDOMAIN PROTEIN
Laszlo Patthy

Proteins that are larger than approximately 200–300 residues usually consist of multiple protein domains. The individual structural domains of multidomain proteins are compact, stable units with a unique three-dimensional structure. The interactions within one domain are more significant than with other domains. The presence of distinct structural domains in multidomain proteins indicates that they fold independently, i.e., the structural domains are also folding domains. Frequently, the individual structural and folding domains of a multidomain protein perform distinct functions.

Many multidomain proteins contain multiple copies of a single type of protein domain, indicating that internal duplication of gene segments encoding a domain has given rise to such proteins. Some multidomain proteins contain multiple types of domain of distinct evolutionary origin. Such chimeric proteins that arose by fusion of two or more gene segments are frequently referred to as mosaic proteins or modular proteins, and the constituent domains are usually referred to as protein modules.

See also Modular Protein, Mosaic Protein, Protein Domain, Protein Module.

MULTIFACTORIAL TRAIT (COMPLEX TRAIT)
Mark McCarthy and Steven Wiltshire

Phenotypes (traits or diseases) for which individual susceptibility is governed by the action of (and interaction between) multiple genetic and environmental factors.

Most of the major diseases of mankind (e.g., diabetes, asthma, many cancers, schizophrenia) are multifactorial traits. While they can be seen to cluster within families, such traits manifestly fail to show classical patterns of Mendelian segregation. Individual susceptibility, it is assumed, reflects variation at a relatively large number of polymorphic sites in a variety of genes, each of which has a relatively modest influence on risk. Depending on the assumptions about the number of genes involved, this may be termed *polygenic* or *oligogenic* inheritance. In addition, the phenotypic expression of this genetic inheritance in any given person will, typically, depend on their individual history of exposure to relevant environmental influences (e.g., food availability in the case of obesity). For such multifactorial traits, no single variant is likely to be either necessary or sufficient for the development of disease. This incomplete correspondence between genotype and phenotype (contrasting with the tight correlation characteristic of Mendelian diseases) provides one of the major obstacles to gene mapping in such traits, because of the consequent reduction in power for both linkage and linkage disequilibrium (association) analyses.

Further Reading

Bennett ST, Todd JA (1996). Human Type 1 diabetes and the insulin gene: Principles of mapping polygenes. *Ann. Rev. Genet.* 30: 343–370.

Cardon LR, Bell JI (2001). Association study designs for complex disease. *Nat. Rev. Genet.* 2: 91–99.

Lander ES, Schork NJ (1994). Genetic dissection of complex traits. *Science* 265: 2037–2048.

Risch N (2000). Searching for genetic determinants in the new millennium. *Nature* 405: 847–856.

Risch N (2001). Implications of multilocus inheritance for gene-disease association studies. *Theor. Popul. Biol.* 60: 215–220.

See also Linkage Analysis, Mendelian Disease, Oligogenic Inheritance, Penetrance, Polygenic Inheritance.

MULTILAYER PERCEPTRON. *SEE* NEURAL NETWORK.

MULTIPLE ANOMALOUS DISPERSION PHASING (ANOMALOUS DISPERSION, MAD PHASING)

Liz P. Carpenter

Multiple anomalous dispersion is a technique used in X-ray crystallography to obtain relative phases for diffraction spots by exploiting the effects of absorption edges on the diffraction of X rays.

The scattering of X rays by an atom generally increases slowly with increasing wavelength. However, there are discontinuities in this profile, where the scattering suddenly drops and then increases slowly again. This is due to the energy of the X ray being close to the energy needed to excite a bound electron into a higher energy level. Electrons are therefore absorbed and emitted at the same or a different wavelength, instead of just being scattered. These discontinuities in the absorption-versus-wavelength spectrum are called absorption edges, and each element has several at different wavelengths, some of which lie within the range of wavelengths that is available at synchrotrons (0.5–2 Å).

These absorption edges give rise to changes in the intensity of reflections. Away from absorption edges Friedel's law applies, whereby $I(h, k, l) = I(-h, -k, -l)$. However, this law no longer applies near the absorption edge, and a phenomenon called anomalous scattering is observed where there are small differences between the Bijvoet pairs, $I(h, k, l)$ and $I(-h, -k, -l)$.

The anomalous scattering can be used to assist in solving the phase problem. In a MAD experiment several data sets are collected at selected wavelengths above and below the absorption edge of an anomalous scattering atom in the crystal. Heavy-atom derivatives such as mercury and seleno-methionine derivatives are good candidates for MAD data collection or protein crystals and brominated DNA is used for DNA–protein complexes.

Related Websites

X-ray absorption edges	http://www.bmsc.washington.edu/scatter/AS_periodic.html
Anomalous scattering	http://www-structure.llnl.gov/Xray/101Index.html

See also Diffraction of X Rays, Phase Problem, X-Ray Crystallography for Structure Determination and associated Further Reading.

MULTIPLE HIERARCHY (POLYHIERARCHY)
Robert Stevens

A hierarchy in which, unlike a tree, a concept may have multiple parents. A multiple hierarchy may be seen in the figure under Directed Acyclic Graph.

See also Directed Acyclic Graph.

MULTIPOINT LINKAGE ANALYSIS
Mark McCarthy and Steven Wiltshire

The determination of evidence for linkage which simultaneously takes into account the information arising from multiple typed loci.

Multipoint linkage analysis can be applied to the analysis of multiple marker loci alone, as in the case in marker map construction, or to the mapping of a putative disease susceptibility locus onto a map of markers, in the case in linkage analysis for Mendelian disease and multifactorial/complex traits. The two most commonly used algorithms in linkage analysis—the Elston–Stewart (E–S) and Lander–Green (L–G) algorithms—calculate the likelihood of the data at each position on a marker map, conditional on the genotypes of all the markers in the map, although the L–G algorithm can handle substantially more markers than the E–S algorithm. A multipoint LOD score is obtained at each position on the map, and the peak in the resulting multipoint linkage profile identifies the position of the maximum evidence for linkage. By combining evidence from several markers, each of which may, in isolation, provide limited information, multipoint linkage analysis can provide enhanced evidence for and localization of susceptibility loci when compared to two-point analyses. Multipoint linkage analysis is now routine in genome scans for complex trait loci.

Examples: Vionnet and colleagues (2000) report the results of a genome scan for type 2 diabetes loci in French pedigrees that employs a multipoint linkage analysis.

Related Websites

LINKAGE	ftp://linkage.rockefeller.edu/software/linkage/
GENEHUNTER	http://www-genome.wi.mit.edu/ftp/distribution/software/genehunter

Merlin	http://www.sph.umich.edu/csg/abecasis/merlin/reference.html
Jurg Ott's linkage page	http://linkage.rockefeller.edu/

Further Reading

Ott J (1999). *Analysis of Human Genetic Linkage.* John's Hopkins University Press, Baltimore, MD, pp. 114–150.

Vionnet N, et al. (2000). Genomewide search for type 2 diabetes-susceptibility genes in French Whites: Evidence for a novel susceptibility locus for early-onset diabetes on chromosome 3q27-qter and independent replication of a type 2-diabetes locus on chromosome 1q21-q24. *Am. J. Hum. Genet.* 67: 1470–1480.

See also Genome Scans; Linkage Analysis, Multifactorial Trait.

MUTATION MATRIX. *SEE* AMINO ACID EXCHANGE MATRIX.

N

National Center for Biotechnology
    Information
Natural Selection
NCBI
NDB
Nearest-Neighbor Methods
Nearly Neutral Theory
Needleman–Wunsch Algorithm
Negentropy
Neighbor-Joining Method
Network
Neural Network
Neutral Theory
Newton-Raphson Minimization
Nit

NIX
NMR
Noise
Noncrystallographic Symmetry
Nonparametric Linkage Analysis
Nonsynonymous Mutation
NOR
NT
N-Terminus
Nuclear Intron
Nuclear Magnetic Resonance
Nucleic Acid Database
Nucleic Acid Sequence Databases
Nucleolar Organizer Region
Nucleotide Base Codes

Dictionary of Bioinformatics and Computational Biology. Edited by Hancock and Zvelebil
ISBN 0-471-43622-4 © 2004 John Wiley & Sons, Inc.

NATIONAL CENTER FOR BIOTECHNOLOGY INFORMATION. *SEE* NCBI.

NATURAL SELECTION
A. Rus Hoelzel

Natural selection is the nonrandom, differential survival and reproduction of phenotypes in a population of organisms of a given species, whereby organisms best suited to their environment contribute a higher proportion of progeny to the next generation.

The idea of natural selection was first mentioned by Charles Darwin in his notebooks in 1838 and later described in detail as the motive force behind evolutionary change in his 1859 volume, *The Origin of Species by Means of Natural Selection, or The Preservation of Favoured Races in the Struggle for Life*.

This process assumes that phenotypically expressed traits that can be selected in natural populations have a heritable basis, later established to be genes encoded by DNA. Selection acts by changing allele frequencies in natural populations, and for a single-locus, two-allele system in Hardy–Weinberg equilibrium for a diploid organism this change can be expressed as

$$\Delta_s p = \frac{pq[p(w_{11} - w_{12}) + q(w_{12} - w_{22})]}{p^2 w_{11} + 2pq w_{12} + q^2 w_{22}}$$

where w_{11}, w_{12}, and w_{22} represent the viabilities of A_1A_1, A_1A_2, and A_2A_2 genotypes, respectively, and p is the allele frequency of A_1, while q is the allele frequency of A_2. These changes led to the adaptation of organisms to their environment. Selection can be directional when one allele is favored, balanced when selection maintains polymorphism in a population (e.g., by overdominance or frequency-dependent selection), or disruptive when homozygotes are favored over the heterozygote condition.

Further Reading
Darwin C (1859). *The Origin of Species by Means of Natural Selection, or The Preservation of Favoured Races in the Struggle for Life*, Penquin Books, London.

Hartl D (2000). *A Primer in Population Genetics*. Sinauer Associates, Sunderland, MA.

See also Adaptation, Kin Selection, Sexual Selection.

NCBI (NATIONAL CENTER FOR BIOTECHNOLOGY INFORMATION)
Guenter Stoesser

The National Center for Biotechnology Information (NCBI) develops and provides information systems for molecular biology, conducts research in computational biology, and develops software tools for analyzing genome data.

Dictionary of Bioinformatics and Computational Biology. Edited by Hancock and Zvelebil
ISBN 0-471-43622-4   © 2004 John Wiley & Sons, Inc.

Analysis resources include BLAST, RefSeq, Electronic PCR, UniGene, OrfFinder, HomoloGene, Single Nucleotide Polymorphisms (dbSNP) database, Human Genome Sequencing, Human MapViewer, Human Mouse Homology Map, Entrez Genomes, Clusters of Orthologous Groups (COG) database, SAGEmap, Gene Expression Omnibus (GEO), Online Mendelian Inheritance in Man (OMIM), and the Molecular Modeling Database (MMDB). Retrieval resources include Entrez, PubMed, LocusLink, and the Taxonomy Browser.

NCBI's mission is to develop information technologies to aid in the understanding of fundamental molecular and genetic processes that control health and disease. NCBI is located and maintained at the National Institute of Health (NIH) in Bethesda, Maryland.

Related Websites

NCBI	http:www.ncbi.nlm.nih.gov
GenBank	http://www.ncbi.nlm.nih.gov/Genbank/
Literature databases	http://www.ncbi.nlm.nih.gov/Literature/
Genome biology	http://www.ncbi.nlm.nih.gov/Genomes/
Data mining tools	http://www.ncbi.nlm.nih.gov/Tools/

Further Reading

Altschul SE, et al. (1990). Basic local alignment search tool. *J. Mol. Biol.* 215: 403–410.

Edgar R, et al. (2002). Gene Expression Omnibus: NCBI gene expression and hybridization array data repository. *Nucleic Acids Res.* 30: 207–210.

Pruitt K, Maglott D (2001). RefSeq and LocusLink: NCBI gene-centered resources. *Nucleic Acids Res.* 29: 137–140.

Schuler GD, et al. (1996). Entrez: Molecular biology database and retrieval system. *Methods Enzymol.* 266: 141–162.

Wheeler DL, et al. (2002). Database resources of the National Center for Biotechnology Information: 2002 update. *Nucleic Acids Res* 30: 13–16.

NDB. *SEE* NUCLEIC ACID DATABASE.

NEAREST-NEIGHBOR METHODS
Patrick Aloy

The *nearest neighbor* is a classification algorithm that assigns a test instance to the class of a "nearby" example whose class is known. When applied to secondary-structure prediction, this means that the secondary-structure state of the central residue of a test segment is assigned according to the secondary structure of the closest homolog of known structure.

A key element in any nearest-neighbor prediction algorithm is the choice of a scoring system for evaluating segment similarity. For such a purpose, a variety of approaches have been used (e.g., similarity matrices, neural networks) with different success.

Some methods that use this approach (e.g., NNSSP) have achieved three-state-per-residue accuracies of 72.2%.

Related Website

NNSSP	http://bioweb.pasteur.fr/seqanal/interfaces/nnssp-simple.html

Further Reading

Salamov AA, Solovyev VV (1995). Prediction of protein secondary structure by combining nearest-neighbor algorithms and multiple sequence alignments. *J. Mol. Biol.* 247: 11–15.

Yi TM, Lander ES (1993). Protein secondary structure prediction using nearest-neighbor methods. *J. Mol. Biol.* 232: 1117–1129.

See also Neural Network, Secondary-Structure Prediction of Protein.

NEARLY NEUTRAL THEORY. *SEE* NEUTRAL THEORY.

NEEDLEMAN–WUNSCH ALGORITHM
Jaap Heringa

The technique to calculate the highest scoring or optimal alignment is generally known as the dynamic programming (DP) technique. While the physicist Richard Bellman first conceived DP and published a number of papers on the topic between 1955 and 1975, Needleman and Wunsch (1970) introduced the technique to the biological community, and their paper remains among the most cited in the area. The technique falls in the class of global alignment techniques, which align protein or nucleotide sequences over their full lengths. The original Needleman–Wunsch algorithm penalizes the inclusion of gaps in the alignment through a single gap penalty for each gap, irrespective of how many amino acids are spanned by the gap, and also penalizes end gaps.

Further Reading

Needleman SB, Wunsch CD (1970). A general method applicable to the search for similarities in the amino acid sequence of two proteins. *J. Mol. Biol.* 48: 443–453.

NEGENTROPY (NEGATIVE ENTROPY)
Thomas D. Schneider

The term *negentropy* was defined by Brillouin (1962, p. 116) as "negative entropy," $N = -S$. Supposedly, living creatures feed on negentropy from the sun. However, it is impossible for entropy to be negative, so negentropy is always a negative quantity. The easiest way to see this is to consider the statistical-mechanics (Boltzmann) form of the entropy equation:

$$S \equiv -k_B \sum_{i=1}^{\Omega} P_i \ln P_i \quad \left(\frac{\text{joules}}{K \cdot \text{microstate}} \right)$$

where k_B is Boltzmann's constant, Ω is the number of microstates of the system, and P_i is the probability of microstate i. Unless one wishes to consider imaginary probabilities, it can be proven that S is positive or zero. Rather than saying "negentropy" or "negative entropy," it is clearer to note that when a system dissipates energy to its surroundings, its entropy decreases. So it is better to refer to $-\Delta S$ (a negative change in entropy).

Further Reading
Brillouin L (1962). *Science and Information Theory*, 2nd ed. Academic Press, New York.

NEIGHBOR-JOINING METHOD
Sudhir Kumar and Alan Filipski

A method for inferring phylogenetic trees based on the minimum-evolution (ME) principle.

Neighbor joining is a stepwise method that begins with a star tree and successively resolves the phylogeny by joining pairs of taxa in each step such that the selected pairing minimizes the sum of branch lengths. Thus, the ME principle is applied locally at each step, and a final tree is produced along with the estimates of branch lengths. This method does not require the assumption of molecular clock and is known to be computationally efficient and phylogenetically accurate for realistic simulated data.

Further Reading
Felsenstein J (1993). *PHYLIP: Phylogenetic Inference Package*. University of Washington, Seattle, WA.

Kumar S, et al. (2001). MEGA2: Molecular evolutionary genetics analysis software. *Bioinformatics* 17: 1244–1245.

Saitou N, Nei M (1987). The neighbor-joining method—a new method for reconstructing phylogenetic trees. *Mol. Biol. Evol.* 4: 406–425.

Studier JA, Keppler KJ (1988). A note on the neighbor-joining algorithm of Saitou and Nei. *Mol. Biol. Evol.* 5: 729–731.

Swofford DL (1998). *PAUP*: Phylogenetic Analysis Using Parsimony (and Other Methods)*. Sinauer Associates, Sunderland, MA.

See also Minimum-Evolution Principle.

NETWORK (GENETIC NETWORK, METABOLIC NETWORK, MOLECULAR NETWORK)
Denis Thieffry

Biological macromolecules [DNA fragments, proteins, RNAs, other (bio)chemical compounds] interact with each other, affecting their respective concentrations, conformations, or activities. A collection of interacting molecular components thus forms a molecular network endowed with emerging dynamical properties.

One speaks of a genetic interaction occurring between two genes when the perturbation (deletion, overexpression) of the expression of the first gene affects that of the second, something which can be experimentally observed at the phenotypic or molecular levels. Such interactions can be direct or indirect, i.e., involving intermediate genes. Turning to the molecular level, one can further distinguish between different types of molecular interaction: protein–DNA interactions (e.g., transcriptional regulation involving a regulatory protein and a DNA cis-regulatory region near the regulated gene), protein–RNA interactions (as in the case of posttranscriptional regulatory interactions), and protein–protein interactions (e.g., phosphorylation of a protein by a kinase, formation of multimeric complexes).

Molecular interactions between biological macromolecules are often represented as a graph, with genes as vertices and interactions between them as edges. Information on genetic interactions can be obtained directly from genetic experiments (e.g., mutations, genetic crosses). Regardless of the types of experimental evidence, one often refers to genetic networks, although the corresponding interactions usually involve various types of molecular components (e.g., proteins, RNA species). The genetic networks described in the literature predominantly encompass transcription regulatory interactions. At the level of the organism, the complete set of regulatory interactions forms the regulome.

Regulatory interactions of various types (e.g., protein–protein interactions, transcriptional regulation) form regulatory cascades. Signaling molecules are sensed by membrane receptors, leading to (cross-) phosphorylation by kinases (or other types of protein–protein interactions), ultimately leading to the translocation or the change of activity of transcription factors, which will in turn affect the transcription of specific subsets of genes in the nucleus. Different regulatory cascades (each forming a graph, typically with a tree topology) can share some elements, thus leading to further regulatory integration, something which can be represented in terms of acyclic graphs. When regulatory cascades ultimately affect the expression of genes coding for some of their crucial components (e.g., membrane receptors), the system is better represented by (cyclic) directed graphs.

In the context of metabolism or (to a lesser extent) gene regulation, one speaks of regulatory feedback when the product of a reaction participates in the regulation of the activity or of the expression of some of the enzymes at work in the reaction (or pathway). Such regulatory feedback can be inhibitory (negative feedback, e.g., the inhibition of a reaction by its product) or activatory (positive feedback). This notion of regulatory feedback has to be distinguished from that of a regulatory (feedback) circuit defined in the context of regulatory graphs. A feedback regulatory circuit often counts a regulatory feedback among its edges.

The term *module* is used in various contexts with varying meanings in biology and bioinformatics. In the context of metabolic or genetic networks, it corresponds to the delineation of interactive structures associated with specific functional processes. From a graph-theoretic perspective, in contrast to regulatory cascades, corresponding to a tree graph topology (devoid of feedback circuits), regulatory modules correspond to (strongly) connected components and thus often to sets of intertwined feedback circuits.

Cell metabolism involves numerous reactions sharing some of their reactants, products, and catalytic enzymes, thus also forming complex networks, including cycles such as the Krebs cycle. The full set of metabolic reactions occurring in one organism is called the metabolic network of the organism, or metabolome.

Information on genetic and metabolic networks is scarcely available in general molecular databases. Furthermore, when the information is present, it lacks the appropriate higher level integration needed to allow queries dealing with multigenic regulatory modules or pathways. As a consequence, a series of research groups or consortia are developing dedicated databases to integrate genetic regulatory data or metabolic information. [See the table below for a list of the prominent (publicly available) genetic databases and metabolic databases.]

Molecular, metabolic, and genetic (sub)graphs may be characterized by different network structures, depending on the pattern of interactions between the genes or molecular species involved. Formally, one can make rigorous distinctions on the basis of graph theory.

When facing a metabolic or genetic network, various biological questions can be phrased in terms of comparison of interactive structures. One can, e.g., check whether a set of genes, each paralogous to a gene of another set, also presents a conserved interaction pattern. When comparing orthologous or paralogous sets of genes across different organisms, one can evaluate the conservation of regulatory networks or metabolic pathways beyond the conservation of the structure or sequence of individual components (genes or molecular species). Formally, these comparisons can be expressed in terms of graph-theoretic notions such as (sub)graph isomorphism (similar interactive structures) or homeomorphism (similar topologies).

A molecular, metabolic, or genetic network can be displayed and analyzed on the basis of the use of various types of representations. The most intuitive description consists simply in using diagrams such as those found in many biological articles. But such a graphical description does not usually conform to strictly defined standards, making difficult any computer implementation or analysis. When turning to more standardized and formal approaches, one can distinguish three main (related) formal representations: the first referring to graph theory, the second referring to the matrix formalism (i.e., the use of a matrix to represent interactions

Table   Examples of Databases Integrating Genetic, Molecular, or Metabolic Interaction Data

Name	Types of data	URL
Amaze	Biochemical pathways	http://www.ebi.ac.uk/research/pfmp/
Ecocyc/ Metacyc	Metabolic pathways	http://biocyc.org
KEGG	Metabolic pathways	http://www.genome.ad.jp/kegg/
TransPath	Signal transduction pathways	http://transpath.gbf.de/
BIND	Protein interactions and complexes	http://www.bind.ca/
GeneNet	Gene networks	http://wwwmgs.bionet.nsc.ru/mgs/systems/genenet/
CSNDB	Cell-signaling networks	http://geo.nihs.go.jp/csndb/

between genes, each regulator corresponding to a specific column and a specific row), and the last referring to dynamical system theory and using different types of (differential) equations.

Molecular, metabolic, and genetic networks can be visualized via dedicated graphical user interfaces (GUIs). These network visualization tools offer various types of functions, such as zooming or editing (selection, deletion, insertion, and displacement of objects).

The variation of metabolic activity or of gene expression across time constitutes what is called *network dynamics*. The term is often used to designate the full set of temporal behavior associated with a network, depending on different signals or perturbations. Formally, this behavior can be represented by time plots or dynamical state transition graphs. Network state usually refers to the levels of activity or concentration of all components of a network at a given time.

In contrast to the derivation of the temporal behavior from prior knowledge of a regulatory (metabolic or genetic) network, the notion of reverse engineering refers to the derivation of the network or of parts of its regulatory structure on the basis of dynamical data (i.e., a set of characterized states of the network). Reverse engineering methods can be developed in the context of various formal approaches, e.g., by fitting a differential model to dynamical data using computer optimization methods.

Further Reading

Bower JM, Bolouri H (2001). *Computational Modeling of Genetic and Biochemical Networks.* MIT Press, Cambridge, MA.

Goldbeter A (1997). *Biochemical Oscillations and Cellular Rhythms: The Molecular Bases of Periodic and Chaotic Behaviour.* Cambridge University Press, Cambridge, UK.

Voit EO (2000). *Computational Analysis of Biochemical Systems: A Practical Guide for Biochemists and Molecular Biologists.* Cambridge University Press, Cambridge, UK.

See also Bayesian Network; Graph Representation of Genetic, Molecular and Metabolic Networks; Mathematical Modeling of Molecular/Metabolic/Genetic Networks.

NEURAL NETWORK (ARTIFICIAL NEURAL NETWORK, BACKPROPAGATION NETWORK, CONNECTIONIST NETWORK, MULTILAYER PERCEPTRON)
Nello Cristianini

A general class of algorithms for machine learning, originally motivated by analogy with the structure of neurons in the brain.

A neural network can be described as a parametrized class of functions specified by a weighted graph (the network's architecture). The weights associated with the edges of the graph are the parameters, each choice of weights identifying a function.

For directed graphs, we can distinguish recurrent architectures (containing cycles) and feed-forward architectures (acyclic). A very important special case of feed-forward networks is given by layered networks, in which the nodes of the graph are organized into an ordered series of disjoint classes (the layers) such that connections are possible only between elements of two consecutive classes and following the natural order. The weight between the unit k and the unit j of a network is indicated with w_{kj}, and we assume that all elements of a layer are connected to all elements of the successive layer (fully connected architecture). In this way, the connections between two layers can be represented by a weight matrix W whose entry jk corresponds to the connection between node j and node k in successive layers (see the accompanying figure). It is customary to call input and output layers the first and the last ones and hidden layers all the remaining.

In a layered network, the function is computed sequentially, assigning the value of the argument to the input layer, then calculating the activation level of the successive layers as described below, until the output layer is reached. The output of the function computed by the network is the activation value of the output unit.

All units in a layer are updated simultaneously, and all the layers are updated sequentially, based on the state of the previous layer. Each unit k calculates its value y_k by a linear combination of the values at the previous layer, x, followed by a nonlinear transformation $t : R \rightarrow R$ as follows: $y_k = t(Wx)$, where t is called the transfer function and for which a common choice is the logistic function

$$f(z) = \frac{1}{1 + e^{-z}}$$

Notice that the input/output behavior of the network is determined by the weights (or, in other words, each neural network represents a class of nonlinear functions parametrized by the weights), and training the network amounts to automatically choosing the values of the weights. A perceptron can be described as a network of this type with no hidden units.

It can also be seen as the building block of complex networks, in that each unit can be regarded as a perceptron (if instead of the transfer function one uses a threshold function, returning Boolean values). Layered feed-forward neural networks are often referred to as multilayer perceptrons.

Given a (labeled) training set of data and a fixed error function for the performance of the network, the training of a neural network can be done by finding those weights that minimize the network's error on such a sample (i.e., by fitting the network to the data). This can be done by gradient descent, if the error function is differentiable, by means of the backpropagation algorithm.

In the parameter space, one can define a cost function that associates each configuration with a given performance on the training set. Such a function is typically nonconvex, so that it can only be locally minimized by gradient descent. Backpropagation provides a way to compute the necessary gradients, so that the network finds a local minimum of the training error with respect to network weights.

The chain rule of differentiation is used to compute the gradient of the error function with respect to the weights. This gives rise to a recursion proceeding from the output unit to the input, hence the name backpropagation.

If y_i is the value of the ith unit, for each w_{kj} connecting it to the previous layer's units, one can write the partial derivative of the error function as

$$\frac{\partial E}{\partial w_{ij}} = \frac{\partial E}{\partial y_i} \frac{\partial y_i}{\partial w_{ij}} = \frac{\partial E}{\partial y_i} t'(x_i) x_i \doteq \varepsilon_i$$

where $t( )$ is the transfer function defined before, and hence the update for such weight will be $\Delta w_{ij} = -\eta \varepsilon_i y_j$, where η is a parameter known as the learning rate. Any differentiable error function can be used.

Among the main problems of neural networks are that this training algorithm is only guaranteed to converge to a local minimum and the solution is affected by the initial conditions since the error function is in general nonconvex. Also problematic

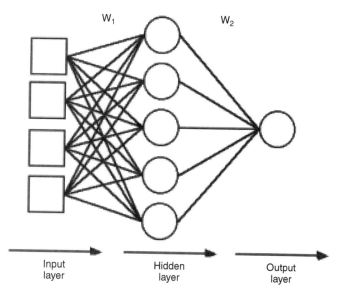

Figure A Feed-Forward Neural Network or Multilayer
Perceptron.

is the design of the architecture, often chosen as the result of trial and error. Some
such problems have been overcome by the introduction of the related method of
support vector machines.

Other types of networks arise from different design choices. For example, radial
basis function networks use a different transfer function; Kohonen networks are
used for clustering problems and Hopfield networks for combinatorial optimization
problems. Also, different training methods exist.

Further Reading

Baldi P, Brunak S (2001). *Bioinformatics: A Machine Learning Approach*. MIT Press,
Cambridge, MA.

Bishop C (1996). *Neural Networks for Pattern Recognition*. Oxford University Press, Oxford,
UK.

Mitchell T (1997). *Machine Learning*. McGraw Hill, New York.

Cristianini N, Shawe-Taylor J (2000). *An Introduction to Support Vector Machines*. Cambridge
University Press.

Duda RO, Hart PE, Stork DG (2001). *Pattern Classification*. Wiley, New York.

NEUTRAL THEORY (NEARLY NEUTRAL THEORY)
A. Rus Hoelzel

Theory of evolution suggesting that most evolutionary change at the molecular level
is caused by the random genetic drift of mutations that are selectively neutral or
nearly neutral.

Proposed by Kimura (1968), the neutral theory was controversial from the start
and has undergone numerous modifications over the years. Although it has been
called "non-Darwinian" because of its emphasis on random processes, it does not
actually contradict Darwin's theory. It proposes that most substitutions have no

influence on survival but does not deny the importance of selection in shaping the adaptive characteristics of many morphological, behavioral, and life history characteristics.

The evidence in support of the neutral theory included the observation that the rate of amino acid substitution in vertebrate lineages was remarkably constant for different loci, which in turn led to the proposal of a molecular clock. The observed rate of change by substitution was also noted to be similar to the estimated mutation rate, consistent with theoretical expectations. One modification to the theory, especially to account for discrepancies between the expected and observed relationship between amino acid substitution rate and generation time, suggested that most mutations are slightly deleterious, as opposed to neutral (e.g., Ohta, 1973).

Further Reading

Kimura M (1968). Evolutionary rate at the molecular level. *Nature* 217: 624–626.

Ohta T (1973). Slightly deleterious mutant substitutions in evolution. *Nature* 246: 96–98.

Ohta T (1992). The nearly neutral theory of molecular evolution. *Annu. Rev. Ecol. Systematics* 23: 263–286.

See also Evolution, Evolutionary Clock, Genetic Drift, Natural Selection.

NEWTON-RAPHSON MINIMIZATION. *SEE* ENERGY MINIMIZATION.

NIT
Thomas D. Schneider

Natural units for information or uncertainty are given in nits.

If there are M messages, then $\ln(M)$ nits are required to select one of them, where ln is the natural logarithm with base e ($= 2.71828\ldots$). Natural units are used in thermodynamics where they simplify the mathematics. However, nits are awkward to use because results are almost never integers. In contrast, the bit unit is easy to use because many results are integer (e.g., $\log_2 32 = 5$) and these are easy to memorize. Using the relationship

$$\frac{\ln(x)}{\ln(2)} = \log_2(x)$$

allows one to present all results in bits.

Related Website

Appendix in primer on information theory gives table of powers of 2 that is useful to memorize	http://www.lecb.ncifcrf.gov/~toms/paper/primer

Further Reading

Schneider TD (1995). *Information Theory Primer*. http://www.lecb.ncifcrf.gov/~toms/paper/primer.

See also Bit, Message.

NIX

John M. Hancock and Martin J. Bishop

A Web tool to aid the analysis of nucleotide sequences.

Similar in philosophy to PIX, the application masks the input sequences for repeats using RepeatMasker. Sequences are then submitted for BLAST database searches. For genomic sequences the following exon-finding programs are run: GRAIL, Genefinder, Genemark, Fex, Hexon, and Fgene. A transfer RNA (tRNA) gene search is also carried out using the program trnascan. If a complementary DNA (cDNA) sequence is input, only GRAIL is used for exon identification and no tRNA search is carried out.

Related Website

| NIX | http://www.hgmp.mrc.ac.uk/Registered/Webapp/nix/ |

See also BLAST, GRAIL, PIX.

NMR (NUCLEAR MAGNETIC RESONANCE)

Liz P. Carpenter

Nuclear magnetic resonance (NMR) is a method for solving the three-dimensional structures of macromolecules and small molecules in solution.

The nuclei of atoms have associated magnetic fields which are referred to as spin states. The spin state of an atom depends on the number of protons and neutrons in the nucleus. ^{12}C and ^{16}O both have $I = 0$, where I is the spin number. Proton, ^{13}C, and ^{15}N have $I = 1/2$ whereas ^{14}N and ^{2}H have $I = 1$. When nuclei with spin greater than zero are placed in a magnetic field, the spin direction aligns with the magnetic field and an equilibrium state is obtained. If radio-frequency pulses are then applied to the samples, higher energy states are produced, and when these revert to the equilibrium state, radio-frequency radiation is emitted. The exact frequency of the radiation emitted depends on the environment around the nucleus, so each nucleus will emit at a different frequency (unless the nuclei are exactly chemically equivalent).

In order to solve the structure of a macromolecule, the types of radio-frequency pulses used are varied and the nature of the nuclei probed. Two useful types of NMR experiments involve COSY (correlation spectroscopy), which gives information about nuclei that are connected through bonds, and the nuclear Overhauser effect (NOE), which gives information about through-space interactions. Since these spectra on individual isotopes of a particular nucleus often give rise to overlapping peaks, it is necessary to perform two-, three-, and four-dimensional experiments using different isotopes of various nuclei. The isotopes of hydrogen, carbon, and nitrogen present in a macromolecule can be varied by producing the protein in bacteria that are grown in media rich in the required isotope. For example, ^{12}C is replaced with ^{13}C and/or ^{13}N is replaced with ^{14}N.

Once spectra have been collected, it is necessary to assign all the peaks to individual atoms and interactions between atoms. This is done using through-bond experiments such as COSY, TOCSY, and three- and four-dimensional experiments. The NOEs (through space) provide information about which parts of the protein that are distant in the sequence are adjacent in the folded structure. Newer techniques exist that can provide longer range interactions based on partial alignment of the molecules in solution, e.g., by use of lipid solutions. Once these distances have been measured, they can be used to generate a series of distance restraints to apply to the protein to form a folded structure. Usually structures are generated by various molecular mechanics techniques, obtaining an ensemble of structures that represents the experimental data.

Related Websites

NMR teaching pages, by David Gorenstein.	http://www.biophysics.org/btol/NMR.html
Gary Trammell's teaching pages on NMR	http://www.uis.edu/~trammell/che425/nmr_theory/
Joseph Hornak's teaching pages	http://www.cis.rit.edu/htbooks/nmr/inside.htm

NOISE
Thomas D. Schneider

Noise is a physical process that interferes with transmission of a message.

Shannon pointed out that the worst kind of noise has a Gaussian distribution. Since thermal noise is always present in practical systems, received messages will always have some probability of having errors.

Further Reading
Shannon CE (1949). Communication in the presence of noise. *Proc. IRE*, 37: 10–21.

See also Error, Thermal Noise.

NONCRYSTALLOGRAPHIC SYMMETRY. *SEE* SPACE GROUP.

NONPARAMETRIC LINKAGE ANALYSIS. *SEE* ALLELE-SHARING METHODS.

NONSYNONYMOUS MUTATION
Laszlo Patthy

Nucleotide substitution occurring in translated regions of protein coding genes that alters the amino acid are called *amino acid changing* or *nonsynonymous mutations*.

NOR. *SEE* NUCLEOLAR ORGANIZER REGION.

NT. *SEE* BASE PAIR.

N-TERMINUS (AMINO TERMINUS)
Roman A. Laskowski

In a polypeptide chain the N-terminus is the end residue that has a free amino group (NH_3). The other end is referred to as the C-terminus. The direction of the chain is defined as running from the N- to the C-terminus. N-terminal refers to being of or relating to the N-terminus.

See also Amino Acid, C-Terminus, Polypeptide.

NUCLEAR INTRON. *SEE* INTRON.

NUCLEAR MAGNETIC RESONANCE. *SEE* NMR.

NUCLEIC ACID DATABASE (NDB)
Guenter Stoesser

The Nucleic Acid Database (NDB) assembles and distributes information about the crystal structures of nucleic acids.

The *Atlas of Nucleic Acid Containing Structures* highlights the special aspects of each structure in the NDB. Archives contain nucleic acid standards, nucleic acid summary information, software programs, coordinate files, and structure factor data.

Data for the crystal structures of nucleic acids are deposited using the AutoDep Input Tool. NDB maintains the mmCIF website; mmCIF (macromolecular Crystallographic Information File) is the IUCr-approved data representation for macromolecular structures. Databases for monomer units and ligands have been created by à la mode, which is "A Ligand And Monomer Object Data Environment" for building models.

NDB is located in the Department of Chemistry and Chemical Biology at Rutgers University in Piscataway, New Jersey.

Related Websites

NDB	http://ndbserver.rutgers.edu/NDB/
NDB Search	http://ndbserver.rutgers.edu/NDB/structure-finder/ndb/
NDB Atlas	http://ndbserver.rutgers.edu/NDB/NDBATLAS/
Structure deposition	http://ndbserver.rutgers.edu/NDB/deposition/

Further Reading
Berman HM, et al. (1992). The Nucleic Acid Database: A comprehensive relational database of three-dimensional structures of nucleic acids. *Biophys. J.* 63: 751–759.
Berman HM, et al. (1996). The Nucleic Acid Database: Present and future. *J. Res. Natl. Inst. Stand. Technol.* 101: 243–257.

Berman HM, et al. (1999). The Nucleic Acid Database: A research and teaching tool. In *Handbook of Nucleic Acid Structure*, S Neidle, Ed. Oxford University Press, New York, pp. 77–92.

Olson WK, et al. (2001). A standard reference frame for the description of nucleic acid base-pair geometry. *J. Mol. Biol.* 313: 229–237.

NUCLEIC ACID SEQUENCE DATABASES
Guenter Stoesser

The principal nucleic acid sequence databases, DDBJ/EMBL/GenBank, constitute repositories of all published nucleotide sequences. These comprehensive archival databases are primary factual databases containing data which can be mined and analyzed by computer analysis. In addition to these primary sequence databases, there exists a large variety of organism- or molecule-specific secondary sequence databases.

While data in primary databases originate from original submissions by experimentalists, sequence data in secondary databases are typically derived from a primary database and are complemented by further human curation, computational annotation, literature research, personal communications, etc. For primary databases the ownership of data lies with the submitting scientist, while secondary databases have editorial power over contents of the database.

Related Websites

EBI database resources	http://www.ncbi.nlm.nih.gov/Database/
NCBI database resources	http://www.ncbi.nlm.nih.gov/Database/

Further Reading
Baxevanis AD (2002). The Molecular Biology Database Collection: 2002 update. *Nucleic Acids Res.* 30: 1–12.

NUCLEOLAR ORGANIZER REGION (NOR)
Katheleen Gardiner

Sites of tandem repeats of ribosomal RNA genes.

In metaphase chromosomes, NORs appear as secondary constrictions. In interphase chromosomes, NORs show attached spherical structures, called nucleoli, where the ribosomal RNAs are transcribed. In the human genome, NORs are located on the short arms of the acrocentric chromosomes 13, 14, 15, 21, and 22.

Further Reading
Miller OJ, Therman E (2000). *Human Chromosomes*. Springer, New York.

Wagner RP, Maguire MP, Stallings RL (1993). *Chromosomes—A Synthesis*. Wiley-Liss, New York.

NUCLEOTIDE BASE CODES. *SEE* IUPAC-IUB CODES.

O

OBF
Observation
Occam's Razor
OilEd
OKBC
Oligogenic Effect
Oligogenic Inheritance
Oligo Selection Program
Omics
Ontolingua
Ontology
Ontology Web Language
Open Bioinformatics Foundation
Open Knowledge Base Connectivity
Open Reading Frame
Open Reading Frame Finder
Open-Source Bioinformatics Organizations

Operational Taxonomic Unit
Operon
OPLS
Optimal Alignment
ORF
ORFan
ORF Finder
Organelle Genome Database
Organismal Classification
Orphan Gene
Ortholog
OSP
OTU
Overdominance
Overfitting
Overtraining
OWL

Dictionary of Bioinformatics and Computational Biology. Edited by Hancock and Zvelebil
ISBN 0-471-43622-4 © 2004 John Wiley & Sons, Inc.

OBF. *SEE* OPEN BIOINFORMATICS FOUNDATION.

OBSERVATION. *SEE* FEATURE.

OCCAM'S RAZOR. *SEE* PARSIMONY.

OILED
Robert Stevens

OilEd is an ontology editor allowing the user to build ontologies using DAML+OIL and OWL. Although OilEd is a somewhat lightweight editor, it has two important characteristics distinguishing it from other similar tools. It provides access to the full functionality of the ontology languages. Thus concept descriptions can use arbitrary Boolean combinations (e.g., and, or, not) and explicit quantification of property fillers is supported. A reasoner can be used to determine concept hierarchies, discovering inferred subsumption relationships and logical inconsistencies within an ontology. OilEd is freely available for download and use, has been used in a number of projects, both industrial and academic, and is also widely used as a tool for teaching.

Related Website

OilEd	http://oiled.man.ac.uk

OKBC. *SEE* OPEN KNOWLEDGE BASE CONNECTIVITY.

OLIGOGENIC EFFECT. *SEE* OLIGOGENIC INHERITANCE.

OLIGOGENIC INHERITANCE (OLIGOGENIC EFFECT)
Mark McCarthy and Steven Wiltshire

A genetic architecture whereby variation in a trait is determined by variation in several genes, each of which has substantial additive effects on the phenotype.

The oligogenic architecture is applicable, as with polygenic inheritance, to both quantitative traits and discrete traits under the liability threshold model. Oligogenicity is to be distinguished from polygenic (many-gene) and Mendelian—(single–gene)

Dictionary of Bioinformatics and Computational Biology. Edited by Hancock and Zvelebil
ISBN 0-471-43622-4   © 2004 John Wiley & Sons, Inc.

inheritance patterns, although the boundaries between these are, in practice, often indistinct.

Further Reading

Falconer D, McKay TFC (1996). *Introduction to Quantitative Genetics*. Prentice-Hall, Harlow, England, pp. 100–183.

Hartl DL, Clark AG (1997). *Principles of Population Genetics*. Sinauer Associates, Sunderland, MA, pp. 397–481.

See also Mendelian Disease, Multifactorial Trait, Polygenic Inheritance, Quantitative Trait.

OLIGO SELECTION PROGRAM. *SEE* OSP.

OMICS
John M. Hancock

Catch-all for fields of study attempting to integrate the study of some feature of an organism.

Recent years have seen an explosion in terms ending in the suffix -*omics*. These all derive from the term *genome* (hence *genomics*), a term invented by Hans Winkler in 1920, although the use of -*ome* is older, signifying the "collectivity" of a set of things (see Web article by Lederberg and McCray). The oldest of these terms, and one that seems due to come back into fashion, may be *biome*. Thus, although the explosion in the use of this terminology may appear fatuous, it does signify a widespread interest in moving toward an integrative, rather than reductionist, approach to biology, following from the early successes of genomics and functional genomics.

Related Websites

Article by Lederberg and McCray	http://www.the-scientist.com/yr2001/apr/comm_010402.html
Omes	http://bioinfo.mbb.yale.edu/what-is-it/omes/omes.html

Further Reading

Winkler H (1920). *Verbreitung und Ursache der Parthenogenesis im Pflanzen- und Tierre-iche*. Verlag Fischer, Jena.

See also Network, System Biology.

ONTOLINGUA
Robert Stevens

Ontolingua provides a distributed collaborative environment to browse, create, edit, modify, and use ontologies. The ontology server architecture provides access to a library of ontologies, translators to other languages, and an editor to create and browse ontologies. Remote users can browse and edit ontologies. Applications can access any of the ontologies in the ontology library using OKBC. The Ontolingua server has extended the original language in two ways: First, it provides explicit support for building ontological modules that can be assembled, extended, and refined in a new ontology. Second, it makes an explicit separation between an ontology's presentation and representation.

Related Website

| Ontolingua | http://www.ksl.stanford.edu/software/ontolingua/ |

Further Reading

Bechhofer S (2002). *Ontology Language Standardisation Efforts; OntoWeb Deliverable D4.0.* http://ontoweb.aifb.uni-karlsruhe.de/About/Deliverables/d4.0.pdf.

ONTOLOGY
Robert Stevens

In computer science an ontology is a way of capturing how a particular community thinks about its field. An ontology attempts to capture a community's knowledge or understanding of a domain of interest. Like any representation, an ontology is only a partial representation of the community's understanding. Ontologies are used to share a common understanding between both people and computer systems. Most ontologies consist of at least concepts, the terms that name those concepts, and relationships between those concepts. Ideally, the concepts have definitions and the collection of terms (lexicon) provides a vocabulary by which a community can talk about its domain. The ontology provides a structure for the domain and constrains the interpretations of the terms the ontology provides. The goal of an ontology is to create an agreed-upon vocabulary and semantic structure for exchanging information about that domain.

Within molecular biology ontologies have found many uses. The most common use is to provide a shared, controlled vocabulary. For example, the Gene Ontology provides a common understanding for the three major attributes of gene products. TAMBIS provides the illusion of a common query interface to multiple, distributed bioinformatics resources and aids in the reconciliation of semantic heterogeneities. The ontologies in RiboWeb and EcoCyc form rich and sophisticated schema for their knowledge bases and offer inferential support not normally seen in a conventional database management system.

Further Reading

Bechhofer S (2002). *Ontology Language Standardisation Efforts; OntoWeb Deliverable D4.0.* http://ontoweb.aifb.uni-karlsruhe.de/About/Deliverables/d4.0.pdf.

Uschold M, et al. (1998). The Enterprise Ontology. *Knowledge Eng. Rev.* 13: 32–89.

ONTOLOGY WEB LANGUAGE (OWL)
Robert Stevens

The successor to DAML+OIL that is to become a W3C standard for the creation and exchange of ontologies. Like DAML+OIL, it is underpinned by an expressive description logic and can be submitted to a reasoning engine to infer taxonomies and check consistency.

OPEN BIOINFORMATICS FOUNDATION (OBF)
John M. Hancock and Martin J. Bishop

An umbrella group that provides a central resource for a variety of open-source bioinformatics projects.

The OBF grew out of the BioPerl project, which was officially organized in 1995. The project was an international association of developers of open-source Perl tools for bioinformatics. The OBF acts as a distribution point for a number of similar or related projects, including biojava.org, biopython.org, Distributed Annotation Server (DAS), bioruby.org, biocorba.org, Ensembl, and EMBOSS.

Related Websites

OBF	http://open-bio.org/
BioCorba	http://biocorba.org
BioJava	http://biojava.org
BioPerl	http://bioperl.org
BioPython	http://biopython.org
BioRuby	http://bioruby.org
Distributed Annotation Server	http://biodas.org
EMBOSS	http://www.emboss.org/
Ensembl	http://www.ensembl.org/

See also EMBOSS, Ensembl, Open-Source Bioinformatics Organizations.

OPEN KNOWLEDGE BASE CONNECTIVITY (OKBC)
Robert Stevens

Open Knowledge Base Connectivity (OKBC) is not an ontology representation language but an application programming interface for accessing knowledge bases stored in knowledge representation systems (KRSs).

OKBC provides a uniform model of KRSs based on a common conceptualization of classes, individuals, slots, facets, and inheritance. OKBC is defined in a programming language–independent fashion and has existing implementations in Common Lisp, Java, and C. The protocol transparently supports networked as well as direct access to KRSs and knowledge bases. OKBC consists of a set of operations that provide a generic interface to underlying KRSs. This interface isolates an application from many of the idiosyncrasies of a specific KRS and enables the development of tools (e.g., graphical browsers, frame editors, analysis tools, inference tools) that operate on many KRSs.

Related Website

OKBC	http://www.ai.sri.com/~okbc/

Further Reading
Bechhofer S (2002). *Ontology Language Standardisation Efforts; OntoWeb Deliverable D4.0.* http://ontoweb.aifb.uni-karlsruhe.de/About/Deliverables/d4.0.pdf.

OPEN READING FRAME (ORF)
Niall Dillon

Set of translation codons spanning the region between a start and a stop codon. In prokaryotes, which have no introns, finding ORFs forms the basis of gene finding. In eukaryotes introns make this approach inapplicable, but defining the largest ORF in a complementary DNA (cDNA) sequence is an important component of identifying the likely protein product of a gene, and of identifying pseudogenes.

OPEN READING FRAME FINDER. *SEE* ORF FINDER.

OPEN-SOURCE BIOINFORMATICS ORGANIZATIONS
John M. Hancock

Bioinformatics organizations dedicated to the open-source model of software development and dissemination.

Although bioinformatics originated as an academic discipline, its increasing importance has led to a plethora of commercial products, some of which have become almost indispensable to their users. The open-source model, familiar from the activities of organizations such as GNU, Linux, and SourceForge, asserts that software and data should be freely available and distributed under nonrestrictive licensing conditions. Organizations operating this model in bioinformatics include bioinformatics.org, openinformatics.org, and the Open Bioinformatics Foundation.

Related Websites

bioinformatics.org	http://bioinformatics.org/
openinformatics.org	http://www.openinformatics.org/
Open Bioinformatics Foundation	http://open-bio.org/
GNU	http://www.gnu.org/
Linux	http://www.linux.org/
Open Source Initiative	http://www.opensource.org/
SourceForge	http://sourceforge.net/

See also Open Bioinformatics Foundation.

OPERATIONAL TAXONOMIC UNIT (OTU)
John M. Hancock

In a phylogenetic tree, the terminal nodes, or leaves, corresponding to observable genes or species.

The term is used to distinguish these observable nodes from internal nodes, which are presumed to have existed earlier in evolution and not to be observable. A further implication of the term is that, in the context of taxonomy, these nodes may correspond to species, subspecies, or some other defined unit of taxonomy. They may, however, correspond to individuals within a population or members of a gene family within an individual species.

See also Phylogenetic Tree.

OPERON
Katheleen Gardiner

A unit of genetic regulation comprising a set of genes under the coordinate regulation of an operator gene.

The term was originally coined to describe bacterial systems and does not apply in higher eukaryotes. The phenomenon has been observed in parasitic eukaryotes such as Leishmania. The occurrence of operons in a species presents special challenges for the prediction of gene regulatory elements.

Further Reading
Lewin B (2000). *Genes VII*. Oxford University Press, New York.

OPLS
Roland L. Dunbrack

A molecular mechanics potential energy function for organic molecules developed by William Jorgensen and colleagues at Yale University.

Related Website

Jorgensen	http://zarbi.chem.yale.edu/

Further Reading
Jorgensen WL (1998). *The Encyclopedia of Computational Chemistry*, Vol. 3. Wiley, Chichester, United Kingdom, pp. 3281–3285.

OPTIMAL ALIGNMENT
Jaap Heringa

The optimal alignment of two sequences refers to the highest scoring alignment out of very many possible alignments, as assessed using a scoring system consisting of a residue exchange matrix and gap penalty values. The most widely applied technique for obtaining an optimal global alignment or local alignment is the dynamic programming algorithm. The fast homology search methods BLAST and FASTA are heuristics to generate approximate optimal alignments, the scores of which are assessed by statistical evaluations, such as E values.

See also Alignment—Multiple, Alignment—Pairwise, Dynamic Programming, Needleman–Wunsch Algorithm.

ORF. *SEE* OPEN READING FRAME.

ORFAN. *SEE* ORPHAN GENE.

ORF FINDER (OPEN READING FRAME FINDER)
John M. Hancock and Martin J. Bishop

A Web-based tool to find and characterize open reading frames within DNA sequences.

Sequences are pasted into the Web interface (or IDs supplied) producing a graphical display of ORFs in all six reading frames. Selected ORFs can then be used for BLAST searches or searching the COG database.

Related Website

ORF Finder	http://www.ncbi.nlm.nih.gov/gorf/gorf.html

See also BLAST, Clusters of Orthologous Groups, Open Reading Frame.

ORGANELLE GENOME DATABASE. *SEE* GOBASE.

ORGANISMAL CLASSIFICATION. *SEE* TAXONOMIC CLASSIFICATION.

ORPHAN GENE (ORFAN)
Jean-Michel Claverie

A (often putative) protein sequence (open reading frame) apparently unrelated to any other previously identified protein sequence from any other organism.

In practice, this notion depends on a similarity threshold (e.g., percentage identity or BLAST score) and/or a given significance level (e.g., Blast *E* value). Orphan genes (often nicknamed ORFans) are more precisely defined as protein sequences which cannot be *reliably* aligned with any other, i.e., with similarity levels in the twilight zone. This definition is time dependent, as it depends on the content of the sequence database. Thus the fraction of ORFans tends to diminish as more genomes get sequenced. The ORFan definition is adapted to the presence of very close species (or bacterial strains) in the databases, but has been used in an evolutionary distance-dependent fashion, meaning "only found in." For instance, *Escherichia coli* ORFans will have clear homologs in the other *E. coli* strain sequences but not elsewhere. The relative proportion of ORFans corresponding to truly unique genes vs. genes that have simply diverged too far from their relative in other genomes is unknown. The determination of the three-dimensional structure of ORFan proteins is one way of addressing this question.

Further Reading
Fischer D, Eisenberg D (1999). Finding families for genomic ORFans. *Bioinformatics* 15: 759–762.

Monchois V, et al. (2001). *Escherichia coli* ykfE ORFan gene encodes a potent inhibitor of C-type lysozyme. *J. Biol. Chem.* 276: 18437–18441.

ORTHOLOG
Dov S. Greenbaum

Similar genes in different species that arose through speciation (vertical descent).

The concept of the ortholog is integral for understanding molecular evolution. The exact definition is important as there are many proteins that are not orthologs (or paralogs for that matter) that have the same function in different organisms. (An inappropriate definition that is sometimes employed defines orthologs as homologs in different species that catalyze the same reaction. Some have declared the term ortholog useless, as it is commonly misused.)

In most cases orthologs share similar functions with their partners and thus allow for transferring of annotation information within groups of orthologs. Orthologs differ from paralogs, which also share similar functions, in that paralogs are deemed to have arisen in the same gene through gene duplication. Koonin (2001) claims that "the 'ortholog meme' ... encapsulates a whole panoply of diverse concepts beyond the strict definition: the existence of discrete gene histories that can be traced back to the last universal common ancestor, at least in principle."

Orthologs differ from paralogs in that orthologs result solely from speciation and mutation while paralogs are the result of gene duplication within a species. Orthologs are usually determined through sequence similarity analyses but are better defined by phylogenetic analysis.

Sets of orthologous genes are collected in the COG (Clusters of Orthologous Groups) and TOAG (TIGR Orthologous Gene Alignment) data sets.

Further Reading

Fitch W (1970). Distinguishing homologous from analogous proteins. *Syst. Zool.* 19: 99–113.

Gogarten JP, Olendzenski L (1999). *Orthologs, paralogs and genome comparisons. Curr. Opin. Genet. Dev.* 9: 630–636.

Koonin EV (2001). An apology for orthologs—or brave new nemes *Genome Biol.* 2: comment 1005–comment 1005.2.

See also Clusters of Orthologous Groups.

OSP (OLIGO SELECTION PROGRAM)
John M. Hancock and Martin J. Bishop

A program to aid primer design for PCR and DNA sequencing.

The program allows the user to specify a range of parameters, including primer and amplified product lengths, GC content, melting temperature, primer 3′ nucleotides, and the different types of primer annealing propensity. Predicted primers are presented to the user in rank order reflecting a quality score derived from the user-defined constraints. The user may supply a database of repetitive or other sequences against which primers may be screened to eliminate false positives.

Related Website

| HGMP OSP | http://www.hgmp.mrc.ac.uk/Registered/Help/osp/ |

Further Reading
Hillier L, Green P (1991). OSP: A computer program for choosing PCR and DNA sequencing primers. *PCR Methods Appl.* 1: 124–128.

OTU. *SEE* OPERATIONAL TAXONOMIC UNIT.

OVERDOMINANCE
A. Rus Hoelzel

Overdominance occurs when the heterozygote condition at a locus is selectively more fit than the homozygotes at that locus.

Considering the relative fitness of genotypes for a two-allele system at a locus in a diploid organism, the following relationship holds:

Genotype:	A_1A_1	A_1A_2	A_2A_2
Relative fitness:	1	$1 - hs$	$1 - s$

where s is the selection coefficient and h is the heterozygous effect. A_1 is dominant when $h = 0$ and A_2 is dominant when $h = 1$, but overdominance occurs when $h < 0$.

Overdominance is also referred to as *heterozygote advantage*. A famous example is the relationship between sickle cell anemia and malarial resistance (Allison, 1954). In this example a mutation for the sickle cell hemoglobin (HbS) causes serious anemia in homozygotes, but the heterozygote exhibits only mild anemia. Because of the "sickle" shape of the red blood cell containing hemoglobin with this mutation (even in the heterozygote), the malarial parasite is less able to infect, and consequently the heterozygote has a selective advantage in locations where malaria is prevalent.

Further Reading
Allison AC (1954). Protection afforded by sickle-cell trait against subtertian malaria. *Br. Med. J.*, February 6.

See also Adaptation, Natural Selection.

OVERFITTING (OVERTRAINING)
Nello Cristianini

In machine learning, the phenomenon by which a learning algorithm identifies relations in the training data that are due to noise or chance and do not reflect the underlying laws governing the given data set.

It occurs mostly in the presence of small and/or noisy training samples, when too much flexibility is allowed to the learning algorithm. When the learning algorithm is overfitting, its performance on the training set appears to be good, but the performance on the test data or in cross-validation is poor. It is addressed by reducing the capacity of the learning algorithm. The size of a neural network or a decision tree and the number of centers in k-means are all rough measures of capacity of a learning algorithm. A precise formal definition is given within the field of statistical learning theory, and is at the basis of the design of a new generation of algorithms explicitly aimed at counteracting overfitting.

Further Reading
Duda RO, Hart PE, Stork DG (2001). *Pattern Classification*. Wiley, New York.
Mitchell T (1997). *Machine Learning*. McGraw-Hill, New York.

See also Cross-Validation, Neural Network, Support Vector Machines.

OVERTRAINING. *SEE* OVERFITTING.

OWL. *SEE* ONTOLOGY WEB LANGUAGE.

P

Pairwise Alignment
PAML
PAM Matrix
PAM Matrix of Nucleotide Substitutions
Paralinear Distance
Paralog
Paraphyletic
Parity Bit
Parsimony
Pattern Analysis
Pattern Discovery
Pattern-of-Change Analysis
Pattern Recognition
PAUP*
PCR
PCR Primer
Penalty
Penetrance
Peptide
Peptide Bond
Percent Accepted Mutation Matrix
Performance Criteria
PFAM
Phantom Indel
Phase
Phase Problem
Phenocopy
Phenylalanine
PHRAP
PHRED
PHYLIP
Phylogenetic Events Analysis
Phylogenetic Footprinting
Phylogenetic Profiles
Phylogenetic Reconstruction
Phylogenetic Shadowing
Phylogenetic Tree
Phylogenomics
Phylogeny, Phylogeny Reconstruction

Physical Mapping
PIPMAKER
PIX
Plasmid
Plesiomorphy
Polar
Polarization
Polyadenylation
Polygenic Effect
Polygenic Inheritance
Polyhierarchy
Polymerase Chain Reaction
Polymorphism
Polypeptide
Polyphyletic
Population Bottleneck
Positional Candidate Approach
Position-Specific Scoring Matrix
Position Weight Matrix
Position Weight Matrix of Transcription
    Factor Binding Sites
Positive Darwinian Selection
Positive Selection
Potential of Mean Force
Power Law
Prediction of Gene Function
Predictive Power
Predictor Variables
Pregap4
Primary Constriction
PRIMEGEN
PRIMER
Primer Generator
Primer Walking
Principal-Components Analysis
PRINTS
Procheck
ProDom
Profile

Dictionary of Bioinformatics and Computational Biology. Edited by Hancock and Zvelebil
ISBN 0-471-43622-4   © 2004 John Wiley & Sons, Inc.

PAIRWISE ALIGNMENT
Jaap Heringa

Many methods have been developed for the calculation of sequence alignments, of which implementations of the dynamic programming algorithm (Needleman and Wunsch, 1970; Smith and Waterman, 1981) are considered the standard in yielding the most biologically relevant alignments. The dynamic programming (DP) algorithm requires a scoring matrix, which is an evolutionary model in the form of a symmetrical 4×4 nucleotide or a 20×20 amino acid exchange matrix. Each matrix cell approximates the evolutionary propensity for the mutation of one nucleotide or amino acid type into another. The DP algorithm also relies on the specification of gap penalties, which model the relative probabilities for the occurrence of insertion/deletion events. Normally, a gap opening and extension penalty is used for creating a gap and each extension respectively (affine gap penalties), so that the chance for an insertion/deletion depends linearly upon the length of the associated fragment. Given an exchange matrix and gap penalty values, which together are commonly called the scoring scheme, the DP algorithm is guaranteed to produce the highest scoring alignment of any pair of sequences—the optimal alignment.

Two types of alignment are generally distinguished: global and local alignment. Global alignment (Needleman and Wunsch, 1970) denotes an alignment over the full length of both sequences, which is an appropriate strategy to follow when two sequences are similar or have roughly the same length. However, some sequences may show similarity limited to a motif or a domain only, while the remaining sequence stretches may be essentially unrelated. In such cases, global alignment may well misalign the related fragments, as these become overshadowed by the unrelated sequence portions that the global method attempts to align, possibly leading to a score that would not allow the recognition of any similarity. If not much knowledge about the relationship of two sequences is available, it is usually better to align selected fragments of either sequence. This can be done using the local alignment technique (Smith and Waterman, 1981). The first method for local alignment, often referred to as the Smith–Waterman (SW) algorithm, is in fact a minor modification of the DP algorithm for global alignment. It basically selects the best scoring subsequence from each sequence and provides their alignment, thereby disregarding the remaining sequence fragments. Later elaborations of the algorithm include methods to generate a number of suboptimal local alignments in addition to the optimal pairwise alignment (e.g., Waterman and Eggert, 1987).

Related Websites

BLAST 2 sequence alignment	http://www.ncbi.nlm.nih.gov/gorf/bl2.html
FASTA program suite	http://fasta.bioch.virginia.edu/fasta/fasta_list.html

Further Reading

Needleman SB, Wunsch CD (1970). A general method applicable to the search for similarities in the amino acid sequence of two proteins. *J. Mol. Biol.* 48: 443–453.

Smith TF, Waterman MS (1981). Identification of common molecular subsequences. *J. Mol. Biol.* 147: 195–197.

Dictionary of Bioinformatics and Computational Biology. Edited by Hancock and Zvelebil
ISBN 0-471-43622-4 © 2004 John Wiley & Sons, Inc.

Waterman MS, Eggert M (1987). A new algorithm for best subsequences alignment with applications to the tRNA-rRNA comparisons. *J. Mol. Biol.* 197: 723–728.

See also Alignment—Multiple.

PAML (PHYLOGENETIC ANALYSIS BY MAXIMUM LIKELIHOOD)
Michael P. Cummings

A package of computer programs for maximum-likelihood-based analysis of DNA and protein sequence data.

The strengths of the programs are in providing sophisticated likelihood models at the level of nucleotides, codons, or amino acids. Specific programs allow for simulating sequences, analyses incorporating rate variation among sites, inferring ancestral sequences, and estimating synonymous and nonsynonymous substitution rates using codon-based models. The programs also allow for likelihood ratio–based hypothesis tests, including those for rate constancy among sites, evolutionary homogeneity among different genes, and the molecular clock.

The programs are written in American National Standards Institute (ANSI) C and are available as source code and executables for some platforms.

Related Website

| PAML | http://abacus.gene.ucl.ac.uk/software/paml.html |

Further Reading

Yang Z (1997). PAML: A program package for phylogenetic analysis by maximum likelihood. *Comput. Appl. Biosci.* 15: 555–556.

Yang Z, Nielsen R (2000). Estimating synonymous and nonsynonymous substitution rates under realistic evolutionary models. *Mol. Biol. Evol.* 17: 32–43.

Yang Z, et al. (2000). Codon-substitution models for heterogeneous selection pressure at amino acid sites. *Genetics* 155: 431–449.

See also Seq-Gen.

PAM MATRIX. *SEE* AMINO ACID EXCHANGE MATRIX, DAYHOFF AMINO ACID SUBSTITUTION MATRIX.

PAM MATRIX OF NUCLEOTIDE SUBSTITUTIONS
Laszlo Patthy

PAM matrices for scoring DNA sequence alignments incorporate the information from mutational analyses which revealed that point mutations of the transition type (A ↔ G or C ↔ T) are more probable than those of the transversion type (A ↔ C, A ↔ T, G ↔ T, G ↔ C).

Further Reading

Li WH, Graur D (1991). *Fundamentals of Molecular Evolution*. Sinauer Associates, Sunderland, MA.

States DJ, et al. (1991). Improved sensitivity of nucleic acid database searches using application specific scoring matrices. *Methods: Companion Methods in Enzymol.* 3: 66–70.

PARALINEAR DISTANCE (LOGDET)

Dov S. Greenbaum

An additive measure of the genealogical distance between two sequences.

Paralinear distance, also known as LogDet (invented simultaneously by Mike Steel and James Lake) is based on the most general models of nucleotide substitution. A paralinear distance is an additive measure of the genealogical distance between two sequences i and j (either DNA or protein). It assumes that all sites in the sequence can vary equally. This assumption of equal variation at all sites is one of the drawbacks of the model.

$$d_{ij} = -\log_e \frac{\det J_{ij}}{(\det D_i)^{1/2}(\det D_j)^{1/2}}$$

where det is the determinant of the matrix

This method does not suffer, to the same extent as other methods, in dealing with unequal rate effects. These effects group together diverse species that, although they are not evolutionarily related, are evolving at a fast pace. In general, it has been found that there is significant heterogeneity in the substitution patterns in different taxa; in the paralinear model base substitutions can vary throughout the tree.

Further Reading

Lake JA (1994). Reconstructing evolutionary trees from DNA and protein sequences: Paralinear distances. *Proc. Natl. Acad. Sci. USA* 91: 1455–1459.

Steel MA (1994). Recovering a tree from leaf colorations it generates under a Markov model. *Appl. Math. Lett.* 7: 19–24.

Székely L, et al. (1993). Fourier inversion formula for evolutionary trees. *Appl. Math. Lett.* 6: 13–17.

PARALOG

Austin L. Hughes, Dov S. Greenbaum, and Laszlo Patthy

Two genes in two different genomes are said to be paralogous (or paralogs) if they are descended from a common ancestral gene with one or more intervening gene duplication events.

When there is a gene duplication event followed by independent mutations and sequence changes within the duplicate genes, the result is a set of paralogous genes, that is, a set of genes within one organism with similar sequence but differing functions. Because they arise from gene duplication events, there is less evolutionary pressure to maintain the sequence in all paralogous genes in the set; thus in some instances, the encoded protein in a paralogous gene may have a totally different function or may decay into pseudogenes.

A commonly cited example of paralogous genes is the *HOX* cluster of transcription factor genes that is found in many organisms. In both mice and humans there are four paralogous clusters of 13 genes, each on different chromosomes.

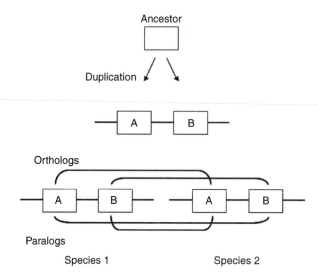

Ancestor

Duplication

Orthologs

Paralogs

Species 1   Species 2

Figure   Illustration of paralogous genes.

In the accompanying figure, gene A in species 1 is paralogous to gene B (in either species 1 or species 2). Likewise gene B is paralogous to gene A.

See also Homology, Ortholog.

PARAPHYLETIC
John M. Hancock

In a phylogenetic tree, a paraphyletic group is a group of nodes which excludes some members which might be natural members of that group.

Fish are strictly an example of a paraphyletic group, as all other vertebrates have fish ancestors. Referring only to fish therefore excludes many groups, such as amphibians, reptiles, birds, and mammals, that are natural members of that group.

See also Phylogenetic Tree, Taxonomic Classification.

PARITY BIT
Thomas D. Schneider

A parity bit determines a code in which one data bit is set to either 0 or 1 so as to always make a transmitted binary word contain an even or odd number of 1's.

The receiver can then count the number of 1's to determine if there was a single error. This code can only be used to detect an odd number of errors but cannot be used to correct any error. Unfortunately for molecular biologists, the now-universal method for coding characters, 7-bit ASCII words, assigns to the symbols for the nucleotide bases A, C, and G only a 1-bit difference between A and C and a 1-bit difference between C and G:

$$A : 101_8 = 1000001_2$$

$$C : 103_8 = 1000011_2$$

$$G : 107_8 = 1000111_2$$

$$T : 124_8 = 1010100_2$$

For example, this choice could cause errors during transmission of DNA sequences to the international sequence repository, GenBank. If we add a parity bit on the front to make an even-parity code (one byte long), the situation is improved and more resistant to noise because a single error will be detected when the number of 1's is odd:

$$A : 101_8 = 01000001_2$$

$$C : 303_8 = 11000011_2$$

$$G : 107_8 = 01000111_2$$

$$T : 324_8 = 11010100_2$$

Related Website

GenBank	http://www.ncbi.nlm.nih.gov/Genbank/

See also Bit Noise, Code.

PARSIMONY. *SEE* MAXIMUM-PARSIMONY PRINCIPLE.

PATTERN ANALYSIS (DATA MINING, INTELLIGENT DATA ANALYSIS, PATTERN DISCOVERY, PATTERN RECOGNITION)
Nello Cristianini

The detection of relations within data sets.

In data analysis, by "pattern" one means any relation that is present within a given data set. Pattern analysis or discovery therefore deals with the problem of detecting relations in data sets. Part of this process is to ensure that (with high probability and under assumptions) the relations found in the data are not the product of chance but can be trusted to be present in future data.

Statistical and computational problems force one to select a priori the type of patterns one is looking for, e.g., linear relations between the features (independent variables) and the labels (dependent variables).

Further Reading
Duda RO, et al. (2001). *Pattern Classification*. Wiley, New York.

See also Machine Learning.

PATTERN DISCOVERY. *SEE* PATTERN ANALYSIS.

PATTERN-OF-CHANGE ANALYSIS. *SEE* PHYLOGENETIC EVENTS ANALYSIS.

PATTERN RECOGNITION. *SEE* PATTERN ANALYSIS.

PAUP* [PHYLOGENETIC ANALYSIS USING PARSIMONY (AND OTHER METHODS)]
Michael P. Cummings

A program for phylogenetic analysis using parsimony, maximum-likelihood, and distance methods.

The program features an extensive selection of analysis options and model choices and accommodates DNA, RNA, protein, and general data types. Among the many strengths of the program is the rich array of options for dealing with phylogenetic trees, including importing, combining, comparing, constraining, rooting, and testing hypotheses.

Versions for the Apple MacIntosh computer, and to a lesser extent Windows, have a graphical user interface; other versions are command line driven. In all versions a user can execute a series of individual commands without interaction by including appropriate commands in the PAUP block (NEXUS file format). The program accommodates several input and output formats for data. Written in American national Standards Institute (ANSI) C, the program is available as executable files for a broad range of platforms.

Related Website

PAUP*	http://paup.csit.fsu.edu/

Further Reading
Swofford DL (2002). *PAUP* 4.0 beta 8 *Phylogenetic Analysis Using Parsimony (and Other Methods)*. Sinauer Associates, Sunderland, MA.

See also Modeltest.

PCR. *SEE* POLYMERASE CHAIN REACTION.

PCR PRIMER

Short single stranded deoxyoligonucleotides used in PCR.

See also Polymerase Chain Reaction.

PENALTY. *SEE* GAP PENALTY.

PENETRANCE
Mark McCarthy and Steven Wiltshire

The conditional probability that an individual with a given genotype expresses a particular phenotype.

The concept of penetrance is most readily understood in relation to Mendelian diseases. Such a disease may, e.g., be described as fully penetrant if susceptibility genotypes are invariably associated with disease development (in other words, the probability associated with those genotypes is 1). In multifactorial traits, where any given variant will characteristically be neither necessary nor sufficient for the development of disease, the penetrance of susceptibility genotypes will be less than 1 ("incomplete penetrance") and the penetrance of nonsusceptibility genotypes greater than zero. In either situation, penetrance may vary with age (most markedly so, for diseases of late onset), gender, or pertinent environmental exposures.

Examples: The penetrance of Huntington's chorea, even in individuals known to have inherited a disease-causing allele, is effectively zero until middle age and climbs to 100% during late middle life.

Further Reading

Ott J (1999). *Analysis of Human Genetic Linkage*, 3rd ed. Johns Hopkins University Press, Baltimore, MD, pp. 151–170.

Peto J, Houlston RS (2001). *Genetics and the Common Cancers. Eur. J. Cancer* 37(Suppl. 8): S88–S96.

Sham P (1998). *Statistics in Human Genetics*. Arnold, London, pp. 88–91.

See also Multifactorial Trait, Oligogenic Inheritance, Phenocopy, Polygenic Inheritance.

PEPTIDE
Roman A. Laskowski

A compound of two or more amino acids joined covalently by a peptide bond between the carboxylic acid group (COOH) of one and the amino group (NH_2) of the other. The amino acids forming a peptide are referred to as amino acid residues as their formation removes an element of water.

Long chains of amino acids joined together in this way are termed polypeptides. All proteins are polypeptides.

Further Reading

Branden C, Tooze J (1998). *Introduction to Protein Structure*, Second Edition, Garland Science, New York.

See also Amide Bond, Amino Acid.

PEPTIDE BOND. *SEE* AMIDE BOND.

PERCENT ACCEPTED MUTATION MATRIX. *SEE* DAYHOFF AMINO ACID SUBSTITUTION MATRIX.

PERFORMANCE CRITERIA. *SEE* ERROR MEASURES.

PFAM
Terri Attwood

A collection of multiple-sequence alignments and hidden Markov models for a range of protein domains and families. Each entry in PFAM is represented by both seed and full alignments. The seed includes representative members of the family, while the full alignment contains all family members detected with a profile hidden Markov model (HMM) constructed from the seed using the HMMer2 software to search SWISS-PROT and TrEMBL.

Each PFAM entry has a unique accession number and a database identifier. The database includes limited biological annotation—in the main, what is provided is taken directly from the PRINTS and PROSITE documentation that feeds InterPro. Some technical information is given on how the HMM was derived, together with both seed and final alignments. The database may be searched either using simple keywords or by means of query sequences (or their identifiers); options are also provided to search the SMART and TIGRFAM HMM collections.

PFAM is made available from the Sanger Institute, Cambridge, United Kingdom. A member of the InterPro consortium, it specializes in characterizing highly divergent domains. As such, it complements PROSITE, which specializes in the identification of functional sites, and PRINTS, which concentrates on hierarchical diagnosis of protein families, from superfamily, through family, down to subfamily levels. In order to increase the coverage of the resource, a fully automatically generated supplement, termed PFAM-B, is also provided, based on sequence clusters identified in ProDom. Results in PFAM-B are not manually verified and carry no annotation.

Related Websites

PFAM	http://www.sanger.ac.uk/Pfam
HMMer2	http://hmmer.wustl.edu/

Further Reading

Bateman A, et al. (2002). The Pfam protein families database. *Nucleic Acids Res.* 30(1): 276–280.

See also Alignment—Multiple, Hidden Markov Model, PRINTS, PROSITE.

PHANTOM INDEL
Jaap Heringa

A phantom indel corresponds to a spurious insertion or deletion in a nucleotide sequence that arises when physical irregularities in a sequencing gel cause the reading software either to call a base too soon or to miss a base altogether. The occurrence of a phantom indel then leads to a frameshift, which is an alteration in the reading sense of DNA resulting from an inserted or deleted base, such that the reading frame for all subsequent codons is shifted with respect to the number of changes made. For example, if a sequence should read GUC–AGC–GAG–UAA (translated into amino acids Val–Ser–Asp as UAA is the stop codon) and a single A is added to the beginning, the new sequence would then read AGU–CAG–CGA–CUA and give rise to a completely different protein product Ser–Gln–Arg–Leu-... as also the stop codon would be changed so that translation would proceed until the next spurious stop codon was encountered. Because the codon table differs between species in some cases, such that the same mRNA can lead to different proteins in those species, the effects of frameshifts can also be different.

PHASE
Mark McCarthy and Steven Wiltshire

Phase describes the relationship between alleles at a pair of loci, most relevantly when an individual is heterozygous at both. For the phase of the alleles in such double heterozygotes to be determined, the haplotypes of their parents need to be known.

Consider the three-generation family in the accompanying figure. The child in the third generation has genotype Aa at one locus and Bb at a second (linked) locus. Based on this information alone, it is not possible to determine the haplotype arrangement of this individual (they could be AB and ab or Ab and aB). However, in the presence of parental (and, as in this case, grandparental) genotypes, which allow the parental haplotypes to be unambiguously defined, the phase of the parental genotypes is clear, and the haplotypes of the child can be deduced. Such information allows us, in this example, to see that (under the assumption of no mutation) there has been a paternal recombination between the two loci (marked with a cross). Linkage analysis relies on the detection of recombinant events, and when phase is known, transmitted haplotypes can readily be separated into those which are recombinant and those which are nonrecombinant. When phase is not known, linkage analysis is still possible and is typically implemented through maximum-likelihood methods using appropriate software. The development of statistical tools (including software listed below) to facilitate phase determination (haplotype construction) in phase-ambiguous individuals (such as unrelated cases and control subjects in whom there is no genetic information from relatives) is becoming increasingly important for linkage disequilibrium analyses as researchers seek to exploit the power of haplotype analyses based on dense maps of single-nucleotide polymorphisms.

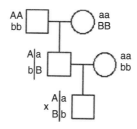

Figure
Three-generation family typed for two loci.

Related Websites
Programs for haplotype (and phase) determination:

PHASE	http://www.stats.ox.ac.uk/mathgen/software.html
HAPLOTYPER	http://www.people.fas.harvard.edu/~junliu/Hapl./docMaintion

| SNPHAP | http://www-gene.cimr.cam.ac.uk/clayton/software/ |
| SIMWALK | http://Lpcid.cit.nih.gov/lserver/SIMWALK2.html |

Further Reading

Niu T, et al. (2002). Bayesian haplotype inference for single-nucleotide polymorphisms. *Am. J. Hum. Genet.* 70: 157–169.

Ott J (1999). *Analysis of Human Genetic Linkage*, 3rd ed. Johns Hopkins University Press, Baltimore, MD, pp. 7 et seq.

Schaid DJ, et al. (2002). Score tests for association between traits and haplotypes when linkage phase is ambiguous. *Am. J. Hum. Genet.* 70: 425–434.

Sham P (1998). *Statistics in Human Genetics*. Arnold, London, pp. 62–67.

Stephens M, et al. (2001). A new statistical method for haplotype reconstruction from population data. *Am. J. Hum. Genet.* 68: 978–989.

See also Haplotype, Linkage Analysis.

PHASE PROBLEM

Liz P. Carpenter

The three-dimensional structure of a macromolecule can be obtained by observation of the diffraction pattern obtained when crystals of the macromolecule are placed in an X-ray beam. In order to calculate the structure of the macromolecule, knowledge of not only the intensity of the X-ray diffraction spots but also their relative phases is necessary. This information cannot be recorded on an X-ray detector, so a single diffraction pattern is not sufficient to derive the structure of an unsolved molecule.

There are a number of techniques that can be used to overcome this problem. In general, a number of data sets are collected under conditions where the diffraction pattern has been perturbed. For a completely unknown structure several data sets can be collected on crystals that have been soaked in heavy-atom compounds containing, e.g., mercury, platinum, or gold. The aim is to obtain a crystal with a few well-ordered binding sites for the heavy atom. Since diffraction is in proportion to the number of electrons on an atom, these high-atomic-weight atoms will shift the diffraction pattern substantially and comparisons with data from crystals of native protein will give information on the phases of the diffraction spots.

A similar technique that has gained popularity recently is the substitution of seleno-methionine for all the methionine residues in a protein. This can be achieved by producing the protein in an organism that cannot synthesize methionine but is given seleno-methionine in the growth medium. The replacement of sulfur with selenium in all methionine residues gives a heavy atom with 36 electrons without perturbing the structure significantly. The use of seleno-methionine is often combined with multiple anomalous dispersion (MAD) techniques in which a number of data sets are collected at different wavelengths around an absorption edge for a heavy atom; e.g., selenium has an absorption edge at 0.9795 Å. The differences in these data sets can then be used to solve the structure.

When a related structure is known, a technique called molecular replacement can be used to find phases for the diffraction pattern.

See also Diffraction of X Rays, Molecular Replacement, Multiple Anomalous Dispersion Phasing.

PHENOCOPY
Mark McCarthy and Steven Wiltshire

Individuals displaying a phenotype (e.g., a disease) which is generally familial and inherited but in whom the disease has been caused not by genetic factors but by environmental exposure.

Under this strict definition, phenocopies equate to sporadic cases. In the context of gene mapping by linkage analysis, the term is often used, more generally, to denote any individual who displays the phenotype of interest but who does not have a susceptibility genotype at the locus under study. This therefore includes cases of disease arising due to mutations in other (unlinked) genes as well as those exclusively due to environmental exposures. In linkage analysis, the presence of phenocopies is represented by ascribing nonzero penetrance to the nonsusceptibility genotype(s). This is typically the case in linkage analysis for multifactorial traits, where the expectation is that variation in any given gene will neither be necessary nor sufficient as a cause of disease. The proportion of phenocopies among all affected individuals is often termed the *phenocopy rate*. Finally, the term is also used in model organisms (such as rodents) to describe animals with similar phenotypes despite disparate genetic (or environmental) causes.

Examples: Guris and colleagues (2001) describe the consequences of knocking out, in mice, the homolog of a human gene (*CKRL*) implicated in the development of DiGeorge syndrome (a developmental abnormality of the heart and cranium associated with a deletion on human chromosome 22q). Since this knockout of a single gene bears many phenotypic similarities to (i.e., it phenocopies) the human disease phenotype which usually follows from contiguous deletion of several genes, these data confirm that loss of *CKRL* function may be one of the key events leading to the DiGeorge phenotype.

Further Reading

Burgess JR, et al. (2000). Phenotype and phenocopy: The relationship between genotype and clinical phenotype in a single large family with multiple endocrine neoplasia type 1 (MEN 1). *Clin. Endocrinol. Oxf.* 53: 205–211.

Guris DL, et al. (2001). Mice lacking the homologue of the human 22q11.2 gene CKRL phenocopy neurocristopathies of DiGeorge syndrome. *Nat. Genet.* 27: 238–240.

Moore RC, et al. (2001). Huntington disease phenocopy is a familial prion disease. *Am. J. Hum. Genet.* 69: 1385–1388.

Ott J (1999). *Analysis of Human Genetic Linkage*, 3rd ed. Johns Hopkins University Press, Baltimore, MD, pp. 151 et seq.

Whittemore AS, Halpern J (2001). Problems in the definition, interpretation, and evaluation of genetic heterogeneity. *Am. J. Hum. Genet.* 68: 457–465.

See also Linkage Analysis, Multifactorial Traits, Penetrance.

PHENYLALANINE
Jeremy Baum

Phenylalanine is a large aromatic amino acid with side chain $-CH_2C_6H_5$ found in proteins. In sequences, written as Phe or F.

See also Amino Acid, Aromatic.

PHRAP
Rodger Staden

A widely used sequence assembly engine.

Related Website

PHRAP	http://www.phrap.org/

PHRED
Rodger Staden

The base-calling program which gave its name to the PHRED scale of base-call confidence values. Defines the confidence $C = -10\log(P_{error})$, where P_{error} is the probability that the base call is erroneous.

Related Website

PHRAP	http://www.phrap.org/

Further Reading
Ewing B, Green P (1998). Base-calling of automated sequencer traces using Phred. II. Error probabilities. *Genome Res.* 8: 186–194.

PHYLIP (PHYLOGENY INFERENCE PACKAGE)
Michael P. Cummings

A set of modular programs for performing numerous types of phylogenetic analyses.

Individual programs are broadly grouped into several categories: molecular sequence methods; distance matrix methods; analyses of gene frequencies and continuous characters; discrete characters methods; and tree drawing, consensus, tree editing, and tree distances. Together the programs accommodate a broad range of data types, including DNA, RNA, protein, restriction sites, and general data types. The programs encompass a broad variety of analysis types, e.g., parsimony, compatibility, distance, invariants, and maximum likelihood, and also include both jackknife and bootstrap resampling methods. The output from one program often forms the input for other programs within the package (e.g., dnadist generates a distance matrix from a file of DNA sequences, which is then used as input for a neighbor to generate a neighbor-joining tree and the tree is then viewed with drawtree). Therefore, for a typical analysis the user makes choices regarding each aspect of an analysis and chooses specific programs accordingly. Programs are run interactively via a text-based interface that provides a list of choices and prompts users for input.

The programs are available as American National Standards Institute (ANSI) C source code. Executables are also available for several platforms.

Related Website

PHYLIP	http://evolution.genetics.washington.edu/phylip.html

Further Reading

Felsenstein J (1981). Evolutionary trees from DNA sequences: A maximum likelihood approach. *J. Mol. Evol.* 17: 368–376.

Felsenstein J (1985a). Confidence limits on phylogenies: An approach using the bootstrap. *Evolution* 39: 783–791.

Felsenstein J (1985b). Phylogenies from gene frequencies: A statistical problem. *Syst. Zool.* 34: 300–311.

PHYLOGENETIC EVENTS ANALYSIS (PATTERN-OF-CHANGE ANALYSIS)

Jamie J. Cannone and Robin R. Gutell

For covariation analysis, a more complete measure of the dependence and independence between the two positions with the same pattern of variation is gauged by (1) the number of base-pair types that covary with one another (i.e., A:U, G:C), (2) the frequency of these base-pair types, and (3) phylogenetic events, or the number of times the base pair was created during the evolution of that base pair. The first two gauges can be measured with a chi-square statistic. The third gauge requires an understanding of the phylogenetic relationships between the sequences that are in the data set.

These three gauges are exemplified by three different base pairs in 16 S rRNA: 9:25, 502:543, and 245:283. The 9:25 base pair has approximately 67% G:C and 33% C:G in the nuclear-encoded rRNA genes in the three primary forms of life, the eucarya, archaea, and bacteria. The minimal number of times these base pairs evolved (phylogenetic events) on the phylogenetic tree is about 4. In contrast, the 502:543 base pair has 27% G:C, 30% C:G, and 42% A:U, with a minimum of 75 phylogenetic events. Lastly, the 245:283 base pair has 38% C:C and 62% U:U in the same set of 16 S rRNA sequences, with approximately 25 phylogenetic events.

The phylogenetic event–counting method can be used to augment the results from the analysis of base-pair types and base-pair frequencies. It can add or subtract support for a base pair predicted with covariation analysis. In some situations, this form of analysis can suggest a base pair that would not have been predicted based on base-pair type and frequencies alone.

Further Reading

Cannone JJ, et al. (2002). The Comparative RNA Web (CRW) Site: An Online Database of Comparative Sequence and Structure Information for Ribosomal, Intron, and Other RNAs. *BioMed Central Bioinformatics* 3(15).

Gutell RR, et al. (1986). Higher order structure in ribosomal RNA. *The EMBO Journal* 5(5): 1111–1113.

Hancock JM, Vogler AP (1998). Modelling secondary structures of hypervariable RNA regions evolving by slippage: The example of the tiger beetle 18S rRNA variable region V4. *Nucleic Acids Res.* 26: 1689–1699.

Related Website

Comparative RNA Web (CRW)	http://www.rna.icmb.utexas.edu/

See also Covariation Analysis, RNA Structure Prediction.

PHYLOGENETIC FOOTPRINTING (PHYLOGENETIC SHADOWING)

James W. Fickett

A technique involving the use of comparative genomics to distinguish transcriptional elements (equivalently, transcription factor binding sites) from surrounding DNA sequence.

It is normally not possible to detect transcriptional elements simply as distinctive patterns. Often, however, the functional element is more highly conserved in evolution than the surrounding DNA, and appropriate alignment techniques can be revealing. In human–mouse comparisons, typically only a few percent of computationally predicted elements are conserved, while most functionally verified elements are indeed found to be conserved (Loots et al., 2002; Wasserman et al., 2000). Phylogenetic footprinting has become a standard approach in unraveling regulatory mechanisms. In a 1-Mb region of the human genome, containing interleukins 4, 13, and 5, only 90 conserved regions were found (compared to an expected many thousands of computationally predicted transcription elements). Analysis of these revealed a new coordinate regulator of the three genes (Loots et al., 2000).

Most commonly, the alignment problem is split into parts. A rough overall alignment of large genomic regions is followed by local refinement and filtering for small regions of high conservation. An informative website and typical tools are described in Schwartz et al. (2000). The choice of a comparison organism is not easy. The degree of conservation varies from region to region, so that no single interspecies evolutionary distance is optimal for all regions (though it is fortunate that human and mouse are at an appropriate distance for many regions; Wasserman et al., 2000). The use of multiple alignments with three or more genomes has been shown to give measurably more reliable information (Miller, 2000; McCue et al., 2002).

Related Websites

Penn State Bioinformatics Group	http://bio.cse.psu.edu
Wadsworth Center Bayesian Bioinformatics	http://www.wadsworth.org/resnres/bioinfo

Further Reading

Blanchette M, et al. (2002). Algorithms for phylogenetic footprinting. *J. Comput. Biol.* 9: 211–223.

Boffelli D, et al. (2003). Phylogenetic shadowing of primate sequences to find functional regions of the human genome. *Science* 299: 1391–1394.

Duret L, Bucher P (1997). Searching for regulatory elements in human noncoding sequences. *Curr. Opin. Struct. Biol.* 7: 399–406.

Loots GG, et al. (2000). *Science* 288: 136–140.

Loots GG, et al. (2002). *Genome Res.* 12: 832–839.

McCue, LA, et al. (2002). *Genome Res.* 12: 1523–1532.

Miller, W. (2000). *Nature Biotechnol.* 18: 148–149.

Schwartz S, et al. (2000). *Genome Res.* 10: 577–586.

Wasserman WW, et al. (2000). *Nature Genet.* 26: 225–228.

See also Comparative Genomics, Profile, Transcription Factor Binding Site.

PHYLOGENETIC PROFILES
Patrick Aloy

Phylogenetic profiles are based on the presence or absence of a certain number of genes in a set of genomes. Systematic presence or absence of a set of proteins (e.g., similarity of phylogenetic profiles) might indicate a common functionality of these proteins. Sometimes this reflects physical interactions between proteins forming a complex but most often implies a related function (e.g., they are components of the same metabolic pathway).

This approach can only be applied to entire genomes (i.e., not individual pairs of proteins) and it cannot be used to test interactions between essential proteins present in most organisms (e.g., ribosomal proteins).

Further Reading

Pellegrini M, Marcotte EM, Thompson MJ, Eisenberg D, Yeates TO (1999). Assigning protein functions by comparative genome analysis: Protein phylogenetic profiles. *Proc. Natl. Acad. Sci. USA* 96: 4285–4288.

See also Rosetta Stone Method.

PHYLOGENETIC RECONSTRUCTION. *SEE* PHYLOGENETIC TREE.

PHYLOGENETIC SHADOWING. *SEE* PHYLOGENETIC FOOTPRINTING.

PHYLOGENETIC TREE (PHYLOGENETIC RECONSTRUCTION, PHYLOGENY, PHYLOGENY RECONSTRUCTION)
Jaap Heringa

Graph in the form of a "tree" representing the evolutionary relationships of sequences or species.

A phylogenetic tree depicts the evolutionary relationships of a target set of sequences. It contains interior and exterior (terminal) nodes. Normally, the input sequences are contemporary and referred to as the operational taxonomic units (OTUs). They correspond with the exterior nodes of the evolutionary tree, whereas the internal nodes represent ancestral sequences which have to be guessed from the OTUs and the tree topology. The length of each branch connecting a pair of nodes may correspond to the estimated number of substitutions between two associated sequences. The minimal evolution hypothesis is that the "true" phylogenetic tree is the rooted tree; i.e., it contains a node ancestral to all other nodes and has the shortest overall length and thus comprises the lowest cumulative number of mutations. The computation of a phylogenetic tree typically involves four steps: (i) selection of a target set of orthologous sequences; (ii) multiple alignment of the sequences; (iii) construction of a pairwise sequence distance matrix by calculating the pairwise distances using the multiple alignment; and (iv) compilation of the tree by applying a clustering technique to the distance matrix. As alternatives to the last two steps, maximum-parsimony or maximum-likelihood methods may also be used to infer a tree from a multiple alignment (see below).

The construction of phylogenetic trees from protein and DNA data was pioneered more than three decades ago by Edwards and Cavalli-Sforza (1963), who reconstructed phylogenies by exploring the concept of minimum evolution based on gene frequency data. Camin and Sokal (1965) first implemented this concept on discrete morphological characters, which they called the parsimony method, a now standard term. Following Camin and Sokal, Eck and Dayhoff (1966) then introduced the parsimony method for molecular sequence data. Other early attempts to reconstruct evolutionary pathways from present-day sequences were based on distance methods which explore a matrix containing all pairwise distances of a set of multiply aligned sequences (Cavalli-Sforza and Edwards, 1967; Fitch and Margoliash, 1967; Sokal and Sneath, 1963). All of these methods, each with different emphasis, try to reconstruct the past using a minimalist approach; i.e., they use as few evolutionary changes as possible. More recently, another important class of computer methods appeared for sequence data based on a stochastic model of evolution and was named the maximum-likelihood technique (Felsenstein, 1981a).

A widely used principle early on for constructing evolutionary or phylogenetic trees from protein sequence data is that of maximum parsimony (Farris, 1970), based on a deterministic model of sequence evolution and requiring as few mutations as possible. Unlike distance methods, parsimony methods are character based and examine each sequence position (character) separately. Multiple alignments of the subject sequences (OTUs) are screened for the occurrence of *phylogenetically informative* alignment positions where at least two residue types are represented more than once. From the phylogenetically informative columns, the number of mutations between each sequence (including hypothetical ancestors) is determined.

The first step of a parsimony method deals with the generation of the interior nodes of a tree. Fitch (1971) proved that it is possible, given an initial alignment of sequences and an initial tree or dendrogram, to construct maximum-parsimony ancestral sequences associated with the interior nodes of the tree such that the minimum number of mutations is obtained. So, for every given tree topology, it is possible to infer the ancestral sequences associated with the internal nodes together with all the branch lengths, yielding the minimum overall length for that tree. The second step is computationally prohibitive since it involves, for a given set of OTUs, exploring all possible tree topologies to find the one with minimum cost. There is no global strategy to infer the most parsimonous tree, which therefore can only be found by exhaustive searching. Unfortunately, the number of trees to be inspected explodes with the number of sequences such that much more than 10 sequences is not feasible. The number of possible unrooted tree topologies for N sequences is $\sum_{i=1}^{N-2}(2i-1)$, yielding about 2×10^6 different topologies for only 10 sequences. Therefore, heuristic search approaches are needed.

Felsenstein (1978, 1981b) and Saitou and Nei (1986) demonstrated that parsimony methods cannot deal well with many biological phenomena, including the occurrence of mutation rates that differ through various lineages, or the fact that not all alignment positions are equally informative given different tertiary structural constraints. Felsenstein (1981b) therefore suggested applying different weights to different alignment positions. Generally, parsimony methods are effective for closely related sequences resulting from divergent evolution.

Another issue is the *multiple alignment* required as input to infer mutations. As Hogeweg and Hesper (1984) and Feng and Doolittle (1987) demonstrated, multiple-alignment accuracy and derivation of a phylogenetic tree are closely interrelated. These authors devised phylogenetic methods that utilize this relationship using distance methods. The construction of a multiple alignment is dependent on an *amino acid exchange matrix* and *gap penalties*, especially when distant sequences are aligned which are likely to acquire many insertions and deletions. An accurate multiple alignment in fact requires an a priori knowledge of evolutionary relationships.

Many different parsimony programs have been described over the years, generally operating on nucleotide sequences. Moore et al. (1973) extended the parsimony method of Fitch (1971) for nucleotide sequences to protein sequences. Also the PHYLIP package (Felsenstein, 1989, 1990) and PAUP (Swofford, 1992) offer various parsimony algorithms for protein sequences. A further class of methods related to maximum-parsimony techniques is that of compatibility analysis (Le Quesne, 1974), where tree branching is made compatible with observed mutations at a maximum number of given multiple-alignment sites; sites that are not compatible are ignored.

Maximum-likelihood methods (Felsenstein, 1981b) present an alternative way to derive the phylogeny of a set of sequences. Here the tree construction and assignment of branch lengths are performed using evolutionary probabilities of nodal connections from which statistical significance is inferred. The probabilities are based on a stochastic model of sequence evolution. Furthermore, alternative tree topologies can be readily evaluated using the associated likelihoods. The method attempts to maximize the probability that the data will fit well onto a tree under a given evolutionary model. Explicit assumptions of sequence evolution are required, such as equal mutation rates, independent evolution of alignment sites, and the like. They can be tailored to a particular data set under study. Golding and Felsenstein (1990) used maximum-likelihood calculations to detect the occurrence of selection in the evolution of restriction sites in *Drosophila*. Unfortunately, likelihood methods are also computer intensive, difficult to apply to large data sets, and can fall into local traps (Golding and Felsenstein, 1990).

Though most maximum-likelihood methods deal with nucleotide sequences (e.g., Felsenstein, 1989, 1990; Olsen et al., 1994), the DNAML program from the PHYLIP package (Felsenstein, 1990) allows codon positions to be independently weighted for protein sequence analysis. Adachi and Hasegawa (1992) devised a maximum-likelihood program for protein phylogeny which can handle larger sets of sequences using a fast star decomposition approach to search through tree topologies (Saitou and Nei, 1987).

Closely related to likelihood methods are the so-called invariant techniques (e.g., Lake, 1987) which compare phylogenetic trees based on their topology only, thereby ignoring branch lengths. Their advantage is insensitivity to different branch lengths and different evolutionary rates at various (alignment) sites, but to date, they can only operate on four-species trees and nucleotide data.

Distance methods derive a tree from a distance matrix which compares pairwise all tree constituents. Distances can be obtained from sequence identities (Fitch and Margoliash, 1967) or pairwise sequence alignment scores (Hogeweg and Hesper, 1984). An agglomerative cluster criterion is utilized reflecting the evolutionary information in the best possible way. One advantage of distance methods over parsimony methods is that they are usually less CPU intensive as they employ a fixed strategy to arrive at a final tree without the need to sample the complete tree space. Many clustering criteria have been introduced over the years, each having an underlying assumption of evolutionary dynamics.

The first cluster method used in molecular sequence phylogeny was the UPGMA or group averaging method (Sokal and Sneath, 1963). It takes the average value over all intergroup distances to measure the evolutionary distance between two groups of sequences and has the underlying assumption of identical mutation rates in all lineages. It is not suitable when there are significant deviations from a unit evolutionary clock. For not too large differences, the method of Fitch and Margoliash (1967) is more appropriate since it infers the branch length for each nearest-neighbor pair with information from the next closest sequence. The Fitch–Margoliash method yields a tree topology identical to that from the UPGMA technique, but the branch lengths may be different.

The method of Saitou and Nei (1987), called neighbor joining (NJ), is acclaimed by many workers in phylogenetic analysis. The NJ method is able to reconstruct evolution properly under a variety of parameters such as different mutation rates, back mutations, and the like (Czelusniak et al., 1990; Huelsenbeck and Hillis, 1993; Saitou and Imanishi, 1989). The method relies on a protocol of progressive pairwise joining of nearest sequences (each sequence being represented by a node) such that each time two nodes are joined, they are represented by an internal node; the two nodes selected at each step for joining keep the overall tree length at a minimum. This process is iterated until all nodes are joined. The NJ method departs from the UPGMA and Fitch–Margoliash techniques in that it does not use the evolutionary (dis)similarity among groups but merely is a strategy to join sequences and calculate branch lengths.

The simplest way to calculate the distance between sequences is by the percent divergence, which for two aligned sequences involves counting the number of nonidentical matches (ignoring positions with gaps) divided by the number of positions considered. When a multiple alignment is used, all positions containing a gap in any of the sequences are commonly ignored. The real evolutionary time between the divergence of two sequences depends on the speed of the evolutionary clock, a matter of ongoing controversy. Even under a uniform clock, sequence identity as a measure of distance underestimates the real number of mutations. Certainly in diverged sequences there is an increasing chance that multiple substitutions have occurred at a site. The greater the divergence, the more the evolutionary times are underestimated. For example, Dayhoff et al. (1978) estimated that a sequence identity of 20% would correspond to about 2.5 mutations per amino acid. The Dayhoff et al. relationship between observed percent identity and percent accepted mutation shows that the estimated evolutionary time rises much faster than the observed sequence distances; the curve is not defined for sequence differences higher than 93%. Kimura (1983) corrected for this effect by curve fitting such that the corrected evolutionary time from the distance K (percent divergence divided by 100) is given by $K_{corrected} = -\ln(1.0 - K - K^2/5.0)$. The formula applies to cases with a reasonably uniform evolutionary clock and sequence identities from 15% and fits the data well from identities higher than 35%. It is good practice to start tree-building routines from such corrected sequence distances. The Dayhoff et al. relationship is valid for a fixed amino acid composition of the data. Whenever deviating compositions occur, the user can resort to the PHYLIP package (Felsenstein, 1989, 1990) to calculate evolutionary distances based on the actual amino acid composition.

Because evolutionary trees may be a result of local traps in the search space, it is important to estimate the significance of a particular tree topology and associated branch lengths. Felsenstein (1985) introduced the concept of bootstrapping, which involves resampling of the data such that the multiple-alignment positions are randomly selected and placed in some order and then a tree is generated by the original method. This process is repeated a statistically significant number of times. Comparison of frequencies at which the $N - 3$ internal branches (N is the number of sequences) occur in the original tree and those from bootstrapping allow probability estimates of significance. It is common practice to consider frequencies higher than or equal to 95% as supportive for the occurrence of an original tree branch in the bootstrapped trees. In this way, bootstrapping tests the stability of groupings given the data set and the method, thus lessening the chance of incorrect tree structures due to conservative and/or back mutations. Bootstrapping can be performed for any method that generates a tree from a multiple alignment, albeit the biological significance of a particular tree is not addressed.

A molecular biologist wishing to derive a phylogeny from a given set of protein sequences should first seek a sensitive multiple-alignment routine. If secondary or tertiary structures are known, the information should be applied to achieve the most accurate alignments. A variety of phylogenetic methods should then be used (preferably including the NJ method) in conjunction with bootstrapping and the results compared carefully for consistency. It should be kept in mind that many wrong trees can be derived from a particular sequence set, often with more seemingly interesting phylogenies than that of the one correct tree.

General-purpose programs used for constructing phylogenies include PHYLIP, PAUP MEGA, and VOSTORG.

Related Website

PHYLIP	http://evolution.genetics.washington.edu/phylip/software.html

Further Reading

Adachi J, Hasegawa M (1992). *MOLPHY: Programs for Molecular Phylogenetics, I. PROTML: Maximum Likelihood Inference of Protein Phylogeny*, Computer Science Monographs no. 27. Institute of Statistical Mathematics, Tokyo, Japan.

Blanken RL, et al. (1982). Computer comparison of new and existing criteria for constructing evolutionary trees from sequence data. *J. Mol. Evol.* 19: 9–19.

Camin JH, Sokal RR (1965). A method for deducing branching sequences in phylogeny. *Evolution* 19: 311–326.

Cavalli-Sforza LL, Edwards AWF (1967). Phylogenetic analysis: Models and estimation procedures. *Am. J. Hum. Genet.* 19: 233–257.

Czelusniak J, et al. (1990). Maximum parsimony approach to construction of evolutionary trees from aligned homologous sequences. *Methods Enz.* 183: 601–615.

Dayhoff MO (1965). *Atlas of Protein Sequence and Structure.* National Biomedical Research Foundation, Silver Spring, MD.

Dayhoff MO, Schwartz RY, Orante BC (1978). A model of evolutionary change in proteins. In *Atlas of Protein Sequence and Structure*, Vol. 5, Suppl. 3, MO Dayhoff, Ed. National Biomedical Research Foundation, Washington, DC, pp. 345–352.

Eck RV, Dayhoff MO (1966). *Atlas of Protein Sequence and Structure 1966.* National Biomedical Research Foundation, Silver Spring, MD. Edwards AFW, Cavalli-Sforza LL (1963). The reconstruction of evolution. *Ann. Hum. Genet.* 27: 104–105.

Farris JS (1970). Methods for computing Wagner trees. *Syst. Zool.* 19: 83–92.

Felsenstein J (1978). Cases in which parsimony or compatibility methods will be positively misleading. *Syst. Zool.* 27: 401–410.

Felsenstein J (1981a). Evolutionary trees from DNA sequences: A maximum likelihood approach. *J. Mol. Evol.* 17: 368–376.

Felsenstein J (1981b). A likelihood approach to character weighting and what it tells us about parsimony and compatibility. *Biol. J. Linn. Soc.* 16: 183–196.

Felsenstein J (1985). Confidence limits on phylogenies: An approach using the bootstrap. *Evolution* 39: 783–791.

Felsenstein J (1989). PHYLIP—phylogeny inference package (version 3.2). *Cladistics* 5: 164–166.

Felsenstein J (1990). *PHYLIP Manual Version 3.3.* University Herbarium, University of California, Berkeley, CA.

Feng DF, Doolittle RF (1987). Progressive sequence alignment as a prerequisite to correct phylogenetic trees. *J. Mol. Evol.* 21: 112–125.

Fitch WM (1971). Toward defining the course of evolution: Minimum change for a specified tree topology. *Syst. Zool.* 20: 406–416.

Fitch WM, Margoliash E (1967). Construction of phylogenetic trees. *Science* 155: 279–284.

Golding B, Felsenstein J (1990). A maximum likelihood approach to the detection of selection from a phylogeny. *J. Mol. Evol.* 31: 511–523.

Hogeweg P, Hesper B (1984). The alignment of sets of sequences and the construction of phyletic trees: An integrated method. *J. Mol. Evol.* 20: 175–186.

Huelsenbeck JP, Hillis DM (1993). Success of phylogenetic methods in the four-taxon case. *Syst. Biol.* 42: 247–264.

Kimura M (1983). *The Neutral Theory of Molecular Evolution.* Cambridge University Press, Cambridge, England.

Lake JA (1987). A rate-independent technique for analysis of nucleic acid sequences. *Mol. Biol. Evol.* 4: 167–191.

Le Quesne WJ (1974). The uniquely evolved character concept and its cladistic application. *Syst. Zool.* 23: 513–517.

Moore GW, et al. (1973). A method for constructing maximum parsimony ancestral amino acid sequences on a given network. *J. Theor. Biol.* 38: 459–485.

Olsen GJ, et al. (1994). fastDNAml: A tool for construction of phylogenetic trees of DNA sequences using maximum likelihood. *Comput. Appl. Biosci.* 10: 41–48.

Saitou N, Imanishi T (1989). Relative efficiencies of the Fitch-Margoliash, maximum parsimony, maximum-likelihood, minimum-evolution, and neighbor-joining methods of phylogenetic tree construction in obtaining the correct tree. *Mol. Biol. Evol.* 6: 514–525.

Saitou N, Nei M (1986). The number of nucleotides required to determine the branching order of three species with special reference to human-chimpanzee-gorilla divergence. *J. Mol. Evol.* 24: 189–204.

Saitou N, Nei M (1987). The neighbor-joining method: A new method for reconstructing phylogenetic trees. *Mol. Biol. Evol.* 4: 406–425.

Sokal RR, Sneath PHA (1963). *Principles of Numerical Taxonomy.* W. H. Freeman, San Fransisco.

Swofford DL (1992). *PAUP: Phylogenetic Analysis Using Parsimony, Version 3.0s,* Illinois Natural History Survey, Champaign, IL.

See also Bayesian Phylogenetic Analysis, Consensus Tree, MacClade, Maximum-Likelihood Phylogeny Reconstruction, Maximum-Parsimony Principle, MEGA, Minimum-Evolution Principle, Neighbor-Joining Method, PAUP, PHYLIP, Tree-Based Progressive Alignment, UPGMA.

PHYLOGENOMICS
Jean-Michel Claverie

An area of research making use of phylogenetic methods (hierarchical dendrogram building) in a large-scale context.

Apparently introduced by Eisen (1998) to designate the use of evolutionary relationship to complement straightforward similarity between homologs to more accurately predict gene functions. The term has then been extended to include a wide range of studies with nothing in common but the computation of trees from a large number of sequences. The database of Clusters of Orthologous Groups (COG) of proteins is one successful example of phylogenomic study. More recently, this term has been used to qualify the phylogenetic classification of organisms based on properties computed from their whole gene content. Accordingly, phylogenomics is part of comparative genomics. Finally, this term is also used to designate whole vs. all other genome analyses to establish relationships between genes of known and unknown functions, such coevolution profiling, "Rosetta stone" gene fusion detection, and genomic context studies.

Related Websites

University of California Museum of Paleontology Phylogenetics Resources	http://www.ucmp.berkeley.edu/subway/phylogen.html
Clusters of Orthologous Groups	http://www.ncbi.nlm.nih.gov/COG/
UCLA-DOE Institute for Genomics and Proteomics	http://www.doe-mbi.ucla.edu/

Further Reading

Eisen JA (1998). Phylogenomics: Improving functional predictions for uncharacterized genes by evolutionary analysis. *Genome Res.* 8: 163–167.

Huynen M, et al. (2000). Predicting protein function by genomic context: Quantitative evaluation and qualitative inferences. *Genome Res.* 10: 1204–1210.

Marcotte EM, et al. (1999). Detecting protein function and protein-protein interactions from genome sequences. *Science* 285: 751–753.

Pellegrini M, et al. (1999). Assigning protein functions by comparative genome analysis: Protein phylogenetic profiles. *Proc. Natl. Acad. Sci. USA.* 96: 4285–4288.

Sicheritz-Ponten T, Andersson SG (2001). A phylogenomic approach to microbial evolution. *Nucleic Acids Res.* 29: 545–552.

Snel B, et al. (1999). Genome phylogeny based on gene content. *Nat. Genet.* 21: 108–110.

Tatusov RL, et al. (1997). A genomic perspective on protein families. *Science* 278: 631–637.

See also Phylogenetic Tree.

PHYLOGENY, PHYLOGENY RECONSTRUCTION. *SEE* PHYLOGENETIC TREE.

PHYSICAL MAPPING. *SEE* CONTIG MAPPING.

PIPMAKER
John M. Hancock

A program to identify evolutionarily conserved regions between genomes.

PIPMAKER carries out alignment between two long genomic sequences using BLASTZ, a variant of the BLAST algorithm. BLASTZ generates a series of local alignments which form the raw material of PIPMAKER. The main novel feature of PIPMAKER is the display of these alignments in the form of a percentage identity plot, or PIP. This displays the local alignments in the context of the first sequence, which may have been annotated to show positions of exons and is processed with RepeatMasker to show positions of repeated sequences, and CpG islands are also identified. The segments of sequence 2 that align with sequence 1 are represented as horizontal lines below sequence 1 and are placed on a vertical axis corresponding to the percentage match in the alignment. In this way regions of strong similarity between genomes are readily visualized. Matches can be constrained so that they have to be in the same order and orientation in the two genomes. MULTIPIPMAKER takes a similar approach but allows comparison of more than two sequences.

Related Website

| PIPMAKER | http://bio.cse.psu.edu/pipmaker/ |

Further Reading
Schwartz S, et al. (2000). PipMaker—A Web server for aligning two genomic DNA sequences. *Genome Res.* 10: 577–586.

See also BLAST.

PIX
John M. Hancock and Martin J. Bishop

A web tool to aid the analysis of protein sequences.
Similar in philosophy to NIX, the program first carries out a categorization of the physical properties of the residues in the sequence. BLAST searches of a number of databases are carried out and protein signature searches are executed using the PFAM, BLOCKS, PRINTS, and PROSITE databases. A number of other analyses are then carried out to predict cell localization, secondary structure, low-complexity regions and long/short globular domains, coiled coils, transmembrane domains, helix–turn–helix domains, signal peptides, antigenic regions, and enzyme digests. In an extended and slower analysis the Predator secondary-structure prediction algorithm is also applied.

Related Website

| PIX | http://www.hgmp.mrc.ac.uk/Registered/Webapp/pix/ |

See also BLOCKS, NIX, PFAM, PRINTS, PROSITE, Secondary-Structure Prediction of Protein.

PLASMID. *SEE* VECTOR

PLESIOMORPHY
A. Rus Hoelzel

An ancestral or unchanged state of an evolutionary character.

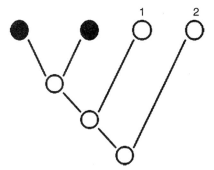

Figure Illustration of plesiomorphy. Open circles (1, 2) represent OTUs with an ancestral state of the character.

As a hypothesis about the pattern of evolution among operational taxonomic units (OTUs), a phylogenetic reconstruction can inform us about the relationship between character states. We can identify ancestral and descendent states and therefore primitive and derived states. A plesiomorphy (meaning "near shape" in Greek) is the primitive state (shown as open circles in the accompanying figure).

A shared primitive state between OTUs, as between 1 and 2 in the figure, is known as a synplesiomorphy.

Further Reading

Maddison DR, Maddison WP (2001). *MacClade 4: Analysis of Phylogeny and Character Evolution.* Sinauer Associates, Sunderland, MA.

See also Apomorphy, Autapomorphy, Synapomorphy.

POLAR
Roman A. Laskowski

A molecule which has an uneven distribution of electrons and a consequent dipole moment is said to be polar. Perhaps the best-known polar molecule is that of water.

In proteins, the amino acid residues having polar side chains are serine, threonine, cysteine, asparagine, tyrosine, histidine, tryptophan, and glutamine.

See also Amino Acid, Side Chain.

POLARIZATION
Roland L. Dunbrack

Changes in electron distribution in a molecule due to intermolecular and intramolecular interactions. Intermolecular interactions include solvent and ions as well as other kinds of molecules. Intramolecular polarization may occur as a flexible molecule changes conformation. Polarization is often not represented explicitly in molecular mechanics potential energy functions, but rather is accounted for by partial charges that are averaged over likely interactions with solvent.

Further Reading

Halgren TA, Damm W (2001). Polarizable force fields. *Curr. Opin. Struct. Biol.* 11: 236–242.

POLYADENYLATION
Niall Dillon

Addition of a poly(A) tail to an mRNA molecule.

The site at which polyadenylation occurs is generated by cleavage of the primary RNA transcript. Almost all eukaryotic mRNAs are posttranscriptionally modified at their 3′ ends by the addition of a poly-A tail. In higher eukaryotes, the site of polyadenylation is specified by the sequence AAUAAA located x bases upstream from the cleavage site. The polyadenylation site is recognized by a multiprotein complex containing at least 12 polypeptides. The C-terminal domain (CTD) of RNA polymerase II is also involved in polyadenylation.

Further Reading

Shatkin AJ, Manley JL (2000). The ends of the affair: Capping and polyadenylation. *Nature Struct. Biol.* 7: 838–842.

POLYGENIC EFFECT. *SEE* POLYGENIC INHERITANCE.

POLYGENIC INHERITANCE (POLYGENIC EFFECT)
Mark McCarthy and Steven Wiltshire

A genetic architecture whereby variation in a trait is determined by variation in many genes, each of which has individually small additive effects on the phenotype.

The term owes its origin to R. A. Fisher, who demonstrated in 1918 that continuous characteristics could be explained in terms of the small additive effects of many genes, each inherited according to Mendel's law of segregation. The aggregate of such genes is sometimes referred to as a polygene. Models for quantitative traits, variance components analysis, and segregation analysis routinely feature polygenic components, often in addition to genes of individually larger measurable effect termed quantitative trait loci. However, use of the term also extends to describing a possible genetic architecture for discrete complex, or multifactorial, traits. Such traits can be thought of as having an underlying continuously-distributed disease liability to which the many genetic and environmental risk factors contribute. An individual manifests the disease when his or her genetic and environmental risk factors together exceed a certain threshold. Many apparently "discrete" traits have underlying measurable quantitative characteristics, as demonstrated, e.g., by the relationship between blood pressure (a continuous trait) and hypertension (a discrete phenotype).

Further Reading

Falconer D, McKay TFC (1996). *Introduction to Quantitative Genetics*. Prentice-Hall, Harlow, England, pp. 100–183.

Fisher RA (1918). The correlation between relatives on the supposition of Mendelian inheritance. *Trans. Royal. Soc. Edinburgh*, 52: 399–433.

Hartl DL, Clark AG (1997). *Principles of Population Genetics*. Sinauer Associates, Sunderland, MA, pp. 397–481.

See also Multifactorial Trait, Oligogenic Inheritance, Quantitative Trait, Variance Components.

POLYHIERARCHY. *SEE* MULTIPLE HIERARCHY.

POLYMERASE CHAIN REACTION (PCR)

A molecular biology technique for amplifying a specific fragment of DNA for further analysis.

PCR combines a simple set of biochemical reactions to amplify a specified region of DNA. It generally employs a pair of short (10–15 nucleotide), single stranded deoxyoligonucleotide primers which can pair with sequences on either side of the region to be amplified, but on opposite strands. The DNA to be amplified is melted at high temperature in the presence of the primers, then cooled to allow the primers to bind to the separated strands. In the presence of deoxynucleotide triphosphates, new DNA is then synthesized by extending these primers. After a set time of synthesis, the reaction is again heated to high temperature to separate the products and the original strands. The reaction is then cycled a number of times. Because the newly synthesized strands are short, extending little beyond the binding site of the second primer, the bulk of the end product of these cycles is a DNA fragment bounded by the two primers. PCR makes use of the ability of DNA polymerases from thermophilic bacteria (classically *Thermus aquaticus*) to tolerate temperatures at which DNA strands separate, allowing a mixture containing DNA, primers, enzymes and deoxynucleotide triphosphates to be mixed at the beginning and put through a series of cycles without further intervention. The technique is widely used because it allows amplification of an original DNA sequence by a factor of 10^9 or 10^{10}.

Further Reading

Mullis KB et al. (1986). Specific enzymatic amplification of DNA in vitro: the polymerase chain reaction. *Cold Spring Harbor Symp. Quant. Biol.* 51: 263–273.

Bartlett JM, Stirling D (2003). *PCR Protocols*. Humana Press, Totowa NJ.

POLYMORPHISM (GENETIC POLYMORPHISM)
Mark McCarthy and Steven Wiltshire

Any genomic position (or locus) at which the DNA sequence shows variation between individuals.

Polymorphic sites are of importance for several reasons. First, by definition, inherited variation in traits of biological and medical importance is due to polymorphic variation. Second, polymorphic loci, if their genomic location is known, represent markers that can be used in linkage and linkage disequilibrium analyses to follow chromosomal segregation, define sites of recombination, and thereby aid susceptibility gene identification. Third, highly polymorphic loci are used in a wide range of legal and forensic situations. Several types of DNA polymorphisms exist, but the most frequently used in genetic analysis are single-nucleotide polymorphisms

(SNPs) and microsatellites (these are usually di-, tri-, or tetra-nucleotide repeats). The degree of polymorphism at a locus is described in terms of either the heterozygosity (the probability that a random individual is heterozygous for any two alleles at the locus) or a related measure termed the polymorphism information content (PIC). Historically, the term polymorphism was restricted to sites at which the minority allele frequency exceeded 1% and carried some imputation that the variant was subject to either neutral or balancing selection. However, in the light of contemporary insights into the development and maintenance of genomic variation in humans, these restrictions have become less relevant.

Related Websites

Databases of polymorphic sites in humans:

Online Mendelian Inheritance in Man	http://www.ncbi.nih.gov/Omim/
DbsNP	http://www.ncbi.nlm.nih.gov/SNP/
HGVBase	http://hgvbase.cgb.ki.se/

Utility programs (including determination of PIC values for polymorphisms):

Jurg Ott's Linkage Utility programs	http://linkage.rockefeller.edu/ott/linkutil.htm

Further Reading

Balding DJ, Donnelly P (1995). Inferring identity from DNA profile evidence. *Proc. Natl. Acad. Sci. USA* 92: 11741–11745.

Gray IC, et al. (2000). Single nucleotide polymorphisms as tools in human genetics. *Hum. Mol. Genet.* 9: 2403–2408.

Sachidanandam R, et al. (2001). A map of human genome sequence variation containing 1.42 million single nucleotide polymorphisms. *Nature* 409: 928–933.

Tabor HK, et al. (2002). Candidate-gene approaches for studying complex genetic traits: Practical considerations. *Nat. Rev. Genet.* 3: 1–7.

Wang D, et al. (1998). Large-scale identification, mapping, and genotyping of single-nucleotide polymorphisms in the human genome. *Science* 280: 1077–1082.

See also Association Analysis, Linkage Analysis, Linkage Disequilibrium, Marker, Single-Neucleotide Polymorphism.

POLYPEPTIDE
Roman A. Laskowski

As defined in the IUPAC Compendium of Chemical Terminology, a polypeptide is any peptide containing 10 or more amino acids. By convention, its direction runs from the N-terminus to the C-terminus.

Related Website

IUPAC	http://www.chemsoc.org/chembytes/goldbook/

Further Reading

Branden C, Tooze J (1991). *Introduction to Protein Structure*. Garland Science, New York.

See also Amino Acid, C-terminus, N-terminus, Peptide.

POLYPHYLETIC
John M. Hancock

In the context of phylogeny or taxonomy, an unnatural assemblage of nodes (species).

A classic example, the warm-blooded vertebrates, consists of mammals plus birds. Although mammals and birds share a common ancestor, the grouping warm-blooded vertebrates excludes many other taxa with the same property—notably reptiles, which are ancestral to birds.

See also Phylogenetic Tree, Taxonomic Classification.

POPULATION BOTTLENECK (BOTTLENECK)
A. Rus Hoelzel

A population bottleneck occurs when a large population quickly reduces in size.

This can impact genetic diversity in the reduced population through the random loss of alleles and through the loss of heterozygosity. The latter is a consequence of the increased chance of recombining alleles that are identical by descent in small populations. If the prebottleneck population was large and highly polymorphic, recessive alleles may have been maintained in the heterozygous state. The expression of these as homozygotes through more frequent consanguineous mating in the bottlenecked population can lead to inbreeding depression (a reduction in fitness resulting from inbreeding). The degree of loss in genetic diversity depends on the size of the bottlenecked population and the intrinsic rate of growth for that population, whereby smaller bottlenecks and lower rates of growth both lead to a greater loss of diversity. The time required to regain lost variation through mutation can be approximated by the reciprocal of the mutation rate (sometimes longer than the life span of the species).

Another possible consequence of a population bottleneck is an impact on quantitative characters that is more controversial than the simple loss of diversity. Given that there are many genes interacting for a given (polygenic) character and the variance resulting from this interaction has both additive and nonadditive components, a disruption of this interaction during a bottleneck event can lead to both an increase in the diversity of that character and the disruption of developmental stability (resulting in increased fluctuating asymmetry).

Further Reading
Hartl DL (2000). *A Primer of Population Genetics*. Sinauer Associates, Sunderland, MA.

See also Evolution, Founder Effect, Genetic Drift.

POSITIONAL CANDIDATE APPROACH
Mark McCarthy and Steven Wiltshire

The dominant strategy for susceptibility gene discovery in multifactorial traits, this is a hybrid approach that seeks to move toward susceptibility gene discovery using information derived from both positional and biological information.

The former will generally be derived through a genome scan for linkage, which may be expected, if adequately powered, to define the approximate genomic position of the most significant susceptibility effects. Further refinement of location may be possible through fine-mapping within these regions using both linkage and linkage disequilibrium analyses. In parallel, transcripts known to map to the region are reviewed to identify those positional candidate genes with the strongest prior claims, on biological grounds, for involvement in the pathogenesis of the disease concerned.

Examples: Two recent examples of this approach have been the identification of *NOD2* as a susceptibility gene for Crohn's disease (Hugot et al., 2001; Ogura et al., 2001) and of *ADAM33* for asthma (van Eerdewegh et al., 2002).

Further Reading

Collins FS (1995). Positional cloning moves from the perditional to traditional. *Nat. Genet.* 9: 347–350.

Hugot J-P, et al. (2001). Association of NOD2 leucine-rich repeat variants with susceptibility to Crohn's disease. *Nature* 411: 599–603.

Ogura Y, et al. (2001). A frameshift mutation in NOD2 associated with susceptibility to Crohn's disease. *Nature* 411: 603–607.

van Eerdewegh P, et al. (2002). Association of the ADAM33 gene with asthma and bronchial hyperresponsiveness. *Nature* 418: 426–430.

See also Candidate Gene, Genome Scan, Multifactorial Trait.

POSITION-SPECIFIC SCORING MATRIX. *SEE* PROFILE.

POSITION WEIGHT MATRIX. *SEE* PROFILE.

POSITION WEIGHT MATRIX OF TRANSCRIPTION FACTOR BINDING SITES
James W. Fickett

A tool to describe the DNA binding specificity of a transcription factor or, equivalently, to describe a prototypical transcription factor binding site.

The most common description of specificity in everyday writing is the consensus sequence, which gives the most commonly occurring nucleotide at each position of the binding site. However, the consensus sequence loses a great deal of information, as it does not distinguish between a highly informative position where T, say, always occurs and an essentially random one where T occurs 28% of the time and A, C, and G each occur 24% of the time. The position weight matrix (PWM) has one column for each position in the binding site and one row for each possible nucleotide. The entries are derived from the relative frequency f_{ij} of each nucleotide i at each position j. If one measures the background frequency p_i of each nucleotide as well and defines a PWM entry as the logarithm of the f_{ij}/p_i, then the matrix can provide statistically and physically meaningful numerical models. Define the PWM score of a potential site as the sum of the corresponding matrix entries. Then, under certain simplifying assumptions, the PWM score of a site is proportional both to the probability that the site is bound by the cognate factor and to the free energy of binding (Stormo and Fields, 1998). The sequence logo (Schneider, 2001) is a graphical device to represent the information in a PWM intuitively. The height of a character at a particular position is proportional to the relative frequency of the corresponding nucleotide, while the overall height of the column of letters is proportional to the information content of that position, in the sense of Shannon (Cover and Thomas, 1991). The PWM score does typically predict in vitro binding fairly well; its greatest limitation is that in vitro binding does not necessarily correlate with biological activity (Tronche et al., 1997). Less frequently, it may be an unrealistic assumption that the different positions in the binding site make independent contributions to the free energy of binding (Bulyk et al., 2002).

The TRANSFAC database covers some of the known transcription factors, binding sites, and derived PWMs (Wingender et al., 2001). To date, binding sites have been laboriously determined one at a time, and only a very small fraction of the total is known (Tupler et al., 2001). However, genomewide localization of DNA-binding proteins is now possible through formaldehyde cross-linking of the protein to the DNA, followed by fragmentation, chromatin immunoprecipitation, and finally hybridization of the resulting fragments to a microarray of genomic fragments (Ren et al., 2000).

The first step in constructing a PWM is alignment of known sites. More generally, one may attempt to discover, and align, a set of binding sites common to a particular transcription factor from a set of regulatory regions thought to contain such sites. For example, putative promoters may be extracted from a set of coordinatedly upregulated genes (as determined by expression array analysis) and these promoters searched for the binding sites of common regulating factors (Hughes et al., 2000). For both these variations on the alignment problem, an iterative algorithm, such as expectation maximization or Gibbs sampling, is often used (Lawrence et al., 1993). If the upstream regions of genes are extensive, with widely placed regulatory regions, as in human, phylogenetic footprinting must be used to limit the search to regions of high promise (Wasserman et al., 2000).

Related Websites

BIOBASE	http://www.gene-regulation.de/
AG BIODV	http://www.gsf.de/biodv
Wadsworth Center Bayesian Bioinformatics	http://www.wadsworth.org/resnres/bioinfo

Further Reading

Bulyk ML, et al. (2002). *Nucleic Acids Res.* 30: 1255–1261.

Cover TM, Thomas JA (1991). *Elements of Information Theory.* Wiley, New York.

Hughes JD, et al. (2000). *J. Mol. Biol.* 296: 1205–1214.

Lawrence CE, et al. (1993). *Science* 262: 208–214.

Ren B, et al. (2000). *Science* 290: 2306–2309.

Schneider TD (2001). *Nucleic Acids Res.* 29: 4881–4891.

Stormo GD, Fields DS (1998). *Trends Biochem. Sci.* 23: 109–113.

Tronche F, et al. (1997). *J. Mol. Biol.* 266: 231–245.

Tupler R, et al. (2001). *Nature* 409: 832–833.

Wasserman, WW, et al. (2000). *Nature Genet.* 26: 225–228.

Wingender E, et al. (2001). *Nucleic Acids Res.* 29: 281–283.

See also Consensus Sequence, Expectation Maximization Algorithm, Phylogenetic Footprinting, Sequence Logo, Transcription Factor, Transcription Factor Binding Site, TRANSFAC.

POSITIVE DARWINIAN SELECTION (POSITIVE SELECTION)
Austin L. Hughes

Positive Darwinian selection is natural selection that acts to favor amino acid replacements.

There is evidence that this type of selection is a relatively rare phenomenon over the course of evolution. One of the most common statistical tests for positive Darwinian selection is to compare the number of synonymous nucleotide substitutions per synonymous site (d_S) with the number of nonsynonymous (i.e., amino acid–altering) nucleotide substitutions per nonsynonymous site (d_N). In most comparisons between homologous protein coding genes, d_S exceeds d_N because most nonsynonymous substitutions are deleterious and thus are eliminated by purifying selection. When d_N is greater than d_S, this is evidence that natural selection has actually favored change at the amino acid level.

Further Reading

Endo T, et al. (1996). Large-scale search for genes on which positive selection may operate. *Mol. Biol. Evol.* 13: 685–690.

Hill RE, Hastie ND (1987). Accelerated evolution in the reactive centre regions of serine protease inhibitors. *Nature* 326: 96–99.

Hughes AL (1999). *Adaptive Evolution of Genes and Genomes*. Oxford University Press, New York.

Hughes AL, Nei M (1988). Pattern of nucleotide substitution at MHC class I loci reveals overdominant selection. *Nature* 335: 167–170.

See also Purifying Selection.

POSITIVE SELECTION. *SEE* POSITIVE DARWINIAN SELECTION.

POTENTIAL OF MEAN FORCE
Roland L. Dunbrack

A potential energy function derived from molecular dynamics simulations or other sources of conformational sampling for a system that represents the energy of some component of the system (e.g., the interaction of a pair of ions) with the effects of the remainder of the system (e.g., solvent water) averaged out. This term is often used to describe statistical potential functions.

See also Statistical Potential Energy.

POWER LAW (ZIPF'S LAW)
Dov S. Greenbaum

A frequency distribution of text strings in which a small number of strings are very frequent.

The most famous example of the power law is that of the usage of words in texts. Zipf's law states that some words–(e.g., *and, or, the*) are used frequently while most other words are used infrequently. When the size of each group of words is plotted against the usage of that word, the distribution can be approximated to a power law function; that is, the number of words (*N*) with a given occurrence (*F*) decays according to the equation $N = aF^{-b}$.

In molecular biology it has been found that short strings of DNA, DNA words, in noncoding DNA followed a similar behavior, suggesting that noncoding DNA resembled a natural language. The power law has also been applicable in contexts such as the connectivity within metabolic pathways; the occurrence of protein families, folds, or protein–protein interactions; and the occurrence of pseudogenes and pseudomotifs.

The finding of power law behavior is nontrivial, as it provides a mathematical description of a biological feature: the dominance, within a larger population, of a few members.

Further Reading

Mantegna RN, et al. (1994). Linguistic features of noncoding DNA sequences. *Phys. Rev. Lett.* 73: 3169–3172.

Qian J, et al. (2001). Protein family and fold occurrence in genomes: Power-law behavior and evolutionary model. *J. Mol. Biol.* 313: 673–681.

Zipf, GK (1949). *Human Behavior and the Principle of Least Effort.* Addison-Wesley, Reading, MA.

PREDICTION OF GENE FUNCTION

Jean-Michel Claverie

The process of interpreting a newly determined sequence in order to classify it into one of the previously characterized families of proteins.

Functional prediction mostly relies on the detection of similarity with proteins of known function. There are numerous levels of increasing sophistication in the recognition of these similarities, leading to functional predictions of decreasing accuracy and confidence. These various levels are:

- Significant global pairwise similarity, associated with the expected positioning of the new sequence in a phylogenetic tree. This is used to define orthologs, i.e., genes with the exact same role in two different species.
- Significant global pairwise similarity. Reciprocal best match throughout whole-genome comparison is used to extend the above definition of orthology.
- Significant global pairwise similarity, associated with the strict conservation of a set of residues at positions defining a function specific motif, is used to define paralogs.
- Significant local pairwise similarity, encompassing a function-specific motif, indicates evolutionarily more distant, but functionally related genes (possibly associated with unrelated domains).
- Significant local similarity solely detected by sequence family-specific scoring schemes (PSI-BLAST, HMM, etc.), indicating an even more distant functional relationship (usually biochemical, e.g., "kinase").
- The occurrence of a set of conserved positions/residues (e.g., PROSITE regular expression) previously associated with a function.

- No sequence similarity, but the evidence of a link with a gene of known function provided by one of the guilt-by-association methods, or experimental protein–protein interaction data (e.g., two-hybrid system).
- No sequence similarity, but the evidence of a link with another gene (e.g., coexpression) or a disease, organ, tissue, or physiological condition (e.g., differential expression) from gene expression measurements and profiling.

three-dimensional structure similarity can also be used to infer a functional relationship between proteins with absolutely no significant sequence similarity.

It should be noted that the operational definitions of orthology and paralogy used here differ from stricter definitions used in the discipline of molecular evolution. For example, in molecular evolution orthology is not taken to imply that two genes play the same role in different species.

Related Websites

NCBI BLAST	http://www.ncbi.nlm.nih.gov/BLAST/
PROSITE	http://www.expasy.ch/prosite/
InterPro	http://www.ebi.ac.uk/interpro/
Blocks	http://www.blocks.fhcrc.org/
PFAM	http://www.sanger.ac.uk/Pfam/
PRINTS	http://bioinf.man.ac.uk/dbbrowser/PRINTS/
ProDom	http://prodes.toulouse.inra.fr/prodom/doc/prodom.html
SMART	http://smart.embl-heidelberg.de/

Further Reading

Altschul SF, et al. (1997). Gapped BLAST and PSI-BLAST: A new generation of protein database search programs. *Nucleic Acids Res.* 25: 3389–3402.

Apweiler R, et al. (2000). InterPro—an integrated documentation resource for protein families, domains and functional sites. *Bioinformatics.* 16: 1145–1150.

Bateman A, et al. (2002). The Pfam protein families database. *Nucleic Acids Res.* 30: 276–280.

Claverie JM (1999). Computational methods for the identification of differential and coordinated gene expression. *Hum. Mol. Genet.* 8: 1821–1832.

Eisen JA (1998). Phylogenomics: Improving functional predictions for uncharacterized genes by evolutionary analysis. *Genome Res.* 8: 163–167.

Enright AJ, et al. (1999). Protein interaction maps for complete genomes based on gene fusion events. *Nature* 402: 86–90.

Marcotte EM, et al. (1999). Detecting protein function and protein-protein interactions from genome sequences. *Science* 285: 751–753.

Tatusov RL, et al. (2001). The COG database: New developments in phylogenetic classification of proteins from complete genomes. *Nucleic Acids Res.* 29: 22–28.

See also Functional Signature, Gene Family, Motif, Ortholog, Paralog.

PREDICTIVE POWER. *SEE* ERROR MEASURES.

PREDICTOR VARIABLES. *SEE* FEATURE.

PREGAP4
Rodger Staden

A widely used sequencing project data preparation program.

Related Website

Pregap4	http://www.mrc-lmb.cam.ac.uk/pubseq/manual/pregap4_unix_toc.html

PRIMARY CONSTRICTION. *SEE* CENTROMERE.

PRIMEGEN (PRIMER GENERATOR)
John M. Hancock and Martin J. Bishop

A program to analyze multiple alignments for the most and least conserved regions within them and aid the design of oligonucleotide probes to them.

The program accepts multiple alignments of DNA or protein sequences in msf format (the format used by the Wisconsin Package for multiple alignments). Identification of sequence type is automatic but may be overridden. Protein sequences are backtranslated for analysis. The program can identify the part or parts of the alignment with the following properties:

Most and least conserved: Conservation is measured by calculating the number of pairwise base identities. This is calculated as the sum over all columns in the window of the number of pairwise identities between sequences for each column of the alignment within this window.

Least and most degenerate: Degeneracy for a particular column in the alignment is effectively defined as the number of different bases occurring in that column. In the case of backtranslation, this is calculated from the respective IUPAC codes. The degeneracy of a window is then the product of the degeneracies of the individual columns.

The output of the program provides sequences of oligonucleotides that can be used to identify clones containing the sequences by hybridization.

Related Website

HGMP PRIMEGEN	http://www.hgmp.mrc.ac.uk/Registered/Option/primegen.html

Further Reading
O'Hara PJ, Venezia D (1991). PRIMEGEN, a tool for designing primers from multiple alignments. *Comput. Appl. Biosci.* 7: 533–534.

See also Alignment—Multiple.

PRIMER (PRIMER3)
John M. Hancock and Martin J. Bishop

A program for the design of PCR primers.

PRIMER takes a DNA sequence as input and, given user-defined constraints, provides a list of the highest quality pairs of PCR primers that will amplify from that sequence consistent with those constraints. Constraints include position and length of the region to be amplified, preferred experimental conditions (of the PCR reaction), product GC content, and potential of the primers to self-anneal or anneal to repetitive elements.

PRIMER3 is a complete rewrite of the original program.

Related Websites

PRIMER	http://www-genome.wi.mit.edu/ftp/distribution/software/primer.0.5/
HGMP PRIMER	http://www.hgmp.mrc.ac.uk/Registered/Help/primer/
PRIMER3	http://www-genome.wi.mit.edu/cgi-bin/primer/primer3_www.cgi

Further Reading

Rozen S, Skaletsky HJ (2000). Primer3 on the WWW for general users and for biologist programmers. In *Bioinformatics Methods and Protocols: Methods in Molecular Biology*, S Krawetz, S Misener, Eds. Humana Press, Totowa, NJ, pp. 365–386.

PRIMER GENERATOR. *SEE* PRIMEGEN.

PRIMER WALKING
Rodger Staden

The technique of using a custom primer to produce a sequence reading from an internal section of a DNA molecule. Used to obtain new sequence starting near the end of a known sequence in order to extend it. Often used to make joins between contigs or to double strand contigs.

Sequence assembly is the process of comparing the individual sequences (readings) to find overlaps and produce their most likely order along the genome. Often the bulk of the sequences generated are forward reads, and a subset reverse reads. In the assembled data, read pairs should have the correct relative orientations and separations. Hence read-pair data can be used to guide the assembly or as a check on its correctness. The assembly process arranges the readings into overlapping sets known as contigs. Using simple voting algorithms or more sophisticated calculations using base-call confidence values, a consensus sequence can be derived for each contig. As the projects progress, new data are used to establish joins between contigs until the whole of the target is covered by a single contig. The new data can be obtained by continuing to sequence clones generated by the shotgun procedure, but more often programs are used to analyze the existing contigs and select suitable sequencing experiments to help join contigs or to confirm poorly determined regions. Despite the increasing sophistication of these programs, the finishing stage of a sequencing project is still a lengthy and labor-intensive process. Major genome centers have detailed finishing criteria to guide these final stages, and the sequence quality annotations which they attach to entries are submitted to the sequence libraries.

See also Base-Call Confidence Values, Consensus Sequence, Contig, Double Stranding, Finishing, Finishing Criteria, Reading.

PRINCIPAL-COMPONENTS ANALYSIS (PCA)
John M. Hancock

A statistical method for identifying a limited set of uncorrelated (orthogonal) variables which explain most of the variability in a sample.

The variables derived by the method are termed the principal components. These are linear combinations of the original variables so that it may be possible to ascribe meaning to them, although this is not always easy to achieve.

Related Website

| PCA | http://www.okstate.edu/artsci/botany/ordinate/PCA.htm |

Further Reading

Hilsenbeck SG, et al. (1999). Statistical analysis of array expression data as applied to the problem of tamoxifen resistance. *J. Natl. Cancer Inst.* 91: 453–459.

Misra J, et al. (2002). Interactive exploration of microarray gene expression patterns in a reduced dimensional space. *Genome Res.* 12: 1112–1120.

PRINTS
Terri Attwood

A database in which protein families are characterized by groups of ungapped conserved motifs, which together form signatures or fingerprints of family membership. Fingerprints are built largely manually from hand-crafted seed alignments. The most conserved motifs are extracted and used to search a SWISS-PROT/TrEMBL composite database using an algorithm that converts the motifs into residue frequency matrices. Sequences that match all the motifs that were not in the seed alignment are assimilated into the process, and the database is searched repeatedly until the scans converge, i.e., until no more new matches can be found that contain all the motifs. The resulting fingerprint is then annotated and deposited in PRINTS.

Each PRINTS entry has a unique accession number and database identifier. The resource includes functional and structural descriptions of each family, plus disease information, literature references, and technical details of how the fingerprint was derived. Links to related entries within PRINTS are provided, as are cross-references to equivalent entries in related resources (PROSITE, InterPro, etc.). All SWISS-PROT and TrEMBL matches are listed, followed by the constituent motifs, which are stored as ungapped minialignments. The database may be searched using keywords or with query sequences via the FingerPRINTScan suite.

PRINTS is made available from the School of Biological Sciences at the University of Manchester, United Kingdom. A founding member of the InterPro consortium, it specializes in hierarchical diagnosis of protein families, from superfamily, through family, down to subfamily levels. The database was originally maintained as a single flat file but has also been migrated to a relational database management system in order to facilitate maintenance and to support more complex queries—the stream-lined version is known as PRINTS-S. To increase the coverage of the resource, an automatically generated supplement, prePRINTS, has been created from sequence clusters in ProDom and is automatically annotated using PRECIS.

Related Websites

PRINTS	http//www.bioinf.man.ac.uk/dbbrowser/PRINTS/
PRINTS-S	http//www.bioinf.man.ac.uk/dbbrowser/sprint/
FingerPRINTScan	http//www.bioinf.man.ac.uk//dbbrowser/fingerPRINTScan/
PRECIS	http//www.bioinf.man.ac.uk/cgi-bin/dbbrowser/precis/ precis.cgi

Further Reading

Attwood TK, Beck ME (1994). PRINTS—a protein motif fingerprint database. *Protein Eng.* 7(7): 841–848.

Attwood TK, Blythe M, Flower DR, Gaulton A, Mabey JE, Maudling N, McGregor L, Mit-chell A, Moulton G, Paine K, Scordis P (2002). PRINTS and PRINTS-S shed light on protein ancestry. *Nucleic Acids Res.* 30(1): 239–241.

Scordis P, Flower DR, Attwood TK (1999). FingerPRINTScan: Intelligent searching of the PRINTS motif database. *Bioinformatics* 15(10): 799–806.

See also Fingerprint, Motif, Protein Family.

PROCHECK
Roland L. Dunbrack

A program for assessing the stereochemical quality of a protein structure, developed by Laskowski and colleagues (1993, 1996). Procheck reports deviations of covalent geometry, planarity, chirality, and side-chain and main-chain dihedral angles from distributions in high-resolution crystal structures.

Related Website

PROCHECK	http://www.biochem.ucl.ac.uk/~roman/procheck/procheck.html

Further Reading

Laskowski RA, et al. (1993). PROCHECK: A program to check the stereochemical quality of protein structures. *J. Appl. Cryst.* 26: 283–291.

Laskowski RA, et al. (1996). AQUA and PROCHECK-NMR: Programs for checking the quality of protein structures solved by NMR. *J. Biomol. NMR* 8: 477–86.

See also Dihedral Angle.

PRODOM
Terri Attwood

A comprehensive collection of domain family alignments, obtained by automatic clus-tering of protein sequences in SWISS-PROT and TrEMBL. By contrast with the "pattern" databases (PROSITE, PRINTS, PFAM, etc.), ProDom is built entirely automatically and does not include hand-crafted familial discriminators. As such, its philosophical basis is rather different, and the populations of a given domain family will depend on the parameters used by the clustering algorithm rather than on the expert knowledge of a biologist or bioinformatician.

Each entry in ProDom is assigned a unique accession number that represents a single domain alignment, which includes all sequences detected in SWISS-PROT and TrEMBL with a level of similarity at or above a defined clustering threshold. By virtue of being an automatically derived resource, ProDom's coverage of the underlying sequence databases is much more comprehensive than that of the pattern databases (i.e., including many thousands of entries rather than hundreds), but it consequently includes no biological annotation. The database may be searched either using keywords via SRS or by means of query sequences via a standard BLAST interface.

ProDom is made available from the Institut National de Recherche Agronomique, Toulouse, France. A member of the InterPro consortium, it specializes in providing graphical interpretations of the domain arrangements in proteins.

Related Websites

ProDom	//www.toulouse.inra.fr/prodom.html
BLAST ProDom	//prodes.toulouse.inra.fr/prodom/doc/blast_form.html

Further Reading
Corpet F, Servant F, Gouzy J, Kahn D (2000). ProDom and ProDom-CG: Tools for protein domain analysis and whole genome comparisons. *Nucleic Acids Res.* 28(1): 267–269.

See also BLAST, SRS.

PROFILE (POSITION-SPECIFIC SCORING MATRIX, POSITION WEIGHT MATRIX, PSSM, WEIGHT MATRIX)
Terri Attwood

A position-specific scoring table that encapsulates the sequence information within sets of aligned sequences. Profiles define which residues are allowed at given positions; which positions are conserved and which degenerate; and which regions can tolerate insertions. In addition to data implicit in the alignment, the scoring system may include evolutionary weights and results from structural studies. Variable penalties are specified to weight against insertions and deletions occurring in secondary-structure elements.

Profiles are used in PROSITE both to provide an alternative to regular expressions whose diagnostic power for particular protein families is not optimal and to offer diagnostic tools for families that are too divergent to be able to derive useful regular expressions. In addition, a collection of profiles is made available from the Swiss Institute for Experimental Cancer Research, Lausanne; in contrast with those included in PROSITE, however, the majority of these are not fully annotated. They may be searched using ProfileScan.

Profiles provide a sensitive means of detecting distant sequence relationships, where only very few residues are well conserved. They therefore provide a useful complement to methods such as fingerprinting, blocks, and regular expressions.

Related Websites

About profiles	http://www.expasy.org/txt/profile.txt
ProfileScan	http://hits.isb-sib.ch/cgi-bin/PFSCAN
PROSITE	http://www.expasy.org/prosite/

Further Reading

Bucher P, Bairoch A (1994). A generalized profile syntax for biomolecular sequence motifs and its function in automatic sequence interpretation. *Proc. Intl. Conf. Intell. Syst. Mol. Biol.* 2: 53–61.

Falquet L, Pagni M, Bucher P, Hulo N, Sigrist CJ, Hofmann K, Bairoch A (2002). The PROSITE database, its status in 2002. *Nucleic Acids Res.* 30(1): 235–238.

See also Alignment, Profile Searching, PROSITE, Sequence Motifs—Prediction and Modeling.

PROFILE, 3D

Andrew Harrison and Christine A. Orengo

A three-dimensional (3D) profile or template encodes information about the structural conservation of positions in the 3D structure or fold adopted by a group of related structures. These structures are typically adopted by proteins related by a common evolutionary ancestor (homologs) or proteins adopting the same fold because of the constraints on the packing of secondary structures. The latter is described as convergent evolution and related structures are then described as analogs.

Different approaches have been developed for capturing information on the structural characteristics of a protein fold and how these are conserved across a structural family. The aim of generating a 3D profile is usually to recognize the most highly conserved positions for a group of structures in order to use this information to recognize other relatives sharing the same structural characteristics for these conserved positions. A 3D profile can therefore also be thought of as a pattern and is analogous to 1D profiles used in sequence alignment, as their performance similarly depends on matching those positions which have been most highly conserved during evolution and will therefore correspond even between very distant relatives.

In constructing 3D profiles, residue positions are usually considered and information on variation in 3D coordinates of the $C\alpha$ or $C\beta$ atom may be determined and described in the profile. However, in order to compare the positions, it is first necessary to multiply align the structures against each other or superpose them in 3D. Thus equivalent positions can be identified across the set and variation in coordinates can be easily calculated. Some approaches also consider variation in the structural environments of residues, where the structural environment may be described by the distances or vectors to all other residues or neighboring residues in the protein. Information on those subsets of residues which have conserved contacts in the fold can also be included. The algorithm for generating the 3D profile usually employs some mechanism for preventing the profile becoming biased toward any subsets of structures in the group which are overrepresented.

When using 3D profiles to recognize relatives adopting a similar 3D structure or fold, it is usual to employ a dynamic programming algorithm which can identify where insertions or deletions occur in the query structure to obtain an optimal alignment of the query against the template. Weights are employed to amplify scores associated with matching highly conserved positions in the 3D profile.

Related Website

Protein Family Cores	http://smi-web.stanford.edu/projects/helix/LPFC

PROFILE SEARCHING
Jaap Heringa

Techniques for searching sequences that are homologous to a given set of related sequences, where the search is carried out using a profile determined from a multiple alignment of these sequences. A natural extension of motif-searching techniques is provided by methods that use information over an entire sequence alignment of a certain protein family to find additional related family members. The earliest conceptually clear technique of this kind of sequence searching was called profile analysis (Gribskov et al., 1987) and combined a full representation of a sequence alignment with a sensitive searching algorithm. The procedure takes as input a multiple alignment of N sequences. First, a profile is constructed from the alignment, i.e., an alignment-specific scoring table which comprises the likelihood of each residue type to occur in each position of the multiple alignment. A typical profile has $L(20 + 2)$ elements, where L is the total length of the alignment, 20 is the number of amino acid types, and the last two columns contain gap penalties (see below). As a measure of similarity between different types of residues, sequences, a residue exchange matrix can be used. Then each element of the profile is calculated for each alignment position r and residue type c as

$$\text{Profile}(r, c) = \sum_{d=1}^{M} W_{d,r} * Comp(\text{Residue}_d, \text{Residue}_c)$$

where $M = 20$ is the number of amino acid types; *Comp* is the comparison value or substitution weight between the residue type c and each possible type of residue d, taken from an amino acid exchange matrix; and $W_{d,r}$ is the weight, which depends on the number of times that each residue type occurs in the position r of the alignment. The two commonly used weighting schemes are linear,

$$W_{d,r} = \frac{\sum_{i=1}^{N} w_i * \delta_d}{\sum_{i=1}^{N} w_i} \quad , \delta_d = \begin{cases} 1 \text{ if Residue}_{i,r} = \text{Residue}_d \\ 0 \text{ if Residue}_{i,r} \neq \text{Residue}_d \end{cases}$$

and logarithmic,

$$W_{d,r} = \frac{\ln\left[1 - \sum_{i=1}^{N} w_i * \delta_d \middle/ \left(1 + \sum_{i=1}^{N} w_i \right) \right]}{\ln\left[1 \middle/ \left(1 + \sum_{i=1}^{N} w_i \right) \right]}$$

where w_i is a weight assigned to each sequence in the alignment, usually 1.0.

Linear weighting simply reflects the fraction of each residue type at the position r while the logarithmic weighting upweights the most frequent of residue types. For any weighting scheme, $W_{d,r} = 0, 1$ if a certain type of residue d does not occur or exclusively occurs in the position r, respectively. Gribskov et al. (1987) used a single extra column in the profile to describe the local weight for both the gap opening and the gap extension penalty. For alignment positions not containing gaps, $P_{open} = P_{extend} = 100$, whereas for positions with insertions/deletions, these values are multiplied by the weighting factor $G_{max}/(1.0 + G_{inc*}L_{gap})$, where G_{max} is the maximum possible multiplier for an alignment position containing gaps, G_{inc} scales the decrease of this quantity as the observed gap length grows, and L_{gap} is the length of the gap crossing a given alignment position. The advantage of such positional gap penalties is that multiple-alignment regions with gaps (loop regions) will be assigned lowered gap penalties, hence will be more likely than core regions to attract gaps in a target sequence during profile searching, consistent with structural considerations. However, the implementation by Gribskov et al. (1987) does not take the frequency of gaps at each alignment position into account for the estimation of gap opening and/or extension penalties. This does not correspond with the expectation that a position rich in gaps would correspond to a loop site such that gaps in a sequence matched with the profile should be accommodated more easily at this position. Many alternative profile implementations therefore reserve the last two columns of the profile for positional gap opening (P_{open}) and gap extension (P_{extend}) penalties, which can be individually determined using protocols that take the above considerations into account. Another issue with the Gribskov et al. implementation is that an alignment column with, e.g., a single glycine and gaps for all other sequences would show the same score for glycine as a column consisting of identically conserved glycine residues.

After calculation of the profile, it is aligned in the Gribskov et al. approach with each sequence in the databank by means of the Smith and Waterman (1981) dynamic programming procedure, which finds the best local alignment path in a search matrix where appropriate profile values are placed into a comparison matrix cell corresponding to each residue in the database sequence and each alignment (or profile) position. Each match between a databank residue and a profile position simply receives the profile propensity for the databank residue type as a score. For each database sequence, the alignment score corresponding to the best local alignment quantifies the degree of similarity of this sequence with the probe profile. The scores are then corrected for sequence length, represented in the form of Z scores, and ranked to create the final list of databank search hits. Top-scoring sequences with scores above some threshold level are then likely to be related to the multiply-aligned sequences used to build the profile.

Another important sequence searching technique based on flexible protein sequence patterns was introduced by Barton and Sternberg (1990). Significant residue positions are selected on the basis of sequence conservation, functional importance, or the presence of secondary structure. These residues, constituting the pattern, can be separated by gaps which serve to exclude variable regions from the analysis. For each gap, a minimal and maximal possible length is derived from the initial sequence set. A look-up table similar to a profile is then calculated, which results in scores to compare each element of the pattern with each residue type. This feature distinguishes the method from regular expression pattern-matching algorithms based on positional match sets, which essentially use a binary exchange matrix. The flexible pattern is subsequently compared to every databank sequence using a modified Needleman–Wunsch technique (1970). Because only partial information contained in the protein family is used, which represents the most essential structural features, the method has high discriminating power. It is especially recommended for sequence alignments in which crucial elements are separated by long noisy stretches.

A technique with somewhat inverted logic compared to that of the profile-related methods was described by Altschul and Lipman (1990). Rather than searching individual databank sequences with the highest degree of similarity to a target set of sequences, their algorithm uses a single query sequence and attempts to find consistent sets of similar sequences in the databank. Similarity to the query sequence is determined by means of alignment of ungapped sequence segments with high individual scores.

The PSI-BLAST method (Altschul et al., 1997) also features a profile-related formalism to describe a set of aligned sequence motifs, called a position-specific scoring matrix (PSSM).

More recently, profile hidden Markov models (profile HMMs) have been developed to do sensitive database searching using statistical descriptions of a sequence family's consensus. Profile HMMs have several advantages over standard profiles. They have a formal probabilistic basis and consistent theory behind gap and insertion scores, in contrast to standard profile methods, which use heuristic methods. HMMs apply a statistical method to estimate the true frequency of a residue at a given position in the alignment from its observed frequency, while standard profiles use the observed frequency itself to assign the score for that residue. This means that a profile HMM derived from only 10–20 aligned sequences can be of equivalent quality to a standard profile created from 40–50 aligned sequences. In general, producing good profile HMMs requires less skill and manual intervention than producing good standard profiles. However, the technique is computationally intensive, which is an issue when integrated in genomic analysis pipelines.

A profile HMM is a linear state machine consisting of a series of nodes, each of which corresponds to a position (column) in the alignment from which it was built. If gaps are not considered, the correspondence is exact, and the profile HMM has a node for each column in the alignment; each node can exist in only one state, a match state. In order to model real sequences, the possibility should be included that gaps might occur when a model is aligned to a sequence. Two types of gaps may arise. A gap may occur when the sequence contains a region that is not present in the model (an insertion in the sequence). The second type occurs when there is a region in the model that is not present in the sequence (a deletion in the sequence). To handle these cases, each node in the profile HMM must now have three states: the match state, an insert state, and a delete state. The model then also needs more types of *transition probabilities*: match → match, match → insert, match → delete, insert → match, insert → insert, etc. In a simple ungapped model, the probability of a transition from one match state to the next match state is 1.0 and the path through the model is strictly linear, moving from the match state of node n to the match state of node $n + 1$. In a gapped model the transition pathway is more complex, but the sum of the probabilities associated with the transitions emanating from each state still add up to 1.0. Profile HMMs furthermore require *emission probabilities* associated with each match state, based on the probability of given residue types at that position in the alignment.

After constructing a profile HMM, the alignment of the model with a database sequence corresponds to following a path through the model to generate a sequence consistent with the model. The probability of any sequence that is generated depends on the transition and emissions probabilities at each node. The most widely used HMM-based profile searching tools currently are SAM-T98 or SAM-T99 (Karplus, 1998) and HMMer2 (Eddy, 1998).

Further Reading

Altschul SF, Lipman DJ (1990). Protein database searches for multiple alignments. *Proc. Natl. Acad. Sci. USA* 87: 5509–5513.

Barton GJ, Sternberg MJE (1990). Flexible protein sequence patterns: A sensitive method to detect weak structural similarities. *J. Mol. Biol.* 212: 389–402.

Eddy SR (1998). Profile hidden Markov models. *Bioinformatics* 14: 755–763.

Gribskov M, et al. (1987). Profile analysis: Detection of distantly related proteins. *Proc. Natl. Acad. Sci. USA* 84: 4355–4358.

Karplus K, Barrett C, Hughey R (1998). Hidden Markov models for detecting remote protein homologies. *Bioinformatics* 14: 846–856.

See also BLAST.

PROGENOTE
Dov S. Greenbaum

A progenote is defined as "a hypothetical stage in the evolution of cells that preceded organisms with typical prokaryotic cellular organization" (Woese and Fox, 1977).

Just as eukaryotes arose from earlier prokaryotes, prokaryotes are thought to have emerged from progenotes. Progenotes allowed for much faster DNA sequence evolution due to their more error prone translation and transcription machinery and their shorter gene sequences.

The inability to rationalize a spontaneously arising system as complex as a prokaryote from random nucleic acids and amino acids necessitated the introduction of this group—an intermediate between an ancestral organism in the RNA world and today's prokaryotes.

Alternatively, contradictory definitions of the word also exist—specifically, the usage of the word to refer to a common ancestor from which all kingdoms in life independently arose. In this sense the term is used similarly to the concept of a last common ancestor: "The two lines of descent, nevertheless shared a common ancestor, that was far simpler than the procaryote. This primitive entity is called a progenote, to denote the possibility that it had not yet completed evolving the link between geno and phenotype" (Woese and Fox, 1977).

Related Website

Progenote	http://www.sp.uconn.edu/~gogarten/progenote/progenote.htm

Further Reading

Dose K (1986). Hypotheses on the appearance of life on Earth (review). *Adv. Space Res.* 6: 181–186.

Doolittle WF, Brown JR (1994). Tempo, mode, the progenote, and the universal root. *Proc. Natl. Acad. Sci. USA* 91: 6721–6728.

Woese CR, Fox GE (1977). The concept of cellular evolution. *J. Mol. Evol.* 10: 1–6.

PROLINE
Jeremy Baum

Proline is a small nonpolar amino acid with side chain —H found in proteins. In sequences, written as Pro or P. It is the only amino acid that occurs in proteins which does not have a main chain —NH hydrogen atom and thus cannot participate in some of the hydrogen bonding found in secondary structures. The ring structure also makes this amino acid much less conformationally flexible.

See also Hydrogen Bond, Main Chain, Secondary Structure of Protein.

PROMOTER
Niall Dillon

Sequence that specifies the site of initiation of transcription.

The term *promoter* is an operational definition describing sequences located close to the transcription start site that specify the site of transcriptional initiation. Polymerase I and II genes have their promoter sequences located immediately upstream and downstream from the transcription initiation site. RNA polymerase III promoters are located upstream from the transcription start site in some genes (snRNA genes) and mainly within the transcribed region in others (5S RNA and tRNA genes). Promoters for all three polymerases have spatially constrained sequences that specify the site or sites of initiation of transcription by binding multiprotein initiation complexes.

Further Reading

Dvir A, et al. (2001). Mechanism of transcription initiation and promoter escape by RNA polymerase II. *Curr. Opin. Genet. Dev.* 11: 209–214.

Geiduschek EP, Kassavetis GA (2000). The RNA polymerase III transcription apparatus. *J. Mol. Biol.* 310: 1–26.

Grummt I (1999). Regulation of mammalian ribosomal gene transcription by RNA polymerase I. *Progr. Nucleic Acid Res. Mol. Biol.* 62: 109–154.

PROMOTERINSPECTOR
John M. Hancock

A program to identify regions likely to contain mammalian promoters.

The program makes use of context-specific features of mammalian promoters identified using a heuristic free learning algorithm. The method expresses the context of promoters in terms of equivalence classes of IUPAC groups, allowing a fuzzy description of the regions. The method has a claimed specificity of 85% and a sensitivity of 48%.

Related Website

PromoterInspector	http://www.genomatix.de/software_services/software/ PromoterInspector/PromoterInspector.html

Further Reading

Scherf M, et al. (2000). Highly specific localization of promoter regions in large genomic sequences by PromoterInspector: A novel context analysis approach. *J. Mol. Biol.* 297: 599–606.

See also ConsInspector, GenomeInspector, MatInspector, ModelInspector.

PROMOTER PREDICTION
James W. Fickett

Computational methods to localize the transcription start site (TSS) in genomic DNA (Ohler and Niemann, 2001) and to interpret transcription factor binding sites to predict under what conditions the gene will be expressed.

Definitions of a promoter differ. The essential thing is that the promoter contains signals necessary to both localize and time the initiation of transcription. If a gene is alternatively spliced, there may be multiple TSSs and, correspondingly, multiple promoters. In prokaryotes, all transcriptional regulation is typically in the promoter. In vertebrates, although the promoter does usually serve both functions of localizing the TSS and transducing control signals, much of the control is typically elsewhere, in so-called enhancers. Together, promoters, enhancers, and repressors constitute transcriptional regulatory regions.

In eukaryotes, computational prediction of the TSS is a difficult and unsolved problem (Fickett and Hatzigeorgiou, 1997). At the same time, it is a key component of gene identification. The core motif in use to recognize eukaryotic promoters is the so-called TATA box (Bucher, 1990), which binds a central protein (TATA binding protein) of the transcription initiation complex. Not all promoters contain a TATA box and, in fact, the great difficulty in promoter recognition is that no single set of motifs is characteristic. The most common components of prediction tools are:

• Scoring for common motifs (like the TATA box)
• Similarity of oligonucleotide frequencies to those of known promoters
• Density of putative transcription factor binding sites
• Phylogenetic footprinting
• Scoring (in vertebrates) for CpG islands in the vicinity

A CpG island is a region of high C + G content and a high frequency of CpG dinucleotides relative to the rest of the genome (Gardiner-Garden and Frommer, 1987). One recent algorithm gains accuracy by redefining the problem as first exon prediction, searching for pairs of promoters and donor splice sites. This and other algorithms seem to perform with acceptable accuracy when a gene occurs in conjunction with a CpG island, but not otherwise (Davuluri et al., 2001). Development of new algorithms depends on a collection of genes with well-characterized TSSs (Praz et al., 2002).

The problem in bacteria is somewhat different. Because operons are common, the ribosome binding site rather than the TSS is used in gene identification to mark the beginning of the gene.

Related Websites

AG BIODV	http://www.gsf.de/biodv
EPD	http://www.epd.isb-sib.ch
Zhang Lab, CSH	http://rulai.cshl.edu/software/index1.htm

Further Reading
Bucher P (1990). *J. Mol. Biol.* 212: 563–578.
Fickett JW, Hatzigeorgiou A (1997). *Genome Res.* 7: 861–878.
Gardiner-Garden M, Frommer, M (1987). *J. Mol. Biol.* 196: 261–282.

Niemann H (2001). *Trends Genet.* 17: 56–60.

Ohler U, Davuluri RV, et al. (2001). *Nature Genet.* 29: 412–417.

Praz V, et al. (2002). *Nucleic Acids Res.* 30: 322–324.

Suzuki Y, et al. (2002). *Nucleic Acids Res.* 30: 328–331.

See also CpG Island, Enhancer; Gene Prediction (Ab Initio, Comparative, Homology Based); Gene Prediction Systems, Integrated; Phylogenetic Footprinting; Promoter; Regulatory Region Prediction; Ribosome Binding Site; TATA Box; Transcriptional Regulatory Region; Transcription Factor; Transcription Start Site.

PROMOTER SCAN (PROSCAN)
John M. Hancock and Martin J. Bishop

A program for the prediction of RNA polymerase II promoters.

PROMOTER SCAN bases its detection of hypothetical promoter sequences on a combination of two factors: matching to a profile representing the relative density of transcription factor binding sites in promoter and nonpromoter sequences (known as the promoter recognition profile) and matching to a weight matrix describing a TATA box published by Bucher (1990).

PROMOTER SCAN II claims to identify 50% of uncharacterized promoters with a false-positive rate of one per 15,000 base pairs.

Related Websites

PROMOTER SCAN at HGMP	http://www.hgmp.mrc.ac.uk/ Registered/Option/proscan.html
PROMOTER SCAN II at Oxford, UK	http://www.molbiol.ox.ac.uk/ documentation/promoter_scan.htm

Further Reading

Bucher P (1990). Weight matrix descriptions of four eukaryotic RNA polymerase II promoter elements derived from 502 unrelated promoter sequences. *J. Mol. Biol.* 212: 563–578.

Prestridge DS (1995). Predicting Pol II promoter sequences using transcription factor binding sites. *J. Mol. Biol.* 5: 923–932.

See also Promoter Prediction, TATA Box, Transcription Factor.

PROSAII
Roland L. Dunbrack

A program for assessing the solvation free energy of a protein model from its atomic coordinates, developed by Manfred Sippl and colleagues. ProsaII reports a residue-by-residue summary of the solvation potential and can be used to identify poorly modeled regions of a protein that may have buried charges or too many exposed hydrophobic residues.

Related Website

ProsaII	http://www.came.sbg.ac.at/Services/prosa.html

Further Reading
Sippl MJ (1993). Recognition of errors in three-dimensional structures of proteins. *Proteins* 17: 355–362.

PROSCAN. *SEE* PROMOTER SCAN.

PROSITE
Terri Attwood

A database of patterns (regular expressions) and profiles used to characterize protein families and functional sites. Constituent regular expressions (regexs) are derived from the single most conserved motif observed in a multiple-sequence alignment of representative family members. The regexs are searched against SWISS-PROT and iteratively refined to determine the full extent of the family. Where a single motif is incapable of capturing the full family, additional expressions are derived until an optimal set of patterns is produced that characterizes all, or most, family members. Profile entries are also hand crafted to produce a discriminator with optimal diagnostic power—they are included in PROSITE to counterbalance the relatively poor diagnostic performance of some regular expressions.

Families in PROSITE are fully documented with, e.g., details of their function, structure, and disease associations. Each entry has a unique identifier and two accession numbers, one for the data file (which lists all the matches to a given regex or profile together with various technical details) and the other for the documentation file. The database may be searched either using simple keywords or by means of query sequences using ScanProsite.

PROSITE is made available from the Swiss Institute of Bioinformatics (SIB), Geneva. A founding member of the InterPro consortium, SIB specializes in characterizing both highly conserved functional sites (via its regular expressions) and divergent domains (via its profile collection).

Related Websites

PROSITE	http://www.expasy.org/prosite/
ScanProsite	http://www.expasy.org/tools/scanprosite/

Further Reading
Falquet L, Pagni M, Bucher P, Hulo N, Sigrist CJ, Hofmann K, Bairoch A (2002). The PROSITE database, its status in 2002. *Nucleic Acids Res.* 30: 235–238.

See also Motif, Profile, Sequence Pattern.

PROTÉGÉ, PROTÉGÉ-2000
Robert Stevens

Protégé (latest version Protégé-2000) is an extensible tool that enables users to build ontologies and use them for developing applications. Protégé uses a frame-based language, including classes and metaclasses arranged into hierarchies with slots representing relationships between these classes and constraining axioms. From such an ontology, the system can automatically generate forms which are used to acquire information about instances of the classes in the ontology and thereby to construct a knowledge base. Protégé can be extended by adding modular plug-ins from a centrally maintained library to enhance the behavior of its knowledge acquisition systems. Developers can easily make new plug-ins to provide new functionality.

Related Website

Protégé	http://protege.stanford.edu/index.html

PROTEIN ARRAY (PROTEIN MICROARRAY)
Jean-Michel Claverie

A newly introduced technique for proteomics and functional genomics based on the immobilization of protein molecules or the in situ synthesis of peptides on a solid support.

The recent explosion of data on gene sequences and expression patterns has generated great interest in the function and interactions of the corresponding proteins. This has triggered the demand for advanced technology offering greater throughput and versatility than two-dimensional gel electrophoresis combined with mass spectrometry. The new technology of protein microarrays incorporates advances in surface chemistries, developments of new capture agents, and new detection methods. As for DNA, the technologies differ by the nature of the physical support and the protocol of protein/peptide production and immobilization. Protein arrays make possible the parallel multiplex screening of thousands of interactions, encompassing protein–antibody, protein–protein, protein–ligand or protein–drug, enzyme–substrate screening, and multianalyte diagnostic assays. In the microarray or chip format, such determinations can be carried out with minimum use of materials while generating large amounts of data. Moreover, since most proteins are made by recombinant methods, there is direct connectivity between results from protein arrays and DNA sequence information.

Related Website

Protein Arrays Resource	http://www.functionalgenomics.org.uk/sections/ resources/protein_arrays.htm

Further Reading

Cahill DJ (2001). Protein and antibody arrays and their medical applications. *J. Immunol. Methods* 250: 81–89.

Fung ET, et al. (2001). Protein biochips for differential profiling. *Curr. Opin. Biotechnol.* 12: 65–69.

Jenkins RE, Pennington SR (2001). Arrays for protein expression profiling: Towards a viable alternative to two-dimensional gel electrophoresis? *Proteomics* 1: 13–29.

Mitchell P (2002). A perspective on protein microarrays. *Nature Biotechnol.* 20: 225–229.

Vaughan CK, Sollazzo M (2001). Of minibody, camel and bacteriophage. *Comb. Chem. High. Throughput Screen.* 4: 417–430.

Walter G, et al. (2000). Protein arrays for gene expression and molecular interaction screening. *Curr. Opin. Microbiol.* 3: 298–302.

Weinberger SR, et al. (2000). Recent trends in protein biochip technology. *Pharmacogenomics* 1: 395–416.

Wilson DS, Nock S (2002). Functional protein biochips. *Curr. Opin. Chem. Biol.* 6: 81–85.

Zhou H, et al. (2001). Solution and chip arrays in protein profiling. *Trends Biotechnol.* 19: S34–S39.

PROTEIN DATA BANK (PDB)

Eric Martz

The single internationally recognized primary repository of all published three-dimensional biological macromolecular structure data, including nucleic acids and carbohydrates as well as proteins.

The PDB archives atomic coordinate files, sometimes accompanied by supporting empirical data (crystallographic structure factors or NMR restraints). Each data file represents a fragment of a molecule, a whole molecule, or a complex of fragments or molecules. Each data file is assigned a unique four-character identification code (PDB ID). The first character is always a numeral (1–9) while the remaining three characters may be numerals or letters. The data files are freely available through the Internet. In some cases, authors do not release data until a year or more after publication of the structure; however, some prominent journals require that the data become publically available upon the date of publication.

The PDB was founded at Brookhaven National Laboratory (Upton, New York) in 1971, at a time when about a dozen protein structures had been solved by crystallography. It was maintained at Brookhaven until 1999, when management changed to a consortium of three geographically separated groups under the umbrella of the Research Collaboratory for Structural Bioinformatics. The number of entries had reached about 15,000 by year 2000. In 2002, about 83% are crystallographic results, 15% are from nuclear magnetic resonance, and 2% are theoretical. About 90% are proteins, while about 10% are nucleic acids or protein–nucleic acid complexes.

There is high redundancy in the PDB data set, however, so the number of sequence-dissimilar proteins is only a few thousand (depending on the stringency of the criteria employed). Furthermore, most of these are single domains rather than the intact naturally expressed forms. Proteins difficult to crystallize, notably integral membrane proteins, are underrepresented. Altogether, empirical structure data are available for less than a few percent of the human genome as of 2002.

Related Websites

Protein Data Bank	http://www.pdb.org
Research Collaboratory for Structural Bioinformatics	http://www.rcsb.org

Further Reading

Berman HM, et al. (2000). The Protein Data Bank. *Nucleic Acids Res.* 28: 235–242.

Bernstein FC, et al. (1977). The Protein Data Bank: A computer-based archival file for macromolecular structures. *J. Mol. Biol.* 112: 535–542.

Meyer EF (1997). The first years of the Protein Data Bank. *Protein Sci.* 6: 1591–1597.

Westbrook J, et al. (2002). The Protein Data Bank: Unifying the archive. *Nucleic Acids Res.* 30: 245–248.

See also Atomic Coordinate File; Models, Molecular; Nuclear Magnetic Resonance; Visualization, Molecular; X-Ray Crystallography.

PROTEIN DATABASES
Rolf Apweiler

Protein databases are a whole range of databases containing information on various protein properties.

The various protein databases hold different types of information on proteins; e.g., sequence data; bibliographical references; description of the biological source of a protein; function(s) of a protein; posttranslational modifications protein families, domains, and sites; secondary, tertiary, and quaternary structure; disease(s) associated with deficiencie(s) in a protein; two-dimensional (2D) gel data; and various other information depending on the type of protein database. The different kinds of protein databases include protein sequence databases, protein structure databases, protein sequence cluster databases, protein family and domain signature databases, 2D gel protein databases, and specialized protein database resources like the proteome analysis database.

Further Reading

Apweiler R (2000). Protein sequence databases. *Adv. Protein Chem.*. 54: 31–71.

Zdobnov EM, Lopez R, Apweiler R, Etzold T (2001). Using the molecular biology data. In *Biotechnology*, Vol. 5b: *Genomics and Bioinformatics*, CW Sernsen, ed. Wiley-VCH, pp. 281–300.

See also Protein Family and Domain Signature Databases, Protein Sequence Cluster Databases, Protein Sequence Databases, Proteome Analysis Database.

PROTEIN DOMAIN
Laszlo Patthy

A structurally independent, compact spatial unit within the three-dimensional structure of a protein.

Distinct structural domains of multidomain proteins usually interact less extensively with each other than do structural elements within the domains. A protein domain of a multidomain protein folds into its unique three-dimensional structure irrespective of the presence of other domains. Different domains of multidomain proteins may be connected by flexible linker segments.

Related Websites

Structure databases of protein domains:

PDB	http://www.rcsb.org/pdb/
DALI	http://www2.embl-ebi.ac.uk/dali/domain/

Sequence databases of protein domains

InterPro	http://www.ebi.ac.uk/interpro

PFAM	http://www.sanger.ac.uk/Pfam
ProDom	http://protein.toulouse.inra.fr/prodom.html
SMART	http://smart.embl-heidelberg.de
SBASE	http://www3.icgeb.trieste.it/~sbasesrv/

See also Multidomain Protein.

PROTEIN FAMILY
Andrew Harrison, Christine A. Orengo, Dov S. Greenbaum, and Laszlo Patthy

Groups of proteins sharing homology, structure, or function.

Given the variety of proteins in nature, it is helpful to pigeonhole proteins into a categorical system. The most common system is that of hierarchical families, akin to a taxonomic classification used for organisms. Protein families were originally conceived as comprising proteins related through sequence identity, although the colloquial use has extended to include also the grouping together of proteins related via similar function. Part of the Dayhoff classification scheme, protein families are more closely related evolutionarily and share a closer ancestor than the more broadly characterized grouping of superfamilies. By contrast, closely related homologous proteins may be grouped into protein subfamilies. Dayhoff originally defined a protein family as a group of proteins of similar function whose amino acid sequences are more than 50% identical; this cutoff value is still used by the Protein Information Resource (PIR). Other databases may use lower cutoff values and may not use functional criteria for the definition of a protein family. In many cases there are specific sequences within each member of the family that are required for the functional or structural features of that protein. The analysis of these conserved portions of the sequence allow researchers to analyze and derive defining signature sequences.

There are now several protein family databases, including PFAM, PRINTS, ProDom, SMART. PFAM is the most comprehensive database to date, containing 3360 protein families (version 7.0). There is usually considerable similarity in the functions of relatives classified within the same protein family. Most protein family databases generate sequence profiles, patterns, or regular expressions from a set of clear relatives within the family and use these to scan the sequence databases, such as SP-TREMBL and GenBank, to recognize further relatives to include in the classification. The comprehensive PFAM database builds hidden Markov models for each family, while the PRINTS database uses a fingerprint of regular expressions encoding conserved patterns at different positions in a multiple alignment of relatives. The PRINTS database contains extensive functional information for each protein family and is manually validated. By contrast, protein families in the ProDom database are identified using completely automated protocols.

Related Websites
Sequence-based classification of protein families:

| PROCLASS | http://www-nbrf.georgetown.edu/gfserver/proclass.html |

PFAM	http://www.sanger.ac.uk/Pfam/
PIR	http://www-nbrf.georgetown.edu/pirwww/pirhome.shtml
PRINTS	http://bioinf.man.ac.uk/dbbrowser/PRINTS/
SMART	http://smart.embl-heidelberg.de/
ProtFam List	http://www.hgmp.mrc.ac.uk/GenomeWeb/prot-family.html
MetaFam	http://metafam.ahc.umn.edu/
ProDom	http://prodes.toulouse.inra.fr/prodom/doc/prodom.html

Structural classification of protein families:

SCOP	http://scop.mrc-lmb.cam.ac.uk/scop
FSSP	http://www2.embl-ebi.ac.uk/dali/fssp/fssp.html
CATH	http://www.biochem.ucl.ac.uk/bsm/cath_new/index.html
PROSITE	http://www.expasy.ch/prosite/
TRANSFAC	http://transfac.gbf.de/TRANSFAC/

Further Reading

Mulder NJ, Apweiler R (2002). Tools and resources for identifying protein families, domains and motifs. *Genome Biol* 3: REVIEWS2001–2001.8.

Silverstein KAT, Shoop E, Johnson JE, Kilian A, Freeman JL, Kunau TM, Awad IA, Mayer M, Retzel EF (2001). The MetaFam Server: A comprehensive protein family resource. *Nucleic Acids Res.* 29: 49–51.

See also Hidden Markov Models, Protein Domain.

PROTEIN FAMILY AND DOMAIN SIGNATURE DATABASES
Rolf Apweiler

Protein family and domain signature databases provide signatures diagnostic for certain protein families or domains.

Protein family and domain signature databases use different sequence motif methodologies and a varying degree of biological information on well-characterized protein families and domains to derive signatures diagnostic for certain protein families or domains. The different protein family and domain signature databases provide varying degrees of annotation for the protein families or domains. These signature databases have become vital tools for identifying distant relationships in novel sequences and hence for inferring protein function. Diagnostically, the signature databases have different areas of optimum application owing to the different strengths and weaknesses of their underlying analysis methods (e.g., regular expressions, profiles, hidden Markov models, fingerprints).

InterPro	http://www.ebi.ac.uk/interpro/
PROSITE	http://www.expasy.org/prosite/
PFAM	http://www.sanger.ac.uk/Pfam/
PRINTS	http://www.biochem.ucl.ac.uk/bsm/dbbrowser/PRINTS/PRINTS.html
SMART	http://SMART.embl-heidelberg.de
Blocks	http://www.blocks.fhcrc.org/

Further Reading

Apweiler R, Attwood TK, Bairoch A, Bateman A, Birney E, Biswas M, Bucher P, Cerutti L, Corpet F, Croning MD, Durbin R, Falquet L, Fleischmann W, Gouzy J, Hermjakob H, Hulo N, Jonassen I, Kahn D, Kanapin A, Karavidopoulou Y, Lopez R, Marx B, Mulder NJ, Oinn TM, Pagni M, Servant F, Sigrist CJ, Zdobnov EM. (2001). The InterPro database, an integrated documentation resource for protein families, domains and functional sites. *Nucleic Acids Res.* 29: 37–40.

Kriventseva EV, Biswas M., Apweiler R (2001). Clustering and analysis of protein families. *Curr. Opin. Struct. Biol.* 11: 334–339.

See also Blocks, InterPro, PFAM, PRINTS, PROSITE, Protein Family.

PROTEIN INFORMATION RESOURCE (PIR)
Rolf Apweiler

The Protein Information Resource is a universal annotated protein sequence database covering proteins from many different species.

The National Biomedical Research Foundation (NBRF) established PIR in 1984 as a successor of the original NBRF Protein Sequence Database. Since 1988 the database has been maintained by PIR-International, a collaboration between the NBRF, the Munich Information Center for Protein Sequences (MIPS), and the Japan International Protein Information Database (JIPID). The PIR release 72 (April 2, 2002) contained 283,175 entries partitioned into four sections. Entries in PIR1 (20,723 entries) are fully classified by superfamily assignment, fully annotated, and fully merged with respect to other entries in PIR1. The annotation content as well as the level of redundancy reduction varies in PIR2 entries. Many entries in PIR2 are merged, classified, and annotated. Entries in PIR3 are not classified, merged, or annotated. PIR3 serves as a temporary buffer for new entries. PIR4 was created to include sequences identified as not naturally occurring or expressed, such as known pseudogenes, unexpressed ORFs, synthetic sequences, and non–naturally occurring fusion, crossover, or frameshift mutations.

Related Website

| PIR | http://pir.georgetown.edu/ |

Further Reading

Wu CH, Huang H, Arminski L, Castro-Alvear J, Chen Y, Hu Z-Z, Ledley RS, Lewis KC, Mewes HW, Orcutt BC, Suzek BE, Tsugita A, Vinayaka CR, Yeh L-S L, Zhang J, Barker WC (2002). The Protein Information Resource: An integrated public resource of functional annotation of proteins. *Nucleic Acids Res.* 30: 35–37.

See also SWISS-PROT, TrEMBL.

PROTEIN MICROARRAY. *SEE* PROTEIN ARRAY.

PROTEIN MODULE
Laszlo Patthy

A protein module is a structurally independent protein domain that has been spread by module shuffling and may occur in different multidomain proteins with different domain combinations.

Related Website

SMART	http://smart.embl-heidelberg.de

Further Reading

Patthy L (1985). Evolution of the proteases of blood coagulation and fibrinolysis by assembly from modules. *Cell* 41: 657–663.

Patthy L (1991). Modular exchange principles in proteins. *Curr. Opin. Struct. Biol.* 1: 351–361.

Patthy L (1996). Exon shuffling and other ways of module exchange. *Matrix Biol.* 15: 301–310.

Schultz J, et al. (2000). SMART: A Web-based tool for the study of genetically mobile domains. *Nucleic Acids Res.* 28: 231–234.

See also Module Shuffling, Multidomain Protein, Protein Domain.

PROTEIN–PROTEIN COEVOLUTION
Laszlo Patthy

Coevolution is a reciprocally induced inherited change in a biological entity in response to an inherited change in another with which it interacts.

Specific protein–protein interactions essential for some biological function (e.g., binding of a protein ligand to its receptor, binding of a protein inhibitor to its target enzyme) are ensured by a specific network of interresidue contacts between the partner proteins. During evolution, sequence changes accumulated by one of the interacting proteins are usually compensated by complementary changes in the other. As a corollary, positions where changes occur in a correlated fashion in the interacting molecules tend to be close to the protein–protein interfaces, permitting the prediction of contacting pairs of residues of interacting proteins.

Coevolution of protein–protein interaction partners is also reflected by congruency of their cladograms. Thus, comparison of phylogenetic trees constructed from multiple-sequence alignments of interacting partners (e.g., ligand families and the corresponding receptor families) shows that protein ligands and their receptors usually coevolve so that each subgroup of ligands has a matching subgroup of receptors.

Bioinformatics approaches have been developed for the prediction of protein–protein interactions based on the correspondence between the phylogenetic trees of interacting proteins and detection of correlated mutations between pairs of proteins.

Further Reading

Goh CS, et al. (2000). Co-evolution of proteins with their interaction partners. *J. Mol. Biol.* 299: 283–293.

Pazos F, Valencia A (2001). Similarity of phylogenetic trees as indicator of protein-protein interaction. *Protein Eng.* 14: 609–614.

Pazos F, et al. (1997). Correlated mutations contain information about protein-protein interaction. *J. Mol. Biol.* 271: 511–523.

See also Coevolution, Alignment—Multiple.

PROTEIN SEQUENCE CLUSTER DATABASES
Rolf Apweiler

Protein sequence cluster databases group related proteins together and are derived automatically from protein sequence databases using different clustering algorithms.

Protein sequence cluster databases use different clustering algorithms to group related proteins together. Since they are derived automatically from protein sequence databases without manual crafting and validation of family discriminators, these databases are relatively comprehensive, although the biological relevance of clusters can be ambiguous and can sometimes be an artifact of particular thresholds.

Related Websites

ProDom	http://www.toulouse.inra.fr/prodom.html
SYSTERS	http://www.dkfz-heidelberg.de/tbi/ services/cluster/systersform
ProtoMap	http://www.protomap.cs.huji.ac.il/
CluSTr	http://www.ebi.ac.uk/clustr/
Clusters of Orthologous Groups (COG) of proteins	http://www.ncbi.nlm.nih.gov/COG

Further Reading

Kriventseva EV, et al. (2001). Clustering and analysis of protein families. *Curr. Opin. Struct. Biol.* 11: 334–339.

See also Clusters of Orthologous Groups, CluSTr.

PROTEIN SEQUENCE DATABASES
Rolf Apweiler

Protein sequence databases are databases containing information on the sequences of proteins and typically also on various other protein properties.

The protein sequence databases are the most comprehensive source of information on proteins. The various protein sequence databases hold different types of information on proteins: e.g., sequence data; bibliographical references; description of the biological source of a protein; function(s) of a protein; posttranslational modifications; disease(s) associated with deficiencie(s) in a protein; and various other information depending on the type of protein sequence database. It is necessary to distinguish between universal protein sequence databases covering proteins from all species, nonredundant protein sequence databases, and specialized data collections storing information about specific families or groups of proteins or about the proteins of a specific organism. The main universal protein sequence databases are the Protein Information Resource (PIR) and SWISS-PROT and TrEMBL.

Related Websites

Protein Information Resource	http://pir.georgetown.edu/
SWISS-PROT	http://www.ebi.ac.uk/swissprot/ http://www.expasy.org/sprot/
TrEMBL	http://www.ebi.ac.uk/swissprot/ http://www.expasy.org/sprot/
Merops	http://merops.iapc.bbsrc.ac.uk/

Further Reading
Apweiler R (2000). Protein sequence databases. *Adv. Protein Chem.* 54: 31–71.

See also MEROPS, Protein Information Resource, SWISS-PROT, TrEMBL.

PROTEIN STRUCTURE
Roman A. Laskowski, Andrew Harrison, and Christine A. Orengo

The final, folded three-dimensional (3D) conformation of the atoms making up a protein, which describes the geometry of that protein in 3D space. The term is loosely applied to the models of the protein structure that are obtained as the result of experiment (e.g., by means of X-ray crystallography or NMR spectroscopy). Such models aim to give as good an explanation for the experimental data as possible but are merely representations of the molecule in question; they may be accurate or inaccurate models, precise or imprecise. Models from X-ray crystallography tend to be static representations of a time-averaged structure, whereas proteins frequently exhibit dynamic behavior.

Nevertheless, knowledge of a protein's structure, as from such an experimentally determined model, can be crucial for understanding its biological function and mechanism. Comparisons of structures can provide insights into general principles governing these complex molecules, the interactions they make, and their evolutionary relationships.

The details of the protein's active site can not only help explain how it achieves its biological function but also assist the design of drugs to target the protein and block or inhibit its activity.

Protein structure is usually represented as a set of 3D coordinates for each atom in the protein. From these coordinates other geometrical relationships and properties can be calculated—e.g., the distances between all Cα atoms within the protein or specific angles (e.g., dihedral angles) which are used to describe the curvature of the polypeptide chain. Protein structures are deposited in the Protein Data Bank (PDB) currently held at the Research Collaboratory of Structural Biology at Rutgers University, New Brunswick, New Jersey. There is also a European node, the Macromolecular Structure Database (MSD), where structures are also deposited, held at the European Bioinformatics Institute near Cambridge, United Kingdom.

Related Websites

RCSB	http://www.rcsb.org/pdb/
MSD	http://www.ebi.ac.uk/msd

Further Reading
Branden C, Tooze J (1991). *Introduction to Protein Structure.* Garland Science, New York.

See also Dihedral Angle, Protein Data Bank.

PROTEIN STRUCTURE CLASSIFICATION DATABASES. *SEE* STRUCTURE—3D CLASSIFICATION.

PROTEOME
Jean-Michel Claverie and Dov S. Greenbaum

The protein complement of a genome, cell, organism, or other biological entity.

The term *proteome*, coined in 1995, has two distinct usages. Colloquially, it is used to broadly define the population of proteins in a cell. This definition is favored by experimentalists, who use procedures such as two-dimensional electrophoresis and matrix-assisted laser desorption ionization time-of-flight (MALDI-TOF) mass spectrometry to measure and quantify the population of proteins in a cell under multiple conditions.

A second definition, favored by computational biologists, is that the proteome refers specifically to the portion of a genome coding for proteins, a parts list of the genome-encoded proteins in a cell, as distinct from the genome which includes also the non–protein coding DNA. This usage does not take into account the level of protein expression in individual cells, but rather is a more general term referring to the hard data encoded in all cells in a given organism at any time point.

It includes the whole spectrum of different molecular forms found after posttranslational modifications (e.g., phosphorylation, glycosylation). Originally defined as all the separable spots on a two-dimensional electrophoresis gel (separation by size and isoelectric point). In a comparative context (similar to gene expression profiling), it is used to refer to the subset of proteins detected in a given cell, organ, or tissue. In a more quantitative way, the "proteome" can also refer to the set of all proteins associated to a measure of their abundance level. More theoretically, it is also used to designate all putative protein sequences encoded in a given genome (predicted proteome), including all the variants generated from alternative transcripts.

Related Websites

Proteome Analysis Database	http://www.ebi.ac.uk/proteome/
Human Proteome Organisation	http://www.hupo.org/
Proteomics Interest Group	http://proteome.nih.gov/
ExPASy (Expert Protein Analysis System)	http://www.expasy.org/
MIPS (cataloging proteins and many of their attributes from multiple genomes)	http://www.mips.biochem.mpg.de/
Proteome-EBI	http://www.ebi.ac.uk/proteome/
MINT	http://cbm.bio.uniroma2.it/mint/

Further Reading

Apweiler R, et al. (2001). Proteome Analysis Database: Online application of InterPro and CluSTr for the functional classification of proteins in whole genomes. *Nucleic Acids Res.* 29: 44–48.

Gygi SP, et al. (2000). Evaluation of two-dimensional gel electrophoresis-based proteome analysis technology. *Proc. Natl. Acad. Sci. USA* 97: 9390–9395.

Patton WF (2002). Detection technologies in proteome analysis. *J. Chromatogr. B Anal. Technol. Biomed. Life Sci.* 771: 3–31.

Rappsilber J, Mann M (2002). Is mass spectrometry ready for proteome-wide protein expression analysis? *Genome Biol.* 3: COMMENT2008.

Zhu H, et al. (2001). Global analysis of protein activities using proteome chips. *Science* 293: 2101–2105.

PROTEOME ANALYSIS DATABASE
Rolf Apweiler

The Proteome Analysis Database is a specialized protein database resource integrating information from a variety of sources that together facilitate the classification of proteins in complete proteome sets. These proteome sets are built from the SWISS-PROT and TrEMBL protein sequence databases that provide reliable, well-annotated data as the basis for the analysis. Proteome analysis data are available for all completely sequenced organisms present in SWISS-PROT and TrEMBL, spanning archaea, bacteria, and eukaryotes. The Proteome Analysis Database provides a broad view of the proteome data classified according to signatures describing particular sequence motifs or sequence similarities and at the same time affords the option of examining various specific details like structure or functional classification. The InterPro, CluSTr, and GO resources have been used to classify the data. The Proteome Analysis Database contained in May 2002 statistical and analytical data for the proteins from more than 70 complete genomes.

Related Website

Proteome Analysis Database	http://www.ebi.ac.uk/proteome/

Further Reading

Apweiler R, Biswas M, Fleischmann W, Kanapin A, Karavidopoulou Y, Kersey P, Kriventseva EV, Mittard V, Mulder N, Phan I, Zdobnov E. (2001). Proteome Analysis Database: Online application of InterPro and CluSTr for the functional classification of proteins in whole genomes. *Nucleic Acids Res.* 29: 44–48.

See also CluSTr, InterPro, SWISS-PROT, TrEMBL.

PROTEOMICS

Jean-Michel Claverie

A research area and the set of approaches dealing with the study of the proteome as a whole (or the components of large macromolecular complexes).

Beside the large-scale aspect, an essential difference with traditional protein biochemistry is that the proteins of interest are not defined prior to the experiment.

The critical pathway of proteome research includes:

1. Sample collection, handling, and storage
2. Protein separation (two-dimensional electrophoresis)
3. Protein identification (peptide mass fingerprinting and mass spectrometry)
4. Protein characterization (amino acid sequencing)
5. Bioinformatics (cross-reference of protein informatics with genomic databases)

Main topics in proteomics are

1. Listing the parts: the detailed identification of all the components of an organism, organ, tissue, organelle, or other intracellular structures
2. Profiling the expression: the analysis of protein abundance across different cell types, organs, tissue, or physiological or disease conditions
3. Finding the function: submitting the proteome (a large subset) to parallel functional assays such as ligand binding, interaction with a macromolecule, etc.

Topics 1 and 2 require good separation and highly sensitive detection, quantitation, and identification techniques. Topic 3 requires protein mass production, robotics, arraying (e.g., protein arrays), and sensitive interaction assays.

Related Websites

ExPASy Proteomics tools	http://us.expasy.org/tools/
Proteomics Interest Group	http://proteome.nih.gov/
Proteome Analysis Database	http://www.ebi.ac.uk/proteome/

Further Reading

Bakhtiar R, Nelson RW (2001). Mass spectrometry of the proteome. *Mol. Pharmacol.* 60: 405–415.

Griffin TJ, Aebersold R (2001). Advances in proteome analysis by mass spectrometry. *J. Biol. Chem.* 276: 45497–45500.

Issaq HJ (2001). The role of separation science in proteomics research. *Electrophoresis* 22: 3629–3638.

Jenkins RE, Pennington SR (2001). Arrays for protein expression profiling: Towards a viable alternative to two-dimensional gel electrophoresis? *Proteomics* 1: 13–29.

Nelson RW, et al. (2000). Biosensor chip mass spectrometry: A chip-based proteomics approach. *Electrophoresis* 21: 1155–1163.

Patton WF (2000). A thousand points of light: The application of fluorescence detection technologies to two-dimensional gel electrophoresis and proteomics. *Electrophoresis* 21: 1123–1144.

Rappsilber J, Mann M (2002). What does it mean to identify a protein in proteomics? *Trends Biochem. Sci.* 27: 74–78.

PSEUDOGENE
Dov S. Greenbaum

Pseudogenes are sequences of genomic DNA which, although they have similarity to normal genes, are nonfunctional.

Pseudogenes are interesting in genomics as they provide a picture of how the genomic sequence has mutated over evolutionary time and can also be used in determining the underlying rate of nucleotide insertion, deletion, and substitution.

Pseudogenes are caused by one of two possible processes. (i) *Duplication to produce nonprocessed pseudogenes.* During duplication any number of modifications—mutations, insertions, deletions, or frameshifts—can occur to the primary genomic sequence. Subsequent to these modifications a gene may lose its function at either or both the translational and transcriptional levels. (ii) *Processed pseudogenes.* These arise when a mRNA transcript is reverse transcribed and integrated back into the genome. Mutations and other disablements may then occur to this sequence over the course of evolution. As they are reverse transcription copies of a mRNA transcript, they do not contain introns.

Pseudogenes are identified through the use of sequence alignment programs. Initially annotated genes are collected into paralog families. These families are then used to survey the entire genome. Potential homologs are determined to be pseudogenes if there is evidence for one of the two methods mentioned above. For each potential pseudogene there are a number of validation steps, including checking for overcounting and repeat elements, overlap on the genomic DNA with other homologs, and cross-referencing with exon assignments from genome annotations.

Related Website

Gerstein Lab Pseudogenes	http://bioinfo.mbb.yale.edu/genome/pseudogene/

Further Reading

Cooper DN (1999). *Human Gene Evolution*. Bios Scientific Publishers, Oxford, UK, pp. 265–285.

Harrison PM, Gerstein M (2002). Studying genomes through the aeons: Protein families, pseudogenes and proteome evolution. *J. Mol. Biol.* 318: 1155–1174.

Mighell AJ, Smith NR, Robinson PA, Markham AF (2000). Vertebrate pseudogenes. *FEBS Lett.* 468: 109–114.

PSI BLAST
Dov S. Greenbaum

Version of the BLAST program employing iterative improvements of an automatically generated profile to detect weak sequence similarity.

"BLAST is a heuristic that attempts to optimize a specific similarity measure. It permits a tradeoff between speed and sensitivity, with the setting of a 'threshold' parameter, T. A higher value of T yields greater speed, but also an increased probability of missing weak similarities" (Altschul, et al., 1997).

Previous versions of BLAST could find only local alignments between proteins or DNA sequences that did not contain gaps. These programs were able to give a value, the E value, that gave some idea of the probability of the results occurring wholly by chance. It is thought, although not mathematically proven, that E values are reliable for gapped sequence alignments as well.

PSI BLAST, Position Specific Iterated Basic Local Alignment Search Tool, is based on the Gapped BLAST program, but after alignments have been found, a profile is constructed from the significant matches. This profile is then compared and locally aligned to proteins in the available protein databases. This is then iterated, if desired, an arbitrarily number of times with the new profiles that are discovered in each successive search.

PSI BLAST is one of the most powerful and popular sequence homology programs available—the papers describing the algorithms for this and related BLASTS some of the most heavily cited.

Related Website

| BLAST | http://www.ncbi.nlm.nih.gov/blast/ |

Further Reading

Altschul SF, et al. (1990). Basic local alignment search tool. *J. Mol. Biol.* 215: 403–410.

Altschul SF, et al. (1997). Gapped BLAST and PSI-BLAST: A new generation of protein database search programs. *Nucleic Acids Res.* 25: 3389–3402.

PSSM. *SEE* PROFILE.

PURIFYING SELECTION
Austin L. Hughes

Natural selection that acts to eliminate mutations that are harmful to the fitness of the organism.

Evidence of purifying selection on a genomic region is evidence that that region, whether coding or not, plays an important function and thus is subject to functional constraint. In protein coding regions, purifying selection eliminates the majority of nonsynonymous (amino acid–altering) mutations. As evidence that such selection has occurred, the number of synonymous nucleotide substitutions per synonymous site (d_S) exceeds the number of nonsynonymous nucleotide substitutions per nonsynonymous site (d_N) in the vast majority of protein coding genes.

Further Reading

Kimura M (1977). Preponderance of synonymous changes as evidence for the neutral theory of molecular evolution. *Nature* 267: 275–276.

Kimura M (1983). *The Neutral Theory of Molecular Evolution*. Cambridge University Press, Cambridge.

Li W-H, et al. (1985). Evolution of DNA sequences. In *Molecular Evolutionary Genetics*, RJ MacIntyre, Ed. Plenum, New York, pp. 1–94.

Q

Dictionary of Bioinformatics and Computational Biology. Edited by Hancock and Zvelebil
ISBN 0-471-43622-4 © 2004 John Wiley & Sons, Inc.

Q INDEX (*Q* HELIX; *Q* STRAND; *Q* COIL; Q3)
Patrick Aloy

Per-residue prediction accuracy is the simplest, although not the best, measure of secondary-structure prediction accuracy. It gives the percentage of residues correctly predicted as helix, strand, or coil or for all three conformational states.

For a single conformational state

$$Q_i = \frac{\text{number of residues correctly predicted in state } i}{\text{number of residues observed in state } i} \times 100$$

where i is helix, strand, or coil.

For all three states

$$Q_3 = \frac{\text{number of residues correctly predicted}}{\text{number of residues}} \times 100$$

where Q_3 is a very generous way of measuring prediction accuracy, as it only considers correctly predicted residues and does not penalize wrong predictions.

Related Website

Accuracy estimators	http://predictioncenter.llnl.gov/local/local.html

Further Reading

Rost B, Sander C (2000). Third generation prediction of secondary structures. *Methods Mol. Biol.* 143: 71–95.

See also Secondary-Structure Prediction of Protein.

QM/MM SIMULATIONS
Roland L. Dunbrack

Molecular mechanics simulations in which part of the system is treated with ab initio or semiempirical quantum mechanics while the rest of the system is treated with standard empirical molecular mechanics energy functions. QM/MM simulations are needed to study inherently quantum-mechanical processes such as breaking and forming covalent bonds. In this case, the active site of an enzyme is treated quantum mechanically, and the rest of the protein and the solvent are treated with empirical energy functions.

Further Reading

Gao J, Truhlar, DG (2002). Quantum mechanical methods for enzyme kinetics. *Annu. Rev. Phys. Chem.* 53: 467–505.

See also Ab Initio, Molecular Mechanics.

Dictionary of Bioinformatics and Computational Biology. Edited by Hancock and Zvelebil
ISBN 0-471-43622-4   © 2004 John Wiley & Sons, Inc.

QSAR (QUANTITATIVE STRUCTURE–ACTIVITY RELATIONSHIP)
Marketa J. Zvelebil

The method of QSAR aims to derive a function connecting biological activity of a set of similar chemical compounds with parameters that describe a structural feature of these molecules which reflects properties of the binding cavity on a protein target. The derived function is used as a guide to select best candidates for drug design.

Basically, the QSAR method puts together statistical and graphical models of biological activity based on molecular structures. The models are then used to make predictions for the activity of untested compounds.

Related Website

QSAR—Tripos	http://www.tripos.com/sciTech/inSilicoDisc/strActRelationship/qsar.html

Further Reading
Kim KH, Greco G, Novellino E (1998). A critical review of recent CoMFA applications. In *3D QSAR in Drug Design*, Vol. 3. H Kubinyi, G Folkers, YC Martin, Eds. Kluwer Academic, Great Britain, p. 257.
Klebe G (1998). Comparative molecular similarity indices: CoMSIA. In *3D QSAR in Drug Design*, Vol. 3. H Kubinyi, G Folkers, YC Martin, Eds. Kluwer Academic, Great Britain, p. 87.

QUANTITATIVE TRAIT (CONTINUOUS TRAIT)
Mark McCarthy and Steven Wiltshire

A trait or characteristic that is measured on a continuous scale.

The quantitative phenotype (P_i) of a particular individual i will reflect the combined effects of those genetic (G_i) and environmental factors (E_i) acting on that individual: $P_i = G_i + E_i$. The architecture of the genetic component, G_i, can vary in terms of the number of genes involved, their frequencies, and the magnitude of their effects on the trait. Any gene influencing the value of a quantitative trait can be termed a quantitative trait locus (QTL), although this term is usually reserved for genes with individually measurable effects on the trait. A trait under the influence of several such genes is often termed oligogenic, whereas a genetic architecture comprising many genes each with individually small additive effects is termed *polygenic*, and the aggregate of such genes is often referred to as the polygene. Many biomedically significant traits (such as obesity, blood pressure) are thought to be under the control of one or several QTLs acting on a polygenic background. The environmental component, E, encompasses all the nongenetic factors acting on the phenotype.

The mean phenotypic value of individuals with a genotype at a given QTL is referred to as the genotype mean, so for a trait influenced by a biallelic QTL (with alleles A and B) there will be three genotypic means: μ_{AA}, μ_{AB}, and μ_{BB}. The overall population mean, μ, represents the weighted sum of these three genotypic means, the weights being the population frequencies of the respective genotypes. An analogous situation applies when the effects of multiple QTLs are considered. Continuous traits are often, although not always, distributed normally in humans.

The phenotype P_i of individual i can be modeled in an alternative fashion. If G_i and E_i, each with mean of zero, are now both defined as deviations from the overall population mean, μ, the phenotype of individual i can be expressed as $P_i = \mu + G_i + E_i$. This model is the basis for partitioning the total phenotypic variance into its genetic and environmental variance components.

It is important to note that the term *quantitative trait* is also widely used to refer to meristic traits—those in which the trait is ordinally distributed—and to discrete traits, in which it is assumed that an underlying continuously distributed liability exists with the disease phenotype expressed when individual liability exceeds some threshold.

Further Reading

Almasy L, Blangero J (1998). Multipoint quantitative-trait linkage analysis in general pedigrees. *Am. J. Hum. Genet.* 62: 1198–1211.

Almasy L, et al. (1999). Human pedigree-based quantitative-trait-locus mapping: localization of two genes influencing HDL-cholesterol metabolism. *Am. J. Hum. Genet.* 64: 1686–1693.

Falconer D, McKay TFC (1996). *Introduction to Quantitative Genetics*. Prentice-Hall, Harlow, United Kingdom.

Flint J, Mott R (2001). Finding the molecular basis of quantitative traits: Successes and pitfalls. *Nat. Rev. Genet.* 2: 437–445.

Hartl Dl, Clark AG (1997). *Principles of Population Genetics*. Sinauer Associates, Sunderland, MA, pp. 397–481.

See also Heritability, Oligogenic Inheritance, Polygenic Inheritance, Variance Components.

QUARTET PUZZLING. *SEE* QUARTET, PHYLOGENETIC.

QUARTETS, PHYLOGENETIC
Sudhir Kumar and Alan Filipski

A phylogenetic quartet is a set of four taxa or groups of taxa.

Quartets are important in phylogenetic analysis because four is the fewest number of taxa to have alternative unrooted topologies. If the configurations for all quartets of a set of taxa can be inferred, then a phylogenetic tree over the entire set of taxa may be created by a process known as quartet puzzling. In quartet puzzling, a phylogenetic tree is built by first starting with three taxa and then adding taxa in such a way as to maximize consistency with the set of quartets. In general, the result is dependent upon the order in which the taxa are added. To minimize the effect of this dependency, we can produce several different trees by quartet puzzling, taking taxa in a different order each time, then take a consensus of the results.

Quartet puzzling requires the rapid evaluation of the topology for many quartets. Maximum-likelihood methods are frequently used. Another way of doing this is known as the relaxed four-point method. In this method, taxa a and b are grouped together to the exclusion of the other two (c and d) if $d_{ab} + d_{cd}$ is less than both $d_{ac} + d_{bd}$ and $d_{ad} + d_{bc}$. If the quantities $d_{ab} + d_{cd}$, $d_{ac} + d_{bd}$, and $d_{ad} + d_{bc}$ are all equal, the topology is unresolved (star tree). Here d_{xy} represents an appropriate distance measure between taxa x and y.

Another way of evaluating the topology of a four-taxon tree is using the four-cluster test of Rzhetsky et al. (1995). Under the minimum-evolution criterion, this test directly compares the three possible alternative configurations for four taxa and estimates the length of the internal branch. Bootstrapping may be used to evaluate the confidence that this branch length is greater than zero. An advantage of this method is that it works not only for four taxa but also for any four clusters of taxa without requiring knowledge of the phylogenetic relationships within each cluster as long as cluster membership is known.

Further Reading

Rzhetsky A, et al. (1995). Four-cluster analysis: A simple method to test phylogenetic hypotheses. *Mol. Biol. Evol.* 12: 163–167.

Schmidt HA, et al. (2002). TREE-PUZZLE: Maximum likelihood phylogenetic analysis using quartets and parallel computing. *Bioinformatics* 18: 502–504.

Strimmer K, von Haeseler A (1996). Quartet puzzling: A quartet maximum-likelihood method for reconstructing tree topologies. *Mol. Biol. Evol.* 13: 964–969.

See also Consensus Tree, Maximum-Likelihood Phylogeny Reconstruction, Minimum-Evolution Principle.

QUATERNARY STRUCTURE
Roman A. Laskowski

Describes the arrangement of two or more interacting macromolecules: for example, how the chains of a multimeric protein pack together and the extent of the contact regions between these chains. Many proteins can only achieve their biological function in concert with one or more other proteins, the functional units ranging in complexity from simple dimers to large multiprotein and RNA assemblies such as the ribosome. The assemblies are held together by hydrogen bonds and van der Waals and coulombic forces.

Related Websites

Quaternary structure predictor	http://www.mericity.com/
Protein quaternary structure (PAS) (server)	http://pqs.ebi.ac.uk

See also Hydrogen Bond.

R

r8s
Ramachandran Plot
Rat Genome Database
Rational Drug Design
Read
Reading
readseq
Reasoning
Reasoner
Recombinant Protein Expression
Recombination
RECOMBINE
Refinement
Regex
Regression Analysis
Regular Expression
Regulatory Region
Regulatory Region Prediction
Regulatory Sequence
Regulome
Relationship
Repeatmasker
Repeats Alignment
Repetitive Sequences
Replication Fork
Replication Origin
Replication Terminus
Replicon
Residue
Resolution

Response
Retrosequence
Retrotransposon
Reverse Complement
R-Factor
Rfrequency
RGD
Ri
Ribosomal RNA
Ribosome Binding Site
RiboWeb
RMSD
RNA
RNA Folding
RNA Splicing
RNA Structure
RNA Structure Prediction
RNA Structure Prediction
RNA Tertiary-Structure Motifs
Robustness
Role
Rooting Phylogenetic Trees
Root-Mean-Square Deviation
Rosetta Stone Method
Rotamer
Rotamer Library
rRNA
Rsequence
Rule

Dictionary of Bioinformatics and Computational Biology. Edited by Hancock and Zvelebil
ISBN 0-471-43622-4 © 2004 John Wiley & Sons, Inc.

r8s

Michael P. Cummings

A program for estimating divergence times and rates of evolution based on a phylogenetic tree with branch lengths using maximum-likelihood and semi- and nonparametric methods.

The program accommodates local, global, and relaxed clock assumptions as well as one or more calibration points.

The program can be used interactively or without interaction by including appropriate commands in the r8s block (NEXUS file format). Written in American National Standards Institute (ANSI) C, the program is available as source code and executable for Linux (x86).

Related Website

r8s	http://ginger.ucdavis.edu/r8s/

Further Reading

Sanderson MJ (2002). Estimating absolute rates of molecular evolution and divergence times: A penalized likelihood approach. *Mol. Biol. Evol.* 19: 101–109.

Sanderson MJ (2003). r8s: Inferring absolute rates of molecular evolution and divergence times in the absence of molecular clock. *Bioinformatics* 19: 301–302.

Sanderson MJ, Doyle JA (2001). Sources of error and confidence intervals in estimating the age of angiosperms from rbcL and 18S rDNA data. *Am. J. Bot.* 88: 1499–1516.

RAMACHANDRAN PLOT

Roman A. Laskowski

A two-dimensional plot of the permitted combinations of phi and psi values in protein structures, where phi and psi are the main-chain dihedral angles. The favored regions were originally computed by Ramakrishnan and Ramachandran in 1965 by taking side chains to be composed of hard-sphere atoms and deriving the regions of phi–psi space that were accessible and those that were "disallowed" on the basis of steric hindrance. Nowadays, the favorable and disallowed regions of the plot are determined empirically using the known protein structures deposited in the PDB.

Different residues exhibit slightly different Ramachandran plots due to the differences in their side chains, the most different being glycine, which, due to its lack of a proper side chain, can access more of the phi–psi space than can other side chains, and proline, whose phi dihedral angle is largely restricted to the range 120° to −30°.

The plot is primarily used as a check on the "quality" of a protein structure. That is, to check whether a given model of a protein structure (whether derived by experiment or other means) appears to be a reasonable one or whether it might be unreliable. By plotting the phi–psi combination for every residue in the protein structure on the Ramachandran plot, one sees whether the values cluster in the favored regions, as expected, or whether any stray into the "disallowed" regions. While it is possible for individual residues to have such disallowed phi–psi values due to strain in the structure, any model of a protein having many such residues has probably been incorrectly determined and is not a reliable representation of the protein in question.

Dictionary of Bioinformatics and Computational Biology. Edited by Hancock and Zvelebil
ISBN 0-471-43622-4 © 2004 John Wiley & Sons, Inc.

Related Websites

Ramachandran plot generators	http://bmc.uu.se/eds/ramachan.html
IISc Ramachandran plot	http://144.16.71.146/rp

Further Reading

Kleywegt GJ, Jones TA (1996). Phi/psi-chology: Ramachandran revisited. *Structure* 4: 1395–1400.

Ramachandran GN, Ramakrishnan C, Sasisekharan V (1963). Stereochemistry of polypeptide chain configurations. *J. Mol. Biol.* 7: 95–99.

Ramachandran GN, Sasisekharan V (1968). Conformations of polypeptides and proteins. *Adv. Prot. Chem.* 23: 283.

Ramakrishnan C, Ramachandran GN (1965). Stereochemical criteria for polypeptide and protein chain conformations for a pair of peptide units. *Biophys. J.* 5: 909–933.

See also Dihedral Angle, Main Chain, Side Chain.

RAT GENOME DATABASE (RGD)
Guenter Stoesser

The Rat Genome Database (RGD) collects and integrates data generated from ongoing rat genetic and genomic research efforts.

RGD develops and provides informatics tools that allow researchers to further explore the data. Besides rat-centric tools in the context of genetic mapping and gene prediction, RGD also provides algorithms which are of use to researchers working with other organisms, particularly human and mouse. RGD is based at the Medical College of Wisconsin (MCW) and is a direct collaboration between the MCW, the Mouse Genome Database (MGD), and the National Center for Biotechnology Information (NCBI).

Related Website

RGD	http://rgd.mcw.edu

Further Reading

Twigger S, et al. (2002). Rat Genome Database (RGD): Mapping disease onto the genome *Nucleic Acids Res.* 30: 125–128.

RATIONAL DRUG DESIGN. *SEE* STRUCTURE-BASED DRUG DESIGN.

READ. *SEE* READING.

READING (GEL READING, READ)
Rodger Staden

A DNA sequence obtained from a single sequencing experiment, typically around 500—800 bp.

A DNA sequence obtained using a primer for the forward strand of the DNA is known as a forward read; a sequence obtained from the reverse strand of the DNA is known as a reverse read. When a pair of sequence readings is obtained from opposite ends of a DNA sequencing template, this is known as a read pair.

READSEQ
Michael P. Cummings

A program for converting sequence files from one format to another.

The program reads and writes a very broad range of file formats. The most recent version of the program is written in Java and is available as source code.

Related Website

| readseq | http://iubio.bio.indiana.edu/soft/molbio/readseq/version2/ |

REASONING
Robert Stevens

Reasoning is the process by which new facts are produced from those already existing. This process is known as inference. Computer-based reasoners can process knowledge encoded in a formal Knowledge Representation Language to produce new facts. In a description logic environment, a reasoner is used to infer the subsumption hierarchy that forms the taxonomy of the ontology. The reasoner will also check that the statements made within the encoding are logically consistent. A computer-based reasoner used for classification is sometimes known as a classifier.

REASONER. *SEE* CLASSIFIER.

RECOMBINANT PROTEIN EXPRESSION
Jean-Michel Claverie

The process of overproducing a protein of interest in an in vivo system (usually the bacteria *Escherichia coli*, the yeast *P. pastoris*, or the baculovirus in insect cells), or more recently in vitro, using cell-free extract.

"Recombinant" indicates the requirement that the coding DNA sequence (e.g., a human cDNA) must be inserted (cloned) into a suitable expression vector (e.g., a plasmid) containing the replication, transcription and translation signals required by the expression host (e.g., *E. coli*). The expression regulatory elements are chosen to be strongly inducible upon addition of a chemical compound, allowing the recombinant protein to become a few percent of the total amount of host proteins. Modern expression vectors are designed to produce proteins with additional N- or C-terminal short sequences (a tag) that are subsequently used for purification on the corresponding affinity column. The most frequently used tag is the His_6 peptide, causing the recombinant proteins to bind strongly onto nickel affinity columns. Recombinant protein expression and purification are two key steps, and well-identified bottlenecks, in functional and structural genomics. Researchers are starting to address the challenge of parallel expression and purification of large numbers of gene products through the principles of high-throughput screening technologies commonly used in pharmaceutical development. The production of considerable quantities of soluble and stable material is the main challenge in the large-scale study of proteins. It must be noted that the current technologies offer no standard protocol for the production of nonglobular, nonsoluble (e.g., membrane) proteins.

Further Reading
Lesley SA. (2001). High-throughput proteomics: Protein expression and purification in the postgenomic world. *Protein Exp. Purif.* 22: 159–164.

Spirin AS (2001). *Cell-Free Translation Systems.* Springer Verlag, New York.

RECOMBINATION
Mark McCarthy and Steven Wiltshire

Recombination describes the rearrangement—or recombining—of alleles between haplotypes on homologous chromosomes during meiosis.

Assume that individual i inherits haplotype A_pB_p from its father and A_mB_m from its mother. If meiosis in individual i produces gametes (and subsequently offspring) with haplotypes A_pB_m or A_mB_p, a recombination event is said to have occurred between loci A and B, and the resulting gametes (and offspring) are recombinants for the two loci. Such offspring, as a result, carry haplotypes of loci A and B that were not inherited by individual i from its parents, i.e., there has been a recombining of alleles in the parental haplotypes inherited by individual i. However, if the gametes produced by individual i are the same as the parental types that it inherited, i.e., A_pB_p and A_mB_m, no recombination has been detected between loci A and B: such gametes (and offspring) contain haplotypes of loci A and B that were inherited by individual i from its parents. (The fact that no recombination has been detected does not necessarily mean that no recombination has occurred, since an even number of recombination events between the loci will also result in nonrecombinant haplotypes.) Recombination occurs when the chromatids of a pair of homologous chromosomes physically cross over at chiasmata during meiosis and exchange DNA. The probability of a recombination between any two loci is termed the *recombination fraction* (θ) and lies between 0 and 0.5. Two loci for which the recombination fraction is <0.5 are said to be in genetic linkage or linked. Recombination fractions may be converted into genetic (or map) distances using map functions. Detecting linkage and estimating the recombination fractions between loci form the basis of linkage analysis.

Example: In a genomewide parametric linkage analysis of a large Australian aboriginal pedigree for loci influencing susceptibility to type 2 diabetes, Busfield et al. (2002) detected linkage to marker *D2S2345* with a maximum two-point LOD score of 2.97 at a recombination fraction of 0.01.

Related Websites
Programs for estimating recombination fractions between loci, for identifying recombinations in individual pedigrees. and/or for performing linkage analysis:

LINKAGE	http://www.nslij-genetics.org/soft
GENEHUNTER2	http.broad.wi.mit.edu/ftp/distribution/software/genehunter/
SIMWALK2	http://hpcio.cit.nih.gov/lserver/SIMWALK2.html
MERLIN	http://www.sph.umich.edu/csg/abecasis/Merlin/index.html

Further Reading
Busfield F, et al. (2002). A genomewide search for type 2 diabetes—susceptibility genes in indigenous Australians. *Am. J. Hum. Genet.* 70: 349–357.

Ott J (1999). *Analysis of Human Genetic Linkage*. Johns Hopkins University Press, Baltimore, MD, pp. 1–23.

Sham P (1998). *Statistics in Human Genetics*. Arnold, London, pp. 51–58.

See also Linkage Analysis, Map Function, LOD Score, Phase.

RECOMBINE. *SEE* LAMARC.

REFINEMENT
Liz Carpenter

When an initial structure is obtained from an experimental electron density map, there will be many inaccuracies in the structure, due to errors in obtaining the intensities and the phases and in positioning the atoms in the electron density map. Refinement techniques are used to improve the crystal structure so that the model agrees with the data more closely while also having reasonable bond lengths and angles.

A number of mathematical techniques are used to improve the agreement between data and the model, including least-squares minimization, simulated annealing, and maximum-likelihood methods.

In general, to define a particular parameter such as the position x, y, z or B factor of an atom, it is necessary to have at least three experimental measurements for each parameter. The resolution of the data will define how many measurements are obtained for each data set. In macromolecular crystallography there are often an insufficient number of measurements to properly define all the parameters. This problem can be overcome by including prior knowledge of the structure, such as information on bond lengths and angles that are known from the structures of small molecules. This information can be included in the form of constraints (where a parameter is set to a certain value) or as restraints (where a parameter is set to vary within a range).

Refinement adjusts the positions of atoms within the unit cell. It can make changes over a small distance, fractions of an angstrom. Refinement protocols will not make large movements, e.g., repositioning an entire loop of a protein that is not in density. This has to be done manually on the graphics or with a structure-building program. Several rounds of refinement and rebuilding of a model are necessary to obtain a final structure.

Related Websites

CNS	http://cns.csb.yale.edu/v1.1/
Refmac	http://www.ccp4.ac.uk/dist/html/refmac5.html
Randy Read's teaching site: refinement and maximum-likelihood methods	http://perch.cimr.cam.ac.uk/Course/ Likelihood/likelihood.html

Further Reading
Murshudov GN, Vagin AA, Dodson EJ (1997). Refinement of macromolecular structures by the maximum-likelihood method. *Acta Cryst.* D53: 240–255.

See also X-Ray Crystallography for Structure Determination.

REGEX. *SEE* REGULAR EXPRESSION.

REGRESSION ANALYSIS
Nello Cristianini

The statistical methodology for predicting values of one or more response (dependent) variables from a collection of predictor (independent) variable values.

In machine learning literature the responses are often called "labels," and the predictors are called "features." This method can also be used to assess the effect of the predictor variables on the responses. When there is more than one predictor variable (feature), it is termed *multivariate regression.* When there is more than one response variable (label), it is called *multiple regression.*

Further Reading
Kleinbaum DG, Kupper LL, Muller KE (1997). *Applied Regression Analysis and Multivariable Methods,* Wadsworth, Belmont, CA.

REGULAR EXPRESSION (REGEX)
Terri Attwood

A single-consensus expression derived from a conserved region (motif) of a sequence alignment and used as a characteristic signature of family membership. Regular expressions (regexes) thus discard sequence data, retaining only the most conserved residue information. Typically, they are designed to be family specific and encode motifs of 15–20 residues. An example regex is

[LIVMW]-[PGC]-x(3)-[SAC]-K-[STALIM]-[GSACNV]-[STACP]-x(2)-[DENF]-[AP]-x(2)-[IY]

Within this expression, the single-letter amino acid code is used: Residues within square brackets are allowed at the specified position in the motif; the symbol x denotes that any residue is allowed at that position; individual residues denote completely conserved positions; and numbers in parentheses indicate the number of residues of the same type allowed sequentially at that location in the motif. Regexs can also be used to disqualify particular residues from occurring at any position; conventionally, these are grouped in curly brackets (e.g., {PG} indicates that proline and glycine are disallowed).

Regexs are tremendously simple to derive and use in database searching. However, they suffer from various diagnostic drawbacks. In particular, in order to be recognized, expressions of this type require sequences to match them exactly. This means that any minor change, no matter whether it is conservative (e.g., V in position 8 of the above motif), will result in a mismatch even if the rest of the sequence matches the pattern exactly. Thus many closely related sequences may fail to match a regex if its allowed residue groups are too strict. Conversely, if the groups are relaxed to include additional matches, the expression may make too many chance matches. Thus a match to a pattern is not necessarily true, and a mismatch is not necessarily false!

Regexs underpin the PROSITE and eMOTIF databases. In the former, expressions are derived exactly from observed motifs in sequence alignments; in the latter, to counteract the limitations of binary (match/no-match) diagnoses, expressions are derived in a more tolerant manner, using a set of prescribed physicochemical groupings (DEQN, HRK, FYW, etc.). Both approaches have drawbacks, and hence caution (and biological intuition) should be used when interpreting matches.

Related Websites

ScanProsite	http://www.expasy.org/tools/scanprosite/
eMOTIF scan	http://motif.stanford.edu/emotif/emotif-scan.html

Further Reading

Attwood TK (2000). The quest to deduce protein function from sequence: The role of pattern databases. *Int. J. Biochem. Cell Biol.* 32(2): 139–155.

Attwood TK, Parry-Smith DJ (1999). *Introduction to Bioinformatics.* Addison-Wesley Longman, Harlow, Essex, United Kingdom.

See also Amino Acid, Conservation, Motif.

REGULATORY REGION. *SEE* TRANSCRIPTIONAL REGULATORY REGION.

REGULATORY REGION PREDICTION
James W. Fickett

The prediction of sequence regions responsible for the regulation of gene prediction.

In eukaryotes, transcriptional regulatory regions (TRRs) are either promoters, at the transcription start site, or enhancers, elsewhere. Enhancers, as well as the regulatory portion of promoters, are typically made up of functionally indivisible cis-regulatory modules (CRM) in which a few transcription factors have multiple binding sites (Yuh and Davidson, 1996). The common computational problem is to recognize functional clusters of transcription factor binding sites. Clusters of sites for a single factor may be part of the key to localizing functional sites for that factor (Wagner, 1999), while coordinate action of multiple factors may hold the key to highly specific expression patterns (Berman et al., 2002). A number of computational methods have been used to characterize the degree of clustering. It is possible that the quality of the position weight matrices used to describe binding sites and the use of phylogenetic footprinting to eliminate false-positive matches are more important, in practical terms, than the particular mathematical treatment of clusters.

There is almost certainly some structure in TRRs beyond mere clustering of the binding sites. It is well known that certain combinations of transcription factors act synergistically (one collection of such cases may be found in the database COMPEL (Kel-Margoulis et al., 2002). It has been shown that physical interaction between pairs of transcription factors is, in at least some cases, reflected in constraints on spacing of their binding sites (Fickett, 1996). Some success has been had, in simple cases, in describing the overall structure of a class of CRMs (Klingenhoff et al., 1999). However, in most cases a formal characterization of any spacing and order constraints remains elusive.

In eukaryotes, algorithms giving practical prediction of regulatory specificity are available for only a few cases. Those algorithms for, e.g., muscle (Wasserman and Fickett, 1998), liver (Krivan and Wasserman, 2001), and anterior–posterior patterning in early *Drosophila* development (Berman et al., 2002) were only developed following years of extensive work, in a broad community, determining the biological function and specificity of the transcription factors involved. It will require further work to deduce the set of relevant transcription factors (or their binding specificities) and develop a prediction algorithm directly from a set of coexpressed genes.

In bacteria, regulatory regions are much simpler, often consisting of the binding site of a single regulator. Because intergenic distances are typically small, the localization problem is one of determining operon groupings. Many bacterial genomes are now available, and comparison of gene and putative transcription element order across many species can often provide insight into both operon structure and the likely regulatory mechanism (Gelfand et al., 2000).

Related Websites

COMPEL	http://compel.bionet.nsc
AG BIODV	http://www.gsf.de/biodv
Wyeth Wasserman	http://www.cmmt.ubc.ca/wasserman/
Cister	http://sullivan.bu.edu/~mfrith/cister.shtml

Further Reading
Berman BP, et al. (2002). Exploiting transcription factor binding site clustering to identify cis-regulatory modules involved in pattern formation in the *Drosophila* genome. *Proc. Natl. Acad. Sci. USA* 99: 757–762.

Fickett JW (1996). *Gene* 172: GC19–GC32.

Gelfand MS, et al. (2000). Prediction of transcription regulatory sites in *Archaea* by a comparative genomic approach. *Nucleic Acids Res.* 28: 695–705.

Kel-Margoulis O, et al. (2002). TRANSCompel: A database on composite regulatory elements in eukaryotic genes. *Nucleic Acids Res.* 30: 332–334.

Klingenhoff A, et al. (1999). Coordinate positioning of MEF2 and myogenin binding sites. *Bioinformatics* 15: 180–186.

Krivan W, Wasserman WW (2001). A predictive model for regulatory sequences directing liver-specific transcription. *Genome Res.* 11: 1559–1566.

Wagner A (1999). Genes regulated cooperatively by one or more transcription factors and their identification in whole eukaryotic genomes. *Bioinformatics* 15: 776–784.

Wasserman WW, Fickett, JW (1998). Identification of regulatory regions which confer muscle-specific gene expression. *J. Mol. Biol.* 278: 167–181.

Yuh C-H, Davidson EH (1996). Modular cis-regulatory organization of Endo16, a gut-specific gene of the sea urchin embryo. *Development* 122: 1069–1082.

See also Cis-Regulatory Module, Enhancer, Phylogenetic Footprinting, Promoter, Transcription Factor Binding Site, Transcription Start Site.

REGULATORY SEQUENCE. *SEE* TRANSCRIPTIONAL REGULATORY REGION.

REGULOME
John M. Hancock

Variously defined as the genomewide regulatory network and the combined transcription factor activities of the cell.

For future studies of systems biology it will be essential to understand the interactions that take place within a given cell type. Characterization of the regulome (effectively the set of interactions taking place between individual gene pairs in a cell type) will provide a necessary catalog of interactions to be built into any such model.

See also Interactome, Network, System Biology.

RELATIONSHIP
Robert Stevens

A relationship is a type of association between two concepts. A relation is the nature of the association and a relationship is the relation plus the concepts it associates together. For example, membership, causality, and part/whole are all types of relations that form relationships within ontologies.

Common relations are:

- Is-kind-of relates subcategories to more general categories. This is one of the most important relationships used to organize categories into a taxonomy.
- Is-instance-of relates individuals to categories.
- Part-of describes the relationship between a part and its whole.

Relations themselves can have properties. A relation can have an inverse; so, "part-of" has an inverse, "has-part." Instead of an inverse, a relation may be symmetric; so "homologous-to" has the same relation in the other direction. This enables the ontology to have the relationships "Protein homologous-to Protein" with the same relation in each direction. Relations may have different kinds of quantification: Existential quantification indicates that there is *some* concept filling the relationship. Similarly, there can be precise cardinality constraints indicating the numbers or range of concepts filling the relation. Universal quantification indicates what type of concept can possibly form the relationship. Similarly, a relation can have domain and range constraints, which determine what is logically allowed to be used on either end of the relation. Relations may also be said to be transitive; this means that the effect of the relation is transferred along a chain of relations of the same type. An "Amino acid" is "part-of" an "Alpha helix," and that class is also "part-of" the class "Protein." If the "part-of" relation is made transitive, the ontology is also stating that the "Amino acid" is "part-of" the "Protein."

REPEATMASKER.

Program for detecting repeated and repetitive sequences in DNA sequences.

Repeatmasker compares input sequences against a library including both interspersed and internally repetitive sequences. It returns a detailed annotation of the repeats found and a version of the input sequence in which repeats are replaced by Ns (a so-called masked sequence). The program uses a modified Smith-Waterman algorithm to align library sequences against the input sequence. Sequence masking by repeatmasker is commonly used before comparing DNA sequences against databases to reduce the chances of spurious hits.

Related Website

Repeatmasker Web Server	http://ftp.genome.washington.edu/cgi-bin/RepeatMasker

See also Smith-Waterman Algorithm.

REPEATS ALIGNMENT. *SEE* PAIRWISE ALIGNMENT.

REPETITIVE SEQUENCES. *SEE* SIMPLE DNA SEQUENCE.

REPLICATION FORK

John M. Hancock

The structure at which DNA replication takes place.

Further Reading

Lewin B (2000). *Genes VII*. Oxford University Press, Oxford, UK.

See also DNA Replication.

REPLICATION ORIGIN (ORIGIN OF REPLICATION)
John M. Hancock

The location or locations at which DNA replication initiates within a genome.

Bacteria typically have a single replication origin while eukaryotic genomes have numerous origins. Bacterial and yeast (*Saccharomyces cerevisiae*) genomes have origins of replication that are well defined at the sequence level, but in many other eukaryotes the sequence determinant of the replication origin is not well defined and its selection is still not well understood.

Further Reading
Lewin B (2000). *Genes VII*. Oxford University Press, Oxford, UK.

See also DNA Replication.

REPLICATION TERMINUS
John M. Hancock

The point or region at which DNA replication terminates.

Bacteria typically have a single replication terminus while eukaryotic genomes have numerous termini. Bacterial genomes have replication termini that are well defined at the sequence level, but in many eukaryotes the sequence determinant of the replication terminus is not well defined and its selection is still not well understood.

Further Reading
Lewin B (2000). *Genes VII*. Oxford University Press, Oxford, UK.

See also DNA Replication.

REPLICON
John M. Hancock

Segment of a genome replicated from a single replication origin.

A replicon spans the region between the replication origin and the terminus. In bacteria this typically corresponds to the whole genome, but in eukaryotes there are numerous replicons.

Further Reading
Lewin B (2000). *Genes VII*. Oxford University Press, Oxford, UK.

See also DNA Replication.

RESIDUE. *SEE* AMINO ACID.

RESOLUTION, X-RAY CRYSTALLOGRAPHY
Liz P. Carpenter

An important parameter in determining the quality of an X-ray crystallographic structure is the resolution of the data. The resolution of a diffraction spot observed in X-ray crystallography is dependent on the spacing between the planes that gave rise to the diffraction spot. Bragg's law states that for diffraction to occur the following equation must be correct: $2d \sin \theta = n\lambda$.

The planes in the crystal are separated by a distance d. For a fixed wavelength, the smaller the value of d, the larger is the angle of diffraction θ. With larger values of θ the intensity of the diffraction decreases, until eventually it becomes indistinguishable from the background. This fall-off in intensity with Bragg angle is due to thermal motion of the atoms in the crystal. On an X-ray diffraction image the spots in the middle of the image with small θ values are strong and those near the edge of the detector, with high θ values are weak or undistinguishable. The maximum resolution of a data set is the smallest value of d for which diffraction spots can be detected in the background.

By analogy with the light microscopy this relates to the smallest distance between two objects that can be separated. So if there are diffraction spots to 2 Å, then atoms 2 Å apart are seen as separate objects, whereas if the data only extends to 3 Å, then only atoms 3 Å apart are separate. This means that in data sets where there are reflections to a resolution of 4 Å only the overall shape of a molecule and a few alpha helices can be distinguished. A 3-Å data set would allow the backbone of a protein to be traced, the helices and sheets to be identified, and some side chains to be positioned, but the detail of interactions would not be visible. At 2 Å the side chains and water structure would be clearly visible. At 1 Å resolution each individual atom is in its own separate density and atoms can be assigned six thermal parameters instead of the usual one B-factor parameter, which indicates the thermal motion of the atoms.

Further Reading

Blundell, T. L. & Johnson, L. N., *Protein Crystallography*, Academic Press (1976).

Drenth, J., *Principles of Protein X-ray Crystallography*, Springer Verlag (1994).

Glusker, J. P., Lewis, M. & Rossi, M., *Crystal Structure Analysis for Chemists and Biologists*, VCH Publishers (1994).

See also Diffraction of X Rays, X-Ray Crystallography for Structure Determination.

RESPONSE. *SEE* LABEL.

RETROSEQUENCE
Austin L. Hughes

A retrosequence is any genomic sequence that originates from the reverse transcription of any RNA molecule and its subsequent integration into the genome, excluding sequences that encode reverse transcriptase.

Usually, the RNA template is the RNA transcript of a gene. Most of the retrosequences derived from protein coding genes are derived from processed mRNAs which thus lack introns. A retrosequence that is functional may be called a retrogene, while one that is nonfunctional is called a retropseudogene. Retropseudogenes derived from processed mRNAs are often called processed pseudogenes.

Further Reading
Weiner AM, et al. (1986). Nonviral retroposons: Genes, pseudogenes, and transposable elements generated by the reverse flow of genetic information. *Annu. Rev. Biochem.* 55: 631–661.

See also Retrotransposon.

RETROTRANSPOSON
Dov S. Greenbaum

Transposable elements that spread by reverse transcription of mRNAs.

Retrotransposons resemble retroviruses in many ways. They contain long terminal repeats that flank two or three reading frames: gag-like, env-like and pol-like. Unlike retroviruses, they are not infectious. They are very abundant within genomes: Over 40% of the human genome is composed of retrotransposons. There are many different families of retrotransposons. Retrotransposons move by first being transcribed into mRNA. These pieces of RNA are transcribed back into DNA and are inserted into the genome. Retrotransposons require the help of the enzymes integrase (for insertion) and reverse transcriptase to convert the mRNA transcript into a DNA sequence that can be inserted back into the genome.

There are two classes of retrotransposons: viral and nonviral. Viral transposons include retrovirus-like transposons such as Ty in *Saccharomyces*, Bs1 in maize, copia in *Drosophila*, and LINE-like elements. Nonviral retrotransposons include SINEs and processed pseudogenes.

Further Reading
Brosius J (1991). Retroposons—Seeds of evolution. *Science* 251: 753. Patience C, et al. (1997). Our retroviral heritage. *Trends Genet.* 13: 116–120.

REVERSE COMPLEMENT
John M. Hancock

As the DNA double helix is made up of two base-paired strands, the sequences of the two strands are not identical but have a strict relationship to one another. For example, a sequence 5′-ACCGTTGACCTC-3′ on one strand pairs with the sequence 5′-GAGGTCAACGGT-3′. This second sequence is known as the reverse complement of the first. Converting a sequence to its complement is a simple operation and may be required, e.g., if a sequence has been read off one strand but a coding region lies on the other strand.

See also Complement, DNA Sequence.

R-FACTOR
Liz P. Carpenter

The *R*-factor is the principal measure of how well the refinement of a crystallographic structure is proceeding. It is a measure of the difference between the observed structure factors $|F_{obs}|$ (proportional to the square root of the measured intensity) and the structure factors calculated from the model of the macromolecular structure, $|F_{calc}|$:

$$R = \frac{\sum\limits_{hkl} ||F_{obs}| - |F_{calc}||}{\sum\limits_{hkl} |F_{obs}|}$$

Values of R-factors are generally expressed as percentages. For a macromolecule a completely random distribution of atoms in the unit cell would give an R-factor of 59%. Values below 24% suggest that the model is generally correct. Near atomic resolution structures (1.3 Å and above) can have R-factors below 15%. The R-factor is sensitive to overrefinement, particularly at lower resolution, worse than say 2.8 Å. Since the refinement programs used to perfect the structure are designed to minimize the R-factor, it tends to decrease even when no real improvement in the structure is occurring. The free R-factor was introduced to overcome this problem.

Further Reading
Glusker, JP, Lewis M, Rossi M (1994). *Crystal Structure Analysis for Chemists and Biologists.* VCH Publishers, Berlin, Germany.

See also Diffraction of X Rays, Free R-Factor, X-Ray Crystallography for Structure Determination.

RFREQUENCY
Thomas D. Schneider

Rfrequency is the amount of information needed to find a set of binding sites out of all the possible sites in the genome. If the genome has G possible binding sites and γ binding sites, then Rfrequency $= \log_2(G/\gamma)$ bits per site. Rfrequency predicts the expected information in a binding site, Rsequence.

Further Reading
Schneider TD (2000). Evolution of biological information. *Nucleic Acids Res.* 28: 2794–2799.
Schneider et al. (1986). Information content of binding sites on nucleotide sequences. *J. Mol. Biol.* 188: 415–431. http://www.lecb.ncifcrf.gov/toms/paper/schneider1986.

See also Information, Rsequence.

RGD. *SEE* RAT GENOME DATABASE.

RI
Thomas D. Schneider

Ri is a shorthand notation for "individual information". Following Claude Shannon, the R stands for "rate of information transmission." For molecular biologists this is usually bits per base or bits per amino acid. The i stands for "individual."

Further Reading
Schneider TD (1997). Information content of individual genetic sequences. *J. Theor. Biol.* 189: 427–441.

See also Binding Site, Bit, Individual Information, Information.

RIBOSOMAL RNA (rRNA)

Jamie J. Cannone and Robin R. Gutell

Ribosomal RNAs (rRNAs) are specialized RNAs that provide the structural and catalytic core of the ribosome, the cellular structure that is the site for protein synthesis.

The three major forms of rRNA are the 5S, small-subunit (SSU, 16S, or 16S-like), and large-subunit (LSU, 23S, or 23S-like) rRNAs. rRNA may comprise up to 90% of a cell's RNA.

Ribosomal RNAs are organized into operons in the genome. In nuclear rRNA operons, the genes are separated by internal spacers, which are often used for phylogenetic analyses. In most (higher) eukaryotes, the large-subunit rRNA is divided into two pieces, the 5.8S rRNA and a larger rRNA (varying in size between 25S and 28S), with a spacer between them in the genomic sequence. Certain organisms extend this theme by dividing their rRNA molecules into fragments that must be assembled correctly to produce a viable rRNA. The sizes of the rRNAs vary over a wide range; the *Escherichia coli* rRNAs, which were the first to be sequenced and serve as a reference organism for comparative analysis of rRNA, are 120, 1542, and 2904 nucleotides for the 5S, SSU, and LSU rRNAs, respectively.

Size variation in rRNA: Approximate ranges of size (in nucleotides) for complete sequences are shown in the table. Unusual sequences (of vastly different) lengths are excluded.

Phylogenetic Domain/Cell Location	rRNA Molecule		
	5 S	SSU	LSU
Bacteria	105–128	1470–1600	2750–3200
Archaea	120–135	1320–1530	2900–3100
Eukaryota nuclear	115–125	1130–3725	2475–5450
Eukaryota chloroplast	115–125	1425–1630	2675–3200
Eukaryota mitochondria	115–125	685–2025	940–4500
Overall	105–135	685–3725	940–5450

Structure models for the rRNAs have been proposed using comparative sequence analysis methods. The majority of the base pairs predicted by these methods are canonical (G:C, A:U, and G:U) base pairs that are consecutive and antiparallel with each other, forming nested secondary-structure helices. Many tertiary-structure interactions were also proposed from comparative analysis. These interactions include base triples, noncanonical base pairs, pseudoknots, and many RNA motifs. The most recent versions of the models for the *Escherichia coli* 16S and 23S rRNAs are shown in the accompanying figure.

More than 20 years after the first 16S and 23S rRNA comparative structure models were proposed, the most recent structure models (see the figure) were evaluated against the high-resolution crystal structures of both the small (Wimberly et al., 2000) and large (Ban et al., 2000) ribosomal subunits that were solved in 2000. The results were affirmative; approximately 97% to 98% of the 16S and 23S rRNA base pairs, including nearly all of the tertiary-structure base pairs, predicted with covariation analysis were present in these crystal structures. In addition, many new motifs have been proposed and characterized based upon the crystal structures.

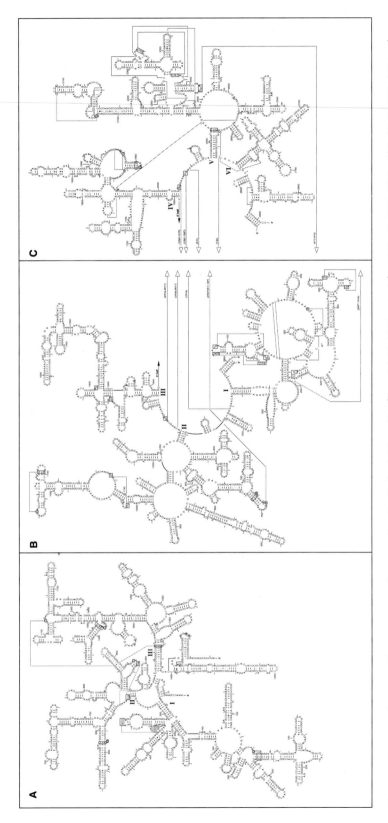

Figure Comparative secondary structure diagrams for the Escherichia coli 16S and 23S rRNAs. A, 16S rRNA; B, 23S rRNA, 5′ half; C, 23S rRNA, 3′ half. Base-pair symbols: line, canonical (G:C or A:U); small filled circle, wobble (G:U); large open circle, large closed circle, all other noncanonical base-pairs. See the Comparative RNA Website for more information.

Related Websites

Comparative RNA Web (CRW) Site (all rRNAs)	http://www.rna.icmb.utexas.edu/
European rRNA Database (SSU and LSU rRNAs)	http://oberon.rug.ac.be:8080/rRNA/
5S rRNA Database	http://biobases.ibch.poznan.pl/5SData/
Ribosomal Database Project II (RDP-II)	http://rdp.cme.msu.edu/html/
Ribosomal Internal Spacer Sequence Collection (RISSC)	http://ulises.umh.es/RISSC/
RNABase ribosomal RNA entries	http://www.rnabase.org/listing/?cat = rrna

Further Reading

Ban N, et al. (2000). The complete atomic structure of the large ribosomal subunit at 2.4 Å resolution. *Science* 289: 905–920.

Gutell RR, et al. (2002). The accuracy of ribosomal RNA comparative structure models. *Curr. Opin. Struct. Biol.* 12: 301–310.

Harms J, et al. (2001). High resolution structure of the large ribosomal subunit from a mesophilic eubacterium. *Cell* 107: 679–688.

Schluenzen F, et al. (2000). Structure of functionally activated small ribosomal subunit at 3.3 Å resolution. *Cell* 102: 615–623.

Wimberly BT, et al. (2000). Structure of the 30S ribosomal subunit. *Nature* 407: 327–339.

Yusupov MM, et al. (2001). Crystal structure of the ribosome at 5.5 Å resolution. *Science* 292: 883–896.

See also RNA Tertiary-Structure Motifs.

RIBOSOME BINDING SITE (RBS)

John M. Hancock

The site on the mRNA that directs its binding to the ribosome.

The identification of an RBS is an important indicator of the position of the true translation start site. The sequence nature of these sites differs between prokaryotes and eukaryotes. The prokaryotic RBS is known as the Shine–Dalgarno sequence; it has the consensus AGGAGG and is located 4–7 bp 5′ of the translation start site. In eukaryotes the equivalent is the Kozak sequence.

Further Reading

Kozak M (1987). An analysis of 5′-noncoding sequences from 699 vertebrate messenger RNAs. *Nucleic Acids Res.* 15: 8125–8148.

Kozak M (1997). Recognition of AUG and alternative initiator codons is augmented by G in position +4 but is not generally affected by the nucleotides in positions +5 and +6. *EMBO J.* 16: 2482–2492.

Shine J, Dalgarno L (1974). The 3′-terminal sequence of *E. coli* 16s ribosomal RNA: Complementarity to nonsense triplets and ribosome binding sites. *PNAS* 71: 1342–1346.

RIBOWEB
Robert Stevens

RiboWeb is an online resource designed to guide researchers in the use of biochemical and biophysical experimental data toward the construction and/or evaluation of molecular models of the ribosome. It consists of a domain-independent part (OWeb), a domain-specific knowledge base of ribosomal structural data, and a set of computational modules written for the data of the knowledge base. This knowledge base is built upon a frame-based ontology of structural biology to which has been added over 20,000 instances, including nearly 11,000 pieces of represented experimental data that yield clues to the structure of the ribosome. The user is guided in the selection of specific data sets and in the conversion of these qualitative data into quantitative distances through series of the computational modules. These distances are subsequently used as constraints in various molecular modeling activities.

Related Website

| RiboWeb | http://riboweb.stanford.edu/riboweb/ |

RMSD. *SEE* ROOT-MEAN-SQUARE DEVIATION.

RNA (GENERAL CATEGORIES)
Jamie J. Cannone and Robin R. Gutell

Ribonucleic acid (RNA) is one of the two major forms of nucleic acids. Each individual element, or nucleotide, of RNA is comprised of three parts: a ribose sugar, a phosphate, and a cyclic base. Four primary bases are found in RNA: adenine, guanine, cytosine, and uracil. Other bases and modified forms of these bases sometimes appear; the McCloskey Lab at the University of Utah has prepared a database of these modified nucleotides.

The major structural difference between RNA and DNA, the other major form of nucleic acid, is the sugar (deoxyribose in DNA, ribose in RNA). The ribose sugar increases the susceptibility to degradation of RNA compared to DNA. Thus, RNA is best suited to (relatively) short-term uses, while DNA is sufficiently stable to be a good medium for genetic inheritance.

While DNA's primary functions are the storage of genetic information and the production of RNA, cellular RNA has several distinct functions. Three major forms of RNA are commonly discussed: messenger RNA, ribosomal RNA, and transfer RNA. (Ribosomal and transfer RNA are discussed separately.) Messenger RNA (mRNA) codes for proteins. mRNA nucleotide sequence is translated to amino acid sequence based on a specific mapping (the "genetic code" for an organism) between sets of three mRNA nucleotides (codons) and amino acids. mRNA is produced from DNA during transcription. mRNAs sometimes contain untranslated regions (UTRs) at their 5′ and 3′ ends that play several key roles in gene regulation and expression. mRNA is both quickly synthesized and degraded as part of the regulation of protein synthesis.

In addition to the three major RNAs, other RNAs that have different properties have been identified. Some RNAs form ribonucleoprotein complexes; others have catalytic activities and, in viruses, carry the genetic information for the organism rather than DNA. Included among these other RNAs are intron RNAs, which must be excised from other genes so that those genes can function, RNase P, and the U RNAs. Many of these interesting RNAs have been characterized, and several examples appear below.

Noncoding RNAs (also referred to as small RNAs) are RNAs that do not code for proteins. (Technically, both ribosomal and transfer RNAs belong to this category.) Certain noncoding RNAs, such as the microRNAs (miRNAs) that have been isolated from plants and animals, have been characterized and implicated in regulatory roles. Other noncoding RNAs have been implicated in the destruction of mRNA via RNA interference (RNAi). Some additional noncoding RNAs that have been characterized are described below.

The bacterial tmRNA (also known as 10Sa RNA or SsrA) is a chimeric molecule with both tRNA-like and mRNA-like characteristics. An incomplete mRNA with a truncated 3' end will cause the ribosome to "stall" with an incomplete protein attached. tmRNA "rescues" the ribosome by binding its tRNA-like portion to the ribosome. This binding positions the mRNA-like portion of tmRNA so that the ribosome can resume translation, using the tmRNA as its template. The mRNA-like portion codes for a signal peptide that will be recognized by bacterial proteases and ensures that the partially produced protein will be degraded.

Small nucleolar RNAs (snoRNAs) are typically 60–300 nucleotide RNAs that are abundant in the nucleolus of a broad variety of eukaryotes. The snoRNAs associate with proteins to form small nucleolar ribonucleoproteins (snoRNPs). snoRNAs come in two major structural forms, one containing the box C and D motifs and the second containing box H and ACA elements. Most snoRNAs are involved in the nucleotide modification process, in either 2'-O-methylation (box C/D snoRNAs) or pseudouridylation (box H/ACA snoRNAs), for a wide range of RNAs by hybridizing to the region of the RNA that needs to be modified. Other snoRNAs are essential for nucleolytic cleavage of precursor RNAs. Two different mechanisms for the synthesis of snoRNAs have been observed; in vertebrates, snoRNAs are processed from previously excised pre-mRNA introns, while in yeast and plants, the sources of snoRNAs are polycistronic snoRNA transcripts. Vertebrate telomerase is a box H/ACA snoRNP.

The signal recognition particle (SRP) RNA is involved in transport of newly translated secretory proteins to the cytosol. The SRP binds to a signal at the N-terminus of a protein as the protein is synthesized. The SRP then binds to a receptor that is bound to the membrane of the endoplasmic reticulum. The protein begins to traverse the membrane *en route* to the cytosol, and protein synthesis continues, with the SRP recycled to assist with another protein–ribosome complex.

Guide RNAs (gRNAs) are a novel class of small noncoding RNA molecules that are transcribed from the maxicircles and minicircles of trypanosome mitochondria. gRNAs contain the necessary information for proper editing (insertion or deletion of uridines) of mitochondrial precursor RNAs in the trypanosomes, resulting in functional RNAs. The 5' end of a gRNA is complementary to its mRNA target sequence that is 3' of the modification site, serving as an "anchor" for the gRNA. The central portion of a gRNA is complementary to the mature, edited mRNA and thus serves as the editing template. The 3' end of a gRNA is a posttranscriptionally added oligo[U] tail; the function of this tail is not presently certain.

Related Websites

| Modified RNA nucleotides | http://medlib.med.utah.edu/RNAmods/ |

Noncoding RNAs	http://biobases.ibch.poznan.pl/ncRNA/
Noncoding RNAs in plants	http://www.prl.msu.edu/PLANTncRNAs/
The RNA World	http://www.imb-jena.de/RNA.html
RNABase	http://www.rnabase.org/
RNAi Database	http://formaggio.cshl.org/%7Emarco/fabio/index.html
RNase P Database	http://jwbrown.mbio.ncsu.edu/RNaseP/home.html
snoRNA Database	http://rna.wustl.edu/snoRNAdb/
SRP Database (Christian Zwieb)	http://psyche.uthct.edu/dbs/SRPDB/SRPDB.html
tmRNA (Kelly Williams)	http://www.indiana.edu/~tmrna/
tmRNA (Christian Zwieb)	http://psyche.uthct.edu/dbs/tmRDB/tmRDB.html
Uridine insertion/deletion RNA editing (gRNA)	http://www.rna.ucla.edu/trypanosome/
UTR Database	http://bighost.area.ba.cnr.it/BIG/UTRHome/
Yeast snoRNA Database	http://www.bio.umass.edu/biochem/rna-sequence/Yeast_snoRNA_Database/snoRNA_DataBase.html

Further Reading

Bachellerie JP, Cavaillé J (1998). Small nucleolar RNAs guide the ribose methylations of eukaryotic rRNAs. In *Modification and Editing of RNA*. H Grosjean, R Benne, Eds. ASM Press, Washington, DC.

Estévez AM, Simpson L (1999). Uridine insertion/deletion RNA editing in trypanosome mitochondria—a review. *Gene* 240: 247–260.

Guthrie C, Patterson B (1988). Spliceosomal snRNAs. *Ann. Rev. Genet.* 22: 387–419.

Hutvagner G, Zamore PD (2002). RNAi: Nature abhors a double-strand. *Curr. Opin. Genet. Dev.* 12: 225–232.

Kiss T (2002). Small nucleolar RNAs: An abundant group of noncoding RNAs with diverse cellular functions. *Cell* 109: 145–148.

Mattick JS (2001). Non-coding RNAs: The architects of eukaryotic complexity. *EMBO Reps.* 2: 986–991.

Reinhart BJ, et al. (2002). MicroRNAs in plants. *Genes Dev.* 16: 1616–1626.

Samarsky DA, Fournier MJ (1999). A comprehensive database for the small nucleolar RNAs from *Saccharomyces cerevisiae*. *Nucleic Acids Res.* 27: 161–164.

RNA FOLDING

Jamie J. Cannone and Robin R. Gutell

The structure and function of any given RNA are dependent on each other. Thus, an understanding of how an RNA folds into its functional form can provide insights on that function. Insight into any potential for fluidity or movement in an RNA structure can also be indicative of functional properties.

The "RNA folding problem" can be summarized as the challenge of folding an RNA's primary structure into its active secondary and tertiary structures. Currently, no complete answer to this question is available, although several approaches have been able to provide insight. RNA-folding algorithms search for secondary-structure helices composed of consecutive, antiparallel canonical base pairs. Another set of constraints comes from thermodynamics, where RNA is expected to fold into its most energetically stable structure. The kinetics of the folding process will also have an impact on the final result.

See also Covariation Analysis, RNA Structure Prediction (Comparative Sequence Analysis), RNA Structure Prediction (Energy Minimization).

RNA SPLICING. *SEE* SPLICING.

RNA STRUCTURE
Jamie J. Cannone and Robin R. Gutell

The arrangement of RNA nucleotides in one-dimensional (primary structure), two-dimensional (secondary structure), and three-dimensional (tertiary structure) space.

For this discussion, we define primary structure to be equivalent to the sequence of an RNA molecule. Secondary structure is two or more consecutive and canonical base pairs that are nested and antiparallel with one another. All other structure, including noncanonical or nonnested base pairs and RNA motifs, is considered to be tertiary structure. Paired nucleotides are involved in interactions in the structure model; all other nucleotides are considered to be unpaired.

Base Pair (Canonical and Noncanonical)
A base pair is formed when two nucleotides hydrogen bond to each other, typically between the bases of the nucleotides. Most RNA base pairs observed to date are oriented with the backbones of the two nucleotides in an antiparallel configuration.

The canonical base pairs G:C and A:U plus the G:U ("wobble") base pair were originally proposed by Watson and Crick. All other base pairs are considered to be noncanonical. While "noncanonical" connotes an unusual or unlikely combination of two nucleotides, a significant number of noncanonical RNA base pairs has been proposed in rRNA comparative structure models and substantiated by the ribosomal subunit crystal structures. A larger number of noncanonical base pairs is present in the crystal structures that were not predicted with comparative analysis.

Stem
A stem is a set of base pairs that are arranged adjacent to and antiparallel with one another. While an RNA helix is a collection of smaller stems connected by loops and bulges, the terms *stem* and *helix* are often used interchangeably.

The base pairs that comprise a stem are nested—i.e., drawn graphically, each base pair either contains or is contained within its neighbors. For two nested base pairs a:a' and b:b', where $a < a'$, $b < b'$, and $a < b$ in the 5' to 3' numbering system for a given RNA molecule, the statement $a < b < b' < a'$ is true. Figure 1 shows nesting for tRNA in two different formats that represent the global arrangement of base

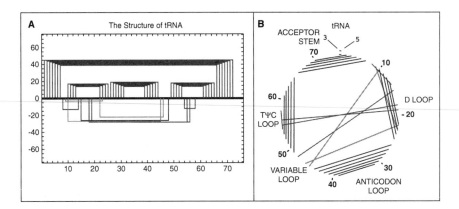

Figure 1 Nested and nonnested base pairs. The comparative model of the secondary and tertiary structure of tRNA is shown in two formats that highlight the helical and nesting relationships. In both panels, the secondary-structure base pairs are shown in blue, tertiary base pairs in red, and base triples in green. Some tertiary base pairs are nested (red lines do not cross blue lines); red lines representing pseudoknot base pairs do cross. A. Histogram format, with the tRNA sequence shown as a "baseline" from left to right (5′ to 3′). Secondary-structure elements are shown above the baseline and tertiary structure elements are shown below the baseline. The distance from the baseline to the interaction line is proportional to the distance between the two interacting positions within the RNA sequence. B. Circular format, with the sequence drawn clockwise (5′ to 3′) in a circle, starting at the top and base–base interactions shown as lines traversing the circle. The tRNA structural elements are labeled.

pairs. Nesting arrangements can be far more complicated in a larger RNA molecule. Most base pairs are nested, and most helices are also nested. Base pairs that are not nested are pseudoknots (see below).

Loop
Unpaired nucleotides in a secondary-structure model are commonly referred to as loops. Many of these loops close one end of an RNA stem and are called "hairpin loops"; phrased differently, the nucleotides in a hairpin loop are flanked by a single stem. Loops that are flanked by two stems come in several forms. A "bulge loop" occurs only in one strand in a stem; the second strand's nucleotides are all forming base pairs. An "internal loop" is formed by parallel bulges on opposing strands, interrupting the continuous base pairing of the stem. Finally, a "multi-stem" loop forms when three or more stems intersect. Figure 2 is a schematic RNA that shows each of these types of loop. The sizes of each type of loop can vary; certain combinations of loop size and nucleotide composition have been shown to be more energetically stable.

Pseudoknots
A pseudoknot is an arrangement of helices and loops where the helices are not nested with respect to each other. Pseudoknots are so named due to the optical illusion of knotting evoked by secondary- and tertiary-structure representations. A simple pseudoknot is represented in Figure 3A. For two nonnested (pseudoknot) base pairs a:a′ and b:b′, where $a < a'$ and $b < b'$ in the 5′ to 3′ numbering system for a given RNA molecule, the following statement will be true: $a < b < a' < b'$. Contrast this situation with a set of nested base pairs (see Stems above), where $a < b < b' <$

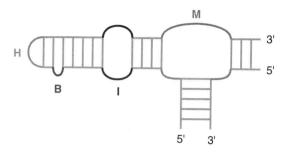

Figure 2 Loop types. This schematic RNA contains one example of each of the four loop types. Labels: B, bulge loop; H, hairpin loop, I, internal loop; M, multistem loop. Stems are shown in gray.

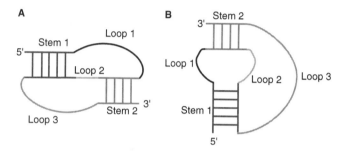

Figure 3 Schematic drawings of a simple pseudoknot. A. Standard format. B. Hairpin loop format (after Hilbers et al., 1998).

a′. Another descriptive explanation (Figure 3B) of pseudoknot formation is when the hairpin loop nucleotides from a stem–loop structure form a helix with nucleotides outside the stem–loop. Figure 1B shows pseudoknot interactions in green; note how these lines cross the blue lines that represent nested helices.

Related Websites

Base-Pair Directory (IMB Jena)	http://www.imb-jena.de/IMAGE_BPDIR.html
Comparative RNA Web Site (CRW) (descriptions and images of structural elements)	http://www.rna.icmb.utexas.edu/
Noncanonical Base-Pair Database	http://prion.bchs.uh.edu/bp_type/
Pseudobase	http://wwwbio.leidenuniv.nl/~Batenburg/PKB.html

Further Reading

Chastain M, Tinoco I Jr (1994). Structural elements in RNA. *Progr. Nucleic Acids Res. Mol. Biol.* 44: 131–177.

Hilbers CW, et al. (1998). New developments in structure determination of pseudoknots. *Biopolymers* 48: 137–153.

Pleij CWA (1994). RNA pseudoknots. *Curr. Opin. Struct. Biol.* 4: 337–344.

ten Dam E, et al. (1992). Structural and functional aspects of RNA pseudoknots. *Biochemistry* 31: 11665–11676.

RNA STRUCTURE PREDICTION (COMPARATIVE SEQUENCE ANALYSIS)

Jamie J. Cannone and Robin R. Gutell

Comparative analysis of RNA sequences is a powerful method for the prediction of RNA structure.

The two underlying principles for comparative sequence analysis are very simple but have a very profound influence on our prediction and understanding of RNA structure. The principles are as follows: (1) Different RNA sequences could have the potential to fold into the same secondary and tertiary structure. (2) The specific structure and function for select RNA molecules are maintained during the evolutionary process of genetic mutations and natural selection. Typically for RNA structure prediction, homologous base pairs that occur at the same positions in all of the sequences in the data set are identified with covariation analysis, resulting in a minimal structure model. Different structural motifs that have characteristic sequences at specific structural elements that are sufficiently conserved in the sequence data set are identified, culminating in a final comparative structure model.

The comparative method has been used for a variety of RNA molecules, including tRNA, the three rRNAs (5S, 16S, and 23S), ITS and IVS rRNAs, group I and II introns, RNase P, tmRNA, SRP RNA, and telomerase RNA. These methods will be more accurate and the predicted structure will have more detail for any one type of RNA when the number of sequences is large and the diversity among these sequences is high.

Related Website

Web (CRW) Site Comparative RNA	http://www.rna.icmb.utexas.edu/METHODS/

Further Reading

Gutell RR, et al. (1994). Lessons from an evolving rRNA: 16S and 23S rRNA structures from a comparative perspective. *Microbiol. Rev.* 58: 10–26.

Gutell RR, et al. (2002). The accuracy of ribosomal RNA comparative structure models. *Curr. Opin. Struct. Biol.* 12: 301–310.

Michel F, et al. (2000). Modeling RNA tertiary structure from patterns of sequence variation. *Methods Enzymol.* 317: 491–510.

Woese CR, Pace NR (1993). Probing RNA structure, function, and history by comparative analysis. In *The RNA World*, RF Gesteland, JF Atkins, Eds. Cold Spring Harbor Laboratory Press, Cold Spring Harbor, NY, pp. 91–117.

See also Covariation Analysis, Phylogenetic Events Analysis, RNA Tertiary-Structure Motifs.

RNA STRUCTURE PREDICTION (ENERGY MINIMIZATION)
Jamie J. Cannone and Robin R. Gutell

Traditionally, molecular biologists search for the most thermodynamically stable structures using energy minimization techniques. Thermodynamic energy values have been experimentally determined for consecutive base pairs and a few other simple structural elements in RNA. The assumption behind energy minimization in RNA folding is that the folding process for an RNA molecule can be determined by summing up the totals of the energy values for its simpler structural elements.

The present set of thermodynamic folding algorithms does not always predict a complete and correct secondary structure for an RNA molecule. This may indicate that either our understanding of all of the thermodynamic parameters is incomplete or the process is based upon a flawed assumption. These algorithms also are unable to predict tertiary-structure base pairs.

Related Websites

Michael Zuker's home page (includes mfold and links to his current research)	http://www.bioinfo.rpi.edu/~zukerm/
Turner Group	http://rna.chem.rochester.edu/

From IMB-JENA (both available from the above link but less direct):

Algorithms and Thermodynamics for RNA Secondary Structure Prediction: A Practical Guide	http://www.bioinfo.rpi.edu/~zukerm/seqanal/
Free-energy and enthalpy tables for RNA folding from the Turner Group	http://www.bioinfo.rpi.edu/~zukerm/rna/energy/

Further Reading

Burkard ME, et al. (1998). The Interactions that shape RNA. In *The RNA World*, 2nd ed., RF Gesteland, JF Atkins, TR Cech, Eds. Cold Spring Harbor Press, pp. 233–264.

Mathews DH (2000). RNA secondary structure prediction. In *Current Protocols in Nucleic Acid Chemistry*, S Beaucage, DE Bergstrom, GD Glick, RA Jones, Eds. Wiley, New York, pp. 11.2.1–11.2.10.

Mathews DH, et al. (1999). Expanded sequence dependence of thermodynamic parameters provides robust prediction of RNA secondary structure. *J. Mol. Biol.* 288: 911–940.

Nagel JHA, Pleij CWA (2002). Self-induced structural switches in RNA. *Biochimie* 84: 913–923.

Zuker M (1989). On finding all suboptimal foldings of an RNA molecule. *Science* 244: 48–52.

Zuker M (2000). Calculating nucleic acid secondary structure. *Curr. Opin. Struct. Biol.* 10: 303–310.

See also Covariation Analysis, Energy Minimization.

RNA TERTIARY-STRUCTURE MOTIFS
Jamie J. Cannone and Robin R. Gutell

Underlying the complex and elaborate secondary and tertiary structures for different RNA molecules is a collection of different RNA building blocks, or structural motifs.

Beyond the abundant G:C, A:U, and G:U base pairs in the standard Watson–Crick conformation that are arranged into regular secondary-structure helices, structural motifs are usually composed of noncanonical base pairs (e.g., A:A) with nonstandard base-pair conformations that are usually not consecutive and antiparallel with one another. Approximately 27 RNA structural motifs have been identified by different research groups with differing methods and criteria; these motifs are listed in alphabetical order below.

- 2'-OH-Mediated Helical Interactions: extensive hydrogen bonding between the backbone of one strand and the minor groove of another in tightly packed RNAs.
- AA.AG@helix.ends: A:A and A:G oppositions exchange at ends of helices.
- Adenosine Platform: two consecutive adenosines form a pseudo–base pair that allows for additional stacking of bases.
- A-Minor: the minor groove faces of adenosines insert into the minor groove of another helix, forming hydrogen bonds with the 2'-OH groups of C:G base pairs.
- A Story: unpaired adenosines in the covariation-based structure models.
- Base Triple: a base pair interacts with a third nucleotide.
- Bulged-G: links a cross-strand A stack to an A-form helix.
- Bulge-Helix-Bulge: Archaeal internal loop motif that is a target for splicing.
- Coaxial Stacking of Helices: two neighboring helices are stacked end to end.
- Cross-Strand Purine Stack: consecutive adenosines from opposite strands of a helix are stacked.
- Dominant G:U Base Pair: G:U is the dominant base pair (50% or greater), exchanging with canonical base pairs over a phylogenetic group, in particular structural locations.
- E-like Loop: a symmetric internal loop that resembles an E loop (with consensus sequence 5'-GHA/GAA-3') forms three noncanonical base pairs.
- E Loop/S Turn: an asymmetric internal or multistem loop (with consensus sequence 5'-AGUA/RAA-3') forms three noncanonical base pairs.
- Kink Turn: named for its kink in the RNA backbone; two helices joined by an internal loop interact via the A-minor motif.
- Kissing Hairpin Loop: two hairpin loops interact to form a pseudocontinuous, coaxially stacked three-stem helix.
- Lone Pair: a base pair that has no consecutive, adjacent base-pair neighbors.
- Lone-Pair Triloop: a base pair with no consecutive, adjacent base-pair neighbors encloses a three-nucleotide hairpin loop.
- Metal Binding: guanosine and uracil residues can bind metal ions in the major groove of RNA.
- Metal Core: specific nucleotide bases are exposed to the exterior to bind specific metal ions.
- Pseudoknots.
- Ribose Zipper: as two helices dock, the ribose sugars from two RNA strands become interlaced.
- Tandem G:A Opposition: two consecutive G:A oppositions occur in an internal loop.
- Tetraloop: four-nucleotide hairpin loops with specific sequences.

- Tetraloop Receptor: a structural element with a propensity to interact with tetraloops, often involving another structural motif.
- Triplexes: stable "triple helix" observed only in model RNAs.
- tRNA D-Loop:T-Loop: conserved tertiary base pairs between the D and T loops of tRNA.
- U Turn: a loop with the sequences UNR or gNRA contains a sharp turn in its backbone, often followed immediately by other tertiary interactions.

Related Websites

Comparative RNA Web (CRW) Site (descriptions and images of structural elements; motif-related publications)	http://www.rna.icmb.utexas.edu/
Distribution of RNA motifs in natural sequences	http://www.centrcn.umontreal.ca/~bourdeav/Ribonomics/
Pseudobase	http://wwwbio.leidenuniv.nl/~Batenburg/PKB.html
RNABase	http://www.rnabase.org/
SCOR (Structural Classification of RNA)	http://scor.lbl.gov/domain_tert.html

Further Reading

Batey RT et al. (1999). Tertiary motifs in RNA structure and folding. *Angewandte Chemie* (international ed. in English) 38: 2326–2343. [review discussing multiple motifs: Adenosine Platform; Base Triple; Coaxial Stacking of Helices; Kissing Hairpin Loops; Pseudoknot; Tetraloop; Tetraloop Receptor; tRNA D-Loop:T-Loop]

Blake RD, et al. (1967). Polynucleotides. 8. A spectral approach to the equilibria between polyriboadenylate and polyribouridylate and their complexes. *J. Mol. Biol.* 30: 291–308. [Triplexes]

Cate JH, Doudna JA (1996). Metal-binding sites in the major groove of a large ribozyme domain. *Structure* 4: 1221–1229.

Cate JH, et al. (1996a). Crystal structure of a group I ribozyme domain: Principles of RNA packing. *Science* 273: 1678–1685. [Metal-Core; Tetraloop Receptor]

Cate JH et al. (1996b). RNA tertiary structure mediation by adenosine platforms. *Science* 273: 1696–1699. [Adenosine Platform; Tetraloop Receptor]

Cate JH, et al. (1997). A magnesium ion core at the heart of a ribozyme domain. *Nature Struct. Biol.* 4: 553–558. [Metal-Core; Tetraloop Receptor]

Chang KY, Tinoco I Jr (1997). The structure of an RNA "kissing" hairpin complex of the HIV TAR hairpin loop and its complement. *J. Mol. Biol.* 269: 52–66. [Kissing Hairpin Loops]

Correll CC, et al. (1997). Metals, motifs, and recognition in the crystal structure of a 5S rRNA domain. *Cell* 91: 705–712. [Cross-Strand Purine Stack; example of Metal Binding]

Costa M, Michel F (1995). Frequent use of the same tertiary motif by self-folding RNAs. *EMBO J.* 14: 1276–1285. [Tetraloop Receptor]

Costa M, Michel F (1997). Rules for RNA recognition of GNRA tetraloops deduced by *in vitro* selection: Comparison with *in vivo* evolution. *EMBO J.* 16: 3289–3302. [Tetraloop Receptor]

Diener JL, Moore PB (1998). Solution structure of a substrate for the archaeal pre-tRNA splicing endonucleases: The bulge-helix-bulge motif. *Mol. Cell.* 1: 883–894. [Bulge-Helix-Bulge]

Dirheimer G, et al. (1995). Primary, secondary, and tertiary structures of tRNAs. In *tRNA: Structure, Biosynthesis, and Function*, D Söll, U RajBhandary, Eds. American Society for Microbiology, Washington, DC, pp. 93–126. [tRNA D-Loop:T-Loop]

Doherty EA, et al. (2001). A universal mode of helix packing in RNA. *Nature Struct. Biol.* 8: 339–343. [A-Minor]

Elgavish T, et al. (2001). AA.AG@Helix.Ends: A:A and A:G base-pairs at the ends of 16S and 23S rRNA helices. *J. Mol. Biol.* 310: 735–753. [AA.AG@helix.ends]

Gautheret D, et al. (1994). A major family of motifs involving G-A mismatches in ribosomal RNA. *J. Mol. Biol.* 242: 1–8. [Tandem G:A Oppositions]

Gautheret D, et al. (1995a). Identification of base-triples in RNA using comparative sequence analysis. *J. Mol. Biol.* 248: 27–43. [Base Triples]

Gautheret D, et al. (1995b). GU base pairing motifs in ribosomal RNAs. *RNA* 1: 807–814. [Dominant G:U Base-Pair]

Gutell RR et al. (1994). Lessons from an evolving ribosomal RNA: 16S and 23S rRNA structure from a comparative perspective. *Microbiol. Rev.* 58: 10–26. [review of several motifs: Base Triple; Coaxial Stacking of Helices; Dominant G:U Base Pair; Lone Pair; Pseudoknot; Tetraloop]

Gutell RR, et al. (2000a). A Story: Unpaired adenosine bases in ribosomal RNA. *J. Mol. Biol.* 304: 335–354. [A Story; Adenosine Platform; E Loop/S Turn; E-like Loop]

Gutell RR, et al. (2000b). Predicting U-turns in ribosomal RNA with comparative sequence analysis. *J. Mol. Biol.* 300: 791–803. [U Turn]

Ippolito JA, Steitz TA (1998). A 1.3-A resolution crystal structure of the HIV-1 trans-activation response region RNA stem reveals a metal ion-dependent bulge conformation. *Proc. Natl. Acad. Sci. USA* 95: 9819–9824. [Metal-Core; Tetraloop Receptor]

Jaeger L, et al. (1994). Involvement of a GNRA tetraloop in long-range RNA tertiary interactions. *J. Mol. Biol.* 236: 1271–1276. [Tetraloop Receptor]

Klein DJ, et al. (2001). The kink-turn: A new RNA secondary structure motif. *EMBO J.* 20: 4214–4221. [Kink-Turn]

Lee JC, et al. (2003). The lonepair triloop: A new motif in RNA structure. *J. Mol. Biol.* 325: 65–83. [Lonepair Triloop]

Leonard GA, et al. (1994). Crystal and molecular structure of r(CGCGAAUUAGCG): An RNA duplex containing two G(anti).A(anti) base pairs. *Structure* 2: 483–494. [2′-OH-Mediated Helical Interactions]

Leontis NB, Westhof E (1998). A common motif organizes the structure of multi-helix loops in 16S and 23S ribosomal RNAs. *J. Mol. Biol.* 283: 571–583. [E Loop/S Turn]

Lietzke SE, et al. (1996). The structure of an RNA dodecamer shows how tandem U-U base pairs increase the range of stable RNA structures and the diversity of recognition sites. *Structure* 4: 917–930. [2′-OH-Mediated Helical Interactions]

Massoulié J (1968). [Associations of poly A and poly U in acid media. Irreversible phenomenon] (French). *Eur. J. Biochem.* 3: 439–447. [Triplexes]

Moore PB (1999). Structural motifs in RNA. *Ann. Rev. Biochem.* 68: 287–300. [review of several motifs: Adenosine Platform; Bulge–Helix–Bulge; Bulged-G; Cross-Strand Purine Stack; Metal Binding; Ribose Zipper; Tetraloop; Tetraloop Receptor; U-Turn]

Nissen P, et al. (2001). RNA tertiary interactions in the large ribosomal subunit: The A-minor motif. *Proc. Natl. Acad. Sci. USA* 98: 4899–4903. [A-Minor]

Pleij CWA (1994). RNA pseudoknots. *Curr. Opin. Struct. Biol.* 4: 337–344. [Pseudoknot]

SantaLucia J Jr, et al. (1990). Effects of GA mismatches on the structure and thermodynamics of RNA internal loops. *Biochemistry* 29: 8813–8819. [Tandem G:A Oppositions]

Tamura M, Holbrook SR (2002). Sequence and structural conservation in RNA ribose zippers. *J. Mol. Biol.* 320: 455–474. [Ribose Zipper]

Traub W, Sussman JL (1982). Adenine-guanine base pairing ribosomal RNA. *Nucleic Acids Res.* 10: 2701–2708. [AA.AG@helix.ends]

Wimberly B (1994). A common RNA loop motif as a docking module and its function in the hammerhead ribozyme. *Nature Struct. Biol.* 1: 820–827. [E Loop/S Turn]

Wimberly B, et al. (1993). The conformation of loop E of eukaryotic 5S ribosomal RNA. *Biochemistry* 32: 1078–1087. [Bulged-G]

Woese CR, et al. (1983). Detailed analysis of the higher-order structure of 16S-like ribosomal ribonucleic acids. *Microbiol. Rev.* 47: 621–669. [AA.AG@helix.ends]

Woese CR, et al. (1990). Architecture of ribosomal RNA: Constraints on the sequence of "tetra-loops." *Proc. Natl. Acad. Sci. USA* 87: 8467–8471. [Tetraloops]

See also RNA Structure.

ROBUSTNESS
Marketa J. Zvelebil

Robustness, in terms of system biology, can be defined as the maintenance of certain functions despite variability in components or the environment.

Robustness allows for complexity of a system as well as its fragility. Minimal cellular life such as mycoplasm survives with only a few hundred genes, but it can only live under specific environmental conditions and is very sensitive to any fluctuations. However, even a slightly more complex life, such as the *Escherichia coli*, has approximately 4000 genes of which only about 300 are classified as essential. The presence of complex regulatory networks for robustness that accommodate various stress situations is thought to be the reason for this additional gene complexity. The *E. coli* organism can live happily in fluctuating environments.

There can be four possible types of mechanisms used to achieve robust behavior: system control, redundancy, structural stability, and modularity. Some papers suggest that robustness is achieved through complexity an in the *E. coli* example.

Further Reading

Kitano H (2002). Systems biology: A brief overview. *Science* 295: 1662–1664.

Lauffenburger DA (2000). Cell signaling pathways as control modules: Complexity for simplicity? *Proc. Natl. Acad. Sci. USA* 97(10): 5031–5033.

See also Modularity, System Biology.

ROLE
Robert Stevens

In description logics, the word *role* is used for "relationships type" or "property". Other knowledge representation and object-oriented programming languages use the word in different ways.

ROOTING PHYLOGENETIC TREES
Sudhir Kumar and Alan Filipski

Rooting is the process of determining placement of the root in a phylogenetic tree.

The usual method is by using an outgroup. An outgroup is a taxon, or clade, in a phylogenetic tree that lies outside of (or is a sister group to) the clade containing the taxa under analysis. An outgroup is commonly used for rooting phylogenetic trees by placing the root between the outgroup and the remaining taxa.

If outgroup knowledge is unavailable, an inferior alternative is midpoint rooting. In midpoint rooting, the root is placed at the midpoint of the longest path between any pair of terminal taxa. If all taxa evolve at the same rate, this will give the correct root, but it may give misleading results in the presence of deviations from a strict molecular clock.

See also Phylogenetic Tree.

ROOT-MEAN-SQUARE-DEVIATION (RMSD)
Roman A. Laskowski

In statistics the RMSD is defined as the square root of the sum of squared deviations of the observed values from the mean value divided by the number of observations. The units of the RMSD are the same as the units used for the observed values.

In protein crystallography it is common to cite the RMSD of bond lengths and bond angles in the final, refined model of the protein structure, these giving the RMSDs of the final bond lengths and angles about the "target values" used during refinement.

The RMSD is also used as a measure of the similarity of two molecular conformations. The two molecules are first superimposed and the distances between equivalent atoms in the two structures calculated. The square root of the sum of the squares of these distances divided by the number of distances gives an indication of how similar or dissimilar the two conformations are. When citing such an RMSD, it is necessary to specify which atoms were used in its calculation. For example, the RMSD between two protein structures might be given as 1.4 Å for equivalenced C-alpha atoms.

In structure determination by NMR spectroscopy it is usual to give an RMSD of all conformers in the ensemble from an averaged set of coordinates.

Further Reading
Norman GR, Steiner DL (1998). *Biostatistics; the Bare Essentials*. BC Decker, Hamilton, Ontario.

See also Cα, Conformation.

ROSETTA STONE METHOD
Jean-Michel Claverie

A sequence analysis method attempting to predict interactions (and functional relationships) between the products of genes A and B of a given organism (e.g., *Escherichia coli*) by the systematic search of gene fusions such as $A_{ortho} - B_{ortho}$ (or $B_{ortho} - A_{ortho}$) in another species (where A_{ortho} and B_{ortho} are orthologous to A and B, respectively).

The name "Rosetta" refers to the "stone of Rosette," the discovery of which allowed a crucial breakthrough in the deciphering of Egyptian hieroglyphs. This basalt slab, found in July 1799 in the small Egyptian village Rosette (Raschid), displays three inscriptions that represent a single text in three different variants of script: hieroglyphs (script of the official and religious texts), Demotic (everyday Egyptian script), and Greek. A three-way comparison allowed the French scholar Jean Francois Champollion (1790–1832) to decipher the hieroglyphs in 1822.

The molecular rationale behind the approach is that protein sequences merging A and B sequences will only be possible if they can fold and form a stable three-dimensional structure where domain A and B interact closely. By extension, isolated proteins constituted of separate A-like and B-like domains are predicted to form a bimolecular complex. The A–B or B–A gene fusion sequence, from which the potential A/B interaction is inferred, is called the Rosetta sequence. The approach is easier to implement for genomes and/or genes without introns. The logic can be extended from a strictly structural interaction to a functional relationship by considering that two enzymes A and B catalyzing different steps of a biochemical pathway could become a bifunctional enzyme where A-like and B-like domains, connected by a flexible linker, might not be required to stably interact. The method, introduced in 1999, has become part of the classical phylogenomics tool chest. The results obtained from the Rosetta stone analysis need to be carefully examined to remove the contribution of pseudogenes and/or split genes found in some bacterial genomes.

Related Website

UCLA-DOE Institute for Genomics and Proteomics	http://www.doe-mbi.ucla.edu/

Further Reading

Eisenberg D, et al. (2000). Protein function in the post-genomic era. *Nature* 405: 823–826.

Enright AJ, et al. (1999). Protein interaction maps for complete genomes based on gene fusion events. *Nature* 402: 86–90.

Huynen M, et al. (2000). Predicting protein function by genomic context: Quantitative evaluation and qualitative inferences. *Genome Res.* 10: 1204–1210.

Marcotte EM, et al. (1999). Detecting protein function and protein-protein interactions from genome sequences. *Science* 285: 751–753.

Ogata H, et al. (2001). Mechanisms of evolution in *Rickettsia conorii* and *R. prowazekii*. *Science* 293: 2093–2098.

Tsoka S, Ouzounis CA (2000). Prediction of protein interactions: Metabolic enzymes are frequently involved in gene fusion. *Nat. Genet.* 26: 141–142.

ROTAMER

Roland L. Dunbrack

Short for "rotational isomer," a single protein side-chain conformation represented as a set of values, one for each dihedral angle degree of freedom. Since bond angles and bond lengths in proteins have rather small variances, they are usually not included in the definition of a rotamer.

A rotamer is usually thought to be a local minimum on a potential energy map or an average conformation over some region of dihedral angle space. However, broad distributions of side-chain dihedral angles (such as amides) may be represented by several rotamers, which may not all be local minima or population maxima or means. *Nonrotameric* is sometimes used to describe side chains that have dihedral angles far from average values or far from a local energy minimum on a potential energy surface.

Further Reading

Dunbrack RL Jr, Cohen FE (1997). Bayesian statistical analysis of protein sidechain rotamer preferences. *Prot. Sci.* 6: 1661–1681.

Bower MJ, Cohen FE, Dunbrack RL Jr (1997). Prediction of protein side-chain rotamers from a backbone-dependent rotamer library: A new homology modeling tool. *J. Mol. Biol.* 267: 1268–1282.

See also Amino Acid; Conformation; Modeling, Molecular.

ROTAMER LIBRARY
Roland L. Dunbrack

A collection of rotamers for each residue type in proteins with side-chain degrees of freedom. Rotamer libraries usually contain information about both conformation and frequency of a certain conformation. Often libraries will also contain information about the variance about dihedral angle means or modes, which can be used in sampling.

Side-chain dihedral angles are not evenly distributed, but for most χ angles occur in tight clusters around certain values. Rotamer libraries therefore are usually derived from statistical analysis of side-chain conformations in known structures of proteins by clustering observed conformations or by dividing dihedral angle space into bins and determining an average conformation in each bin. This division is usually on physicochemical grounds, as in the divisions for rotation about $sp^3 - sp^3$ bonds into three 120° bins centered on each staggered conformation (60°, 180°, −60°).

Rotamer libraries can be backbone -independent, secondary-structure dependent, or backbone dependent. The distinctions are made depending on whether the dihedral angles for the rotamers and/or their frequencies depend on the local backbone conformation or not. *Backbone-independent* rotamer libraries make no reference to backbone conformation and are calculated from all available side chains of a certain type. *Secondary-structure-dependent* libraries present different dihedral angles and/or rotamer frequencies for α-helix, β-sheet, or coil secondary structures. *Backbone-dependent* rotamer libraries present conformations and/or frequencies dependent on the local backbone conformation as defined by the backbone dihedral angles ϕ and ψ, regardless of secondary structure. Finally, a variant of backbone-dependent rotamer libraries exists in the form of *position-specific* rotamers, those defined by a fragment usually of five amino acids in length, where the central residue's side chain conformation is examined.

Related Websites

Backbone-Dependent Rotamer Library	http://www.fccc.edu/research/labs/dunbrack/bbdep.html
Penultimate Rotamer Library	http://kinemage.biochem.duke.edu/databases/rotamer.php

Further Reading

Dunbrack RL Jr, Cohen FE (1997). Bayesian statistical analysis of protein sidechain rotamer preferences. *Prot. Sci.* 6: 1661–1681.

Dunbrack RL Jr, Karplus M (1994). Conformational analysis of the backbone-dependent rotamer preferences of protein sidechains. *Nature Struct. Biol.* 1: 334–340.

Lovell SC, Word JM, Richardson JS, Richardson DC (2000). The penultimate rotamer library. *Proteins: Struct. Funct. Genet.* 40: 389–408.

Ponder JW, Richards FM (1987). Tertiary templates for proteins: Use of packing criteria in the enumeration of allowed sequences for different structural classes. *J. Mol. Biol.* 193: 775–792.

rRNA. *SEE* RIBOSOMAL RNA.

RSEQUENCE
Thomas D. Schneider

Rsequence is the total amount of information conserved in a binding site, represented as the area under the sequence logo.

Further Reading

Schneider TD, et al. (1986). Information content of binding sites on nucleotide sequences. *J. Mol. Biol.* 188: 415–431. Schneider TD (2000). Evolution of biological information. *Nucleic Acids Res.* 28: 2794–2799.

See also Binding Site, Information, Rfrequency, Sequence Logo, Uncertainty.

RULE
Terri Attwood

A short regular expression (typically three to six residues in length) used to identify generic (non-family-specific) patterns in protein sequences. Rules tend to be used to encode particular functional sites: e.g., sugar attachment sites, phosphorylation, glycosylation, hydroxylation, sulfation sites. Example rules are shown below:

Functional Site	Rule
Protein kinase C phosphorylation site	[ST]-*x*-[RK]
N-glycosylation site	N-{*P*}-[ST]-{*P*}

Their small size means that rules do not provide good discrimination—in a typical sequence database, matches to expressions of this type number in the thousands. They therefore cannot be used to show that a particular functional site exists in a protein sequence, but rather give a guide as to whether such a site *might* exist. Biological knowledge must be used to confirm whether such matches are likely to be meaningful.

Rules are used in PROSITE to encode functional sites, such as those mentioned above and many others. For routine searches of the database, however, the Scan-Prosite tool provides an option to exclude patterns with a high probability of occurrence so that outputs should not be flooded with spurious matches.

Related Website

ScanProsite	http://www.expasy.org/tools/scanprosite/

Further Reading
Attwood TK (2000). The quest to deduce protein function from sequence: The role of pattern databases. *Int. J. Biochem. Cell Biol.* 32(2): 139–155.

Attwood TK, Parry-Smith DJ (1999). *Introduction to Bioinformatics.* Addison-Wesley Longman, Harlow, Essex, United Kingdom.

See also Regular Expression.

S

Safe Zone
SAGE
SAM
SAR
Satellite DNA
Scaffold Attachment Region
SCF
Schematic (Ribbon, Cartoon) Models
Score
Scoring Matrix
SCWRL
Search by Signal
Secondary Structure of Protein
Secondary-Structure Prediction of Protein
Second Law of Thermodynamics
Secretome
Segregation Analysis
Selenomethionine Derivatives
Self-Consistent Mean-Field Algorithm
Self-Organizing Map
Semantic Network
SeqCount
Seq-Gen
Sequence Alignment
Sequence Assembly
Sequence Complexity
Sequence Conservation
Sequence Distance Measures
Sequence Logo
Sequence Motif
Sequence Motifs—Prediction
  and Modeling
Sequence of Proteins
Sequence Pattern
Sequence Retrieval System
Sequence Similarity
Sequence Similarity–Based
  Gene Prediction
Sequence Similarity Search

Sequence Simplicity
Sequence-Tagged Site
Sequence Walker
Serial Analysis of Gene Expression
Serine
Sex Chromosome
Sexual Selection
Shannon Entropy
Shannon Sphere
Shannon Uncertainty
Short-Period Interspersion, Short-Term
  Interspersion
Shuffle Test
Side Chain
Side-Chain Prediction
Signal-to-Noise Ratio
Signature
Silent Mutation
SIMPLE
SIMPLE34
Simple DNA Sequence
Simple Repeat
Simple Sequence Repeat
Simulated Annealing
SINE
Single-Nucleotide Polymorphism
Site
Sites
Small-Sample Correction
Smith—Waterman Algorithm
SNP
Solvation Free Energy
SOV
Space Group
Space-Filling Models
Specificity
Spliced Alignment
Splicing
Spotted cDNA Microarray

Dictionary of Bioinformatics and Computational Biology. Edited by Hancock and Zvelebil
ISBN 0-471-43622-4 © 2004 John Wiley & Sons, Inc.

SAFE ZONE
Patrick Aloy

The *safe zone* is the region of the sequence–structure space where sequence alignments can unambiguously distinguish between protein pairs of similar and different structure. This is for high values of the pairwise sequence identity (i.e., >40%).

The safe zone is occupied by homologous proteins that have evolved from a common ancestor. If the alignment of a sequence of unknown structure and one with known three-dimensional conformation falls into the safe zone, homology modeling techniques can be successfully applied to obtain accurate models for our sequence of interest.

Further Reading
Rost B (1999). Twilight zone of protein sequence alignments. *Protein Eng.* 12: 85–94.

See also Alignment—Multiple, Homology.

SAGE (SERIAL ANALYSIS OF GENE EXPRESSION)
Jean-Michel Claverie

Method for the detection and quantitation of transcripts.

The SAGE method is based on the isolation of unique sequence tags from individual transcripts and concatenation of tags serially into long DNA molecules (see accompanying figure). Rapid sequencing of concatemer clones reveals individual tags and allows quantitation and identification of cellular transcripts. Two principles underlie the SAGE methodology: (1) A short sequence tag (10–14 bp) contains enough information to uniquely identify a transcript provided that the tag is obtained from a unique position within each transcript. (2) Counting the number of times a particular tag is observed provides an estimate of the expression level of the corresponding transcript. Sequencing one insert (from a SAGE library) provides information on the expression of up to 30 different genes rather than on a single one with standard complementary DNA (cDNA) libraries used to generate ESTs. The same number of sequencing reactions thus provides a much larger set of tags that can then be used, like ESTs, in the "gene discovery" or "gene profiling" modes. While average EST sequencing projects are set up to generate 5000 tags, SAGE projects routinely generate on the order of 50,000 tags. The same statistical analysis (i.e., tag counting) applies to both methods, but SAGE experiments allow more accurate and more sensitive expression estimates to be made. SAGE studies have been particularly successful in the characterization of various cancers.

Related Websites

SAGE	http://www.sagenet.org/
Cancer Genome Anatomy Project	http://www.ncbi.nlm.nih.gov/ncicgap/

Further Reading
Audic S, Claverie J-M (1997). The significance of digital gene expression profiles. *Genome Res.* 7: 986–995.

Dictionary of Bioinformatics and Computational Biology. Edited by Hancock and Zvelebil
ISBN 0-471-43622-4   © 2004 John Wiley & Sons, Inc.

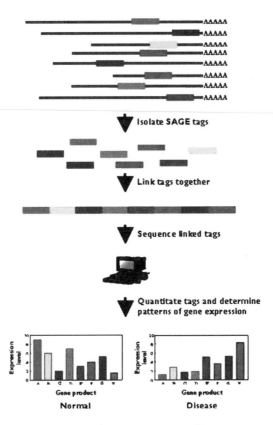

Figure   The SAGE concept (from
http://www.sagenet.org/).

Claverie JM (1999). Computational methods for the identification of differential and coordinated gene expression. *Hum. Mol. Genet.* 8: 1821–1832.

Lal A, et al. (1999). A public database for gene expression in human cancers. *Cancer Res.* 59: 5403–5407.

Velculescu VE, et al. (1995). Serial analysis of gene expression. *Science* 270: 484–487.

Velculescu VE, et al. (2000). Analysing uncharted transcriptomes with SAGE. *Trends Genet.* 16: 423–425.

SAM (SYSTEM FOR ASSEMBLING MARKERS)
John M. Hancock and Martin J. Bishop

Program for assembling markers on individual physical mapping clones onto a map.

SAM accepts data on the marker content of a set of clones and organizes them into a map representing the likely order of those markers in the genome. It also displays the order of the clones on that map it produces. A graphical user interface allows the user to interact with the clone map.

Related Website

| SAM | http://www.sanger.ac.uk/Software/sam/ |

Further Reading
Soderlund C, Dunham I (1995). SAM: A system for iteratively building marker maps. *Comput. Appl. Biosci.* 11: 645–655.

See also Marker.

SAR. *SEE* MATRIX ATTACHMENT REGION.

SATELLITE DNA
Katheleen Gardiner

Arrays of tandemly repeated simple sequence DNA; found within and adjacent to centromeres in heterochromatin.

Multiple different families include the human centromeric alphoid sequence. Different repeat families are found within a single array that may exceed megabases in size.

Further Reading
Cooper DN (1999). *Human Gene Evolution*. Bios Scientific Publishers, Oxford, UK, pp. 265–285.

Ridley M (1996). *Evolution*. Blackwell Science, Cambridge, MA, pp. 265–276.

SCAFFOLD ATTACHMENT REGION. *SEE* MATRIX ATTACHMENT REGION.

SCF
Rodger Staden

The most widely used file format for storing DNA trace data.

The files store the chromatogram amplitudes for the four base types along the length of the sequence, the base calls, a mapping of the base calls to the amplitudes, and the base-call confidence values. In the future it may be replaced by ZTR, which is more flexible and compact.

Related Website

Manual	http://www.mrc-lmb.cam.ac.uk/pubseq/manual/formats_unix_toc.html

Further Reading
Dear S, Staden R (1992). A standard file format for data from DNA sequencing instruments. *DNA Sequence* 3: 107–110.

SCHEMATIC (RIBBON, CARTOON) MODELS
Eric Martz

Smoothed backbone models that indicate secondary structure schematically.

Alpha helices are represented as helical ribbons or further simplified to cylinders. Beta strands are represented as relatively straight ribbons, with arrowheads designating the carboxy ends. In some cases, the cylinders representing alpha helices may have their carboxy ends pointed. Backbone regions that have neither alpha nor beta secondary structure may be represented as thin ribbons or ropes, following a smoothed backbone trace.

Hand-drawn schematic depictions first appeared in the mid-1960s. Schematic ribbon drawings were perfected and popularized by Jane Richardson in the early 1970s. At that time, some visualization software packages were already being adapted to display schematic models. Schematic models are designated "cartoons" in the popular visualization freeware RasMol and its derivatives.

Further Reading
Lesk AM, Hardman KD (1982). Computer-generated schematic diagrams of protein structures. *Science* 216: 539–540.

Lesk AM, Hardman KD (1985). Computer-generated pictures of proteins. *Methods Enzymol.* 115: 381–390.

Richardson JS (1981). The anatomy and taxonomy of protein structure. *Adv. Protein Chem.* 34: 167–339.

Richardson JS (1985). Schematic drawings of protein structures. *Methods Enzymol.* 115: 359–380.

Richardson JS (2000). Early ribbon drawings of proteins. *Nature Struct. Biol.* 7: 624–625.

See also Backbone Models; Models, Molecular and accompanying illustration; Secondary Structure of Protein; Visualization, Molecular.

Eric Martz is grateful for help from Eric Francoeur, Peter Murray-Rust, Byron Rubin, and Henry Rzepa.

SCORE
Thomas D. Schneider

Many methods in bioinformatics give results as scores.

It is worth noting that scores can be multiplied by an arbitrary constant and remain comparable. In contrast, information is measured in bits, and this cannot be multiplied by an arbitrary constant and still retain the same units of measure. Scores cannot be compared between different binding sites, whereas it is reasonable to compare bits. For example, it is interesting that donor splice sites do not have the same information as acceptor splice sites (Stephens and Schneider 1992).

Further Reading
Stephens RM, Schneider TD (1992). Features of spliceosome evolution and function inferred from an analysis of the information at human splice sites. *J. Mol. Biol.* 228: 1124–1136. http://www.lecb.ncifcrf.gov/toms/paper/splice/.

See also Acceptor Splice Site, Bit, Donor Splice Site, Information.

SCORING MATRIX (SUBSTITUTION MATRIX)
Terri Attwood

A table of pairwise values encoding relationships between either nucleotides or amino acid residues that is used to calculate alignment scores in sequence comparison methods. In the simplest type of scoring matrix (a unitary or identity matrix), the pairwise value for identities (e.g., guanine pairing with guanine or leucine pairing with leucine) is 1, while the value given for nonidentical pairs (say, guanine with adenine or leucine with valine) is 0, as illustrated below:

	A	**C**	**T**	**G**
A	1	0	0	0
C	0	1	0	0
T	0	0	1	0
G	0	0	0	1

The unitary matrix is sparse, meaning that most of its elements are 0. Its diagnostic power is therefore relatively poor, because all identical matches carry equal weighting. To improve diagnostic performance, the aim is to enhance the scoring potential of weak, but biologically significant signals without also amplifying noise. To this end, scoring matrices have been derived for protein sequence comparisons that, e.g., weight matches between nonidentical residues according to observed substitution rates across large evolutionary distances. The most commonly used series are the Dayhoff mutation data (MD) and Blocks substitution (BLOSUM) matrices. Here, the greater the value for an amino acid pair, the more related or substitutable those amino acids are likely to be (e.g., tyrosine pairing with phenylalanine has a higher value than alanine pairing with tryptophan).

When using scoring matrices, it should be appreciated that they are inherently noisy, because they indiscriminately weight relationships that may be inappropriate in the context of a particular sequence comparison—in other words, scores of random matches are boosted along with those of weak biological signals. The biological significance of all matches should therefore be evaluated with this in mind.

Further Reading

Dayhoff MO, Schwartz RM, Orcutt BC (1978). A model of evolutionary change in proteins. In *Atlas of Protein Sequence and Structure*, Vol. 5, Suppl. 3, MO Dayhoff, Ed. NBRF, Washington, DC, pp. 345–352.

Henikoff S, Henikoff JG (1992). Amino acid substitution matrices from protein blocks. *Proc. Natl. Acad. Sci. USA* 89(22): 10915–10919.

Schwartz RM, Dayhoff MO (1978). Matrices for detecting distant relationships. In *Atlas of Protein Sequence and Structure*, Vol. 5, Suppl. 3, MO Dayhoff, Ed. NBRF, Washington, DC, pp. 353–358.

See also Alignment—Multiple, BLOSUM Matrix, Dayhoff Amino Acid Substitution Matrix, Substitution Process.

SCWRL

Roland L. Dunbrack

A program for predicting side-chain conformations given a backbone model and sequence, used primarily in comparative modeling. SCWRL uses a backbone-dependent rotamer library to build the most likely conformation for each side chain for the backbone conformation. Steric conflicts with the backbone and between side chains are removed with a combination of dead-end elimination and branch-and-bound algorithms. SCWRL allows the user to specify conserved side chains whose Cartesian coordinates can be preserved from the input structure. SCWRL also allows the user to provide a second input file of atomic coordinates for ligands (small molecules, ions, DNA, other proteins) that can act as a background against which the predicted side chains must fit without significant steric conflicts.

Related Websites

SCWRL	http://www.fccc.edu/research/labs/dunbrack/scwrl/
Backbone-Dependent Rotamer Library	http://www.fccc.edu/research/labs/dunbrack/ sidechain.html

Further Reading

Bower MJ, et al. (1997). Prediction of protein side-chain rotamers from a backbone-dependent rotamer library: A new homology modeling tool. *J. Mol. Biol.* 267: 1268–1282.

Dunbrack RL Jr (1999). Comparative modeling of CASP3 targets using PSI-BLAST and SCWRL. *Proteins* 3(Suppl.): 81–87.

Dunbrack RL Jr, Cohen FE (1997). Bayesian statistical analysis of protein sidechain rotamer preferences. *Protein Sci.* 6: 1661–1681.

See also Comparative Modeling, Dead-End Elimination Algorithm, Rotamer Library.

SEARCH BY SIGNAL. *SEE* SEQUENCE MOTIFS—PREDICTION AND MODELING.

SECONDARY STRUCTURE OF PROTEIN
Roman A. Laskowski

Describes the local conformation of a protein's backbone. The secondary structure can be regular or irregular. The regular conformations are stabilized by hydrogen bonds between main-chain atoms. The most common of these conformations are the alpha helix and beta sheet first proposed by Linus Pauling in 1951. Others include the 3/10, or collagen, helix, and various types of turns. Regions of backbone that have an irregular conformation are said to have random-coil, or simply coil, conformation.

The regular conformations are characterized by specific combinations of the backbone torsion angles, phi and psi, and map onto clearly defined regions of the Ramachandran plot. In secondary-structure prediction the different types are usually simplified into three categories: helix, sheet, and coil.

Various computer algorithms exist for automatically computing the secondary structure of each part of a protein, given the three-dimensional coordinates of the protein structure. The best known is the DSSP algorithm of Kabsch and Sander (1983).

Related Website

DSSP	http://bioweb.pasteur.fr/seqanal/interfaces/dssp.html

Further Reading

Kabsch W, Sander C (1983). Dictionary of protein secondary structure: Pattern recognition of hydrogen-bonded and geometrical features. *Biopolymers* 22: 2577–2637.

See also Alpha Helix, Backbone, Beta Sheet, Dihedral Angle, Hydrogen Bond, Main Chain, Ramachandran Plot.

SECONDARY-STRUCTURE PREDICTION OF PROTEIN
Patrick Aloy

Protein *secondary-structure prediction* is the determination of the regions of secondary structure in a protein (e.g., α-helix, β-strand, coil) from its amino acid sequence.

The first automated methods for protein secondary-structure prediction appeared during the 1970s and were mainly based on residue preferences derived from very limited databases of known three-dimensional (3D) structures. They reached three-state-per-residue accuracies of about 50%.

During the 1980s a second generation of prediction algorithms appeared, the main improvements coming from the use of segment statistics (putting single residues in context) and larger databases. These methods achieved three-state-per-residue accuracies slightly higher than 60%.

A third generation of secondary-structure prediction methods emerged during the 1990s. They combined evolutionary information extracted from multiple-sequence alignments with "intelligent" algorithms, such as neural networks, to predict at 75% accuracy.

Analyses of 3D structures of close homologs and NMR experiments suggest an upper limit for secondary-structure prediction accuracy of 80%. Recent studies have estimated the accuracy of current methods to be $76 \pm 10\%$, so we might have already reached this upper limit, and only more entries in structural databases can improve the accuracies of the current methods.

Related Websites

PredictProtein	http://cubic.bioc.columbia.edu/predictprotein/
Jpred	http://www.compbio.dundee.ac.uk/~www-jpred/submit.html
A review of methods	http://cmgm.stanford.edu/WWW/www_predict.html

Further Reading

Rost B, Sander C (2000). Third generation prediction of secondary structures. *Methods Mol. Biol.* 143: 71–95.

Zvelebil MJ, Barton GJ, Taylor WR, Sternberg MJE (1987). Prediction of protein secondary structure and active site using the alignment of homologous sequences. *J. Mol. Biol.* 195: 957–967.

See also Web-Based Secondary-Structure Prediction Programs.

SECOND LAW OF THERMODYNAMICS
Thomas D. Schneider

The second law of thermodynamics is the principle that the disorder of an isolated system (entropy) increases to a maximum.

The second law appears in many surprisingly distinct forms. Transformations between these forms were described by Jaynes (1988). The relevant form for molecular information theory is

$$E_{\min} = K_b T \ln(2) = -q/R \text{ (joules per bit)}$$

where K_b is Boltzmann's constant, T is the absolute temperature, $\ln(2)$ is a constant that converts to bits, $-q$ is the heat dissipated away from the molecular machine, and R is the information gained by the molecular machine (Schneider, 1991).

Supposedly Maxwell's demon violates the second law, but if we approach the question from the viewpoint of modern molecular biology, the puzzles go away (Schneider, 1994).

Related Websites

Probability theory as extended logic	http://bayes.wustl.edu/
A light-hearted introduction to the "Mother of all Murphy's Laws" which clarifies an issue that is often incorrectly presented in discussions about entropy	http://www.secondlaw.com http://jchemed.chem.wisc.edu/Journal/Issues/Current/abs1385.html
An equation for the second law of thermodynamics	http://www.lecb.ncifcrf.gov/toms/paper/secondlaw/html/index.html
Information is not entropy, information is not uncertainty!	http://www.lecb.ncifcrf.gov/toms/information.is.not.uncertainty.html
Rock candy: an example of the second law of thermodynamics	http://www.lecb.ncifcrf.gov/toms/rockcandy.html

Further Reading

Jaynes ET (1988). The evolution of Carnot's principle. In *Maximum-Entropy and Bayesian Methods in Science and Engineering*, Vol. 1, GJ Erickson, CR Smith, Eds. Kluwer Academic, Dordrecht, The Netherlands, pp. 267–281. http://bayes.wustl.edu/etj/articles/ccarnot.ps.gz, http://bayes.wustl.edu/etj/articles/ccarnot.pdf.

Lambert FL (1999). Shuffled cards, messy desks, and disorderly dorm rooms—examples of entropy increase? Nonsense! *J. Chem. Ed.* 76: 1385–1387.

Schneider TD (1991). Theory of molecular machines. II. Energy dissipation from molecular machines. *J. Theor. Biol.* 148: 125–137. http://www.lecb.ncifcrf.gov/~toms/paper/edmm/.

Schneider TD (1994). Sequence logos, machine/channel capacity, Maxwell's demon, and molecular computers: A review of the theory of molecular machines. *Nanotechnology* 5: 1–18. http://www.lecb.ncifcrf.gov/~toms/paper/nano2/.

See also Bit, Entropy, Information, Molecular Information Theory.

SECRETOME
Dov S. Greenbaum

The secretome is the population of protein products that are secreted from the cell.

First used in reference to the secreted proteins of *Bacillus subtilis*. The analysis of the secretome covers the protein transport system within the cell and specifically the secretory pathways. The secretome also represents a subsection of the proteome/translatome, i.e., those proteins that are exported.

The secretome can be measured experimentally via biochemical identification of all secreted proteins. High-throughput proteomic methods such as two-dimensional electrophoresis to separate the proteins in the extracellular environment of the organism followed by mass spectrometry identification of each of the individual proteins are often used. The secretome can also be determined computationally via dedicated algorithms designed to determine which protein sequences signal the cell to excrete that specific protein. Presently computational methods suffer from the inability to detect lipoproteins and those proteins which, although they contain no recognizable secretion signals, are nevertheless secreted from the cell.

In particular, the exploration of secretomes is of commercial use with regard to the exploitation of so-called industrial bacteria such as *B. subtilis* that are used to produce, through their secretory pathways, industrially important products such as synthetics and drugs. Furthermore, the secretome provides an easily and biochemically well described population which is informative with regard to intercellular interaction both within and between organisms.

Further Reading

Antelmann H, et al. (2001). A proteomic view on genome-based signal peptide predictions. *Genome Res.* 11: 1484–1502.

Tjalsma H, et al. (2000). Signal peptide-dependent protein transport in *Bacillus subtilis*: A genome-based survey of the secretome. *Microbiol. Mol. Biol. Rev.* 64: 515–547.

See also Proteome, Translatome.

SEGREGATION ANALYSIS
Mark McCarthy and Steven Wiltshire

Analysis of the distribution and segregation of phenotypes within pedigrees designed to characterize the genetic architecture of the trait concerned.

In its simplest form, segregation analysis of Mendelian traits in simple nuclear pedigrees amounts to nothing more than comparing the observed segregation ratio of affected/unaffected offspring arising from a particular parental mating type with that expected under different possible disease models (dominant, recessive, codominant, and sex linked). For large pedigrees and traits displaying incomplete penetrance, computer-implemented likelihood ratio–based methods are used. The parameters of the genetic model that need to be specified and/or estimated from the data include putative disease allele frequency, penetrance probabilities, ascertainment correction, and transmission probabilities. The more challenging problem of segregation analysis of complex traits—both qualitative and quantitative—usually proceeds by one of two likelihood ratio–based approaches. Mixed-model segregation analysis explicitly models the quantitative trait in terms of a major gene effect, polygenic effect, and environmental deviation. For discrete traits, a continuous underlying liability function is specified, and the model includes penetrance probabilities rather than genotypic means. Regressive models instead express the quantitative trait in terms of the effects of a segregating major gene together with a residual familial correlation, achieved by regressing each offspring on its predecessors in the pedigree. For both methods, hypothesis testing proceeds by comparing model likelihoods to identify the most parsimonious, statistically significant model. The value of segregation analysis methods for the dissection of multifactorial trait architecture is limited by two main factors: (1) It is difficult to describe the appropriate correction to account for the ascertainment scheme by which the families were collected. (2) Data sets are rarely large enough to allow discrimination between the large number of alternative models involved in any realistic description of the architecture of a multifactorial trait (which is likely to involve several interacting susceptibility loci).

Examples: An et al. (2000) used mixed-model segregation analysis to detect the presence of a recessive major gene influencing baseline resting diastolic blood pressure as part of the HERITAGE study. Thein et al. (1994) used regressive-model segregation analysis to detect the presence of a dominant or codominant major gene influencing hereditary persistence of fetal hemoglobin. Rice et al. (1999) used mixed-model segregation analysis to examine body mass index in Swedish pedigrees: Although they found evidence for transmission for a major gene effect, transmission probabilities were not consistent with Mendelian segregation.

Related Websites
Programs for performing mixed-model or regressive-model segregation analysis:

PAP	http://hasstedt.genetics.utah.edu/
POINTER	http://cedar.genetics.soton.ac.uk/pub/PROGRAMS/pointer
SAGE	http://darwin.cwru.edu/sage/index.htm

Further Reading

An P, et al. (2000). Complex segregation analysis of blood pressure and heart rate measured before and after a 20-week endurance exercise training program: The HERITAGE Family Study. *Am. J. Hypertens.* 13: 488–497.

Ott J (1990). Cutting a Gordian knot in the linkage analysis of complex human traits. *Am. J. Hum. Genet.* 46: 219–221.

Rice T, et al. (1999). Segregation analysis of body mass index in a large sample selected for obesity: The Swedish Obese Subjects study. *Obes. Res.* 7: 246–255.

Sham P (1998). *Statistics in Human Genetics.* Arnold, London, pp. 13–50, 254–261.

Thein SL, et al. (1994). Detection of a major gene for heterocellular hereditary persistence of fetal hemoglobin after accounting for genetic modifiers. *Am. J. Hum. Genet.* 54: 214–228.

Weiss KM (1995). *Genetic Variation and Human Disease.* Cambridge University Press, Cambridge, pp. 52–116.

See also Mendelian Disease, Multifactorial Trait, Quantitative Trait, Variance Components.

SELENOMETHIONINE DERIVATIVES. *SEE* HEAVY-ATOM DERIVATIVE, MULTIPLE ANOMALOUS DISPERSION PHASING, PHASE PROBLEM, X-RAY CRYSTALLOGRAPHY FOR STRUCTURE DETERMINATION.

SELF-CONSISTENT MEAN-FIELD ALGORITHM
Roland L. Dunbrack

An algorithm for protein side-chain placement developed by Patrice Koehl and Marc Delarue (1996). Each rotamer of each side chain has a certain probability, $p(r_i)$. The total energy is a weighted sum of the interactions with the backbone and interactions of side chains with each other:

$$E_{\text{tot}} = \sum_{i=1}^{N} \sum_{r_i=1}^{n_{\text{rot}}(i)} p(r_i)E_{\text{bb}}(r_i) + \sum_{i=1}^{N-1} \sum_{r_i=1}^{n_{\text{rot}}(i)} \sum_{j=i+1}^{N} \sum_{r_j=1}^{n_{\text{rot}}(j)} p(r_i)p(r_j)E_{\text{sc}}(r_i, r_j)$$

In this equation, $p(r_i)$ is the density or probability of rotamer r_i of residue i, $E_{\text{bb}}(r_i)$ is the energy of interaction of this rotamer with the backbone, and $E_{\text{sc}}(r_i, r_j)$ is the interaction energy (van der Waals, electrostatic) of rotamer r_i of residue i with rotamer r_j of residue j. Some initial probabilities are chosen for the p's and the energies calculated. New probabilities $p'(r_i)$ can then be calculated with a Boltzmann distribution based on the energies of each side chain and the probabilities of the previous step:

$$E(r_i) = E_{\text{bb}}(r_i) + \sum_{j=1, j\neq i}^{N} \sum_{r_j=1}^{n_{\text{rot}}(j)} p(r_j)E_{\text{sc}}(r_i, r_j)$$

$$p'(r_i) = \frac{\exp(-E(r_i)/kT)}{\sum_{r_i=1}^{n_{\text{rot}}(i)} \exp(-E(r_i)/kT)}$$

Alternating steps of new energies and new probabilities can be calculated from these expressions until the changes in probabilities and energies in each step become smaller than some tolerance.

Further Reading
Koehl P, Delarue M (1996). Mean-field minimization methods for biological macromolecules. *Curr. Opin. Struct. Biol.* 6: 222–226.

See also Rotamer.

SELF-ORGANIZING MAP (KOHONEN MAP, SOM)
John M. Hancock

Self-organizing maps are unsupervised machine learning algorithms which are designed to learn associations between groups of inputs.

The classical output of a SOM algorithm is a one- or two-dimensional representation. Because of this, SOMs can also be considered as a means of visualizing complex data, as they effectively project a set of data onto a lower dimension representation.

A SOM takes as input a set of sample vectors. It then carries out an iterative matching between these vectors and a set of weight vectors. The weight vectors carry values corresponding to the values in the sample vectors plus coordinates that represent their position on the map. After initializing the weight vectors (which may be carried out randomly or in some other way), the algorithm selects a sample vector and identifies the weight vector most similar to it. Once this best fit is identified, weight vectors adjacent to the winning weight vector are adjusted to make them more similar to the winning matrix. Over tens of thousands of iterations, this process should result in a map in which adjacent weight vectors link a cluster of sample vectors.

Related Websites

Introductions to SOMs	http://www.ucl.ac.uk/oncology/MicroCore/HTML_resource/ SOM_Intro.htm
	http://davis.wpi.edu/~matt/courses/soms/

Further Reading
Wang J, et al. (2002). Clustering of the SOM easily reveals distinct gene expression patterns: Results of a reanalysis of lymphoma study. *BMC Bioinformatics*. 3: 36.

See also Clustering.

SEMANTIC NETWORK
Robert Stevens

Any representation of knowledge based on nodes representing concepts and arcs representing binary relations between those concepts. Early semantic networks were relatively informal and intended to indicate associations. Formalization led to knowledge representation languages and conceptual graphs.

SEQCOUNT. *SEE* ENCPRIME/SEQCOUNT.

SEQ-GEN
Michael P. Cummings

A program that simulates nucleotide sequences for a phylogenetic tree using Markov models.

The program allows for a choice of models and model parameter values. Written in American National Standards Institute (ANSI) C, the program is available as source code and executable files for some platforms.

Related Website

Seq-Gen	http://evolve.zoo.ox.ac.uk/software/Seq-Gen/main.html

Further Reading
Rambaut A, Grassly NC (1997). Seq-Gen: An application for the Monte Carlo simulation of DNA sequence evolution along phylogenetic trees. *Comput. Appl. Biosci*. 13: 235–238.

See also PAML.

SEQUENCE ALIGNMENT. *SEE* ALIGNMENT—MULTIPLE, ALIGNMENT—PAIRWISE.

SEQUENCE ASSEMBLY
Rodger Staden

The process of arranging a set of overlapping sequence readings into their correct order along the genome.

SEQUENCE COMPLEXITY (SEQUENCE SIMPLICITY)
Katheleen Gardiner

Information content of a DNA sequence related to repeat sequence content.

Complex sequences lack repeat motifs and are also considered as unique sequence (one copy per genome) and as having high information content. A pool of all the coding regions of the human genome thus represents a much more complex sequence than a similar pool of all intronic sequences, which contain repetitive sequence such as Alus and L1s.

Originally measured by reassociation kinetics of hybridization, complexity or unique sequence content can now be identified from direct analysis of genomic DNA sequence.

Further Reading
Cooper DN (1999). *Human Gene Evolution*. Bios Scientific, Oxford, UK, pp. 265–285.
Ridley M (1996). *Evolution*. Blackwell Science, Cambridge, MA, pp. 265–276.
Wagner RP, Maguire MP, Stallings RL (1993). *Chromosomes—A Synthesis*. Wiley-Liss, New York.

See also Low-Complexity Region, SIMPLE, Simple DNA Sequence.

SEQUENCE CONSERVATION. *SEE* CONSERVATION.

SEQUENCE DISTANCE MEASURES
Sudhir Kumar and Alan Filipski

A sequence distance measure is a numerical representation of the phylogenetic dissimilarity or divergence between two molecular (DNA or protein) sequences.

The simplest example of a sequence distance measure is the p distance, which is simply the fraction of sites that differ between two aligned DNA or protein sequences. A Poisson correction may be applied to account for multiple substitutions at a site. In this case the corrected distance is given by $d = -\log(1 - p)$.

More complex distance measures are available for DNA sequences. The Jukes–Cantor distance is a simple distance measure for DNA which corrects for multiple substitutions under the assumption that there are exactly four states (nucleotides) at any site. It also assumes that the probability of change from any nucleotide to another is the same and that all sites evolve at the same rate. The Kimura two-parameter distance is based on an extension of the Jukes–Cantor model to allow for different rates for transitional and transversional substitutions.

Gamma distances are a class of distance measures in which the rate variation among sites is taken into account in estimating the sequence distance. In this case, the evolutionary rates at different sites are assumed to come from a gamma distribution with a shape parameter α. A value of $\alpha = 1$ corresponds to a negative exponential distribution of rates with a large number of sites evolving with slow rates and a few sites evolving with faster rates. A value of $\alpha < 1$ represents a larger skew in rates whereas a value >1 shows more homogeneous rates among sites. A very large value of α (theoretically infinite but practically >5) corresponds to the case of uniform evolutionary rates among sites.

Sequence distances may be calculated using all sites in a set of sequences or, for protein coding DNA sequences, using exclusively synonymous or nonsynonymous sites, in which case synonymous or nonsynonymous sequence distances are obtained. Nonsynonymous distances can be obtained by codon-by-codon or protein sequence analysis. Synonymous distances can be estimated by using codon-by-codon approaches or by extracting neutral or fourfold degenerate sites and conducting site-by-site analysis.

Further Reading

Felsenstein J (1993). *PHYLIP: Phylogenetic Inference Package.* University of Washington, Seattle, WA.

Johnson NL, Kotz S (1970). *Continuous Univariate Distributions.* Hougton Mifflin, New York.

Kumar S, et al. (2001). MEGA2: Molecular evolutionary genetics analysis software. *Bioinformatics* 17: 1244–1245.

Li W-H (1997). *Molecular Evolution.* Sinauer Associates, Sunderland, MA.

Nei M, Kumar S (2000). *Molecular Evolution and Phylogenetics.* Oxford University Press, Oxford, UK.

Yang Z (1996). Among-site rate variation and its impact on phylogenetic analyses. *Trends. Ecol. Evol.* 11: 367–372.

See also Evolutionary Distance.

SEQUENCE LOGO
Thomas D. Schneider

A sequence logo is a graphic representation of an aligned set of binding sites (see the accompanying figure). A logo displays the frequencies of bases at each position as the relative heights of letters, along with the degree of sequence conservation as the total height of a stack of letters, measured in bits of information. Subtle frequencies are not lost in the final product as they would be in a consensus sequence. The vertical scale is in bits, with a maximum of 2 bits possible at each position. Note that sequence logos are an average picture of a set of binding sites (which is why logos can have several letters in each stack) while sequence walkers are the individuals that make up that average (which is why walkers have only one letter per position). More examples are in the Sequence Logo Gallery in the discussion of binding site symmetry.

Related Websites

Sequence Logo Gallery	http://www.lecb.ncifcrf.gov/~toms/sequencelogo.html
How to make sequence logos	http://www.lecb.ncifcrf.gov/~toms/logoprograms.html

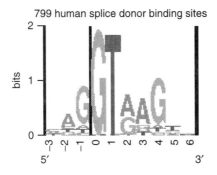

799 human splice donor binding sites

Figure Sequence logo of human splice donor sites (Schneider and Stephens, 1990).

Web-based server	http://weblogo.berkeley.edu
Structure logos	http://www.cbs.dtu.dk/gorodkin/appl/slogo.html

Further Reading

Blom N, et al. (1999). Sequence and structure-based prediction of eukaryotic protein phosphorylation sites. *J. Mol. Biol.* 294: 1351–1362.

Gorodkin J, et al. (1997). Displaying the information contents of structural RNA alignments: The structure logos. *Comput. Appl. Biosci.* 13: 583–586.

Schneider TD (1996). Reading of DNA sequence logos: Prediction of major groove binding by information theory. *Methods Enzymol.* 274: 445–455. http://www.lecb.ncifcrf.gov/~toms/paper/oxyr/.

Schneider TD, Stephens RM (1990). Sequence logos: A new way to display consensus sequences. *Nucleic Acids Res.* 18: 6097–6100. http://www.lecb.ncifcrf.gov/~toms/paper/logopaper/.

See also Binding Site, Binding Site Symmetry, Consensus Sequence, Conservation, Information, Sequence Walker.

SEQUENCE MOTIF. *SEE* MOTIF.

SEQUENCE MOTIFS—PREDICTION AND MODELING (SEARCH BY SIGNAL)

Roderic Guigó

Approach to identifying the function of a sequence by searching for patterns within it.

Given an alphabet (e.g., that of nucleic acids A, C, G, T), a motif is an object denoting a set of sequences on this alphabet, either in a deterministic or probabilistic way. Given a sequence S and a (deterministic) motif m, we will say that the motif m occurs in S if any of the sequences denoted by m occurs in S. We will use here interchangeably the terms motif, pattern, signal, etc., although these terms may be used with different meaning. The simplest motif is just one string or sequence in the alphabet, often a so-called exact word. Exact words may encapsulate biological functions, often when in the appropriate context. For instance, the sequence TAA denotes, under the appropriate circumstances, a translation stop codon.

Often, however, biological functions are carried out by related, but not identical, sequences. Usually these sequences can be aligned. For instance the sequences

```
CTAAAAATAA
TTAAAAATAA
TTTAAAATAA
CTATAAATAA
TTATAAATAA
CTTAAAATAG
TTTAAAATAG
```

are all known to bind the *MEF2* (myocyte enhancer factor 2) transcription factor (Fickett, 1996; Yu et al., 1992). Sets of functionally related aligned sequences can be described by consensus sequences. The simplest form of a consensus sequence is obtained by picking the most frequent base at each position in the set of aligned sequences. More information on the underlying sequences can be captured by extending the alphabet with additional symbols that allow alternative possibilities that may occur at a given position to be denoted. For instance, using the IUB (International Union of Biochemistry and Molecular Biology) nucleotide codes, the sequences above could be represented by the motif YTWWAAATAR, where Y = [CT], W = [AT], and R = [AG].

Regular expressions extend the alphabet further. For instance, the pattern C..?[STA]..C[STA][^P]C denotes the following amino acid sequences: cysteine (C); any amino acid (.); any amino acid that may or may not be present (.?); serine, threonine, or alanine ([STA]); any amino acid (.); cysteine (C); serine, threonine, or alanine ([STA]); any amino acid but proline ([^P]); and cysteine (C). This pattern is the iron–sulfur binding region signature of the 2Fe–2 S ferredoxin, from the PROSITE database (Falquet et al., 2002). In its early days PROSITE was a database mostly of regular expressions.

Information about the relative occurrence of each symbol at each position is lost in the motifs above. For instance in the alignment of *MEF2* sites, both A and G are possible at the last position, but A appears in five sequences, while G appears only in two. This may reflect some underlying biological feature, e.g., that the affinity of the binding is increased when adenosine instead of guanine appears at this position. We can capture explicitly this information by providing the relative frequency or probability of each symbol at each position in the alignment. These probabilities conform to the so-called position weight matrices (PWMs) or position-specific scoring matrices (PSSMs). Below is the PSSM derived from a set of aligned canonical vertebrate donor sites:

	-3	-2	-1	+1	+2	+3	+4	+5	+6
A	35.1	59.6	8.7	0.0	0.0	50.7	72.1	7.0	15.8
C	34.8	13.3	2.7	0.0	0.0	2.8	7.6	4.7	17.2
G	18.5	13.2	80.9	100.0	0.0	43.9	12.2	83.1	18.8
T	11.6	13.9	7.7	0.0	100.0	2.5	8.1	5.2	48.3
	C/A	A	G	G	T	A	A	G	T

This matrix, M, allows us to compute the probability of a given query under the hypothesis that it is a donor site. Indeed, the probability of sequence $s = s_1, \ldots, s_l$ if s is a donor site is $P(s) = \prod_{i=1}^{l} M_{s_i i}$, where M_{ij} is the probability of symbol i at position j. (In the case of the example, length $l = 9$). We can compute as well the background probability $Q(s)$ of the sequence s. The logarithm of the ratio of these two probabilities, $R(s) = \log[P(s)/Q(s)]$, is often used to score the query sequence. If the score is positive, the sequence occurs in donor sites more often than at random, while if the score is negative, the sequence occurs less often in donor sites than randomly expected. The score $R(s)$ can be easily computed if the individual coefficients of the matrix are themselves converted into log-likelihood ratios: $L_{ij} = \log(M_{ij}/f_i)$, where f_i is the background probability of nucleotide i. Assuming equiprobability as the background probability of individual nucleotides, the frequency matrix above becomes

```
        -3     -2     -1     +1     +2     +3     +4     +5     +6
A      0.34   0.87  -1.05  -inf   -inf   0.71   1.06  -1.27  -0.46
C      0.33  -0.63  -2.22  -inf   -inf  -2.17  -1.19  -1.68  -0.38
G     -0.30  -0.64   1.17   1.39  -inf   0.56  -0.72   1.20  -0.29
T     -0.77  -0.59  -1.18  -inf   1.39  -2.29  -1.13  -1.58   0.66
       C/A     A      G      G      T      A      A      G      T
```

Now, the score $R(s)$ is simply $\sum_{i=1}^{l} L_{s_i i}$.

While a regular expression m denotes a subset of all sequences in the range of lengths l_1, l_2, a PSSM m is indeed a probability distribution over the set of all sequences of length l. Thus, while finding matches to regular expression m in sequence S is finding all subsequences of S of length $l_1 \leq l \leq l_2$ that belong to the set denoted by m, finding matches to the PSSM m is scoring each subsequence of S of length l according to matrix m—in practice, sliding a window of length l along the sequence. Usually, however, only subsequences scoring over a predefined threshold are considered matches to the motif described by the PSSM.

Sequence logos are useful graphical representations of PSSMs, while sequence walkers are useful to visually scan a sequence with a PSSM.

In PSSMs, adjacent positions are assumed to be independent. This is often unrealistic, e.g., in the case of the donor sites above. The frequencies of the donor site PSSM reflect the complementarity between the precursor RNA molecule at the donor site and the 5′ end of the U1 small nuclear (snRNA). The stability of this interaction is affected by the stacking energies, which depend on nearest-neighbor arrangements along the nucleic acid sequences. Positions along the donor site sequence, thus, do not appear to be independent. If this is the case, the donor site motif would be better described by the conditional probability of each nucleotide at each position, depending on the nucleotide at the precedent position. Below are these conditional probabilities on the tree exon positions for the vertebrate donor sites above.

	position -3				position -2				position -1			
	A	C	G	T	A	C	G	T	A	C	G	T
A	29.2	31.9	25.5	13.4	62.4	9.5	15.2	12.9	7.0	1.7	86.2	5.1
C	48.6	32.5	6.2	12.7	69.2	11.6	6.4	12.8	19.1	7.1	55.2	18.5
G	38.8	36.2	17.7	7.3	62.6	15.8	12.3	9.3	12.3	2.4	79.1	6.2
T	16.4	41.3	29.5	12.9	17.7	25.6	29.5	27.2	2.9	3.3	84.4	9.4
	35.1	34.8	18.5	11.6	59.6	13.3	13.2	13.9	8.7	2.7	80.9	7.7

As it is possible to see, although the overall probability of G at position -3 is 0.18, this probability is 0.30 if the nucleotide at -4 is T, while it is only 0.06 if this nucleotide is C. These conditional (or transition) probabilities conform to a first-order nonhomogeneous Markov model. These models are called nonhomogeneous because the transition probabilities may change along the sequence. Higher order Markov models can also be used.

PSSMs and nonhomogeneous Markov models are particularly useful to model protein–protein and protein–nucleic acid interactions (Berg and von Hippel, 1987; Schneider, 1997) and thus have been used to describe a variety of elements: splice sites, translation initiation codons, promoter elements, amino acid motifs, etc.

One limitation of Markov models is that they can model only interactions between adjacent positions. In some cases, however, strong dependencies appear to exist between nonadjacent positions. Burge and Karlin (1997) introduced maximal dependence decomposition, a decision tree–like method to model distant dependencies in donor sites. These dependencies can be interpreted in terms of the thermodynamics of RNA duplex formation between U1 snRNA and the 5' splice site region of the pre-mRNA. A number of methods have been further developed to infer all relevant dependencies between positions in a set of aligned sequences (see, e.g., Agarwal and Bafna, 1998; Cai et al., 2000).

More complex patterns, e.g., including gaps or heterogeneous domains, can be modeled with more complex structures.

Related Websites
Here we list a number of severs for splice site prediction. Splice site motifs have been modeled using a wide variety of methods and they are, thus, a good case study. Interested readers can obtain information on the particular methods at the addresses below.

GENESPLICER	http://www.tigr.org/tdb/GeneSplicer/gene_spl.html
SPLICEPREDICTOR	http://bioinformatics.iastate.edu/cgi-bin/sp.cgi
BRAIN	ftp://ftp.ebui.ac.uk/software/dos/
NETPLANTGENE	http://www.cbs.dtu.dk/services/NetPGene/
NETGENE	http://www.cbs.dtu.dk/services/NetGene2/
SPL	http://www.softberry.co'm/ berry.phtml?topic=gfind&prg=SPL
NNSPLICE	http://www.fruitfly.org/seq_tools/splice.html
GENIO/splice	http://genio.informatik.uni-tuttgart.de/GENIO/splice/
SPLICEVIEW	l25.itba.mi.cnr.it/~webgene/wwwspliceview.html

Further Reading
Agarwal P, Bafna V (1998). Detecting non-adjoining correlations within signals in DNA. In *RECOMB 98: Proceedings of the Second Annual International Conference on Computational Molecular Biology*. ACM Press, New York, pp. 2–8.

Berg O, von Hippel P (1987). Selection of DNA binding sites by regulatory proteins, statistical-mechanical theory and application to operators and promoters. *J. Mol. Biol.* 193: 723–750.

Bucher P, et al. (1994). A flexible search technique based on generalized profiles. *Comput. Chem.* 20: 3–24.

Burge CB (1998) Modeling dependencies in pre-mRNA splicing signals. In *Computational Methods in Molecular Biology*, S Salzberg, D Searls, S Kasif, Eds. Elsevier Science, Amsterdam, pp. 127–163.

Burge CB, Karlin S (1997). Prediction of complete gene structures in human genomic DNA. *J. Mol. Biol.* 268: 78–94.

Cai D, et al. (2000). Modeling splice sites with bayes networks. *Bioinformatics*, 16: 152–158.

Falquet L et al. (2002). The prosite database, its status in 2002. *Nucleic Acids Res.* 30: 235–238.

Fickett CW (1996). Quantitative discrimination of MEF2 situations. *Mol. Cell. Bio.* 16: 437–441.

Schneider T (1997). Information content of individual genetic sequences. *J. Theor. Biol.* 189: 427–441.

Yu Y, et al. (1992). Human myocyte-specific enhancer factor 2 comprises a group of tissue-restricted MADs box transcription factors. *Genes Dev.* 6: 1783–1798.

See also Donor Splice Site, Hidden Markov Models, Motif Searching, Neural Network, Sequence Pattern, Sequence Logo, Sequence Walker.

SEQUENCE OF PROTEINS
Roman A. Laskowski

The order of neighboring amino acids in a protein, listed from the N- to the C-terminus. This is the primary structure of the protein. It is most simply represented by a string of letters, such as NEGDAAKGEFN, where each letter codes for one of the 20 amino acids. The sequence uniquely determines how the protein folds into its final three-dimensional conformation.

The sequence is relatively straightforward to determine experimentally, either directly using automated techniques or by derivation via the genetic code, from regions of DNA known to encode for the protein. Protein sequence databases nowadays contain the sequences of many hundreds of thousands of proteins from a wide variety of organisms.

Many databases exist that contain protein sequence data, such as SWISS-PROT and NCBI.

Related Websites

SWISS-PROT	http://ca.expasy.org/sprot/
NCBI	http://www.ncbi.nlm.nih.gov/

See also Amino Acid, Conformation, C-Terminus, N-Terminus.

SEQUENCE PATTERN
Thomas D. Schneider

A sequence pattern is defined by the nucleotide sequences of a set of aligned binding sites or by a common protein structure.

In contrast, consensus sequences, sequence logos, and sequence walkers are only models of the patterns found experimentally or in nature. Models do not capture everything in nature. For example, there might be correlations between two different positions in a binding site (Gorodkin et al., 1997; Stephens and Schneider,

1992). A more sophisticated model might capture these but still not capture three-way correlations. It is impossible to make the more detailed model if there is not enough data.

Further Reading

Gorodkin J, et al (1997). Displaying the information contents of structural RNA alignments: The structure logos. *Comput. Appl. Biosci.* 13: 583–586.

Durves D, et al. (2002). Why we see what we do. *Amer. Sci.* 90; 236–243. http://americanscientist.org/articles/02artictes/purves.html

Schneider TD (2002) Consensus sequence zen. *Appl. Bioinformatics* 1: 111–119. http://www.lecb.ncifcrf.gov/~toms/paper/zen.

Stephens RM, Schneider TD (1992). Features of spliceosome evolution and function inferred from an analysis of the information at human splice sites. *J. Mol. Biol.* 228: 1124–1136. http://www.lecb.ncifcrf.gov/~toms/paper/splice/.

See also Binding Site, Consensus Sequence, Sequence Logo, Sequence Walker.

SEQUENCE RETRIEVAL SYSTEM. *SEE* SRS.

SEQUENCE SIMILARITY
Jaap Heringa

Numerous studies into protein sequence relationships evaluate sequence alignments using a simple binary scheme of matched positions being identical or nonidentical. Sequence identity is normally expressed as the percentage of identical residues found in a given alignment, where normalization can be performed using the length of the alignment or the shorter sequence. The scheme is simple and does not rely on an amino acid exchange matrix. However, if two proteins are said to share a given percentage in sequence identity, this is based on a sequence alignment which will have been almost always constructed using an amino acid exchange matrix and gap penalty values, so that sequence identity cannot be regarded as independent from sequence similarity. Using sequence identity as a measure, Sander and Schneider (1991) estimated that if two protein sequences are longer than 80 residues, they could relatively safely be assumed to be homologous whenever their sequence identity is 25% or more. Despite its popularity, using sequence identity percentages is not optimal in homology searches (Abagyan and Batalov, 1997), and as a result, no major sequence analysis method employs sequence identity scores in deriving statistical significance estimates.

Sequence alignment methods are essentially pattern-search techniques, leading to an alignment with a similarity score even in the absence of any biological relationship. In some database search engines, such as PSI BLAST (Altschul et al., 1997), sequences are scanned for the presence of so-called low-complexity regions (Wootton and Federhen, 1996), which are then excluded from alignment. Although similarity scores of unrelated sequences are essentially random, they can behave like "real" scores and, e.g., like the latter are correlated with the length of the sequences compared. Particularly in the context of database searching, it is therefore important to know what scores can be expected by chance and how scores that deviate from random expectation should be assessed. If within a rigid statistical framework a sequence similarity is deemed statistically significant, this provides confidence in deducing that the sequences involved are in fact biologically related. As a result of the complexities of protein sequence evolution and distant relationships observed in nature, any statistical scheme will invariably lead to situations where a sequence is assessed as unrelated while it is in fact homologous (false negative), or the inverse, where a sequence is deemed homologous while it is in fact biologically unrelated (false positive). The derivation of a general statistical framework for evaluating the significance of sequence similarity scores has been a major task. However, a rigid framework has not been established for global alignment and has only partly been completed for local alignment.

Further Reading

Abagyan RA, Batalov S (1997). Do aligned sequences share the same fold? *J. Mol. Biol.* 273: 355–368.

Altschul SF, et al. (1997) Gapped BLAST and PSI-BLAST: A new generation of protein database search programs. *Nucleic Acids Res.* 25: 3389–3402.

Sander C, Schneider R (1991). Database of homology derived protein structures and the structural meaning of sequence alignment. *Proteins Struct. Funct. Evol.* 9: 56–68.

Wootton JC, Federhen S (1993). Statistics of local complexity in amino acid sequences and sequence databases. *Comput. Chem.* 17: 149–163.

See also Conservation.

SEQUENCE SIMILARITY–BASED GENE PREDICTION. *SEE* GENE PREDICTION, HOMOLOGY BASED.

SEQUENCE SIMILARITY SEARCH
Laszlo Patthy

Related genes, proteins, and protein domains usually retain a significant degree of sequence similarity that can be measured by comparing their aligned sequences. Sequence similarity, however, does not necessarily reflect homology since it may also result from convergence or may simply occur by chance.

Various types of sequence similarity searches may be performed with a query sequence to identify genes or proteins that share sequence similarity with a significance that may imply common ancestry. The search attempts to align the query sequence with all sequences in the database and calculates the similarity scores. The search provides a list of database sequences that can be aligned with the query sequence, ranked in the order of decreasing similarity scores.

Sequence similarity searches may be performed with a single query sequence, with multiple alignments, patterns, motifs, or position-specific scoring matrices.

Related Websites

FASTA	http://fasta.bioch.virginia.edu/fasta/
MAST	http://meme.sdsc.edu/meme/website/mast.html
BLAST	http://www.ncbi.nlm.nih.gov/BLAST/
PROSITE	http://www.expasy.ch/prosite
INTERPRO	http://www.ebi.ac.uk/interpro

CDD/IMPALA	http://www.ncbi.nlm.nih.gov/Structure/cdd/cdd.shtml
PSI-BLAST	http://www.ncbi.nlm.nih.gov/BLAST/
PROFILESEARCH	ftp.sdsc.edu/pub/sdsc/biology
PFAM	http://www.sanger.ac.uk/Pfam
FASTA	http://www.ebi.ac.uk/fasta33/
SW	http://www.ebi.ac.uk/bic_sw/

See also Homology.

SEQUENCE SIMPLICITY. *SEE* SEQUENCE COMPLEXITY.

SEQUENCE-TAGGED SITE (STS)
John M. Hancock and Rodger Staden

A unique sequence-defined landmark in the genome which can be detected by a specific PCR reaction.

An STS is a short (200–500-bp) DNA sequence that occurs uniquely in a genome and whose location and base sequence are known. STSs are useful for localizing and orienting mapping and sequence data and serve as landmarks on the physical map of a genome.

SEQUENCE WALKER
Thomas D. Schneider

Measure to quanify the evolutionary watchness of sequences. A sequence walker is a graphic representation of a single possible binding site, with the height of letters indicating how bases match the individual information weight matrix at each position. Bases that have positive values in the weight matrix are shown right-side up; bases that have negative values are shown upside down and below the "horizon." As in a sequence logo, the vertical scale is in bits; the maximum is 2 bits and the minimum is negative infinity. Bases that do not appear in the set of aligned sequences are shown negatively and in a black box. Bases that have negative values lower than can fit in the space available have a purple box. The zero coordinate is inside a rectangle which (in this case) runs from -3 to $+2$ bits in height. If the background of the rectangle is light green, the sequence has been evaluated as a binding site, while if it is pink it is not a binding site. The accompanying figure shows sequence walkers for human acceptor splice sites at intron 3 of the iduronidase synthetase gene (IDS, *L35485*). An A-to-G mutation decreases the information content of the normal site while simultaneously increasing the information content of a cryptic site, leading to a genetic disease (Rogan et al., 1998).

Related Websites

Ri	http://www.lecb.ncifcrf.gov/~toms/paper/ri/
Walker	http://www.lecb.ncifcrf.gov/~toms/paper/walker/
Web page on walkers	http://www.lecb.ncifcrf.gov/~toms/walker/
Try sequence walkers yourself with the Delila Server:	http://www.lecb.ncifcrf.gov/~toms/delilaserver.html

Further Reading

Allikmets R, et al. (1998). Organization of the ABCR gene: Analysis of promoter and splice junction sequences. *Gene* 215: 111–122. http://www.lecb.ncifcrf.gov/~toms/paper/abcr/.

Arnould I, et al. (2001). Identifying and characterizing a five-gene cluster of ATP-binding cassette transporters mapping to human chromosome 17q24: A new subgroup within the ABCA subfamily. *GeneScreen* 1: 157–164.

Emmert S, et al. (2001). The human XPG gene: Gene architecture, alternative splicing and single nucleotide polymorphisms. *Nucleic Acids Res.* 29: 1443–1452.

Hengen PN, et al. (1997). Information analysis of Fis binding sites. *Nucleic Acids Res.* 25: 4994–5002. http://www.lecb.ncifcrf.gov/~toms/paper/fisinfo/.

Kahn SG, et al. (1998). Xeroderma pigmentosum group C splice mutation associated with mutism and hypoglycinemia—a new syndrome? *J. Invest. Dermatol.* 111: 791–796.

Rogan PK, et al. (1998). Information analysis of human splice site mutations. *Hum. Mut.* 12: 153–171. http://www.lecb.ncifcrf.gov/~toms/paper/rfs/.

Schneider TD (1997a). Information content of individual genetic sequences. *J. Theor. Biol.* 189: 427–441. http://www.lecb.ncifcrf.gov/~toms/paper/ri/.

Schneider TD (1997b). Sequence walkers: A graphical method to display how binding proteins interact with DNA or RNA sequences. *Nucleic Acids Res.* 25: 4408–4415. http://www.lecb.ncifcrf.gov/~toms/paper/walker/.

Schneider TD (1999). Measuring molecular information. *J. Theor. Biol.* 201: 87–92. http://www.lecb.ncifcrf.gov/~toms/paper/ridebate/.

Schneider TD, Rogan PK (1999). Computational analysis of nucleic acid information defines binding sites, United States Patent 5867402. http://www.lecb.ncifcrf.gov/~toms/paper/flexrbs/.

Shultzaberger RK, Schneider TD (1999). Using sequence logos and information analysis of Lrp DNA binding sites to investigate discrepancies between natural selection and SELEX. *Nucleic Acids Res.* 27: 882–887. http://www.lecb.ncifcrf.gov/~toms/paper/lrp/.

Shultzaberger RK, et al. (2001). Anatomy of *Escherichia coli* ribosome binding sites. *J. Mol. Biol.* 313: 215–228.

Stephens RM, Schneider TD (1992). Features of spliceosome evolution and function inferred from an analysis of the information at human splice sites. *J. Mol. Biol.* 228: 1124–1136. http://www.lecb.ncifcrf.gov/~toms/paper/splice/.

Svojanovsky SR, et al. (2000). Redundant designations of BRCA1 intron 11 splicing mutation; c. 4216-2A>G; IVS11-2A>G; L78833, 37698, A>G. *Hum. Mut.* 16: 264. http://www3.interscience.wiley.com/cgi-bin/abstract/73001161/START.

Wood TI, et al. (1999). Interdependence of the position and orientation of SoxS binding sites in the transcriptional activation of the class I subset of *Escherichia coli* superoxide-inducible promoters. *Mol. Microbiol.* 34: 414–430.

Zheng M, et al. (1999). OxyR and SoxRS regulation of fur. *J. Bacteriol.* 181: 4639–4643. http://www.lecb.ncifcrf.gov/~toms/paper/oxyrfur/.

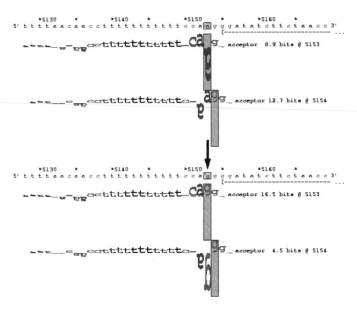

Figure Sequence walkers for human acceptor splice sites at intron 3 of the iduronidase synthetase gene (IDS, *L35485*).

Zheng M, et al. (2001). Computation-directed identification of OxyR-DNA binding sites in *Escherichia coli. J. Bacteriol.* 183: 4571–4579.

See also Acceptor Splice Site, Individual Information, Sequence Logo, Weight Matrix, Zero Coordinate.

SERIAL ANALYSIS OF GENE EXPRESSION. *SEE* SAGE.

SERINE
Jeremy Baum

Serine is a small polar amino acid with side chain —CH$_2$OH found in proteins. In sequences, written as Ser or S.

See also Amino Acid, Polar.

SEX CHROMOSOME (X CHROMOSOME, Y CHROMOSOME)
Katheleen Gardiner

A member of a chromosome pair that is dissimilar in the heterogametic sex.

In mammals the heterogametic sex is the male, with one X and one Y chromosome; the homogametic sex is the female, with two X chromosomes. Unlike autosome pairs, the sex chromosomes do not have identical gene contents. For example, the human X chromosome is larger than the Y chromosome and contains unique, essential genes. This leads to the possibility of sex-linked (or X-linked, in human) inheritance of some traits.

Further Reading

Miller OJ, Therman E (2000). *Human Chromosomes*. Springer, Vienna.

Scriver CR, Beaudet AL, Sly WS, Valle D, Eds. (1995). *The Metabolic and Molecular Basis of Inherited Disease*, McGraw-Hill, New York.

Skaletsky H, et al. (2003). The male-specific region of the human Y chromosome is a mosaic of discrete sequence classes. *Nature* 423: 825–837.

Wagner RP, Maguire MP, Stallings RL (1993). *Chromosomes—A Synthesis*. Wiley-Liss, New York.

SEXUAL SELECTION

A. Rus Hoelzel

Sexual selection is a special case of natural selection, where the characteristics being selected have to do with mating success.

Some traits best explained as a consequence of female mate choice, such as the male peacock's famous tail, may appear to be costly in terms of natural selection if this characteristic made the male more susceptible to predation or put him at an energetic disadvantage. However, it is lifetime reproductive success that matters in terms of fitness, and longevity is only important if this contributes to greater reproduction.

Sexual selection can be divided into two types. In the peacock example, selection is for a display character that attracts females. This could be physical (coloration, plumage, etc.) or behavioral (vocal displays such as bird song, visual displays such as the dance of the bird of paradise, etc.). For this type of sexual selection the evolutionary process is dependent on mate choice. The other type of sexual selection depends instead on competition among males for access to females. In this case competition over access to mates leads to the evolution of secondary sexual characteristics that facilitate domination over competitors. For example, in various species of pinnipeds and cervids there is strong sexual dimorphism between males and females such that the males are larger. Males may also evolve physical attributes that facilitate their domination over other males through visual or vocal displays.

Further Reading

Ridley M (1996). *Evolution*. Blackwell Science, Cambridge, MA, pp. 265–276.

See also Evolution, Natural Selection.

SHANNON ENTROPY (SHANNON UNCERTAINTY)

Thomas D. Schneider

Information theory measure of uncertainty.

The story goes that Shannon did not know what to call his measure and so asked the famous mathematician von Neumman. Von Neumann said he should call it the entropy because nobody knows what that is and so Shannon would have the advantage in every debate (the story is paraphrased from Tribus and McIrvine, 1971). This has led to much confusion in the literature because entropy has different units than uncertainty. It is the latter which is usually meant. Recommendation: When making computations from symbols, always use the term *uncertainty*, with recommended units of bits per symbol. If referring to the entropy of a physical system, then use the term *entropy*, which has units of joules per kelvin (energy per temperature). It is also important to note that information is not entropy and that information is not uncertainty!

Related Websites

| Reference | http://www.lecb.ncifcrf.gov/~toms/information.is.not.uncertainty.html |
| Pitfalls | http://www.lecb.ncifcrf.gov/~toms/pitfalls.html |

Further Reading
Tribus M, McIrvine EC (1971). Energy and information. *Sci. Am.* 225: 179–188. (Note: the table of contents in this volume incorrectly lists this as volume 224).

See also Entropy, Uncertainty.

SHANNON SPHERE
Thomas D. Schneider

A sphere in a high-dimensional space which represents either a single message of a communications system (after sphere) or the volume that contains all possible messages (before sphere) could be called a Shannon sphere, in honor of Claude Shannon, who recognized its importance in information theory.

The radius of the smaller after spheres is determined by the ambient thermal noise, while that of the larger before sphere is determined by both the thermal noise and the signal power (signal-to-noise ratio) measured at the receiver. The logarithm of the number of small spheres that can fit into the larger sphere determines the channel capacity (Shannon, 1949). The high-dimensional packing of the spheres is the coding of the system.

There are two ways to understand how the spheres come to be. Consider a digital message consisting of independent voltage pulses. The independent voltage values specify a point in a high-dimensional space since independence is represented by coordinate axes set at right angles to each other. Thus three voltage pulses correspond to a point in a three-dimensional space and 100 pulses correspond to a point in a 100-dimensional space. The first, "non-Cartesian" way to understand the spheres is to note that thermal noise interferes with the initial message during transmission of the information such that the received point is dislocated from the initial point. Since noisy distortion can be in any direction, the set of all possible dislocations is a sphere. The second, "Cartesian" method is to note that the sum of many small dislocations to each pulse, caused by thermal noise, gives a Gaussian distribution at the receiver. The probability that a received pulse is disturbed a distance x from the initial voltage is of the form $p(x) \approx e^{-x^2}$. Disturbance of a second pulse will have the same form, $p(y) \approx e^{-y^2}$. Since these are independent, the probability of both distortions is multiplied, $p(x, y) = p(x)p(y)$. Combining equations yields

$$p(x, y) \approx e^{-(x^2+y^2)} = e^{-r^2}$$

where r is the radial distance. If $p(x, y)$ is a constant, the locus of all points enscribed by r is a circle. With more pulses the same argument holds, giving spheres in high-dimensional space. Shannon used this construction in his channel capacity theorem.

For a molecular machine containing n atoms there can be as many as $3n - 6$ independent components (degrees of freedom) so there can be $3n - 6$ dimensions. The velocity of these components corresponds to the voltage in a communication system and they are disturbed by thermal noise. Thus the state of a molecular machine can also be described by a sphere in a high-dimensional velocity space.

Further Reading

Schneider TD (1991). Theory of molecular machines. I. Channel capacity of molecular machines. *J. Theor. Biol.* 148: 83–123. http://www.lecb.ncifcrf.gov/~toms/paper/ccmm/

Schneider TD (1994). Sequence logos, machine/channel capacity, Maxwell's demon, and molecular computers: A review of the theory of molecular machines. *Nanotechnology* 5: 1–18. http://www.lecb.ncifcrf.gov/~toms/paper/nano2/.

Shannon E (1949). Communication in the presence of sense. *Proc. IRE* 37: 10–21.

See also Before State, Channel Capacity, Coding, Gumball Machine, Information Theory, Molecular Machine, Message, Thermal Noise, Signal-to-Noise Ratio.

SHANNON UNCERTAINTY. *SEE* SHANNON ENTROPY.

SHORT-PERIOD INTERSPERSION, SHORT-TERM INTERSPERSION. *SEE* INTERSPERSED SEQUENCE.

SHUFFLE TEST
David T. Jones

A method for estimating the significance of matches in fold recognition, most commonly for threading methods.

One major problem in fold recognition is how to determine the statistical significance of matches. In the absence of any theoretical methods for calculating the significance of matches in a fold recognition search, empirical methods are typically used. The most commonly used approach is sequence shuffling, which is often used to assess the significance of global sequence alignments. The test is fairly straightforward to implement. The score for a test sequence threaded onto a particular template is first calculated. Then the test sequence is shuffled; i.e., the order of the amino acids is randomized. This shuffling ensures that random sequences are generated with identical amino acid composition to the original protein. This shuffled sequence is then also threaded onto the given template, and the energy for the shuffled sequence is compared to the energy for the unshuffled sequence. In practice, a large number of shuffled threadings are carried out (typically 100–1000) and the mean and standard deviation of the energies calculated. The unshuffled threading score can then be transformed into a Z score by subtracting the mean and then dividing by the standard deviation of the shuffled scores. As a rule of thumb, Z scores of 4 or more correspond to significant matches.

See also Threading.

SIDE CHAIN
Roman A. Laskowski

In a peptide or protein chain, all atoms other than the backbone (or main-chain) atoms are termed side-chain atoms. For most of the amino acids, the side-chain atoms spring off from the backbone Cα, the exception being glycine, which has no side chain (other than a single hydrogen atom), and proline, whose side chain links back onto its main-chain nitrogen.

In proteins, interactions between side chains help determine the structure and stability of the folded state. Hydrophobic–hydrophobic contacts are instrumental in forming the folded protein's core. Side chains on the surface of the protein play a crucial role in recognition of and interaction with other molecules e.g., DNA, other proteins, metabolites, metals) and for performing the protein's biological function (e.g., the side chains that carry out the biochemical reactions in enzymes—one example being the catalytic triad).

Related Website

Atlas of Protein Side-Chain Interactions	http://www.biochem.ucl.ac.uk/bsm/sidechains

Further Reading
Singh J, Thornton J M (1992). *Atlas of Protein Side-Chain Interactions*, Vols. I and II, IRL Press, Oxford.

See also Backbone, Main Chain.

SIDE-CHAIN PREDICTION
Roland L. Dunbrack

A step in comparative modeling of protein structures in which side chains for the residues of the target sequence are built onto the backbone of the template or parent structure. For residues that are conserved in the target–parent alignment, the Cartesian coordinates of the parent side chain can be preserved in the model. But residues that are different in the two sequences require new side-chain coordinates in the model.

Most side-chain prediction methods are based on choosing conformations from rotamer libraries and an optimization scheme to choose conformations with low energy, as defined by some potential energy function. Side chains are usually built with standard bond lengths and bond angles, and dihedral angles are obtained from the rotamer library used. However, Cartesian coordinate rotamer libraries can also be used to account for side-chain flexibility. Alternatively, flexibility around rotamer dihedral angles can improve results in some cases.

Related Websites

SCWRL	http://www.fccc.edu/research/labs/dunbrack/scwrl/
SCAP	http://trantor.bioc.columbia.edu/~xiang/jackal/

Further Reading
Bower MJ, et al. (1997). Prediction of protein side-chain rotamers from a backbone-dependent rotamer library: A new homology modeling tool. *J. Mol. Biol.* 267: 1268–1282.

Holm L, Sander C (1991). Database algorithm for generating protein backbone and sidechain coordinates from a Cα trace: Application to model building and detection of coordinate errors. *J. Mol. Biol.* 218: 183–194.

Liang S, Grishin NV (2002). Side-chain modeling with an optimized scoring function. *Protein Sci.* 11: 322–331.

Mendes J, et al. (2001). Incorporating knowledge-based biases into an energy-based side-chain modeling method: Application to comparative modeling of protein structure. *Biopolymers* 59: 72–86.

See also Empirical Potential Energy Function, Rotamer Library, Target, Template.

SIGNAL-TO-NOISE RATIO
Thomas D. Schneider

The signal-to-noise ratio is the ratio between the received signal power and the noise at the receiver of a communications system. In molecular biology, the equivalent is the energy dissipated divided by the thermal noise.

See also Channel Capacity, Noise.

SIGNATURE. *SEE* FINGERPRINT.

SILENT MUTATION. *SEE* SYNONYMOUS MUTATION.

SIMPLE (SIMPLE34)
John M. Hancock

Program for detecting simple sequence regions in DNA and protein sequences.

SIMPLE was originally developed to analyze DNA sequences for potential substrates for DNA slippage that were not obviously tandemly repeated. The program detects both tandem repeats and these less obvious "cryptically simple" sequences by a heuristic algorithm that counts the frequency of repeats of the central motif of a 65-bp window and compares it to frequencies observed in random simulated copies of that sequence with the same length and base or (latterly) dinucleotide composition. It thus attempts to distinguish repetition that is beyond the chance expectation for the composition of a given test sequence. More recent versions of the program identify repeated motifs and apply the algorithm to protein sequences.

Related Website

| SIMPLE | http://www.biochem.ucl.ac.uk/bsm/SIMPLE/index.html |

Further Reading

Albà MM, et al. (2002). Detecting cryptically simple protein sequences using the SIMPLE algorithm. *Bioinformatics* 18: 672–688.

Hancock JM, Armstrong JS (1994). SIMPLE34: An improved and enhanced implementation for VAX and SUN computers of the SIMPLE algorithm for analysis of clustered repetitive motifs in nucleotide sequences. *Comput. Appl. Biosci.* 10: 67–70.

Tautz D, et al. (1986). Cryptic simplicity in DNA is a major source of genetic variation. *Nature* 322: 652–656.

See also Low-Complexity Region, Simple DNA Sequence.

SIMPLE34. *SEE* SIMPLE.

SIMPLE DNA SEQUENCE (SIMPLE REPEAT, SIMPLE SEQUENCE REPEAT)
John M. Hancock and Katheleen Gardiner

DNA sequence composed of repeated identical or highly similar short motifs.

Simple sequences may be polymorphic in length. The term encompasses both minisatellites and microsatellites and may also include nontandem arrangements of motifs. Sequences containing overrepresentations of short motifs that are not tandemly arranged are known as cryptically simple sequences. Simple repeats are widely dispersed in the human genome and account for at least 0.5% of genomic DNA.

Further Reading
Cooper DN (1999). *Human Gene Evolution*. Bios Scientific Publishers, Oxford, UK, pp. 265–285.
Ridley M (1996). *Evolution*. Blackwell Science, Oxford, UK, pp. 265–276.
Rowold DJ, Herrera RJ (2000). Alu elements and the human genome. *Genetics* 108: 57–72.
Tautz D, Trick M, Dover GA (1986). Cryptic simplicity in DNA is a major source of genetic variation. *Nature*. 322: 652–656.

See also Microsatellite, Minisatellite, SIMPLE.

SIMPLE REPEAT. *SEE* SIMPLE DNA SEQUENCE.

SIMPLE SEQUENCE REPEAT. *SEE* SIMPLE DNA SEQUENCE.

SIMULATED ANNEALING
Roland L. Dunbrack

A Monte Carlo simulation used to find the global minimum energy of a system, in which the temperature is slowly decreased in steps, with the system equilibrated at each temperature step. The name is derived from the analogous process of heating and slowly cooling a substance to produce a crystal. If the cooling is done too quickly, the substance will freeze into a disordered glasslike state. But if cooling is slow, the substance will crystallize.

Further Reading
Kirkpatrick S et al. (1983). Optimization by simulated annealing. *Science* 220: 671–680.

See also Energy Minimization, Monte Carlo Simulations.

SINE (SHORT INTERSPERSED NUCLEAR ELEMENT)
Dov S. Greenbaum

A class of nonviral retrotransposons less than ~500 bp in length.

SINEs constitute approximately 10% of the human genome, the majority of these belonging to the Alu family. These short sequences, while transcribed by RNA *PolIII*, do not encode any functional proteins. Most SINEs have three components: a 5′ tRNA related region, a tRNA unrelated region, and a 3′ AT-rich region.

In addition to the Alu repeat, many of the other SINEs are copies of tRNAs or small nuclear RNAs. For example, there are almost half a million copies of the tRNA-derived, MIR SINE within the human genome.

SINEs have had a profound effect in shaping the human genome and are thought to be causal agents of disease in humans. SINEs are useful in diagnosing and determining ancestry among populations.

Further Reading

Shedlock AM, Okada N (2000). SINE insertions: Powerful tools for molecular systematics. *Bioessays* 22: 148–160.

Weiner AM (2002). SINEs and LINEs: The art of biting the hand that feeds you. *Curr. Opin. Cell Biol.* 14: 343–350.

SINGLE-NUCLEOTIDE POLYMORPHISM (SNP)
Dov S. Greenbaum

Single-nucleotide polymorphisms are DNA variations, i.e., differences in a single base within a genomic sequence within a population. There is 99.9% identity between most humans; the remaining variation can be partly attributed to SNPs. They are evolutionarily stable and occur frequently in both coding and noncoding regions—approximately every 100–300 bases within the human genome. Studies have shown that the occurrence of SNPs is lowest in exons (i.e., coding regions) and higher in repeats, introns, and pseudogenes. The majority of these polymorphisms have a cytosine replaced by a thymine.

These variations do not, for the most part, produce phenotypic differences. SNPs are important in biomedical research as they can be used as markers for disease-associated mutations and they allow population geneticists to follow the evolution of populations. Additionally, these minimal variations within the genome may have significant effects on an organism's response to environment, bacteria, viruses, toxins, disease, and drugs.

To mine the human genome for SNPs, the SNP Consortium, a unique association of both commercial and academic institutions, was set up in 1999. SNPs can be found through random sequencing using consistent and uniform methods.

Related Websites

HAPMAP	http://snp.cshl.org/
NCBI SNP	http://www.ncbi.nlm.nih.gov/SNP/

Further Reading

Gray IC, Campbell DA, Spurr NK (2000). Single nucleotide polymorphisms as tools in human genetics. *Hum. Mol. Genet.* 9: 2403–2408.

Kwok P-Y, Gu Z (1999). Single nucleotide polymorphism libraries: Why and how are we building them? *Mol. Med. Today* 5: 538–543.

SITE. *SEE* CHARACTER.

SITES
Michael P. Cummings

A program for the comparative analysis of polymorphism in DNA sequence data.

Sites is typically applied to sets of similar sequences (i.e., multiple alleles of the same locus within a species). The program provides summary statistics for synonymous sites, replacement sites, noncoding sites, codons, transitions, transversions, GC content, and other features. The program also estimates several population genetic parameters ($\theta[= 4N\mu]$, π, γ, $4Nc$, population size, linkage disequilibrium) and provides hypothesis tests for departures from neutral theory expectations.

The program can be used in an interactive fashion whereby the program provides prompts and lists of choices. Alternatively, the program can be invoked with command line options to perform analyses. Written in American National Standards Institute (ANSI) C, the program is available as source code and as executables for several platforms.

Related Website

Jody Hey software	http://lifesci.rutgers.edu/% 7 Eheylab/Heylab software.htm

Further Reading

Hey J, Wakeley J (1997). A coalescent estimator of the population recombination rate. *Genetics* 145: 833–846.

Wakeley J, Hey J (1997). Estimating ancestral population parameters. *Genetics* 145: 847–855.

SMALL-SAMPLE CORRECTION
Thomas D. Schneider

A correction to the Shannon uncertainty measure to account for the effects of small sample sizes.

Related Website

The Small Sample Correction for Uncertainty and Information Measures	http://www.lecb.ncifcrf.gov/~toms/small.sample.correction.html

Further Reading

Basharin GP (1959). On a statistical estimate for the entropy of a sequence of independent random variables. *Theory Probability Appl.* 4: 333–336.

Miller GA (1955). Note on the bias of information estimates. In *Information Theory in Psychology*, H Quastler, Ed. Free Press, Glencoe, IL, pp. 95–100.

Schneider TD, et al. (1986). Information content of binding sites on nucleotide sequences. *J. Mol. Biol.* 188: 415–431. http://www.lecb.ncifcrf.gov/~toms/paper/schneider1986/

See also Uncertainty.

SMITH–WATERMAN ALGORITHM
Jaap Heringa

A technique for *local alignment* based upon the *dynamic programming* algorithm (Smith and Waterman, 1981).

The Smith–Waterman technique is a minor modification of the global dynamic programming algorithm (Needleman and Wunsch, 1970). The algorithm selects and aligns similar segments of two sequences and needs an amino acid exchange matrix with negative values; if a nonnegative matrix is used, the algorithm will function as a *global alignment* method.

Further Reading

Needleman SB, Wunsch CD (1970). A general method applicable to the search for similarities in the amino acid sequence of two proteins. *J. Mol. Biol.* 48: 443–453.

Smith TF, Waterman MS (1981). Identification of common molecular subsequences. *J. Mol. Biol.* 147: 195–197.

See also Dynamic Programming, Global Alignment, Local Alignment.

SNP. *SEE* SINGLE-NUCLEOTIDE POLYMORPHISM.

SOLVATION FREE ENERGY
Roland L. Dunbrack

The free-energy change involved in moving a molecule from vacuum to solvent. The solute is assumed to be in a fixed conformation, and the free-energy change includes terms for interaction of the solute with the solvent (electrostatic and dispersion forces, including polarization of the solvent and solute) averaged over time and changes in solvent–solvent interactions. The contribution of hydrophobicity to solvation free energy is usually obtained from terms that are proportional to the exposed surface area of solute atoms.

Further Reading

Lazaridis T, Karplus M (1999). Effective energy function for proteins in solution. *Proteins* 35: 133–152.

Wesson L, Eisenberg D (1992). Atomic solvation parameters applied to molecular dynamics of proteins in solution. *Prot. Sci.* 1: 227–235.

SOV
Patrick Aloy

Segment overlap (SOV) is a measure for the evaluation of secondary-structure prediction methods based on secondary-structure segments rather than individual residues.

SOV is a more stringent measure of prediction accuracy than Q3 and discriminates reasonably well between good and mediocre or bad predictions, as it penalizes wrongly predicted segments.

For a single conformational state

$$S_{ov}(i) = 100 * \frac{1}{N(i)} \sum_{S(i)} \left[\frac{minov(S_1, S_2) + \delta(S_1, S_2)}{maxov(S_1, S_2)} * len(S_1) \right]$$

with the normalization value $N(i)$ defined as

$$N(i) = \sum_{S(i)} len(S_1) + \sum_{S'(i)} len(S_1)$$

where S_1 and S_2 correspond to the secondary-structure segments being compared; $S(i)$ is the set of all the overlapping pairs of segments (S_1, S_2); $S'(i)$ is the set of all segments S_1 for which there is no overlapping segment S_2 in state i; $len(S_1)$ is the number of residues in the segment S_1; $min\, ov(S_1, S_2)$ is the length of the overlap between S_1 and S_2 for which both segments have residues in state i; and $maxov(S_1, S_2)$ is the total extent for which either of the segments S_1 or S_2 has a residue in state i. Then $\delta(S_1, S_2)$ is defined as

$$\delta(S_1, S_2) = min\{(max\, ov(S_1, S_2)$$
$$- min\, ov(S_1, S_2)); min\, ov(S_1, S_2); int(len(S_1)/2); int(len(S_2)/2)\}$$

where $min\{x_1; x_2; \dots; x_n\}$ is the minimum of n integers.

For all three states

$$S_{ov} = 100 * \left[\frac{1}{N} \sum_{i \in (H,E,C)} \sum_{S(i)} \frac{min\, ov(S_1, S_2) + \delta(S_1, S_2)}{max\, ov(S_1, S_2)} * len(S_1) \right]$$

where the normalization value N is the sum of $N(i)$ over all three conformational states

$$N = \sum_{i \in \{H,E,C\}} N(i)$$

The normalization ensures that all S_{ov} values are within the 0–100 range and thus can be used as percentage accuracy and compared to other measures, such as Q3.

Related Website

Accuracy estimators	http://predictioncenter.llnl.gov/local/local.html

Further Reading
Zemla A, Venclovas C, Fidelis K, Rost B (1999). SOV, a segment-based measure for protein secondary structure prediction assessment. *Proteins* 34: 220–223.

See also Q Index, Secondary-Structure Prediction of Protein.

SPACE GROUP (CRYSTAL SYMMETRY, NONCRYSTALLOGRAPHIC SYMMETRY)

Liz P. Carpenter

The space group of a crystal is a description of the crystal symmetry relating the molecules in the unit cell (the repeat unit of the crystal).

A crystal consists of a series of stacked parallelepipeds, called unit cells, which are related to each other by simple translations parallel to the side of the unit cell and of the same dimensions as the unit-cell edges. Within these unit cells there can be several molecules related by strict symmetry operators. A symmetry operator is an operation which, when applied a sufficient number of times, will place the asymmetric unit on top of itself or on top of an asymmetric unit in the same place and orientation in an adjacent unit cell. The volume to which these operators are applied to construct the unit cell is called the asymmetric unit.

Three types of symmetry operators occur are in macromolecular crystals: rotations, screw axes, and centering. Rotations involve rotating the asymmetric unit about an axis by 60°, 90°, 120°, or 180°. These are called two-, three-, four-, or sixfold rotations since repetition of these operators two, three, four, or six times places the asymmetric unit in its original position. Screw axes combine a rotation with a translation of some fraction of a unit-cell dimension; e.g., a 3_1 screw axis consists of a rotation of 120° combined with a translation of one-third of a unit-cell dimension. When this operation is repeated three times, the asymmetric unit will be superimposed upon an asymmetric unit in the next unit cell. The third form of symmetry which is observed in macromolecular crystals is centering. This occurs when there is a second copy of the asymmetric unit either on the middle of a face of the unit cell or in the center of the unit cell. If, e.g., the centering is on the face made by the a and b axes, then it is referred to as C centered. If the centering is in the middle of the unit cell, it is called F centered. If there is no centering, then the cell is termed primitive or P.

The space group of a crystal consists of a description of all the symmetry operators that apply within the crystal: e.g., the space group $P2_12_12$ is primitive, with no centering, two 2_1 screw axes, and one twofold axis. All three axes are perpendicular to each other. Macromolecules are chiral, having some centers which are either left or right handed. They cannot therefore be related by mirrors, and so the number of possible space groups for macromolecules is only 65. Small molecules can have mirror planes and consequently can be in any of the 230 possible space groups. In general, if a symmetry operator is applicable throughout the crystal, the symmetry is called crystallographic symmetry and is described in the space group. In some cases, however, there is local symmetry within the asymmetric unit. An example of this would be a threefold axis relating three molecules in a trimer. This type of symmetry is referred to as noncrystallographic symmetry (NCS) and is not included in the space group description.

There are also cases where there are several copies of a molecule in the asymmetric unit which are not related to each other by any proper symmetry operator. This is also a case of noncrystallographic symmetry.

Related Website

Bernhard Rupp on space groups and list of 65 groups	http://www-structure.llnl.gov/Xray/tutorial/spcgrps.htm

Further Reading

Hahn Th (1995). *International Tables for Crystallography*, Vol. A: *Space Group Symmetry*, 4th ed. Kluwer Academic, New York.

See also Binding Site Symmetry, Diffraction of X Rays, Unit Cell, X-Ray Crystallography for Structure Determination and associated Further Reading.

SPACE-FILLING MODELS
Eric Martz

A space-filling model (or surface model) depicts the molecular volume into which other atoms cannot move, barring conformational or chemical changes. Space-filling models are three-dimensional depictions of molecular structures in which atoms are rendered as spheres with radii proportional to van der Waals radii or, less frequently, covalent or ionic radii.

Space-filling models originated with physical models designed to represent steric constraints—the atoms in the model cannot come unrealistically close together, occupying the same space. In the 1950s, Corey and Pauling used hard wood and plastic models at a scale of 1 or 0.5 in./Å for their work on the folding of the polypeptide chain. In the 1960s, a committee of the U.S. National Institutes of Health improved the design of these models, especially through a connector designed by Walter L. Koltun, increasing their suitability for macromolecules. Commercial production of these Corey–Pauling–Koltun, or CPK, models was implemented by the American Society of Biological Chemists, with financial support from the U.S. National Science Foundation. CPK models are associated with a commonly used color scheme for the elements.

Most atomic coordinate files for proteins lack hydrogen atoms. In these cases, some visualization programs (e.g., RasMol and its derivatives) increase the spherical radii of carbon, oxygen, nitrogen, and sulfur atoms to "united atom" radii approximating the sums of the volumes of these atoms with their bonded hydrogens. Because of the ease and speed with which they can produce accurate space-filling models, computer visualization programs have supplanted the construction of physical space-filling models for most purposes.

Related Websites

Physical molecular models	http://www.netsci.org/Science/Compchem/feature14b.html
Harvard Apparatus, distributors of CPK models	http://www.harvardbioscience.com
Radii used in space-filling models of RasMol and its derivatives	http://www.umass.edu/microbio/rasmol/rasbonds.htm

Further Reading

Corey RB, Pauling L (1953). Molecular models of amino acids, peptides and proteins. *Rev. Sci. Instrum.* 24: 621–627.

Koltun WL (1965). Precision space-filling atomic models. *Biopolymers* 3: 665–679.

Platt JR (1960). The need for better macromolecular models. *Science* 131: 1309–10.

Yankeelov JA Jr, Coggins JR (1971). Construction of space-filling models of proteins using dihedral angles. *Cold Spring Harbor Symp. Quant. Biol.* 36: 585–587.

See also Models, Molecular and accompanying figure; Surface Models; Visualization, Molecular.

Eric Martz is grateful for help from Eric Francoeur, Peter Murray-Rust, Byron Rubin, and Henry Rzepa.

SPECIFICITY
Thomas D. Schneider

The term *specificity* is often ill defined. It has been used to refer to livers (tissue specificity), energy, binding patterns, and other mutually inconsistent concepts. It is often more useful to use an appropriate precise term (*energy, bits, information*, etc.) instead.

See also Error Measures.

SPLICED ALIGNMENT
Roderic Guigó

Given a genomic sequence, the spliced alignment algorithms find a legal chain of exons predicted along this sequence with the best fit to a related target protein, or cDNA.

The program PROCRUSTES (Gelfand et al., 1996) introduced the spliced alignment concept. In PROCRUSTES the set of candidate exons is constructed by selection of all blocks between candidate acceptor and donor sites and further gentle statistical filtration. PROCRUSTES considers all possible chains of candidate exons in this set and finds a chain with the maximum global similarity to a given target protein sequence. In the Las-Vegas version (Sze and Pevzner, 1997), PROCRUSTES generates a set of suboptimal spliced alignments and uses it to assess the confidence level for the complete predicted gene or individual exons. Other examples of spliced alignment algorithms are GENEWISE (Birney and Durbin, 1997) for DNA–protein alignment and SIM4 (Florea et al., 1998) and EST_GENOME (Mott, 1997) for DNA–cDNA alignment.

Most spliced alignment algorithms use dynamic programming to obtain the optimal chain of exons in the query sequence. Therefore, running them in database search mode to identify potential genes in large anonymous genomic sequences may become computationally prohibitive.

Spliced algorithms are quite accurate in recovering the correct set of exons when the target protein is very similar to the protein encoded in the genomic sequence, but the accuracy drops as the similarity between query and target sequences decreases (Guigó et al., 2000).

Related Websites

PROCRUSTES	http://www-hto.usc.edu/software/procrustes/
SIM4 documentation	http://globin.cse.psu.edu/html/docs/sim4.html
EST_GENOME	http://www.well.ox.ac.uk/~rmott/est_genome.shtml

Further Reading

Birney E, Durbin R (1997). Dynamite: A flexible code generating language for dynamic programming methods used in sequence comparison. In *Proceedings of the Fifth International Conference on Intellegent Systems for Molecular Biology*, AAP Press Menlo Park, CA, pp. 179–186.

Florea L, et al. (1998). A computer program for aligning a cDNA sequence with a genomic DNA sequence. *Genome Res.* 8: 967–974.

Gelfand MS, et al. (1996). Gene recognition via spliced alignment. *Proc. Natl. Acad. Sci. USA* 93: 9061–9066.

Guigó R, et al. (2000). Sequence similarity based gene prediction. In *Genomics and Proteomics: Functional and Computational Aspects*, S Suhai, Ed. Kluwer Academic/Plenum Publishing, Dordrecht, The Netherlands, pp. 95–105.

Mott R (1997). EST_GENOME: A program to align spliced DNA sequences to unspliced genomic DNA. *Comput. Appl. Biosci.* 13: 477–478.

Sze SH, Pevzner P (1997). Las Vegas algorithms for gene recognition: Suboptimal and error-tolerant spliced alignment. *J. Comput. Biol.* 4: 297–309.

See also Dynamic Programming, Gene Prediction Accuracy, GENEWISE, Homology-Based Gene Prediction.

SPLICING (RNA SPLICING)
John M. Hancock

The process of removal of introns from RNAs to produce the mature RNA molecule.

The term is commonly used to describe the removal of nuclear introns from the primary transcript to produce the mature mRNA, but it may also be applied to other types of RNA processing.

Further Reading
Lewin B (2000). *Genes VII*. Oxford University Press, Oxford, UK.

See also Intron.

SPOTTED cDNA MICROARRAY
Stuart M. Brown

Microarray constructed by spotting DNA onto glass microscope slides.

The microarray technique is an extension of the pioneering work done by Southern (1975) on the hybridization of labeled probes to nucleic acids attached to a filter. This technology evolved into filter-based screening of clone libraries (dot blots), but a key innovation by Schena and colleagues (1995) was to use automated pipetting robots to spot tens of thousands of DNA probes onto glass microscope slides. The probes were derived from PCR amplification of inserts from plasmids containing known cDNA sequences. After developing methods for printing and hybridizing microarrays, Schena and colleagues published detailed plans for constructing array spotters on the Internet as the "Mguide."

These "spotted arrays" generally use full-length cDNA clones as probes and hybridize each array with a mixture of two cDNA samples (an experimental and a control) labeled with two different fluorescent dyes. The hybridized chip is then scanned at the two fluorescent wavelengths and ratios of signal intensities are calculated for each gene. By using full-length cDNA probes, this system has the advantage of high sensitivity and the ability to detect virtually all alternatively spliced transcripts. Its disadvantages include the potential for cross-hybridization from transcripts of genes with similar sequences (members of multigene families) and the logistical difficulty of maintaining large collections of cDNA clones and doing large-scale PCR amplifications.

| Pat Brown's MGuide to cDNA microarrays | http://cmgm.stanford.edu/pbrown/mguide/index.html |
| DeRisi Lab microarray resources | http://www.microarrays.org/ |

Further Reading

Phimister B (1999). Going global. *Nat. Genet. Suppl.* 21: 1.

Schena M, et al. (1995). Quantitative monitoring of gene expression patterns with a complementary DNA microarray. *Science* 270: 467–470.

Southern EM (1975). Detection of specific sequences among DNA fragments separated by gel electrophoresis. *J. Mol. Biol.* 98: 503–517.

SRS (SEQUENCE RETRIEVAL SYSTEM)
Guenter Stoesser

The SRS has become an integration system for both data retrieval and applications for data analysis. It provides capabilities to search multiple databases by shared attributes and to query across databases. The SRS server at the EBI integrates and links a comprehensive collection of specialized databanks along with the main nucleotide and protein databases. This server contains more than 130 biological databases and integrates more than 10 data analysis applications. Complex querying and linking across all available databanks can be executed.

Related Website

| Sequence Retrieval System (SRS) | http://srs.ebi.ac.uk/ |

Further Reading

Etzold T, et al. (1996). SRS: Information retrieval system for molecular biology data banks. *Methods Enzymol.* 266: 114–128.

Zdobnov EM, et al. (2002). The EBI SRS server—recent developments. *Bioinformatics* 18: 368–373.

See also Nucleic Acid Sequence Databases, Protein Databases.

STADEN
John M. Hancock and Martin J. Bishop

Package of software for the treatment of sequencing data and for sequence analysis.
The STADEN package is one of the longest established sequence analysis packages and has gone through a series of revisions. Its functions can be classified under the following general headings: sequence trace and reading file manipulation, sequence assembly (including sequencing trace viewing and preparation and contaminant screening), mutation detection, and sequence analysis. The sequence

analysis tool *spin* carries out a wide variety of analyses, including base and dinucleotide composition, codon usage, finding open reading frames, mapping restriction sites, searching for subsequences or motifs, and gene finding using a number of statistics. It also provides sequence comparison functions. Since Version 1.4 (February 2004) the package is open source.

Related Website

STADEN	http://staden.source.forge.net

Further Reading

Staden R, et al. (1998). The Staden Package, 1998. In *Computer Methods in Molecular Biology*, Vol. 132: *Bioinformatics Methods and Protocols*, S Misener, SA Krawetz, Eds. Humana, Totowa, pp. 115–130.

See also Chimeric DNA Sequence, DNA Sequencing Trace, Open Reading Frame, Sequence Assembly.

STANDARD GENETIC CODE. *SEE* GENETIC CODE.

STANFORD HIV RT AND PROTEASE SEQUENCE DATABASE (HIV RT AND PROTEASE SEQUENCE DATABASE)

Guenter Stoesser

The HIV RT and Protease Sequence Database is a curated database containing a compilation of most published HIV RT and protease sequences, including submissions from international collaboration databases (DDBJ/EMBL/GenBank), and sequences published in journal articles. The database catalogs evolutionary and drug-related sequence variation in HIV reverse transcriptase (RT) and protease enzymes, which are the molecular targets of antiretroviral drugs. Sequence changes in the genes coding for these enzymes are directly responsible for phenotypic resistance to RT and protease inhibitors. The database project constitutes a major resource for studying evolutionary and drug-related variation in the molecular targets of anti-HIV therapy. The HIV RT and Protease Sequence Database is maintained at the Stanford University Medical Center.

Related Websites

Stanford HIV RT and Protease Sequence Database	http://hivdb.stanford.edu/
Sequence analysis programs	http://hivdb.stanford.edu/seqAnalysis.html

Further Reading

Kantor R, et al. (2001). Human Immunodeficiency Virus Reverse Transcriptase and Protease Sequence Database: An expanded data model integrating natural language text and sequence analysis programs. *Nucleic Acids Res.* 29: 296–299.

Shafer RW, et al. (2000). Human immunodeficiency virus type 1 reverse transcriptase and protease mutation search engine for queries. *Nat. Med.* 6: 1290–1292.

Shafer RW, Deresinski S (2000). Human immunodeficiency virus information on the web: A guided tour. *Clin. Infect. Dis.* 31: 568–77.

START CODON. *SEE* GENETIC CODE.

STATISTICAL MECHANICS
Patrick Aloy

Statistical mechanics is the branch of physics in which statistical methods are applied to the microscopic constituents of a system in order to predict its macroscopic properties.

The earliest application of this method was Boltzmann's attempt to explain the thermodynamic properties of gases on the basis of the statistical properties of large assemblies of molecules. Since then, statistical mechanics theories have been applied to a variety of biological problems with great success. For example, statistical mechanics theories applied to protein folding rely on a series of major simplifications about the energy as a function of conformation and/or simplifications of the representation of the polypeptide chain, such as one point per residue on a cubic lattice, that allow us to obtain a tractable model of a very complex system.

STATISTICAL POTENTIAL ENERGY
Roland L. Dunbrack

Any potential energy function derived from statistical analysis of experimentally determined structures. Such energy functions are usually of the form $E(x) = -k \log(p(x))$, where x is some conformational degree of freedom and k is some constant. When $p(x) = 0$, E is technically infinity, but often some large noninfinite value is arbitrarily chosen. The value of k chosen is often $k_B T$, where k_B is the Boltzmann constant and T is the temperature in degrees kelvin. However, there is no theoretical justification for this choice, since the ensemble of structural features used to derive the statistical energy function is not a statistical mechanical ensemble, such as the canonical ensemble. A value of k can be chosen by plotting energy differences from the statistical analysis, as $\log(p(x_1)) - \log(p(x_2))$, versus the same energy differences, $(E(x_2) - E(x_1))$, from a different source, such as molecular mechanics. The slope of the line is $-k$. Statistical potential energy functions are used in side-chain prediction and in threading.

Further Reading

Sippl MJ (1990). Calculation of conformational ensembles from potentials of mean force: An approach to the knowledge-based predictions of local structures in globular proteins. *J. Mol. Biol.* 213: 859–883.

Thomas PD, Dill KA (1996). Statistical potentials extracted from protein structures: How accurate are they? *J. Mol. Biol.* 257: 457–469.

Thomas P. D, Dill KA (1996). An iterative method for extracting energy-like quantities from protein structures. *Proc. Natl. Acad. Sci. USA* 93: 11628–11633.

Dunbrack RL Jr (1999). Comparative modeling of CASP3 targets using PSI-BLAST and SCWRL. *Proteins* 3(Suppl.): 81–87.

See also Side-Chain Prediction, Threading.

STEEPEST-DESCENT METHOD. *SEE* GRADIENT DESCENT.

STOP CODON. *SEE* GENETIC CODE.

STRICT CONSENSUS TREE. *SEE* CONSENSUS TREE.

STRUCTURAL ALIGNMENT
Andrew Harrison and Christine A. Orengo

A structural alignment identifies equivalent positions in the three-dimensional (3D) structures of the proteins being compared. More than 30 different methods have been developed for aligning protein structures, although most are based on comparing similarities in the properties and/or relationships of the residues or secondary structures. One of the earliest approaches was pioneered by Rossman and Argos, who developed a method for superposing two protein structures in 3D by translating both proteins to a common coordinate frame and then rotating one protein relative to the other to maximize the number of equivalent positions superposed within a minimum distance from each other.

Other approaches (e.g., COMPARER, STAMP, SSAP) usually employ dynamic programming algorithms to optimize the alignment by efficiently determining where insertions or deletions of residues occur between the structures being compared. Structural environments of residues can be compared to identify equivalent positions, comprising, e.g., information on distances from a residue to neighboring residues. Some approaches (DALI, COMPARER) use sophisticated optimization protocols like simulated annealing and Monte Carlo optimization. Some multiple-structure alignment methods have also been developed (STAMP, SSAPM, CORA) for aligning groups of related structures.

Once the set of equivalent positions has been determined, many approaches superpose the structures on these equivalent positions to measure the root-mean-square deviation (RMSD). This is effectively the average distance between the superposed positions calculated by taking the square root of the average squared distances between all equivalent positions measured in angstroms. Related protein structures possessing similar folds typically possess an RMSD less than 4 Å, although this depends on the size of the domain. Some researchers prefer to calculate the RMSD over those positions which are very structurally similar and superpose within 3 Å, in which case it is usual to quote the number of positions over which the RMSD has been measured.

Related Websites

DALI	http://www2.ebi.ac.uk/dali/
CATH	http://www.biochem.ucl.ac.uk/bsm/cath_new
CE	http://cl.sdsc.edu/ce.html

Further Reading

Orengo CA, Michie AD, Jones S, Jones DT, Swindells MB, Thornton JM (1997). CATH—A hierarchic classification of protein domain structures. *Structure* 5(8): 1093–1108.

Taylor WR, Orengo CA (1989). Protein structure alignment. *J. Mol. Biol.* 208: 1–22.

See also Alignment—Multiple, Alignment—Pairwise, Dynamic Programming, Indel, Monte Carlo Simulations, Simulated Annealing.

STRUCTURAL GENOMICS
Jean-Michel Claverie

A research area and the set of approaches dealing with the determination of protein three-dimensional (3D) structures in a high-throughput fashion.

The concept of structural genomics was first introduced in reference to the project of determining the 3D structure of all gene products in a given (small) bacterial genome (hence the use of "genomics"). Given the high redundancy of protein structures (i.e., many different protein sequences fold in similar ways), such projects are now seen as a waste of resources and effort. Structural genomics projects are now developing in two different directions: *fundamental* projects aimed at determining the complete set of basic protein folds (estimates range from 1000 to 40,000) and *applied* structural genomics aimed at determining the structure of many proteins of a given family or sharing common properties usually of biomedical interest (e.g., evolutionary conserved pathogen proteins of unknown function). In the latter context, structural genomics can be seen as part of functional genomics. Given that protein structures evolve much slower than sequences, 3D structure similarities allow a fraction of proteins of unknown function to be linked with previously described families, when no sequence relationships can be established with confidence (i.e., in the "twilight zone" of sequence similarity). The knowledge of a representative 3D structure is also invaluable for the interpretation of conservation patterns identified by multiple-sequence alignments. Purely computational studies attempting to predict protein folds and perform homology modeling from all protein coding sequences in a given genome have also been referred to as "structural genomics."

Structural Genomics projects are made possible by a combination of recent advances in:

- Recombinant protein expression (in vivo and in cell-free extract robotized screening)
- Crystallogenesis (robotized large-scale screening of conditions)
- X-ray beam of higher intensity and adjustable wavelength (synchrotron radiation facilities) allowing single-crystal-based "phasing protocols"
- Crystal handling (cryo-cooling)
- Software for structure resolution (noise reduction, phasing, density map interpretation)
- NMR technology

However, structural genomics projects are presently limited to proteins that are not associated with membranes due to the extreme difficulty in expressing (and crystallizing) nonsoluble proteins.

Related Websites

Midwest Center for Structural Genomics	http://www.mcsg.anl.gov/
New York Structural Genomics Research Consortium	http://www.nysgrc.org/
Joint Center for Structural Genomics	http://www.jcsg.org/scripts/prod/home.html

Further Reading

Baker D, Sali A. (2001). Protein structure prediction and structural genomics. *Science* 294: 93–96.

Eisenstein E, et al. (2000). Biological function made crystal clear—annotation of hypothetical proteins via structural genomics. *Curr. Opin. Biotechnol.* 11: 25–30.

Kim, SH (1998). Shining a light on structural genomics. *Nat. Struct. Biol.* 5(Suppl): 643–645.

Moult J, Melamud E (2000). From fold to function. *Curr. Opin. Struct. Biol.* 10: 384–389.

Orengo CA, et al. (1999). From protein structure to function. *Curr. Opin. Struct. Biol.* 9: 374–382.

Simons KT, et al. (2001). Prospects for ab initio protein structural genomics. *J. Mol. Biol.* 306: 1191–1199.

Skolnick J, et al. (2000). Structural genomics and its importance for gene function analysis. *Nat. Biotechnol.* 18: 283–287.

Zarembinski TI, et al. (1998). Structure-based assignment of the biochemical function of a hypothetical protein: A test case of structural genomics. *Proc. Natl. Acad. Sci.* USA 95: 15189–15193.

STRUCTURAL MOTIF

Andrew Harrison and Christine A. Orengo

A structural motif typically refers to a small fragment of a three-dimensional structure, usually smaller than a domain, which has been found to recur either within the same structure or within other structures possessing different folds and architectures. Most structural motifs which have been characterized comprise fewer than five secondary structures and are common to a particular protein class. For example, the $\alpha\beta$ motif, split $\alpha\beta$ motif, and β-plait are found within many structures in the alpha–beta class, while the β meander, α-hairpin, greek roll, and jelly roll motifs are found in the mainly-α class. Some structural motifs are found to recur in different protein classes, e.g., the α-hairpin which occurs in some mainly-α structures and some alpha–beta structures. Recurrent structural motifs may correspond to particularly favored folding arrangements of the polypeptide chain due to thermodynamic or kinetic factors. Some researchers have suggested that some structural motifs may correspond to evolutionary modules which have been extensively duplicated and combined in different ways with themselves and other modules to give rise to the variety of different folds observed. The PROMOTIF database contains information on the structural characteristics of many structural motifs reported in the literature.

Related Website

| PROMOTIF | http://www.biochem.ucl.ac.uk/~gail/promotif/promotif.html |

See also Protein Domain, Secondary Structure of Protein.

STRUCTURE—3D CLASSIFICATION

Andrew Harrison and Christine A. Orengo

Structure classifications were established in the mid-1990s. In these, protein structures are classified according to phylogenetic relationships and phonetic (i.e., purely geometric) relationships. Although multidomain structures can be classified into evolutionary families and superfamilies, most structural classifications also distinguish the individual domain structures within them and domain structures are grouped into further hierarchical levels (see below). The largest, most complete classifications accessible over the Web are the SCOP, CATH, and DALI domain databases (see below). Other resources such as HOMSTRAD, CAMPASS, and 3Dee are also valuable Web-accessible resources.

Although the various resources employ different algorithms and protocols for recognizing related structures, most group domain structures exist at the following levels: at the top of the hierarchy, class, groups domains depending on the percentage and packing of α-helices and β-strands in the structure. Within each class, architecture groupings are sometimes used to distinguish structures having different arrangements of the secondary structures in 3D, while topology or fold group cluster structures have similar topologies regardless of evolutionary relationships. That is, there are similarities in both the orientations of the secondary structures in 3D space and their connectivity. Within each fold group, domains are clustered according to evolutionary relationships into homologous superfamilies. Families are groups of closely related domains within each superfamily, usually sharing significant sequence and functional similarity.

Related Websites

SCOP	http://scop.mrc-lmb.cam.ac.uk/scop/
DALI	http://www2.ebi.ac.uk/dali/fssp/fssp.html
CATH	http://www.biochem.ucl.ac.uk/bsm/cath_new/index.html
HOMSTRAD	http://www-cryst.bioc.cam.ac.uk/data/align/
CAMPASS	http://www-cryst.bioc.cam.ac.uk/~campass/
3Dee	http://www.compbio.dundee.ac.uk/3Dee

Further Reading

Murzin AG, Brenner SE, Hubbard T, Chothia C (1995). SCOP: A structural classification of proteins database for the investigation of sequences and structures. *J. Mol. Biol.* 247: 536–540.

Orengo CA (1994). Classification of protein folds. *Curr. Opin. Struct. Biol.* 4: 429–440.

See also Phylogeny, Phylogeny Reconstruction; Protein Domain; Superfamily.

STRUCTURE-BASED DRUG DESIGN (RATIONAL DRUG DESIGN)
Marketa J. Zvelebil

Structure-based drug design is based on the identification of candidate molecules that will bind selectively to a specific enzyme and thus (usually) inhibit its activity. The identification and docking of such ligand are two of the most important steps in rational drug design. Once a binding or active site has been identified, it is possible to model a known ligand of interest into the binding site or to find a potential ligand from a database of small molecules. The ligand is then fitted into its binding site. Analyzing the interactions between ligand and protein leads to a much better understanding of the protein's function and therefore to ways that the function could be modified. Binding is generally determined by the shape, chemical properties of both the binding site and substrate, and orientation of the ligand.

Related Websites

DOCK	http://dock.compbio.ucsf.edu/
Network Science	http://www.netsci.org/Science/Compchem/feature14i.html

Further Reading
Ladbury JE, Connelly PR, Eds. *Structure-Based Drug Design: Thermodynamics, Modeling and Strategy.* Springer-Verlag, New York.

See also DOCK, Docking.

STRUCTURE FACTOR
Liz P. Carpenter

Structures of molecules can be derived from observing the diffraction of X rays by crystals of the molecule.

The X-ray diffraction experiment gives a set of indexed diffraction spots each of which has a measured intensity, I. Each diffraction spot is the sum of all the X rays scattered in that direction by all the electrons in all the atoms of the crystal. The regular array of molecules in a crystal gives rise to constructive interference in some directions. The sum of all the X rays in a particular direction is a wave that is described by a vector called the structure factor **F**. This vector has both amplitude $|F|$ and a relative phase α. The observed structure factor amplitude, $|F_{obs}|$, is proportional to the square root of the intensity:

$$|F_{obs}| = K\sqrt{|I_{obs}|}$$

where K is a scaling constant.

A diffraction pattern can be calculated from any molecule placed in the unit cell using the equation

$$F_{calc}(hkl) = \sum_{j=1}^{\text{all atoms}} f_j \exp(2\pi i[hx + ky + lz])$$

where
$F_{calc}(hkl)$ is the calculated structure factor for the diffraction spot with indices hkl
f_j is the atomic scattering factor for the jth atom
x, y, z are the coordinates of the jth atom in the unit cell

The atomic scattering factor is the scattering from a single atom in a given direction, and it can be calculated from the atomic scattering factor equation. In general:

$$f_j\left(\sin\frac{\theta}{\lambda}\right) = \sum_{1}^{4} a_i \exp\left[-b_i\left(\sin\frac{\theta}{\lambda}\right)^2\right] + c$$

where a_i, b_i, and c are known coefficients.

See also Diffraction of X Rays, X-Ray Crystallography for Structure Determination and associated Further Reading.

STS. *SEE* SEQUENCE-TAGGED SITE.

SUBFUNCTIONALIZATION
Austin L. Hughes

Subfunctionalization is hypothesized to occur when, after duplication of an ancestral gene with broad tissue expression, mutations occur causing each of the daughter genes to be expressed in a subset of the tissues in which the ancestral gene was expressed.

This process is most easily visualized by considering a gene with two different regulatory elements (A and B in the accompanying figure) which control expression in two different tissues. After gene duplication, a mutation in one gene copy knocks out regulatory element A, while in the other copy a mutation knocks out regulatory element B. As a result, each of the daughter genes is necessary if the organism is to continue to express the gene product in all tissues in which it was expressed before gene duplication. A similar process might occur with regard to protein functions if the ancestral gene encoded a protein performing two separate functions, which could be performed by separate genes after gene duplication.

Further Reading

Force A, et al. (1999). Preservation of duplicate genes by complementary degenerative mutations. *Genetics* 151: 1531–1545.

Hughes AL (1994). The evolution of functionally novel proteins after gene duplication. *Proc. R. Soc. Lond. B* 256: 119–124.

Lynch M, Force A (2000). The probability of duplicate gene preservation by subfunctionalization. *Genetics* 154: 459–473.

Orgel LE (1977). Gene-duplication and the origin of proteins with novel functions. *J. Theor. Biol.* 67: 773.

See also Gene Sharing.

SUBSTITUTION PROCESS
Sudhir Kumar and Alan Filipski

This refers to the stochastic parameters of the process by which one nucleotide or amino acid replaces another in evolutionary time.

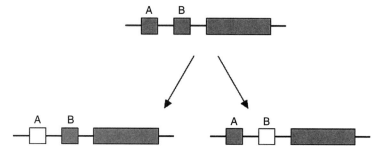

Figure Illustration of subfunctionalization. Of the regulatory elements A and B, a different one is inactivated in two copies of the gene, resulting in differential expression in different tissues.

Usually this substitution process is taken to be a Markov process; i.e.. the transition probabilities depend only on the current state, not on any prior history of the site. In DNA sequences, an important parameter governed by the substitution process is nucleotide composition, or relative numbers of the four nucleotide bases. Of particular biological interest is G+C content, i.e., the proportion of nucleotide bases in a sequence that are either guanine or cytosine

The homogeneity of a substitution process refers to the presence of the same process of substitution (at DNA or protein sequence level) over time in a given lineage. Stationarity may refer to composition or process; in the former case, it refers to the property of transition probabilities not changing over time; in the latter case, it refers to a constant base composition (except for stochastic variation) over time.

Substitution pattern refers to the pattern of relative probabilities of state changes. For example, if we accelerate the process by doubling all substitution rates of one nucleotide to another, the substitution pattern itself stays unchanged.

Further Reading

Kumar S, Gadagkar SR (2001). Disparity index: A simple statistic to measure and test the homogeneity of substitution patterns between molecular sequences. *Genetics* 158: 1321–1327.

Nei M, Kumar S (2000). *Molecular Evolution and Phylogenetics*. Oxford University Press, Oxford, UK.

Tavare S (1986). Some probabilistic and statistical problems on the analysis of DNA sequences. *Lect. Math. Life Sci.* 17: 57–86.

Yang Z (1994). Estimating the pattern of nucleotide substitution. *J. Mol. Evol.* 39: 105–111.

SUPERFAMILY

Dov S. Greenbaum

Superfamilies fit within the Dayhoff protein classification system that uses a hierarchical structure with reference to the protein relatedness.

While originally coined to refer to evolutionarily related proteins (the term is traditionally used only in reference to full sequences and not only domains), it has also come to signify groups of proteins related through functional or structural similarities independent of evolutionary relatedness.

The PIR (International Protein Sequence Database), classifying over 33,000 superfamilies as of January 2002, divides proteins into distinct superfamilies using an automated process that takes into account many criteria, including domain arrangement and percent identity between sequences. In general, proteins in the same superfamily do not differ significantly in length and have similar numbers and placements of domains. SCOP (Structural Classification of Proteins) defines *superfamily* more rigidly as a classification grouping the most distantly related proteins having a common evolutionary ancestor.

Homeomorphic superfamilies are defined as those families containing proteins that can be lined up end to end, i.e., containing the same overall domain architecture in the same order. All members of the family are deemed to have shared a similar evolutionary history.

Related Websites

Superfamily	http://supfam.mrc-lmb.cam.ac.uk/SUPERFAMILY/index.html
PIR	http://pir.georgetown.edu/PIR

Further Reading

Gough J, Chothia C (2002). SUPERFAMILY: HMMs representing all proteins of known structure. SCOP sequence searches, alignments and genome assignments. *Nucleic Acids Res.* 30: 268–272.

Lo Conte L, et al. (2002). SCOP database in 2002: Refinements accommodate structural genomics. *Nucleic Acids Res.* 30: 264–267.

SUPERFOLD

Andrew Harrison and Christine A. Orengo

Superfold is the name given to those frequently occurring domain structures, also known as FODs, which have been found to recur in proteins which do not appear to be related by divergent evolution. Although analyses of structural classifications have shown that most fold groups contain only proteins related by a common evolutionary ancestor (homologs), some folds are adopted by proteins which are not homologs. For example, more than 20 different homologous superfamilies adopt the TIM-barrel fold. Although these may be very distantly related proteins where all trace of sequence or functional similarity has disappeared during evolution, they may also be the consequence of convergence of different superfamilies to a favored folding arrangement. Superfolds may have been selected due to the thermodynamic stability of the fold or kinetic factors associated with the speed of folding.

Related Websites

DALI	http://www2.ebi.ac.uk/dali/
SCOP	http://scop.mrc-lmb.cam.ac.uk/scop/
CATH	http://www.biochem.ucl.ac.uk/bsm/cath_new

Further Reading

Baker D (2000). A surprising simplicity to protein folding. *Science* 405: 39–42.

Holm L, Sander C (1998). Dictionary of recurrent domains in protein structures. *Proteins: Struct. Funct. Genet.* 33: 88–96.

Murzin AG, Brenner SE, Hubbard T, Chothia C (1995). SCOP: A structural classification of proteins database for the investigation of sequences and structures. *J. Mol. Biol.* 247: 536–540.

See also Homologous Genes.

SUPERSECONDARY STRUCTURE

Roman A. Laskowski

Assemblies of commonly occurring secondary-structure elements which constitute a higher level of structure than secondary structures but a lower level than structural domains.

Examples include beta–alpha–beta motifs, Greek keys, alpha and beta hairpins, and helix–turn–helix motifs.

Further Reading

Branden C, Tooze J (1991). *Introduction to Protein Structure*, Garland, New York.

See also Motif, Secondary Structure of Proteins.

SUPERTREE. *SEE* CONSENSUS TREE.

SUPERVISED AND UNSUPERVISED LEARNING
Nello Cristianini

One can distinguish two main general classes of machine learning settings in data analysis applications: supervised and unsupervised. In the first case, the data (independent features) are "labeled" by specifying to which category they belong (dependent feature), and the algorithm is requested to predict the labels on new unseen data. In the second case they are not labeled, and the algorithm is required to discover patterns (such as cluster structures) in the data.

Further Reading

Duda RO, Hart PE, Stork DG (2001). *Pattern Classification*, Wiley, New York.

Mitchell T (1997). *Machine Learning*. McGraw-Hill, New York.

See also Classification, Clustering, Feature, Label, Machine Learning, Regression Analysis.

SUPPORT VECTOR MACHINES (MAXIMAL MARGIN CLASSIFIERS, SVMS)
Nello Cristianini

Algorithms for learning complex classification and regression functions belonging to the class of kernel methods.

In the binary classification case, SVMs work by embedding the data into a feature space by means of kernels and by separating the two classes with a hyperplane in such a space. For statistical reasons, the hyperplane sought has maximal margin or maximal distance from the nearest data point. The presence of noise in the data, however, requires a slightly adapted algorithm.

The maximal margin hyperplane in the feature space can be found by solving a convex (quadratic) optimization problem, and this means that the training of SVMs is not affected by local minima. This hyperplane in the feature space can correspond to a complex (nonlinear) decision function in the original input domain.

The final decision function can be written as $f(x) = \langle w, \phi(x) \rangle + b = \sum y_i \alpha_i K(x_i, x) + b$, where the pairs (x_i, y_i) are labeled training points, x is a generic test point, K is a kernel function, and the α_i, are parameters tuned by the training algorithm. There is one such parameter per training point, and only points whose parameter α_i is nonzero affect the solution. It turns out that only points lying nearest to the hyperplane have nonzero coefficient, and those are called "support vectors" (see accompanying figure).

Generalizations of this algorithm to deal with noisy data, regression problems, unsupervised learning, and a number of other important cases have been proposed in recent years. First introduced in 1992, SVMs are now one of the major tools in pattern recognition applications, mostly due to their computational efficiency and statistical stability.

Related Websites

Kernel Machines	www.kernel-machines.org
Support Vector Net	www.support-vector.net

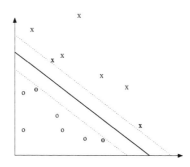

Figure   A maximal margin.

Further Reading
Cristianini N, Shawe-Taylor J (2000). *An Introduction to Support Vector Machines.* Cambridge University Press, Cambridge, UK.
Vapnik V (1995). *The Nature of Statistical Learning Theory.* Springer Verlag, Berlen, Germany.

See also Kernel Function, Kernel Methods, Separating Hyperplane.

SURFACE MODELS
Eric Martz

Models that represent the surfaces of molecules, i.e., the volume into which other atoms cannot penetrate, barring conformational or chemical changes.

Most commonly, the surface is a solvent-accessible surface and is generated by rolling a spherical probe over the space-filling model of the molecule. For a probe representing a water molecule, a sphere of radius 1.4 Å is most commonly used. The concept of the solvent-accessible surface was introduced by Lee and Richards in 1971 (see Connolly, 1983) to quantitate the burial of hydrophobic moieties in folded proteins. Electron density maps of molecules, resulting from X-ray crystallography, are another representation of molecular surfaces. For early low-resolution electron density maps, physical models were constructed with layers of balsa wood that represented the molecular surface.

The surface can be defined either by the path of the probe's center (the original Lee and Richards "solvent-accessible" surface) or by the path of the probe's proximal surface (Richmond and Richards, 1978—the "contact surface" see Connolly, 1983). The volume enclosed by the solvent-accessible surface is larger than the unified van der Waals radius by the radius of the probe and has been termed "solvent-excluded volume" by some (Richmond, 1984; see Connolly, 1983), while others use that term to designate the volume of the contact surface (Connolly, 1985). The volume enclosed by the contact surface represents the portion of the van der Waals surface that can be contacted by solvent and includes the van der Waals volume plus the interstitial volume, the latter being the spaces between atoms that are too small to admit solvent.

Molecular surfaces are often colored to represent molecular electrostatic potential or molecular lipophilicity potential. A popular and highly regarded surface rendering and coloring software program is GRASP.

Related Websites

Molecular surfaces	http://www.netsci.org/Science/Compchem/feature14.html
GRASP	http://trantor.bioc.columbia.edu/grasp/

Further Reading

Connolly ML (1983). Solvent-accessible surfaces of proteins and nucleic acids. *Science* 221: 709–713.

Connolly ML (1985). Computation of molecular volume. *J. Am. Chem. Soc.* 107: 1118–1124.

See also Models, Molecular and accompanying figure; Space-Filling Model; van der Waals Radius; Visualization, Molecular.

Eric Martz is grateful for help from Eric Francoeur, Peter Murray-Rust, Byron Rubin, and Henry Rzepa.

SURPRISAL
Thomas D. Schneider

The surprisal is how surprised one would be by a single symbol in a stream of symbols. It is computed from the probability of the ith symbol, P_i, as $u_i = -\log_2 P_i$. For example, late at night, as I write this, the phone rarely rings, so the probability of silence is close to 1 and the surprisal for silence is near zero. (If the probability of silence is 99%, then $u_{silence} = -\log_2(0.99) = 0.01$ bits per second, where the phone can ring only once per second.) On the other hand, a ring is rare so the surprisal for ringing is very high. (For example, if the probability of ringing is 1% per second, then $u_{ring} = -\log_2(0.01) = 6.64$ bits per second). The average of all surprisals over the entire signal is the uncertainty ($0.99 * 0.01 + 0.01 * 6.64 = 0.08$ bits per second in this example). The term comes from Myron Tribus' book.

Further Reading

Tribus M (1961). *Thermostatics and Thermodynamics.* van Nostrand, Princeton, NJ.

See also Uncertainty.

SVMS. *SEE* SUPPORT VECTOR MACHINES.

SWISSMODEL
Roland L. Dunbrack

A program for comparative modeling of protein structure developed by Guex and Peitsch (1997). SwissModel is intended to be a complete modeling procedure accessible via a Web server that accepts the sequence to be modeled and then delivers the model by electronic mail. SwissModel follows the standard protocol of homolog identification, sequence alignment, determining the core backbone, and modeling loops and side chains. SwissModel determines the core backbone from the alignment of the target sequence to the parent sequence(s) by averaging the structures according to their local degree of sequence identity with the target sequence. The program builds new segments of backbone for loop regions by a database scan of the PDB using anchors of four Cα atoms on each end. Side chains are now built for those residues without information in the parent structure by using the most common (backbone-independent) rotamer for that residue type. If a side chain cannot be placed without steric overlaps, another rotamer is used. Some additional refinement is performed with energy minimization with the GROMOS program.

Related Website

SwissModel	http://www.expasy.org/swissmod/

Further Reading

Guex N, Peitsch MC (1997). SWISS-MODEL and the Swiss-PdbViewer: An environment for comparative protein modeling. *Electrophoresis* 18: 2714–2723.

See also Comparative Modeling.

SWISS-PROT
Rolf Apweiler

The SWISS-PROT protein sequence knowledgebase is a universal annotated protein sequence database covering proteins from many different species.

SWISS-PROT is an annotated protein sequence knowledgebase established in 1986 and maintained collaboratively by the Swiss Institute of Bioinformatics (SIB) and the European Bioinformatics Institute (EBI). It provides a high level of annotation, a minimal level of redundancy, a high level of integration with other biomolecular databases, and extensive external documentation. Each entry in SWISS-PROT is thoroughly analyzed and annotated by biologists to ensure a high standard of annotation and maintain the quality of the database. SWISS-PROT contains data that originate from a wide variety of organisms; in May 2002 the database release 40.16 contained around 108,159 annotated sequence entries from more than 7000 different species.

In SWISS-PROT two classes of data can be distinguished: the core data and the annotation. For each sequence entry the core data consist of the sequence data, the citation information (bibliographical references), and the taxonomic data (description of the biological source of the protein), while the annotation describes the following:

- Function(s) of the protein
- Posttranslational modification (e.g., carbohydrates, phosphorylation, acetylation, GPI-anchor)
- Domains and sites (e.g., calcium binding regions, ATP binding sites, zinc fingers, homeobox, kringle)
- Secondary structure
- Quaternary structure (homodimer, heterotrimer, etc.)
- Similarities to other proteins
- Disease(s) associated with deficiencie(s) in the protein
- Sequence conflicts, variants, etc.

In SWISS-PROT annotation is mainly found in the comment lines (CC), in the feature table (FT), and in the keyword lines (KW). Most comments are classified by "topics"; this approach permits the easy retrieval of specific categories of data from the database.

SWISS-PROT tries to merge separate entries corresponding to different literature reports to minimize the redundancy of the database. Any conflicts between various sequencing reports are indicated in the feature table of the corresponding entry. SWISS-PROT provides the user with a high degree of integration between different data collections. In May 2002 it was cross-referenced with 45 different databases through the use of more than 803,000 pointers to information found in data collections other than SWISS-PROT.

Related Websites

SWISS-PROT	http://www.ebi.ac.uk/swissprot/
ExPASy	http://www.expasy.org/sprot/

Further Reading

Apweiler R (2000). Protein sequence databases. *Adv. Protein Chem.* 54: 31–71. Apweiler R (2001). Functional information in SWISS-PROT: The basis for large-scale characterisation of protein sequences. *Brief. Bioinform.* 2: 9–18.

Bairoch A, Apweiler R (2000). The SWISS-PROT protein sequence database and its supplement TrEMBL in 2000. *Nucleic Acids Res.* 28: 45–48.

See also Protein Information Resource, TrEMBL.

SYBYL
Roland L. Dunbrack

A molecular modeling and visualization computer program package marketed by Tripos.

Related Website

Sybyl	http://www.tripos.com/sciTech/inSilicoDisc/moleculeModeling/sybase.html

See also Comparative Modeling.

SYMMETRY. *SEE* BINDING SITE SYMMETRY.

SYMMETRY PARADOX
Thomas D. Schneider

Specific individual sites may not be symmetrical (i.e., completely self-complementary) even though the set of all sites is bound symmetrically.

This raises an experimental problem: How do we know that a site is symmetric when bound by a dimeric protein if each individual site has variation on the two sides? If we assume that the site is symmetrical then we would write DELILA instructions for both the sequence and its complement (see accompanying figure). The resulting sequence logo will, by definition, be symmetrical. If, on the other hand, we write the instructions so as to take only one orientation from each sequence, perhaps arbitrarily, then by definition the logo will be asymmetrical.

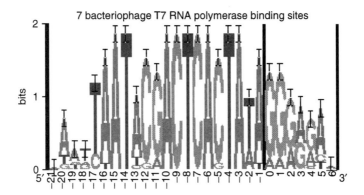

Figure A sequence logo for T7 RNA polymerase binding sites (Schneider and Stephens, 1990; Schneider and Stormo 1989).

That is, one gets the output of what one puts in. This is a serious philosophical and practical problem for creating good models of binding sites. One solution would be to use a model that has the maximum information content (Schneider and Mastronarde, 1996), although this may be difficult to determine in many cases because of small sample sizes. Another solution is to orient the site by some biological criteria such as direction of transcription controlled by a motivator.

Related Websites

DELILA	http://www.lecb.ncifcrf.gov/~toms/delila/instshift.html
DELILA instructions	http://www.lecb.ncifcrf.gov/~toms/delilainstructions.htmlsymmetry

Further Reading

Schneider TD, Mastronarde D (1996). Fast multiple alignment of ungapped DNA sequences using information theory and a relaxation method. *Discr. Appl. Math.* 71: 259–268. http://www.lecb.ncifcrf.gov/~toms/paper/malign.

Schneider TD, Stephens RM (1990). Sequence logos: A new way to display consensus sequences. *Nucleic Acids Res.* 18: 6097–6100. http://www.lecb.ncifcrf.gov/~toms/paper/logopaper/.

Schneider TD, Stormo GD (1989). Excess information at bacteriophage T7 genomic promoters detected by a random cloning technique. *Nucleic Acids Res.* 17: 659–674.

See also Binding Site Symmetry, DELILA Instructions, Information, Sequence Logo.

SYMPATRIC EVOLUTION (SYMPATRIC SPECIATION)
A. Rus Hoelzel

Sympatric evolution occurs when populations of the same species diverge genetically within the same geographic location (having partially or fully overlapping geographic ranges).

Such differentiation, together with the development of pre- or postzygotic reproductive isolation, can result in sympatric speciation. The relative importance of this mode of speciation remains controversial, but there have been a number of possible mechanisms proposed. Perhaps the best known is disruptive selection (Maynard Smith, 1966). Under this theory a stable polymorphism in a heterogeneous environment could lead to speciation if there is strong selection to different niches for different phenotypes expressed at the same locus. Another plausible mechanism involves the shifting of hosts in parasitic species.

A mechanism of sympatric speciation known to occur in nature is polyploidy, especially in plants. This involves the duplication of whole sets of chromosomes, and there are two types. *Allopolyploids* are formed by the hybridization between two species followed by a doubling in the number of chromosomes, and this is the most common form [e.g., resulting in speciation among *Brassica* sp. (e.g., broccoli) and *Triticum* sp. (wheat)]. *Autopolyploids* are formed by the doubling of chromosome number without the involvement of hybridization.

Further Reading
Maynard Smith J (1966). Sympatric speciation. *Am. Nat.* 100: 637–650.

See also Allopatric Evolution, Evolution.

SYMPATRIC SPECIATION. *SEE* SYMPATRIC EVOLUTION.

SYNAPOMORPHY
A. Rus Hoelzel

Two evolutionary characters sharing a derived state.

As a hypothesis about the pattern of evolution among operational taxonomic units (OTUs), a phylogenetic reconstruction can inform us about the relationship between character states. We can identify ancestral and descendant states and therefore identify primitive and derived states. An apomorphy (meaning "near shape" in Greek) is a derived state (shown as filled circles in the figure):

A shared derived state between OTUs, as between 1 and 2 in the figure, is known as a synapomorphy. This is an instance of homology, since 1 and 2 share derived characters from a common ancestor.

Further Reading
Maddison DR, Maddison WP (2001). *MacClade 4: Analysis of Phylogeny and Character Evolution*, Sinauer Associates, Sunderland, MA.

See also Apomorphy, Autapomorphy, Plesiomorphy.

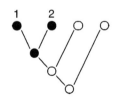

Figure   Example of synapomorphy. The two OTUs 1 and 2 share a common derived state.

SYNCHROTRON
Liz P. Carpenter

A synchrotron is a particle accelerator which emits electromagnetic radiation in highly focused beams of defined wavelength and high intensity necessary for experiments such as X-ray crystallography with small and weakly diffracting crystals. A synchrotron consists of a linear accelerator and one or more storage rings. The linear accelerator produces electrons or positrons traveling at great speed, and the storage rings consist of circular tunnels several hundred meters in diameter under high vacuum in which a current of electrons or positrons is forced to travel in a circular path due to powerful magnetic fields. When particles are forced to change direction under the influence of a magnetic field, radiation is emitted tangentially. Single wavelengths are selected by taking only the radiation that is reflected from one plane in a crystal and a highly parallel beam is obtained using a series of shutters and tubes called collimators. Samples and detectors are positioned several meters from the storage ring.

Synchrotrons have the advantage of providing very intense beams, and on some experimental stations the wavelength of the beam can be selected and adjusted as required so that data can be collected at a number of defined wavelengths.

Examples of synchrotrons in Europe include the synchrotron light source (SRS) at Daresbury, United kingdom (http://www.srs.ac.uk/srs/), and the European Synchrotron Radiation facility in Grenoble, France (http://www.esrf.fr/).

Related Websites

SRS	http://www.srs.ac.uk/srs/
ESRF	http://www.esrf.fr/

SYNONYMOUS MUTATION (SILENT MUTATION)
Laszlo Patthy

Nucleotide substitutions occurring in translated regions of protein coding genes are synonymous (or silent) if they cause no amino acid change. This results as the altered codon codes for the same amino acid as the original one.

SYNTENY
Katheleen Gardiner

Refers to genes that reside on the same chromosome.

Conserved synteny is used for genes that are found on a single chromosome in different species. Conserved synteny does not imply that gene order or proximity is conserved between species. Regions of a chromosome harboring syntenic genes can be considered to be homologous between two species. In popular usage, a homologous region is increasingly often referred to as a syntenic region.

Further Reading

Barbazuk WB, Korf I, Kadavi C, Heyen J, Tate S, Wun E, Bedell JA, McPherson JD, Johnson SL (2000). The syntenic relationship of the zebrafish and human genomes. *Genome Res.* 10: 1351–1358.

Kuroiwa A, Tsuchiya K, Matsubara K, Namikawa T, Matsuda Y (2001). Construction of comparative cytogenetic maps of the Chinese hamster to mouse, rat and human. *Chromosome Res.* 9: 641–648.

Trachtlec Z, Forejt J (2001). Synteny of orthologous genes conserved in mammals, snake, fly, nematode and fission yeast. *Mamm. Genome* 12: 227–231.

SYSTEM BIOLOGY

Marketa J. Zvelebil

System biology is a field of science that studies the organization and dynamic interactions of all individual elements in systems that cannot be fully understood by analysis of their respective components in isolation.

The general aim of system biology is to develop predictive computer models of the integrated networks that explain the many functional elements found in biological systems. Such pathway modeling enables the investigation of how, e.g., extracellular signals are processed to produce functional responses. Similar modeling work can also be used to investigate larger scale multicellular systems such as whole organs or even whole organisms.

To use the method of system biology, mathematical descriptions of biological processes are used together with data from engineering, physics, and the power of modern computers. All these methods are used to obtain a detailed description of the parts (components) and their interactions and then to reconstruct them into an interconnected whole. In other words, mathematical models are applied to biological processes to identify rules concerned with molecular or cellular associations or dependencies (*causal dependencies*). This enables the simulation of pathway behavior.

A number of simulation programs are available to mathematically model biological pathways; some example are Jarnac and Gepasi. In addition, a markup language has been developed for use solely in system biology (called SBML) and an intercommunication workbench (SBW).

Related Websites

Jarnac	http://www.cds.caltech.edu/~hsauro/Jarnac.htm
JDesigner	http://www.cds.caltech.edu/~hsauro/JDesigner.htm
SBW	http://www.sbw-sbml.org/index.html
Institute for System Biology	http://www.systemsbiology.org/
Gepasi	http://www.gepasi.org/
SysBio Links	http://www.sbw-sbml.org/links.html

Further Reading

Chong L, Ray LB (2002). Whole-istic biology. *Science* 295: 1661.

Csete ME, Doyle JC (2002). Reverse engineering of biological complexity. *Science* 295: 1664.

Davidson EH, Rast JP, Oliveri P, Ransick A, Calestani C, Yuh CH, Minokawa T, Amore G, Hinman V, Arenas-Mena C, Otim O, Brown CT, Livi CB, Lee PY, Revilla R, Rust AG, Pan ZJ,

Schilstra MJ, Clarke PJC, Arnone MI, Rowen L, Cameron RA, McClay DR, Hood L, Bolouri HA (2002). Genomic regulatory network for development. *Science* 295: 1669.

Kitano H (2002). Systems biology: A brief overview. *Science* 295: 1662.

Noble D (2002). Modeling the heart—from genes to cells to the whole organ. *Science* 295: 1678.

Trachtlec Z, Forejt J (2001). Synteny of orthologous genes conserved in mammals, snake, fly, nematode and fission yeast. *Mamm. Genome* 12: 227–231.

See also Robustness.

SYSTEM FOR ASSEMBLING MARKERS. *SEE* SAM.

T

TAMBIS
Tandem Repeat
Target
TATA Box
Taxonomic Classification
Taxonomy
Telomere
Template
Template Gene Prediction
Term
Terminology
THEATRE
Thermal Noise
Thesaurus
THREADER
Threading
Threonine
TIM-Barrel
Transcriptional Regulatory Region
Transcription Factor
Transcription Factor Binding Site
Transcription Start Site

Transcriptome
TRANSFAC
Transfer RNA
Translation
Translation Start Site
Translatome
Transposable Element
Transposon
Tree
Tree-Based Progressive Alignment
Tree of Life
Tree Puzzle
TreeView
TrEMBL
Trev
Trinucleotide Repeat
tRNA
Tryptophan
Turn
TWEAK algorithm
Twilight Zone
Two-Dimensional Gel Electrophoresis
Tyrosine

Dictionary of Bioinformatics and Computational Biology. Edited by Hancock and Zvelebil
ISBN 0-471-43622-4 © 2004 John Wiley & Sons, Inc.

TAMBIS
Robert Stevens

TAMBIS (Transparent Access to Multiple Bioinformatics Information Sources) uses an ontology of molecular and bioinformatics tasks to provide a user with the illusion of a common query interface to multiple, distributed, heterogeneous bioinformatics resources. TAMBIS uses the description logic (DL) GRAIL to encode its ontology. The ontology captures knowledge about what classes can exist within its domain. TAMBIS accesses its ontology through a terminology server (TeS) that allows dynamic creation and classification of concept descriptions. Thus, new terms can be built from those classes and relationships present within the ontology, sent to the DL reasoner, and placed within the current taxonomy. A class represents a set of instances in the world (the class Protein represents all proteins), so retrieving the instances of a class is in effect answering a query. TAMBIS allows users to create conceptual queries; these source-independent queries are resolved against the concrete resources by consulting mappings from the concepts and relationships to the functions over and values within resources.

Related Website

TAMBIS	http://img.cs.man.ac.uk/tambis

Further Reading
Goble CA, et al. (2001). Transparent access to multiple bioinformatics information resources. *IBM Syst. J.* 40: 532–551.

TANDEM REPEAT
Katheleen Gardiner

Identical DNA sequences immediately adjacent to each other and in the same orientation. Identical adjacent sequences in opposite orientations are termed *inverted repeats*.

Further Reading
Cooper DN (1999). *Human Gene Evolution*. Bios Scientific Publishers, Oxford, UK, pp. 265–285.
Ridley M (1996). *Evolution*. Blackwell Science, Cambridge, MA, pp. 265–276.

See also Gene Family, Microsatellite, Minisatellite.

TARGET
Roland L. Dunbrack

In comparative modeling, the protein of unknown structure for which a model will be constructed, based on a homologous protein of known structure, referred to as the template (or parent). In some situations such as sequence database searching, the target is also referred to as the query.

See also Comparative Modeling, Template.

Dictionary of Bioinformatics and Computational Biology. Edited by Hancock and Zvelebil
ISBN 0-471-43622-4 © 2004 John Wiley & Sons, Inc.

TATA BOX
Niall Dillon

Consensus sequence located 30–35 bases upstream from the transcription start site of many RNA polymerase II transcribed genes.

Two different elements have been identified that are involved in specifying the transcriptional initiation site of polymerase II transcribed genes in eukaryotes. These are the TATA box and the initiator. Promoters can contain one or both of these elements, and a small number have neither. The TATA box is located 30–35 bp upstream from the site of transcriptional initiation and binds the multiprotein complex *TFIID*, which is involved in specifying the initiation site.

Further Reading
Roeder RG (1996). The role of general initiation factors in transcription by RNA polymerase II. *Trends Biochem. Sci.*, 21: 327–335.

TAXONOMIC CLASSIFICATION (ORGANISMAL CLASSIFICATION)
Dov S. Greenbaum

A hierarchically structured terminology for organisms.

There are seven nested taxa (plural of taxon) or levels in the present classification (or phylogeny or organismal diversity) system:

1. Kingdom, of which there are conventionally five; proposed (in 1969) by Robert Whittaker:
   a. Plantae: multicellular eukaryotes which derive energy from photosynthesis and whose cells are enclosed by cellulose cell walls
   b. Animalia: multicellular, mobile, heterotrophic eukaryotes whose cells lack walls
   c. Fungi: eukaryotic, heterotrophic, and (usually) multicellular organisms that have multinucleated cells enclosed in cells with cell walls
   d. Protista: eukaryotes that are not plants, fungi, or animals
   e. Monera: unicellular prokaryotic organisms lacking membrane-bound organelles but surrounded by a cell wall
   It has been suggested that this be compressed into three: archaea, bacteria, and eukarya.
2. Phylum or division (In the animal kingdom, the term *phylum* is used instead of *division.*)
3. Class
4. Order
5. Family
6. Genus
7. Species

The hierarchical format has kingdom as the largest and most encompassing group and species as the most narrow. As such, the closer a set of organisms are evolutionarily, the more groups they share.

The species name, designed by Carolus Linnaeus, an eighteenth-century Swedish botanist, and ratified by the International Congress of Zoologists, is, by convention, binomial. Each species is given a two-part Latin name, formed by appending a specific epithet to the genus name, e.g., *Homo sapiens* for the human species. Other general rules of classification include the fact that all taxa must belong to a higher taxonomic group.

Related Websites

NCBI taxonomy	http://www.ncbi.nlm.nih.gov/Taxonomy/tax.html/
Taxonomy	http://www.ucmp.berkeley.edu/help/taxaform.html (
2004 International Congress of Zoology	http://icz.ioz.ac.cn/
Species list	http://species.enviroweb.org/
ICBN—St. Louis Code	http://www.bgbm.fu-berlin.de/iapt/nomenclature/code/ SaintLouis/0000St.Luistitle.htm

Further Reading

Cavalier-Smith T (1993). Kingdom protozoa and its 18 phyla. *Microbiol. Rev.* 57: 953–994.

Lake JA (1991). Tracing origins with molecular sequences: Metazoan and eukaryotic beginnings. *Trends Biochem. Sci.* 16: 46–50.

Whittaker RH (1969). New concepts of kingdoms of organisms. *Science* 163: 150–160.

Whittaker RH, Margulis L (1978). Protist classification and the kingdoms of organisms. *BioSystems* 10: 3–18.

TAXONOMY
Robert Stevens

1. Generically, the grouping of concepts into hierarchies based solely on the subtype relationships.
2. A classification of organisms which may correspond to their phylogenetic relationships
3. The science and art of developing 2 above.

See also Taxonomic Classification.

TELOMERE
Katheleen Gardiner

Specialized DNA sequences found at the ends of eukaryotic (linear) chromosomes.

Telomeres are composed of simple repeats—TTAGGG is common in vertebrates—reiterated several hundred times. A telomere stabilizes the chromosome ends, prevents fusion with other DNA molecules, and allows DNA replication to proceed to the end of the chromosome without loss of DNA material.

Further Reading

Blackburn EH (2001). Switching and signaling at the telomere. *Cell* 106: 661–673.

Miller OJ, Therman E (2000). *Human Chromosomes*. Springer, New York.

Wagner RP, Maguire MP, Stallings RL (1993). *Chromosomes—A Synthesis*. Wiley-Liss, New York.

TEMPLATE (PARENT)
Roland L. Dunbrack

In comparative modeling, the protein of known structure used as a template for building a model of a protein of unknown structure, referred to as the target. The template and target proteins are usually homologous to one another. The template is sometimes referred to as the parent.

See also Comparative Modeling, Target.

TEMPLATE GENE PREDICTION. *SEE* GENE PREDICTION, AB INITIO.

TERM
Robert Stevens

A label or name that is used to refer to a concept in some language.

The concept Protein may have terms such as *Polypeptide, Protein, Folded Protein,* etc. The concept and its label are formally independent but often confused, because it is impossible to refer to concepts without the use of terms to name them. Ontology developers should adopt conventions for their terms. For instance, it is usual to use the singular form of a noun and use initial capitals. As far as is possible, the terms used within an ontology should follow a community's conventional understanding of a domain as well as assist in controlling usage of terms within a domain.

TERMINOLOGY
Robert Stevens

The concepts of an ontology together with their lexicon form a terminology.

See also Lexicon, Ontology.

THEATRE
John M. Hancock and Martin J. Bishop

Program for identifying conserved transcription factor binding sites and coding regions.

THEATRE carries out a number of analyses and presents them in a single output file in postscript format. It accepts aligned DNA sequences (in a number of formats) or unaligned sequences (in which case it carries out an alignment using CLUSTALW). Sequences are subjected to a BLAST search of the databases as well as searches for ORFs (using Genemark), repeats (using RepeatMasker), and CpG islands (using Cpgplot). Transcription factor binding sites are identified using MatInspector and the EMBOSS program tfscan. Results are presented in relation to the multiple alignment to allow identification of conserved sites.

Related Website

THEATRE	http://www.hgmp.mrc.ac.uk/Registered/Webapp/theatre/

See also BLAST, CLUSTAL, Coding Region Prediction, CpG Island, EMBOSS, MatInspector, Open Reading Frame, RepeatMasker, Transcription Factor.

THERMAL NOISE
Thomas D. Schneider

Thermal noise is caused by the random motion of molecules at any temperature above absolute zero kelvin.

Since the third law of thermodynamics prevents one from extracting all heat from a physical system, one cannot reach absolute zero and so cannot entirely avoid thermal noise. In 1928 Nyquist worked out the thermodynamics of noise in electrical systems, and in a back-to-back paper Johnson demonstrated that the theory was correct.

Further Reading
Johnson JB (1928). Thermal agitation of electricity in conductors. *Phys. Rev.* 32: 97–109.
Nyquist H (1928). Thermal agitation of electric charge in conductors. *Phys. Rev.* 32: 110–113.

See also Noise.

THESAURUS
Robert Stevens

A thesaurus is a hierarchy or multihierarchy of index terms organized on the relationships of "broader than" (parents) and "narrower than" for children: "The vocabulary of a controlled indexing language, formally organized so that the a priori relationships between concepts (for example as 'broader' and 'narrower') are made explicit" (ISO 2788, 986:2).

The terms within a thesaurus are traditionally used to index a classification of documents, usually in a library: "A controlled set of terms selected from natural language and used to represent, in abstract form, the subjects of documents" (ISO 2788, 1986:2). Thus, the terms of a thesaurus are used to find documents within a classification.

THREADER. *SEE* THREADING.

THREADING
David T. Jones

The process of replacing the side chains of one protein with the side chains of another protein while keeping the main-chain conformation constant.

The side-chain replacement is carried out according to a specific sequence-to-structure alignment. Threading is typically used as one of a number of approaches to recognizing protein folds. For fold recognition, threading methods consider the three-dimensional structure of a protein and evaluate the compatibility of the target sequence with the template by means of pair potentials and solvation potentials. A variety of algorithms can be used to find the optimum threading alignment given a particular energy or scoring function, including double dynamic programming, simulated annealing, branch-and-bound searching, and Gibbs sampling.

Related Websites

GenThreader	http://bioinf.cs.ucl.ac.uk/psipred/index.html
TOPITS	http://www.embl-heidelberg.de/predictprotein/predictprotein.html
UCLA-DOE	http://fold.doe-mbi.ucla.edu/
3D-PSSM	http://www.sbg.bio.ic.ac.uk/~3dpssm/

Further Reading

Bryant SH, Lawrence CE (1993). An empirical energy function for threading protein sequence through the folding motif. *Proteins* 16(1): 92–112.

Jones DT, Miller RT, Thornton JM (1995). Successful protein fold recognition by optimal sequence threading validated by rigorous blind testing. *Proteins* 23(3): 387–397.

Miller RT, Jones DT, Thornton JM (1996). Protein fold recognition by sequence threading: Tools and assessment techniques. *Faseb. J.* 10(1): 171–178.

See also Fold, Fold Library, Fold Recognition.

THREONINE
Jeremy Baum

Threonine is a polar amino acid with side chain —$CH(OH)CH_3$ found in proteins. In sequences, written as Thr or T.

See also Amino Acid, Polar.

TIM-BARREL
Roman A. Laskowski

A beta barrel consisting of eight parallel beta strands with each pair of adjacent strands connected by a loop containing an alpha helix springing from the end of the first strand and looping around the outside of the barrel before connecting to the start of the next strand. The name comes from the name of the first protein in which such a barrel was observed: triose phosphate isomerase (TIM). It is one of the most common tertiary folds observed in high-resolution protein crystal structures.

Related Websites

| DATE, a database of TIM barrels in enzymes | http://mbu.iisc.ernet.in/~pbgrp/date/ |
| TIM-DB, a database for TIM barrel proteins | http://www.quimica.urv.es/~pujadas/TIM/ |

See also Alpha Helix, Beta Barrel, Beta Strand, Fold.

TRANSCRIPT
John M. Hancock

The product of transcription.

In eukaryotes, transcripts may be mature (mRNA) or immature (hnRNA).

Further Reading
Lewin B (2000). *Genes VII*. Oxford University Press, Oxford, UK.

See also hnRNA, mRNA, Transcription.

TRANSCRIPTION
John M. Hancock

The process of generating an RNA molecule from a gene.

Transcription is initiated at the transcription start site under the control of the promoter and other regulatory elements. It proceeds until termination occurs, which may take place at a specific site or may be probabilistic in nature. The product of transcription is an immature RNA which then undergoes a variety of processing reactions to produce a mature message.

Further Reading
Lewin B (2000). *Genes VII*. Oxford University Press, Oxford, UK.

See also hnRNA, mRNA.

TRANSCRIPTIONAL REGULATORY REGION (REGULATORY SEQUENCE)
Niall Dillon and James W. Fickett

Sequence that is involved in regulating the expression of a gene (e.g., promoter, enhancer, LCR).

In eukaryotes, transcriptional regulatory regions (TRRs) are either promoters, at the transcription start site, or enhancers or repressors elsewhere. Enhancers and repressors as well as the regulatory portion of promoters are typically made up of functionally indivisible cis-regulatory modules (CRMs) in which a few transcription factors have multiple binding sites.

Further Reading
Lewin B (2000). *Genes VII*. Oxford University Press, Oxford, UK.

See also Enhancer, Promoter, Transcription Factor, Transcription Factor Binding Site, Transcription Start Site.

TRANSCRIPTION FACTOR
James W. Fickett

Generally speaking, any protein that acts in conjunction with an RNA polymerase to modify the initiation of transcription from the genome (Kuras and Struhl, 1999).

Transcription factors are sometimes divided into general factors, which are associated with RNA polymerase II in the transcription of all or most genes, and specific factors, which limit the expression of a gene to a particular cell type, developmental stage, or signal response process. Specific transcription factors are also called activators and repressors. The distinction between *general* and *specific* has broken down to some extent with the discovery that some transcription factors thought to be general can be, in fact, an important control point for large subsets of the transcriptome (Lee et al., 2000). There are many classes of transcription factors. Some bind DNA and some mediate signal transduction, without DNA contact, to the so-called transcription initiation complex (RNA polymerase II and associated proteins). Some modify the DNA, either chemically or architecturally, while others bind sequence specifically and act on other proteins. The DNA binding site of a transcription factor is sometimes called a transcription element. For characterization of the DNA-binding specificity of a transcription factors, see the discussion of position weight matrix.

Related Website

| BIOBASE | http://www.gene-regulation.de/ |

Further Reading

Kuras L, Struhl K (1999). Binding of TBP to promoters in vivo is stimulated by activators and requires Pol II holoenzyme. *Nature* 399: 609–613.

Lee TI, et al. (2000). Redundant roles for the TFIID and SAGA complexes in global transcription. *Nature* 405: 701–704.

See also Profile, Transcription, Transcription Factor Binding Site, Transcriptome.

TRANSCRIPTION FACTOR BINDING SITE
John M. Hancock

Sites on the genome to which transcription factors bind.

Transcription factors often (but not always) regulate gene transcription by binding to specific sequences in their promoters or other regulatory regions. They then often participate in multicomponent protein complexes which affect the transcription level of the gene in question.

Transcription factor binding sites are important indicators of the location of promoters or regulatory regions near genes. Unfortunately, these sites are relatively small (generally less than 10 bp) and can occur many times by chance in a typical genome. A number of approaches have been taken in an attempt to overcome the problem of distinguishing between real and artifactually identified binding sites, but this is still a major open problem in bioinformatics.

Related Websites

| TRANSFAC | http://www.generegulation.de/pub/databases.html#transfac |
| TESS | http://www.cbil.upenn.edu/tess/ |

See also Transcription Factor.

TRANSCRIPTION START SITE (TSS)
John M. Hancock

The point at the 5′ end of a gene at which RNA polymerase initiates transcription.

The TSS will be adjacent to the promoter and is the place at which the RNA polymerase complex associates with the DNA. Details of the context of the TSS will depend on the gene, what organism it is in, what type of polymerase is involved, and other factors.

Further Reading
Lewin B (2000). *Genes VII*. Oxford University Press, Oxford, UK.

TRANSCRIPTOME
Dov S. Greenbaum and Jean-Michel Claverie

The complete set of all transcripts that can be generated from a given genome. This includes all alternative transcripts (splice variant, alternative 3′ and 5′ UTR). In an experimental context, it is also used to designate the subset of transcripts found in a given cell, organ, or tissue. In a more quantitative way, the "transcriptome" can also refer to the set of all transcripts indexed by (associated with) a measurement of their abundance level.

The transcriptome connects the genome to the translatome and is usually measured, on a high-throughput scale, using cDNA microarray, Affymetrix GeneChip, or SAGE technologies. While there has been an effort to quantify the protein content of the cell, this does not diminish the importance of quantifying and understanding the mRNA population in the cell. While the genome is generally static, the ability of the transcriptome machinery to produce alternatively processed forms of transcripts provides much of the cellular heterogeneity and adaptability.

Transcriptome analysis has been used, among other uses, to compare disease states, predict cancer types, diagnose diseases, and cluster genes into functional categories for functional annotation transfer. In addition to comparing cell states, whole organisms can be compared with each other based on the degree of similarity or specific differences in their mRNA populations.

There still remain many uncertainties and inaccuracies in the mRNA data due to differences in experimental techniques between laboratories and the delicate and sensitive nature of the experimental procedure. As such, researchers must use caution whenever analyzing an mRNA data set. A more robust approach might include creating comprehensive data sets assembling multiple similar experiments into a more reliable data set.

Related Websites

Yeast transcriptome analysis	http://bioinfo.mbb.yale.edu/expression/transcriptome/
SGD Expression Connection	http://db.yeastgenome.org/cgi-bin/SGD/expression/expressionConnection.pl
NCBI Gene Expression Omnibus	http://www.ncbi.nlm.nih.gov/geo/

Further Reading

Saha S, et al. (2002). Using the transcriptome to annotate the genome. *Nat. Biotechnol.* 20: 508–512.

Velculescu VE, et al. (1997). Characterization of the yeast transcriptome. *Cell* 88: 243–251.

TRANSFAC

John M. Hancock

A database of transcription factors and binding sites.

The database contains data on transcription factors, their target genes, and regulatory binding sites. Information on expression patterns have recently been introduced for human and mouse. The database can be queried on cell type of origin, factor classification or name, regulated gene, binding site, or binding site profile. Sequences can be searched for binding sites using the program Match.

Related Websites

TRANSFAC	http://www.generegulation.de/pub/databases.html#transfac
TESS	http://www.cbil.upenn.edu/tess/

TRANSFER RNA (tRNA)

Jamie J. Cannone and Robin R. Gutell

Transfer RNAs (tRNAs) are typically 70–90 nucleotide in length in nuclear and chloroplast genomes and are directly involved in protein synthesis.

The carboxyl terminus of an amino acid is specifically attached to the 3' end of the tRNA (aminoacylated). These aminoacylated tRNAs are substrates in translation interacting with a specific mRNA codon to position the attached amino acid for catalytic transfer to a growing polypeptide chain. Thus, tRNAs decode (or translate) the nucleotide sequence during protein synthesis.

tRNAs have a characteristic "cloverleaf" structure that was initially determined with comparative analysis (see accompanying figure). Crystal structures of tRNA substantiated this secondary structure and revealed that different tRNAs formed very similar tertiary structures, underscoring the key underlying principle of comparative analysis. The "variable loop" of tRNA is primarily responsible for length variation among tRNAs; some of the mitochondrial tRNAs are smaller than the typical tRNA, shortening or deleting the D or T ψ C helices.

Nonmitochondrial tRNAs come in two types: 1 and 2. Structurally, the major difference between the two types is the addition of a stem-loop structure in the variable loop of type 2 tRNAs. The tRNA types do not correlate with the two classes of aminoacyl-tRNA synthetases, where class 1 synthetases attach amino acids to the 2'-OH and class 2 synthetases attach amino acids to the 3'-OH of the terminal nucleotide of the tRNA.

Over 50 modified nucleotides have been observed in different tRNA molecules.

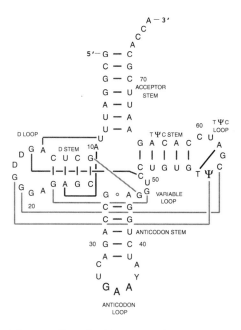

Figure Transfer RNA secondary structure
(*Saccharomyces cerevisiae* phenylalanine
tRNA). Structural features are labeled.

Related Websites

Aminoacyl-tRNA Synthetases Database	http://rose.man.poznan.pl/aars/index.html
Genomic tRNA	http://lowelab.ucsc.edu/GtRNAdb/
Mattias Sprinzl's tRNA compilation	http://www.uni-bayreuth.de/departments/biochemie/sprinzl/trna/
Modified RNA nucleotides	http://medlib.med.utah.edu/RNAmods/

Further Reading

Holley RW, et al. (1965). Structure of a ribonucleic acid. *Science* 147: 1462–1465.

Kim SH (1979). Crystal structure of yeast tRNAphe and general structural features of other tRNAs. In *Transfer RNA: Structure, Properties, and Recognition*, PR Schimmel, D Soll, JN Abelson, Eds. Cold Spring Harbor Laboratory Press, Cold Spring Harbor, NY, pp. 83–100.

Kim SH, et al. (1974). Three-dimensional tertiary structure of yeast phenylalanine transfer RNA. *Science* 185: 435–440.

Levitt M (1969). Detailed molecular model for transfer ribonucleic acid. *Nature* 224: 759–763.

Marck C, Grosjean H (2002). tRNomics: Analysis of tRNA genes from 50 genomes of Eukarya, Archaea, and Bacteria reveals anticodon-sparing strategies and domain-specific features. *RNA* 8: 1189–1232.

Quigley GJ, Rich A (1976). Structural domains of transfer RNA molecules. *Science* 194: 796–806.

Robertus JD, et al. (1974). Structure of yeast phenylalanine tRNA at 3Å resolution. *Nature* 250: 546–551.

TRANSLATION
John M. Hancock

The process of reading information encoded in a mature mRNA and converting it into a protein molecule.

Translation takes place on the ribosome and involves a number of cytoplasmic factors as well as the ribosomal components (rRNAs and ribosomal proteins). It consists of a series of complex chemical processes, including the selection and binding of cognate tRNA molecules, the movement of mRNA and nascent protein chains between binding sites on the ribosomal surface, and the sequential formation of peptide bonds between the growing protein chain and amino acids on incoming tRNA molecules.

Proteins may be subject to a variety or posttranslational modifications, including cutting by specific proteases, phosphorylation, and glycosylation before the mature protein is formed. These processes my be cell-type specific in eukaryotes and give rise to additional variation seen in proteomic experiments.

Further Reading
Lewin B (2000). *Genes VII*. Oxford University Press, Oxford, UK.

TRANSLATION START SITE
Roderic Guigó

Codon on an mRNA sequence at which the translation by the ribosome into an amino acid sequence is initiated.

The start codon is usually AUG (which codes for the amino acid methionine), but in prokaryotes it can also be GUG (coding for leucine). In prokaryotes, binding of the ribosome to the start AUG is mediated by the so-called Shine–Dalgarno motif (Shine and Dalgarno, 1974), a sequence complementary to the 3′ of the 16S rRNA of the 30 S subunit of the ribosome. The consensus Shine–Dalgarno sequence is UAAG-GAG, but the motif is highly variable.

In eukaryotes, ribosomes do not bind directly to the region of the messenger RNA that contains the AUG initiation codon. Instead, the 40 S ribosomal subunit starts at the 5′ end and "scans" down the message until it arrives at the AUG codon (Kozak, 1999). A short recognition sequence, the Kozak motif, usually surrounds the initiation codon. The consensus of this motif is ACCAUGG. If this sequence is too degenerate, the first AUG may be skipped, and translation is initiated at a further downstream AUG.

Related Websites

NetStart Prediction	http://www.cbs.dtu.dk/services/NetStart/
ATGpr Prediction	http://www.hri.co.jp/atgpr/
AUG_EVALUATOR Prediction	http://l25.itba.mi.cnr.it/webgene/wwwaug_help.html

Further Reading

Kozak M (1999). Initiation of translation in prokaryotes and eukaryotes. *Gene* 234: 187–208.

Shine J, Dalgarno L (1974). The 3′-terminal sequence of *E. coli* 16s ribosomal RNA: Complementary to nonsense triplets and ribosome binding sites. *Proc. Natl. Acad. Sci. USA* 71: 1342–1346.

See also Gene Prediction, Ab Initio; Motif Searching.

TRANSLATOME

Dov S. Greenbaum

An alternative term to *proteome* which refers specifically to the complement of translated products in a given cell or cell type.

The translatome refers to any quantifiable population of proteins within a cell at any given moment or under specific cellular circumstances. The translatome can be measured through multiple high-throughput technologies. Basically this requires that each protein be isolated, measured, and then identified. The isolation usually occurs through two-dimensional electrophoresis and protein identification occurs primarily through mass spectrometry methods.

Quantification can be experimentally determined, e.g., through radioactive labeling or computationally (using specific software such as Melanie, Z3, and similar programs) to quantify the pixel representation for individual proteins on a two-dimensional gel image.

While there has been significant work on the elucidation of the transcriptome, much work has yet to be done with regard to the translatome, the analysis of the end product of gene expression. Although there is some correlation between mRNA expression and protein abundance and mRNA expression clustering has been very useful in research, the translatome is still a superior picture of the cell and its processes.

Related Website

| Translatome | http://bioinfo.mbb.yale.edu/expression/translatome/ |

Further Reading

Greenbaum D, et al. (2002). Analysis of mRNA expression and protein abundance data: An approach for the comparison of the enrichment of features in the cellular population of proteins and transcripts. *Bioinformatics* 18: 585–596.

Gygi SP, et al. (2000). Measuring gene expression by quantitative proteome analysis. *Curr. Opin. Biotechnol.* 11: 396–401.

See also Proteome, Proteomics.

TRANSPOSABLE ELEMENT (TRANSPOSON)

Dov S. Greenbaum and Katheleen Gardiner

DNA sequences that can move from one genomic site to another. First described in maize by Barbara McClintock.

This insertion of new pieces of DNA into a sequence can increase or decrease the amount of DNA and possibly cause mutations if they jump into a coding region of a gene.

There are three distinct classes of transposable element:

Class I: Retrotransposons transcribe mRNA into DNA and insert this new piece of DNA into the genome. Retrotransposons are often flanked by long terminal repeats (LTRs) that can be over 1000 bases in length. About 40% of the entire human genome consists of retrotransposons.

Class II are pieces of DNA that move directly from one position to another. They require the activity of an enzyme, transposase, to cut and paste them into new positions within the genome. The transposase enzyme is often encoded within the transposon itself. Class II transposons can be site specific, i.e., they insert into specific sequences within the genome; others insert randomly.

Class III MITEs (miniature inverted repeat transposable elements), which are too small to encode any protein, have been found in the genome of humans, rice, apples+, and *Xenopus*.

Transposons can be mutagens, i.e., they can cause mutations within the genome of a cell. This is accomplished in four ways:

 (i) Insertion into a gene or its flanking regions can either enhance or prevent gene expression depending on where the transposon is inserted.
 (ii) When a transposon leaves its original site, there may be a failure of the cell to repair the gap and this may cause mutations.
(iii) Long strings of repeats caused by transposon insertion can interfere in pairing during meiosis.
(iv) Poly(A) tails carried by some transposable elements may evolve into microsatellites, in some cases causing disease, as in the case of the GAA repeat associated with an Alu element in the Friedreich's ataxia gene.

Further Reading

Cooper DN (1999). *Human Gene Evolution*. Bios Scientific Publishers, Oxford, UK, pp. 265–285.

Finnegan DJ (1992). Transposable elements. *Curr. Opin. Genet. Dev.* 2: 861–867.

McDonald JF (1993). Evolution and consequences of transposable elements. *Curr. Opin. Genet. Dev.* 3: 855–864.

Ridley M (1996). *Evolution*. Blackwell Science, Cambridge, MA, pp. 265–276.

Rowold DJ, Herrera RJ (2000). Alu elements and the human genome. *Genetics* 108: 57–72.

See also Alu Repeat, L1 Element, LINE, SINE.

TRANSPOSON. *SEE* TRANSPOSABLE ELEMENT.

TREE
Robert Stevens

A strict hierarchy.

See also Hierarchy, Phylogenetic Tree.

TREE-BASED PROGRESSIVE ALIGNMENT
Jaap Heringa

The most commonly used heuristic multiple-sequence alignment methods are based on the progressive alignment strategy (Feng and Doolittle, 1987; Hogeweg and Hesper, 1984; Taylor, 1988), with CLUSTALW (Thompson et al., 1994) being the most widely used implementation. The idea is to establish an initial order for joining the sequences and to follow this order in gradually building up the alignment. Many implementations use an approximation of a phylogenetic tree between the sequences as a so-called guide tree that dictates the alignment order. Although appropriate for many alignment problems, the progressive strategy suffers from its greediness. Errors made in the first alignments during the progressive protocol cannot be corrected later as the remaining sequences are added in ("Once a gap, always a gap"; Feng and Doolittle, 1987).

Further Reading

Feng DF, Doolittle RF (1987). Progressive sequence alignment as a prerequisite to correct phylogenetic trees. *J. Mol. Evol.* 21: 112–125.

Hogeweg P, Hesper B (1984). The alignment of sets of sequences and the construction of phyletic trees: An integrated method. *J. Mol. Evol.* 20: 175–186.

Taylor WR (1988). A flexible method to align large numbers of biological sequences. *J. Mol. Evol.* 28: 161–169.

Thompson JD, et al. (1994). CLUSTAL W: Improving the sensitivity of progressive multiple sequence alignment through sequence weighting, positions-specific gap penalties and weight matrix choice. *Nucleic Acids Res.* 22: 4673–4680.

See also Alignment—Multiple.

TREE OF LIFE
Sudhir Kumar and Alan Filipski

A phylogenetic framework depicting the evolutionary relationships among and within the three domains of extant life: eukaryotes, eubacteria, and archaea.

At present, many aspects of deep evolutionary relationships remain controversial or unknown as lateral gene transfer appears to have been common in the early history of life. For this reason, a traditional phylogenetic tree model may not be an appropriate representation of relationships among genomes of these primitive organisms.

The last common ancestor of all extant life forms on Earth is denoted the last universal common ancestor (LUCA). The characteristics and the time of origin of this organism is usually inferred by examining genes and biochemical mechanisms common to all extant life forms. It is possible that no such organism ever existed and that several lineages arose from prebiotic precursors.

Related Websites

Tree of Life Web Project	http://tolweb.org/tree/phylogeny.html
Looking for LUCA	http://www-archbac.u-psud.fr/Meetings/LesTreilles/LesTreilles_e.html

Further Reading

Brown JR, Doolittle WF (1995). Root of the universal tree of life based on ancient aminoacyl-tRNA synthetase gene duplications. *Proc. Natl. Acad. Sci. USA* 92: 2441–2445.

Hedges SB, et al. (2001). A genomic timescale for the origin of eukaryotes. *BMC Evol. Biol.* 1: 4.

Woese C (1998). The universal ancestor. *Proc. Natl. Acad. Sci. USA* 95: 6854–6859.

See also Phylogenetic Tree.

TREE PUZZLE. *SEE* QUARTETS, PHYLOGENETIC.

TREEVIEW
Michael P. Cummings

A program that displays and prints phylogenetic trees.

There are several options for the form of displayed trees (rectangular cladogram, slanted cladogram, phylogram, radial). The program accommodates a number of commonly used input file formats.

Executables are available for several platforms.

Related Website

| TreeView | http://taxonomy.zoology.gla.ac.uk/rod/treeview.html |

Further Reading

Page RDM (1996). TreeView: An application to display phylogenetic trees on personal computers. *Comput. Appl. Biosci.* 12: 357–358.

See also MacClade, PHYLIP.

TREMBL
Rolf Apweiler

The TrEMBL protein sequence database, a supplement to SWISS-PROT, is a universal computer-annotated protein sequence database covering proteins from many different species.

The TrEMBL protein sequence database was created in 1996 as a supplement to SWISS-PROT to make new protein sequences available as quickly as possible. TrEMBL (Translation of EMBL nucleotide sequence database), consists of computer-annotated entries derived from the translation of all coding sequences (CDS) in the EMBL nucleotide sequence database, except for those already included in SWISS-PROT, and peptide sequences directly submitted or found in the literature. TrEMBL is split in two main sections, SP-TrEMBL and REM-TrEMBL. SP-TrEMBL (SWISS-PROT TrEMBL) contains the entries (623,159 in release 20 of March 2002) which should be eventually incorporated into SWISS-PROT. REM-TrEMBL (REMaining TrEMBL) contains the entries that will not get included in SWISS-PROT (77,594 in release 20 of March 2002).

TrEMBL follows the SWISS-PROT format and conventions as closely as possible. The production of TrEMBL starts with the translation of coding sequences (CDS) in the EMBL nucleotide sequence database. At this stage all annotation in a TrEMBL entry comes from the corresponding EMBL entry. A first postprocessing step is to reduce redundancy by merging separate entries corresponding to different literature reports. Another postprocessing step is the automated enhancement of the TrEMBL annotation to bring TrEMBL entries closer to SWISS-PROT standards.

Related Websites

SWISS-PROT	http://www.ebi.ac.uk/swissprot/
ExPASy	http://www.expasy.org/sprot/

Further Reading

Apweiler R (2000). Protein sequence databases. *Adv. Protein Chem.* 54: 31–71.

Bairoch A, Apweiler R (2000). The SWISS-PROT protein sequence database and its supplement TrEMBL in 2000. *Nucl. Acids Res.* 28: 45–48.

Fleischmann W, Kretschmann E, Fleischmann W, Apweiler R (2001). Automatic rule generation for protein annotation with the C4.5 data mining algorithm applied on SWISS-PROT. *Bioinformatics* 17: 920–926.

Moeller S, Gateau A, Apweiler R (1999). A novel method for automatic and reliable functional annotation. *Bioinformatics* 15: 228–233.

O'Donovan C, Martin MJ, Glemet E, Codani J-J, Apweiler R (1999). Removing redundancy in SWISS-PROT and TrEMBL. *Bioinformatics* 15: 258–269.

See also Protein Information Resource, SWISS-PROT.

TREV
Rodger Staden

A widely used DNA trace viewing program.

Related Website

Trev	http://www.mrc-lmb.cam.ac.uk/pubseq/manual/trev_unix_toc.htm

TRINUCLEOTIDE REPEAT
Katheleen Gardiner

A tandem repeat of three nucleotides.

Repeats of this class have become of special interest because many trinucleotide repeats starting with C and ending with G (CAG, CGG, CTG, CCG) and present normally in ~5 to ~50 copies tend to be intergenerationally unstable. Repeats of this kind have been found to be associated with human genetic diseases when they expand beyond a critical copy number. At least 14 neurological disease associations have been identified, including fragile X syndrome, Huntington's disease, and myotonic dystrophy. Repeats may be located within coding regions (e.g., polyglutamine tracts) or in regulatory regions where they may alter methylation status and disrupt gene expression.

Further Reading

Cummings CJ, Zoghbi HY (2000). Trinucleotide repeats: Mechanisms and pathophysiology. *Annu. Rev. Genomics Hum. Genet.* 1: 281–328.

Ferro P, dell'Eva R, Pfeffer U (2001). Are there CAG repeat expansion-related disorders outside the central nervous system? *Brain Res. Bull.* 56: 259–264.

Grabczyk E, Kumari D, Usdin K (2001). Fragile X syndrome and Friedreich's ataxia: Two different paradigms for repeat induced transcript insufficiency. *Brain Res. Bull.* 56: 367–373.

tRNA. *SEE* TRANSFER RNA.

TRYPTOPHAN
Jeremy Baum

Tryptophan is a large aromatic amino acid with side chain —H found in proteins. In sequences, written as Trp or W.

See also Amino Acid, Aromatic.

TURN
Roman A. Laskowski

A reversal in the direction of the backbone of a protein that is stabilized by hydrogen bonds between backbone NH and CO groups and which is not part of a regular secondary-structure region such as an alpha helix.

Turns are classified into various types, the most common being beta and gamma turns, which themselves have further subclasses. In a beta turn, which consists of four residues, the CO group of residue i is usually, but not always, hydrogen bonded to the NH group of residue $i+3$. A gamma turn consists of three residues and has a hydrogen bond between residues i and $i+2$.

Related Website

| BTPRED | http://www.biochem.ucl.ac.uk/bsm/btpred/index.html |

See also Alpha Helix, Backbone, Secondary Structure of Proteins.

TWEAK ALGORITHM
Roland L. Dunbrack

A method developed by Fine and colleagues used in loop modeling to adjust the conformation of a loop so that it connects to the anchor points. TWEAK is designed to make the minimal changes needed in backbone dihedral angles from the starting configuration to close the loop. This is accomplished with Lagrange multipliers to minimize the change in dihedral angles, subject to the constraint that the loop be closed. TWEAK depends on a matrix inversion and because of this suffers from numerical instability when the matrix becomes singular.

Further Reading
Shenkin PS, et al. (1987). Predicting antibody hypervariable loop conformation. I. Ensembles of random conformations for ringlike structures. *Biopolymers* 26: 2053–2085.

See also Anchor Points.

TWILIGHT ZONE
Terri Attwood

A zone of identity (in the range ~0–20%) within which sequence alignments may appear plausible to the eye but are not statistically significant (in other words, the same alignment could have arisen by chance).

Many different sequence analysis methods have been devised to detect evolutionary relationships and ultimately penetrate deeper into the twilight zone: Some rely on pairwise sequence comparisons, others use characteristics of multiple alignments (i.e., consensus approaches); some use complex weighting schemes (e.g., exploiting mutation or structural information), others use only observed amino acid sequence data. Examples of the types of method used to detect evolutionary relationships and their sensitivity ranges are shown in the accompanying figure.

Each method offers a slightly different perspective, depending on the type of information used in the search; none should be regarded as giving the right answer or the full picture—none is infallible. For best results, a combination of approaches should be used.

It should be noted that there is a theoretical limit to the effectiveness of sequence analysis techniques because some sequences have diverged to such an extent that their relationships are only apparent at the level of shared structural features. Such

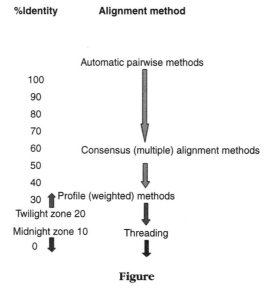

Figure

characteristics cannot be detected even using the most sensitive profile methods because no significant sequence similarity remains. To detect relationships in this "midnight zone'," threading algorithms tend to be employed to determine whether a sequence is likely to be compatible with a given fold.

Further Reading

Doolittle RF (1986). *Of URFs and ORFs: A Primer on How to Analyse Derived Amino Acid Sequences.* University Science Books, Mill Valley, CA.

Rost B (1998). Marrying structure and genomics. *Structure* 6(3): 259–263.

TWO-DIMENSIONAL GEL ELECTROPHORESIS (2DE)
Dov S. Greenbaum

Technique used in proteomics to separate protein molecules.

Two-dimensional gel electrophoresis is a relatively old technology. Essentially a protein population is run through a gel in two dimensions: by charge (pI) through isoelectric focusing (this is done using a pH gradient) and by size via sodium dodecyl sulfate—polyacrylamide gel electrophoresis (SDS–PAGE). SDS–PAGE uses SDS as a detergent to denature the proteins. Finally, the proteins are visualized either through radioactive labeling or through staining (e.g., Coomasie blue).

Spots/proteins of interest are excised from the gel, digested with trypsin, and then run though mass spectrometry. The masses resulting from this fragmentation can be compared with a database of proteins and the spot can then be identified.

Once a population of spots is identified, gels can be compared and relative abundances of proteins can be determined under different conditions. Presently there are many programs that are designed to measure and compare the intensity of a spot within an image of the 2D gel. These include PDQuest, Melanie, and Phoretix 2D.

The method allows for the separation of possibly thousands of proteins; each one can be seen as an individual spot on the gel. Additionally, the procedure is fast, relatively easy, and cheap.

Still there are many problems with the method as it does not resolve low-mass and rare proteins easily. Additionally, some protein types, such as membrane or hydrophobic proteins, have been shown to be problematic.

Related Website

2D Gel Electrophoresis for Proteomics Tutorial	http://www.aber.ac.uk/~mpgwww/Proteome/Tut_2D.html

Further Reading

Görg A, et al. (2000). The current state of two-dimensional electrophoresis with immobilized pH gradients. *Electrophoresis* 21: 1037–1053.

TYROSINE
Jeremy Baum

Tyrosine is a large aromatic amino acid with side chain —$CH_2C_6H_4OH$ found in proteins. In sequences, written as Tyr or Y.

See also Amino Acid, Aromatic.

U

Uncertainty

Ungapped Threading Test B

UniGene/LocusLink

Unit Cell

Universal Genetic Code

Unknome

Unsupervised Learning

Untranslated Region

UPGMA

Upstream

UTR

Dictionary of Bioinformatics and Computational Biology. Edited by Hancock and Zvelebil
ISBN 0-471-43622-4   © 2004 John Wiley & Sons, Inc.

UNCERTAINTY
Thomas D. Schneider

Uncertainty is a logarithmic measure of the average number of choices that a receiver or a molecular machine has available.

The uncertainty is computed as

$$H = -\sum_{i=1}^{M} P_i \log_2 P_i \text{ bits/symbol}$$

where P_i is the probability of the ith symbol and M is the number of symbols. Uncertainty is the average surprisal. The information is the difference between the uncertainty before and after symbol transmission.

Related Website

Information is not entropy, information is not uncertainty!	http://www.lecb.ncifcrf.gov/~toms/ information.is.not.uncertainty.html

See also After State, Before State, Entropy, Information, Negentropy, Surprisal, Small-Sample Correction.

UNGAPPED THREADING TEST B (SIPPL TEST)
David T. Jones

A method for evaluating the efficacy of a particular set of potentials or scoring function for threading.

Sometimes called the Sippl test (after Manfred Sippl, who popularized the test). The idea is to try to locate the native conformation for a particular protein sequence among a large set of decoys. These decoys are generated by taking fragments of others proteins, where the fragment length is equal to the length of the test protein. For example, a template protein of length 105 can provide six different decoy structures for a 100-residue protein in an ungapped threading test. Because the length of the template fragment is equal to the length of the test protein, there is no need to insert gaps in either protein; hence the test is labeled "ungapped."

A minimum requirement for a useful threading potential is that the energy calculated for the native structure of a protein should be lower than the calculated energy for all of the decoy conformations. Usually potentials are tested by performing the ungapped threading test for a large number of test proteins and calculating what fraction of the test proteins pass the test.

Despite the simplicity of the test, passing it must be considered a minimum requirement for a useful set of threading potentials. Many quite poor threading potentials achieve good results on the ungapped threading test.

UNIGENE/LOCUSLINK
Dov S. Greenbaum

The UniGene database provides for the clustering of overlapping GenBank sequences into cohesive (although many clusters do not represent whole contigs)

Dictionary of Bioinformatics and Computational Biology. Edited by Hancock and Zvelebil
ISBN 0-471-43622-4 © 2004 John Wiley & Sons, Inc.

unigene clusters, each representing a single gene. The dynamic nature of the automated processes which create the UniGene database (i.e., the relocation of sequences from one cluster to another) required the establishment of a more stable database, LocusLink. This database provides curated information regarding specific loci within a genome.

UniGene and LocusLink, both projects of the National Center for Biotechnology Information (NCBI), provide additional genetic annotation for each entry, including functional data, homologs, mapping, expression, and sequence information.

Presently, UniGene is limited to the human, rat, mouse, cow, zebrafish, clawed frog, wheat, rice, barley, maize, and cress, while LocusLink provides information regarding human, mouse, rat, fly, and zebrafish.

Related Websites

UniGene	www.ncbi.nlm.nih.gov/UniGene
LocusLink	www.ncbi.nlm.nih.gov/LocusLink/

Further Reading

Schuler GD (1997). Pieces of the puzzle: Expressed sequence tags and the catalog of human genes *J. Mol. Med.* 75: 694–698.

Pruitt KD, Maglott DR (2001). RefSeq and LocusLink: NCBI gene-centered resources. *Nucleic Acids Res.* 29: 137–140.

UNIT CELL
Liz P. Carpenter

The unit cell of a crystal is the building block from which a crystal is made. The whole crystal can be constructed from the unit cell by simple translations along the directions of the unit-cell edges by repeats of the unit-cell lengths. Crystals of macromolecules are necessary so that the three-dimensional structure of the macromolecule can be obtained by X-ray crystallography.

See also Diffraction of X Rays, X-Ray Crystallography for Structure Determination.

UNIVERSAL GENETIC CODE. *SEE* GENETIC CODE.

UNKNOME
Dov S. Greenbaum

That part of the genome whose function is unknown.

The unknome is presently the largest "ome." It represents the majority of genetic information, a population of sequences for which we have no annotation.

While the unknome presents no useful scientific information, its size is a constant reminder of how much more information must still be collected to fully understand and compare distinct genomes.

Related Websites

Omes	http://bioinfo.mbb.yale.edu/what-is-it/omes/
More Omes	http://www.genomicglossaries.com/content/omes.asp

UNSUPERVISED LEARNING. *SEE* SUPERVISED AND UNSUPERVISED LEARNING.

UNTRANSLATED REGION (UTR, 5′ UTR, 3′ UTR)
Niall Dillon

Regions of an mRNA molecule that do not code for a protein.

The 5′ untranslated region contains the Kozak sequence which is involved in recognition of the translation start site. Both 5′ and 3′ untranslated regions can also contain sequences that mediate posttranscriptional regulation by affecting RNA stability.

Further Reading
Lewin B (2000). *Genes VII*. Oxford University Press, Oxford, UK.

See also Kozac Sequence.

UPGMA
Sudhir Kumar and Alan Filipski

A simple clustering method for reconstructing a phylogeny when rates of evolution are assumed constant over different lineages.

This unweighted pair-group method using arithmetic averaging constructs a phylogenetic tree in a stepwise manner under the molecular clock assumption. It is a distance-based hierarchical clustering method in which the closest pair of taxa is clustered together in each step. The distance between any two clusters is defined to be the average of the pairwise distances between sequences, one from each cluster.

If the data are not known to obey the molecular clock, it may still be possible to apply UPGMA by first estimating a topology (using a method such as the neighbor-joining method), performing statistical tests to identify and eliminate sequences whose evolutionary rate differs significantly from the average, and finally reconstructing the tree with UPGMA based on a constant-evolutionary-rate assumption. The result is a phylogenetic tree based on partial data in which branch lengths are proportional to time. This is known as a linearized tree.

Further Reading
Nei M (1987). *Molecular Evolutionary Genetics*. Columbia University Press, New York.
Sneath PHA, Sokal RR (1973). *Numerical Taxonomy; the Principles and Practice of Numerical Classification*. W. H. Freeman, San Francisco.

See also Evolutionary Distance, Molecular Clock.

UPSTREAM
Niall Dillon

Describes a sequence distal to a specific point in the direction opposite to the direction of transcription (i.e., in a $5'$ direction on the strand being transcribed).

UTR. *SEE* UNTRANSLATED REGION.

V

Valine
Variance Components
VAST
VC
VecScreen
Vector
Vector Alignment Search Tool

VectorNTI
Verlet Algorithm
VISTA
Visualization, Molecular
Visualization of
    Multiple-Sequence Alignments

Dictionary of Bioinformatics and Computational Biology. Edited by Hancock and Zvelebil
ISBN 0-471-43622-4   © 2004 John Wiley & Sons, Inc.

VALINE
Jeremy Baum

Valine is a nonpolar amino acid with side chain —$CH(CH_3)_2$ found in proteins. In sequences, written as Val or V.

See also Amino Acid, Polar.

VARIANCE COMPONENTS (COMPONENTS OF VARIANCE, VC)
Mark McCarthy and Steven Wiltshire

Describes the decomposition of the total variance of a quantitative trait into its various genetic and environmental components.

The phenotype P of individual i can be described according to the following standard model:

$$P_i = \mu + G_i + E_i$$

where μ is the overall population mean of the trait and G_i and E_i are random variables representing the independent deviations from μ caused by individual i's genotype and nonshared (micro) environment, respectively: both G_i and E_i have a mean of zero and variances σ_G^2 and σ_E^2, respectively, in the population as a whole. Optionally, terms modeling a common family environment shared by relatives, genetic-environmental interactions, and the fixed effects of a set of measured covariates can be incorporated into this model. The overall variance of the trait, σ_P^2, can be written as the sum of these components of variance. For the basic model described above,

$$\sigma_P^2 = \sigma_G^2 + \sigma_E^2$$

The genetic variance component (σ_G^2) can in turn be further partitioned into additive (σ_A^2), dominance (σ_D^2), and epistasis (σ_I^2) components:

$$\sigma_G^2 = \sigma_A^2 + \sigma_D^2 + \sigma_I^2$$

These three variance components reflect the effects on the phenotypic variation of the trait, of the individual susceptibility alleles at a given locus, of combining different alleles at the same locus, and of interactions (or *epitasis*) between different alleles at different loci, respectively.

This framework is the basis of a widely used from of quantitative trait *linkage analysis*, in which the total phenotypic variance is modeled as the sum of the effects of a *quantitative trait* locus (QTL), σ_Q^2 (the location of which is to be determined), a residual *genetic* component, σ_{PG}^2, which contains the effects of all other genes influencing the trait, and the nonshared environment component, σ_E^2:

$$\sigma_P^2 = \sigma_Q^2 + \sigma_{PG}^2 + \sigma_E^2$$

As above, the two genetic components may themselves be partitioned into additive, dominance and epistasis components, and terms for familial (common)

Dictionary of Bioinformatics and Computational Biology. Edited by Hancock and Zvelebil
ISBN 0-471-43622-4 © 2004 John Wiley & Sons, Inc.

environment, gene-environment interactions and covariates can be included. The covariance between pedigree members' trait values is expressed in terms of the variance components attributable to the QTL and residual polygenic effect, as described above (usually restricting this to the additive components of each, for simplicity) and measures of the genetic similarity between the relatives, namely the proportion of alleles shared *identical by descent* at each individual genetic *marker* (for the QTL) and across the genome as a whole (for the residual genetic term). The likelihood of the pedigree at each position on the *marker* map is calculated from these variance and covariance terms, the trait values and the population mean assuming multivariate normality. The null hypothesis of no linkage (i.e., $\sigma_Q^2 = 0$) can be tested by comparing the log-likelihood of the data given $\sigma_Q^2 = 0$ with that given $\sigma_Q^2 > 0$ (i.e., linkage), a significant likelihood ratio being evidence for linkage. This likelihood ratio statistic can be expressed as a LOD (or logarithm of odds) score.

Examples: Using variance components analysis, Hirschhorn et al. (2002), Perola et al. (2002), Wiltshire et al. (2002) and Xu et al. (2002) observed evidence for QTLs on several chromosomes influencing adult stature in Scandinavian, Quebecois, British, and Dutch populations.

Related Website
Programs for variance components linkage analysis of quantitative traits include

GENEHUNTER 2	http://www.broad.mit.edu/ftp/distribution/software/genehunter/
SOLAR	http://www.sfbr.org/solar/

Further Reading

Almasy L, Blangero J (1998). Multipoint quantitative trait linkage analysis in general pedigrees. *Am J Hum Genet* 62: 1198–1211.

Amos CI (1994). Robust variance-components approach for assessing genetic linkage in pedigrees. *Am J Hum Genet* 54: 535–543.

Falconer DS, Mackay TFC (1996). *Introduction to quantitative genetics.* Prentice-Hall, Englewood Cliffs, NJ, pp. 100–159.

Hirschhorn JN, Lindgren CM, Daly MJ, Kirby A, Schaffner SF, Burtt NP, Altshuler D, Parker A, Rioux JD, Platko J, Gaudet D, Hudson TJ, Groop LC, Lander ES (2002). Genomewide linkage analysis of stature in multiple populations reveals several regions with evidence of linkage to adult height. *Am J Hum Genet* 69: 106–116.

Perola M, Ohman M, Hiekkalinna T, Leppavuori J, Pajukanta P, Wessman M, Koskenvuo M, Palotie A, Lange K, Kaprio J, Peltonen L (2002). Quantitative-trait-locus analysis of body-mass index and of stature, by combined analysis of genome scans of five Finnish study groups. *Am J Hum Genet* 69: 117–123.

Pratt SC, Daly MJ, Kruglyak L (2000). Exact multipoint quantitative-trait linkage analysis in pedigrees by variance components. *Am J Hum Genet* 66: 1153–1157.

Wiltshire S, Frayling TM, Hattersley AT, Hitman GA, Walker M, Levy JC, O'Rahilly S, Groves CJ, Menzel S, Cardon LR, McCarthy MI (2002). Evidence for linkage of stature to chromosome 3p26 in a large U.K. Family data set ascertained for type 2 diabetes. *Am J Hum Genet* 70: 543–546.

Xu J, Bleecker ER, Jongepier H, Howard TD, Koppelman GH, Postma DS, Meyers DA (2002). Major recessive gene(s) with considerable residual polygenic effect regulating adult height: confirmation of genomewide scan results for chromosomes 6, 9, and 12. *Am J Hum Genet* 71: 646–650.

See also Identity by Descent; Linkage Analysis; LOD Score; Multifactorial/Complex Trait; Quantitative Traits.

VAST (VECTOR ALIGNMENT SEARCH TOOL)

John M. Hancock and Martin J. Bishop

A program to identify structurally similar proteins based on statistical criteria.

VAST attempts to identify "surprising" similarities between protein structures using the common statistical approach that the similarity would be expected to occur by chance with less than a certain threshold likelihood. It compares units of tertiary structure, which are defined as pairs of secondary structural elements within a protein. It bases a measure of similarity between structural elements on the type, relative orientation, and connectivity of these pairs of secondary structural elements and derives expectations by drawing these properties at random. Results of the analysis are precompiled and made available via the NCBI website. *VAST Search* allows the VAST algorithm to be used to search a new protein structure against the NCBI's protein structure database MMDB.

Related Websites

VAST	http://www.ncbi.nlm.nih.gov/Structure/VAST/vast.shtml
VAST search	http://www.ncbi.nlm.nih.gov/Structure/VAST/vastsearch.html

Further Reading

Gibrat J-F, et al. (1996). Surprising similarities in structure comparison. *Curr. Opin. Struct. Biol.* 6: 377–385.

Madej T, et al. (1995). Threading a database of protein cores. *Protein Struct. Funct. Genet.* 23: 356–369.

See also DALI, NCBI, Protein Data Bank.

VC. *SEE* VARIANCE COMPONENTS.

VECSCREEN

John M. Hancock and Martin J. Bishop

A program to screen sequences for contaminating vector sequences.

VecScreen uses BLAST to screen sequences for vector sequences held in the Uni-Vec database of vector sequences. Because contaminant sequences are likely to be identical to known vector sequences, or almost so, parameters for the BLAST search are set accordingly. VecScreen returns three types of positive match, according to the length and quality of the match (strong, moderate, weak), and segment of suspect origin.

Related Website

VecScreen	http://www.ncbi.nlm.nih.gov/VecScreen/VecScreen.html

See also BLAST, Chimeric DNA Sequence, Vector.

VECTOR

Molecular biology tool for the cloning of fragments of DNA.

Various vectors are used in molecular biology, from plasmids, which are modified versions of circular molecules that naturally infect bacterial cells, to bacterial and yeast artificial chromosomes (BACs, YACs). Vectors are distinguished by the size of the foreign DNA fragment that can be inserted in them (largest for artificial chromosomes) and the nature of the other sequences into them, which may include specific types of promoter or other regulatory sequence, and genes, which may serve to mark cells in which the vector is present.

Related Websites

The UniVec Database	http://www.ncbi.nlm.nih.gov/VecScreen/UniVec.html
Vectordb	http://genome-www2.stanford.edu/vectordb/

Further Reading
Lewin B (2000) *Genes VII.* Oxford University Press, Oxford, UK.

VECTOR ALIGNMENT SEARCH TOOL. *SEE* VAST.

VECTORNTI
John M. Hancock

A desktop package that provides a limited set of sequence manipulation and analysis facilities.

Unlike other desktop bioinformatics packages, VectorNTI is based around a database which allows data and results to be stored and shared within a work group. The package supports the following classes of function: clone design; PCR primer design; oligoanalysis; restriction fragment analysis; mutagenesis analysis; about 60 types of graphical protein, DNA, and RNA analyses; multiple-sequence analysis; and production of an animated virtual gel.

Related Website

Commercial site	http://www.informaxinc.com/solutions/vectornti/index.html

See also LaserGene, MacVector, Wisconsin Package.

VERLET ALGORITHM
Roland L. Dunbrack

The most common algorithm used to integrate Newton's equations of motion in molecular dynamics simulations. By writing Taylor series expansions for atom positions at times $t + \Delta t$ and $t - \Delta t$ in terms of the current positions $\mathbf{r}$, velocity, and acceleration $\mathbf{a}$ and subtracting, a formula for $\mathbf{r}(t + \Delta t)$ is obtained:

$$\mathbf{r}(t + \Delta t) = 2\mathbf{r}(t) - \mathbf{r}(t - \Delta t) + \mathbf{a}(t)\Delta t^2 + O(\Delta t^4)$$

in which the error in positions in fourth order in the time step Δt. The acceleration is obtained from Newton's equation, $\mathbf{F} = m\mathbf{a}$, and the force $\mathbf{F}$ is obtained from the energy function $\mathbf{F} = -\nabla U(\mathbf{r})$, where U is the potential energy function.

Related Website

Molecular dynamics primer	http://www.fisica.uniud.it/~ercolessi/md/md/md.html

See also Molecular Dynamics Simulations.

VISTA
John M. Hancock

Package for aligning and investigating long genomic sequences.

VISTA (Visualization Tools for Alignments) consists of three core programs: AVID, mVISTA, and rVISTA. AVID is a sequence alignment program designed to rapidly align long segments of genomic sequence. It approaches the problem of alignment by finding a set of maximal repeated substrings between two sequences. These are found by concatenating the two sequences and identifying repeated substrings using suffix trees. Alignment is then carried out by defining anchor sequence pairs and aligning between them using the Needleman–Wunsch algorithm. mVISTA allows visualization of AVID alignments while rVISTA identifies transcription factor binding sites that are conserved in an AVID alignment.

Related Website

VISTA	http://www-gsd.lbl.gov/vista/

Further Reading

Bray N, et al. (2003). AVID: A global alignment program. *Genome Res.* 13: 97–102.

Loots GG, et al. (2002). rVista for comparative sequence-based discovery of functional transcription factor binding sites. *Genome Res.* 12: 832–839.

See also Multiple-Sequence Alignment, Needelman–Wunsch Algorithm, Transcription Factor.

VISUALIZATION, MOLECULAR
Eric Martz

Looking at a model of a molecule in order to grasp its three-dimensional structure.

Molecular visualization involves generation of three-dimensional images or physical models of molecules that can be examined from multiple perspectives. Computer renderings can be rotated and viewed from varied distances ("zoomed") with views targeted to selected focal points. Computer visualization facilitates rendering of the model in different modes [e.g., backbone, schematic ("cartoon"), space filled, surface] and with different color schemes. Optionally, some parts may be hidden to avoid obscuring moieties of interest. Some visualization software can display animations ("movies" or morphs) of conformational changes in proteins (see Database of Molecular Motion in Related Websites).

"Molecular modeling" includes visualization, but "modeling" may also include the ability to change the conformation or covalent structure of the molecule, while visualization, strictly speaking, may be limited to displaying a model using software tools that are unable to modify its structure. Some popular modeling software packages include DeepView (free), WhatIf, Insight, Quanta, and for crystallographic modeling, O.

Visualization of macromolecules requires that data for a three-dimensional model be available, either from empirical or theoretical sources. Such data are typically represented as atomic coordinate files that specify the positions of each atom in space. Most empirical macromolecular structures are obtained by X-ray crystallography, which produces an electron density map. The amino acid sequence of the protein is usually known in advance, and real-time interactive visualization software is used to assist in visualizing the electron density map and fitting the amino acids into it. One of the most popular computer systems used by crystallographers for this purpose in the 1980s was manufactured by Evans and Sutherland and cost about a quarter of a million U.S. dollars at that time.

By the end of the 1980s, powerful computers were much less expensive. The first widely popular software that brought macromolecular visualization to ordinary personal computers was the free program MAGE released in 1992 by David C. Richardson. This supported presentations of the authors' viewpoints called Kinemages as well as interactive visualization. In 1993, Roger A. Sayle released RasMol, with excellent support for self-directed interactive visualization, which also became widely popular. In the mid-1990s, a team at GlaxoWellcome released a powerful visualization and modeling program called Swiss-PDBViewer (recently renamed DeepView). In 1996, RasMol's open-source code was adapted to a Web browser plugin, Chime ("Chemical MIME"), initially used for Web-delivered presentations of the authors' viewpoints. RasMol was also used as part of the foundation of WebLab Viewer. In 1999, Chime was employed as the basis for Protein Explorer, a Web-based user interface that facilitates self-directed interactive visualization while freeing the user from having to learn the RasMol command language. About the same time, the U.S. National Center for Biotechnology Information developed a completely independent self-directed interactive macromolecular visualization program, Cn3D. Galleries of sample images as well as visualization software itself can be obtained through the World Index of Molecular Visualization Resources.

Software that allows interactive rotation of the model compromises resolution and image quality in order to achieve adequate rotation speed in real time. Most figures of three-dimensional structures published in scientific journals are generated by software that emphasizes image quality and resolution at the expense of speed of image generation. The most popular is probably Molscript, a program that emits an image description metalanguage that serves as input to rendering software such as Raster3D or POVRay.

Related Websites

World Index of Molecular Visualization Resources	http://molvisindex.org
Protein Explorer (easiest and most powerful for beginners)	http://proteinexplorer.org

Chime (methods, templates, examples)	http://www.umass.edu/microbio/chime
Chime (program)	http://www.mdlchime.com
National Center for Biotechnology Information's Cn3D	http://www.ncbi.nlm.nih.gov/Structure/CN3D/cn3d.shtml
RasMol (help for beginners)	http://www.umass.edu/microbio/rasmol
RasMol (latest version)	http://openrasmol.org
Mage	http://kinemage.biochem.duke.edu/
WebLab	http://www.accelrys.com/viewer/
Physical Molecular Models	http://www.netsci.org/Science/Compchem/feature14b.html
History of visualization of biological macromolecules	http://www.umass.edu/microbio/rasmol/history.htm
Database of Molecular Motions	http://molmovdb.mbb.yale.edu/MolMovDB/
MolScript	http://www.avatar.se/molscript/
Raster3D	http://www.bmsc.washington.edu/raster3d/raster3d.html
POV-Ray	http://www.povray.org/
List of modeling software	http://molvisindex.org.

Further Reading

Bernstein, HJ (2000). Recent changes to RasMol, recombining the variants. *Trends Biochem. Sci.* 25: 453–455.

Guex N, Diemand A, Peitsch MC (1999). Protein modelling for all. *Trends Biochem. Sci.* 24: 364–367.

Martz E (2002). Protein Explorer: Easy yet powerful macromolecular visualization. *Trends Biochem. Sci.* 27: 107–109.

Richardson DC, Richardson JS. (1992). The kinemage: A tool for scientific communication. *Protein Sci.* 1: 3–9.

Wang Y, Geer LY, Chappey C, Kans JA, Bryant SH (2000). Cn3D: Sequence and structure views for Entrez. *Trends Biochem. Sci.* 25: 300–302.

See also Backbone; Models, Molecular; Protein Data Bank.

Eric Martz is grateful for help from Eric Francoeur, Peter Murray-Rust, Byron Rubin, and Henry Rzepa.

VISUALIZATION OF MULTIPLE-SEQUENCE ALIGNMENTS
Terri Attwood

The use of suitable tools for visualizing and manually editing alignments is therefore essential. Particularly important is the choice of an appropriate coloring scheme (an example is given below), which can immediately highlight highly conserved residues and physicochemically related residue groups.

Property	Residue	Color
Acidic	Asp, Glu	Red
Basic	His, Lys, Arg	Blue
Polar uncharged	Ser, Thr, Asn, Gln	Green
Hydrophobic aliphatic	Ala, Val, Leu, Ile, Met	White
Hydrophobic aromatic	Phe, Tyr, Trp	Purple
Disulfide bond former	Cys	Yellow
Structural properties	Pro, Gly	Brown

Related Websites

CINEMA	http://www.bioinf.man.ac.uk/dbbrowser/CINEMA2.1/
JalView	http://www.ebi.ac.uk/~michele/jalview/contents.html

Further Reading
Parry-Smith DJ, Payne AWR, Michie AD, Attwood TK (1998). CINEMA—A novel Colour INteractive Editor for Multiple Alignments. *Gene* 221(1): GC57–GC63.

See also Amino Acid.

W

Dictionary of Bioinformatics and Computational Biology. Edited by Hancock and Zvelebil
ISBN 0-471-43622-4 © 2004 John Wiley & Sons, Inc.

WEB-BASED SECONDARY-STRUCTURE PREDICTION PROGRAMS

Patrick Aloy

DSC

DSC is a rule-based method to predict secondary structure in proteins. It uses an expert system that combines residue information such as conformational propensities, hydrophobicity, and position of indels in homologous sequences to achieve a three-state-per-residue accuracy of 70.1%.

Jpred

Jpred is an Internet Web server for predicting protein secondary structure from single sequences or multiple-sequence alignments. It works by combining a number of modern, high-quality prediction methods to form a consensus.

This method showed three-state-per-residue prediction accuracies of 74.8% and 72.9% on data sets of 126 and 396 proteins, respectively.

PHD (PHDpsi)

PHD is probably the most widely used method for predicting protein secondary structure. It is based on a two-layered feed-forward neural network able to predict three states per residue (helix, beta, and coil). PHD represented a real breakthrough in the field as it used, for the first time, evolutionary information from multiple alignments. This resulted in an increase of 6–8 percentage points in the prediction accuracy, reaching 70.8% when tested on a nonredundant set of 130 protein chains.

Recently, a version of PHD that uses PSI-BLAST (PHDpsi) has achieved a three-state-per-residue accuracy of 75%.

PSIPRED

PSIPRED is a method for protein secondary-structure prediction from sequence information. It is based on two feed-forward neural networks that compute and analyze the output obtained from PSI-BLAST.

The latest version of PSIPRED includes a new algorithm that averages the output from four separate neural networks in the prediction process, achieving a three-state-per-residue accuracy of between 76.5% and 78.3%. The method was cross-validated on a set of 187 unique folds and also performed very well at CASP blind tests.

Related Websites

Jura	http://jura.ebi.ac.uk:8888/
PredictProtein	http://cubic.bioc.columbia.edu/predictprotein/
PSIPRED	http://bioinf.cs.ucl.ac.uk/psipred/

Further Reading

Cuff JA, Barton GJ (1999). Evaluation and improvement of multiple sequence methods for protein secondary structure prediction. *Proteins* 34: 508–519.

King RD, Sternberg MJ (1996). Identification and application of the concepts important for accurate and reliable protein secondary structure prediction. *Protein Sci.* 5: 2298–2310.

See also Secondary-Structure Prediction of Protein.

Dictionary of Bioinformatics and Computational Biology. Edited by Hancock and Zvelebil
ISBN 0-471-43622-4 © 2004 John Wiley & Sons, Inc.

WEIGHT MATRIX. *SEE* SEQUENCE MOTIFS—PREDICTION AND MODELING.

WHATCHECK
Roland L. Dunbrack

A program for assessing the stereochemical quality of a protein structure, developed by Gert Vriend and colleagues. Whatcheck reports deviations in covalent geometry, dihedral angles, steric clashes, and hydrogen bonding.

Related Website

| Whatcheck | http://www.cmbi.kun.nl/swift/whatcheck/ |

Further Reading
Hooft RW, et al. (1996). Errors in protein structures. *Nature* 381: 272.
Hooft RW, et al. (1997). Objectively judging the quality of a protein structure from a Ramachandran plot. *Comput. Appl. Biosci.* 13: 425–30.

WHATIF
Roland L. Dunbrack

A multifunctional program for protein structure analysis, comparative modeling, and visualization developed by Gert Vriend and colleagues. WhatIfhas a graphical interface for menu-driven commands, including access to a database of protein fragments for loop and side-chain modeling.

Related Website

| WhatIf | www.cmbi.kun.nl/whatif |

Further Reading
Rodriguez R, et al. (1998). Homology modeling, model and software evaluation: Three related resources. *Bioinformatics* 14: 523–528.
Vriend G (1990). WHAT IF: A molecular modeling and drug design program. *J. Mol. Graphics* 8: 52–56.

See also Comparative Modeling, Side-Chain Prediction.

WIRE-FRAME MODELS (STICK MODELS, SKELETAL MODELS)
Eric Martz

Models of molecules in which covalent bonds are represented by lines. The lines can be thickened to cylinders, or in physical models, represented by wire rods or sticks of wood or plastic. In contrast to *ball and stick models*, atoms are represented only by the intersections of the lines. Double and triple bonds may be represented by multiple lines, and resonant bonds with dotted lines. (See illustration at *Models, Molecular.*)

Wire-frame models, augmented with representations of bases in sheet metal, were used by Watson and Crick in the early 1950s to construct their original model of the DNA double helix. In the early 1960s, Kendrew et al. obtained with *crystallography* the first atomic-resolution structures for proteins, and constructed models using brass wire rods in a framework of metal and wire supports, hence the name "wire-frame". Initially, these models were scaled at 5 cm/Å, then at 2 cm/Å. As larger proteins and nucleic acids were solved, the scale was reduced to 1 cm, then 0.5 cm/Å. (Some of the early models had to be partially disassembled when it was necessary to transport them through doorways into a new location.) Wire-frame models were constructed to fit crystallographic electron density maps drawn on transparent lucite sheets using a half-silvered mirror to align the wire-frame model with the maps. This apparatus was developed by Fred Richards, and is known as a Richards' box, or "Fred's Folly."

Wire-frame renderings were the earliest computer graphic molecular models, since they were easiest to accomplish. Later, as hardware and software became more powerful, other styles of molecular *models* were added.

Related Websites

Physical Molecular Models	http://www.netsci.org/Science/Compchem/feature14b.html
History of Visualization of Biological Macromolecules	http://www.umass.edu/microbio/rasmol/history.htm

Further Reading

Crick FHC, Watson JD (1954). The complementary structure of deoxyribonucleic acid. *Proc. Roy. Soc. A* 223: 80–96. (Figures 5 and 6 are photographs of a wire-frame model augmented with sheet-metal bases.)

Watson JD (1980). *The Double Helix*, W.W. Norton & Co., New York. (Chapters 27 and 28 discuss model-building and show photographs).

Kendrew JC et al. (1960). Structure of Myoglobin. A Three-Dimensional Fourier Synthesis at 2 Å Resolution. *Nature* 185: 422–27.

Kendrew JC (1961). The Three-Dimensional Structure of a Protein Molecule. *Scientific American* 205(6): 96–110.

Richards FM (1968). The matching of physical models to three-dimensional electron-density maps: A simple optical device. *J. Mol. Biol.* 37: 225–230.

See also Models, Molecular; Visualization, Molecular.

Eric Martz is grateful for help from Eric Francoeur, Peter Murray-Rust, Byron Rubin, and Henry Rzepa.

WISCONSIN PACKAGE (GCG)
John M. Hancock

A package of sequence analysis utilities that has become a standard for the general user.

Originally released in 1982 as an academic package by the Genetics Computer Group (GCG) at the University of Wisconsin and subsequently distributed on a commercial basis, the Wisconsin Package contains a wide range of utility programs for DNA and protein sequence manipulation and analysis. Programs classically run on the command line in a Unix environment although X Windows (SeqLab) and Web (SeqWeb) interfaces are also available. Version 10.3 (2001) of the package includes 114 programs under 15 general areas: Editing, Fragment Assembly, Mapping, Sequence Comparison, Database Searching, Multiple-Sequence Analysis, Pattern Recognition and Compositional Analysis, RNA Secondary Structure, Protein Sequence Analysis, Evolutionary Analysis, Translation, Manipulation, Display and Publication, Sequence Exchange, and Miscellaneous.

Related Websites

| Commercial site | http://www.accelrys.com/products/gcg_wisconsin_package/index.html |
| Unofficial guide | http://www.cbc.umn.edu/MBsoftware/GCG/Unofficial_Guide/MolBio_man.html |

Further Reading
Womble DD (2000). GCG: The Wisconsin Package of sequence analysis programs. *Methods Mol. Biol.* 132: 3–22.

See also EMBOSS.

X

X Chromosome

Xenolog

XML

X-Ray Crystallography for Structure Determination

X CHROMOSOME. *SEE* SEX CHROMOSOME.

XENOLOG
Laszlo Patthy

Homologous genes acquired through horizontal transfer of genetic material between different species are called xenologs.

See also Homology.

XML (EXTENSIBLE MARKUP LANGUAGE)
Eric Martz

XML is a markup language that identifies data elements within a document. It is widely considered the universal format for structured documents and data on the Web. Its power is that the recipient of a document can recognize the contained data elements following a worldwide standard and can therefore more easily utilize and display the data in any desired manner.

XML annotates content, while, in contrast, HTML (HyperText Markup Language) annotates primarily appearance. HTML has a fixed set of tags, while XML (the X stands for extensible) is designed to accommodate an ever-growing set of discipline-specific tags. XML is an emerging standard, whereas HTML is a standard already in universal use for annotating information on the Web.

HTML specifies a static appearance for a document when displayed in a Web browser. HTML alone does not enable the recipient Web browser to identify the types of data elements contained in the document. For example, HTML can specify that a particular phrase be displayed centered in a large, red, bold font but tells nothing about the kind of information in the phrase. In contrast, XML identifies the type of information. For example, XML tags could identify a certain portion of a document as the amino acid sequence for a specified protein. Presently, the content of a document that is of interest to a user must usually be selected by nonstandard methods that are often inadequately automated. Once the desired information is extracted in computer-readable form, the format employed is often understood by only a subset of the programs in the world that deal with that type of information. XML provides a worldwide standard for content identification.

XML itself is set of rules for writing discipline-specific implementations. The specifications for XML are managed by the World Wide Web Consortium in the form of W3C recommendations. Each subject discipline must develop its own a document type definition (DTD) or XML schema (see examples below). Discipline-specific content within a document is identified by a namespace. A document can contain mixtures of types of content, i.e., multiple namespaces. Within each namespace, the allowed content-identifying tags and the nesting relationships of the tags are specified by the corresponding schema or DTD. For example, in the Chemical Markup Language (CML) schema or DTD, one element is cml:molecule. It can have child elements such as cml:atomArray and cml:bondArray. The cml:bondArray

Dictionary of Bioinformatics and Computational Biology. Edited by Hancock and Zvelebil
ISBN 0-471-43622-4   © 2004 John Wiley & Sons, Inc.

element has cml:atom elements as its children. The XML schema for a particular discipline, such as CML, is enforced by a validation process. Each schema has a primary website, where the discipline-specific tags and rules for validation are defined.

Examples of well-defined XML languages for common generic use are

1. XHTML (a general document markup language derived from HTML),
2. SVG (scalable vector graphics) for expressing diagrams and visual information, and
3. MathML (mathematical markup language).

Specific scientific, technical, and medical (STM) namespaced applications of XML include

1. STMML (for description of scientific units and quantities),
2. CML (Chemical Markup Language) for expressing molecular content,
3. CellML for computer-based biological models,
4. GAME (genome annotation markup elements),
5. MAGE-ML (MicroArray and Gene Expression Markup Language),
6. Molecular Dynamics [Markup] Language (MoDL),
7. StarDOM—transforming scientific data into XML,
8. Bioinformatic Sequence Markup Language (BSML),
9. BIOpolymer Markup Language (BIOML),
10. Gene Expression Markup Language (GEML),
11. GeneX Gene Expression Markup Language (GeneXML),
12. Genome Annotation Markup Elements (GAME),
13. XML for Multiple Sequence Alignments (MSAML),
14. Systems Biology Markup Language (SBML), and
15. Protein Extensible Markup Language (PROXIML).

Related Websites

World Wide Web Consortium managing the specifications for XML	http://www.w3c.org
Chemical Markup Language	http://www.xml-cml.org

Further Reading

Bosak J, Bray T (1999). XML and the second generation web. *Sci. Am.*, May 6. http://www.sciam.com.

Ezzell C (2000). Hooking up biologists. *Sci. Am.*, 283: 22. November 19.

Murray-Rust P, Rzepa HS (2002). Scientific publications in XML—towards a global knowledge base. *Data Sci.* 1: 84–98.

See also Atomic Coordinate File, MIME Types.

Eric Martz is grateful for help from Eric Francoeur, Peter Murray-Rust, Byron Rubin, and Henry Rzepa.

X-RAY CRYSTALLOGRAPHY FOR STRUCTURE DETERMINATION

Liz P. Carpenter

X-ray crystallography is a technique for studying the structures of macromolecules and small molecules in the crystalline state.

Crystals of the molecule under study are exposed to a beam of X rays usually of one wavelength. X-rays are scattered by the electrons in the atoms of the crystal. The regular array of atoms in the crystal results in constructive interference in some directions. This gives rise to diffraction spots which are recorded on a detector. Since there are no lenses available for X rays, the image of the molecule in the crystal cannot be obtained directly from the diffraction pattern. An image of the molecule in the crystal can, however, be obtained computationally. The density of electrons at each point in the repeat unit of the crystal (the unit cell) can be calculated given a knowledge of the intensities of the diffraction spots and their phase. Detectors can record intensity but not phase information, so a variety of techniques (molecular replacement, heavy-atom derivatives, MAD, SAD, SIRAS) are used to obtain information on the phase of the diffraction spots. An initial electron density map can then be calculated, from which the positions of the atoms in the protein structure can be obtained. X-ray crystallography has the advantage that it can be used to obtain three-dimensional structures of any molecule from a few atoms to millions of atoms. The disadvantage of X-ray crystallography is that well-ordered crystals are required for data collection.

Related Websites

Bernard Rupp's teaching site	http://www-structure.llnl.gov/Xray/101index.html
Cambridge structural medicine course	http://perch.cimr.cam.ac.uk/course.html

Further Reading

Blundell TL, Johnson LN (1999). *Protein Crystallography*. Academic, New York.

Drenth J (1999). *Principles of Protein X-Ray Crystallography*, 2nd ed. Springer-Verlag, New York.

Glusker JP, Lewis M, Rossi M (1994). *Crystal Structure Analysis for Chemists and Biologists*. VCH Publishers, Berlin, Germany.

Ladd MFC, Palmer RA (1994). *Structure Determination by X-Ray Crystallography*. Plenum, New York, UK.

Rhodes G (1993). *Crystallography Made Crystal Clear*. Academic, New York.

See also Crystallization of Macromolecules; Crystal, Macromolecular; Diffraction of X Rays, Electron Density Map; Heavy-Atom Derivative; Phase Problem; Refinement; *R*-Factor; Space Group.

Y

Y Chromosome
YAC

Y CHROMOSOME. *SEE* SEX CHROMOSOME.

YAC. *SEE* VECTOR.

Dictionary of Bioinformatics and Computational Biology. Edited by Hancock and Zvelebil
ISBN 0-471-43622-4   © 2004 John Wiley & Sons, Inc.

Z

Zero Base
Zero Coordinate
Zero Position
Zeta Virtual Dihedral Angle
Zipf's Law

ZERO BASE. *SEE* ZERO COORDINATE.

ZERO COORDINATE (ZERO BASE, ZERO POSITION)
Thomas D. Schneider

The zero coordinate is the position by which a set of binding sites is aligned. Not having a zero as part of a coordinate system is a disadvantage because it makes computations tricky.

For consistency, one can place the zero coordinate on a binding site according to its symmetry and some simple rules:

Asymmetric sites: at a position of high sequence conservation or the start of transcription or translation.
Odd-symmetry site: at the center of the site.
Even-symmetry site: for simplicity, the suggested convention is to place the zero base on the 5' side of the axis so that the bases 0 and 1 surround the axis.

See also Binding Site Symmetry, Alignment—Pairwise, Alignment—Multiple.

ZERO POSITION. *SEE* ZERO COORDINATE.

ZETA VIRTUAL DIHEDRAL ANGLE
Roman A. Laskowski

Defined by a protein's four main-chain atoms: $C\alpha$—N—C'—$C\beta$. Its absolute value lies around 33.9°. A positive value indicates a normal L amino acid, whereas a negative value identifies the very rare D conformation.

See also Amino Acid, Main Chain.

ZIPF'S LAW. *SEE* POWER LAW.

Dictionary of Bioinformatics and Computational Biology. Edited by Hancock and Zvelebil
ISBN 0-471-43622-4 © 2004 John Wiley & Sons, Inc.

SYNONYM LIST

Dictionary of Bioinformatics and Computational Biology. Edited by Hancock and Zvelebil
ISBN 0-471-43622-4 © 2004 John Wiley & Sons, Inc.

AUTHOR INDEX

Aloy, Patrick

Apweiler, Rolf

Attwood, Terri

Baum, Jeremy

Dictionary of Bioinformatics and Computational Biology. Edited by Hancock and Zvelebil
ISBN 0-471-43622-4   © 2004 John Wiley & Sons, Inc.

Guigó, Roderic

Gutell, Robin R.

Hancock, John M.

INDEX OF ENTRIES

Dictionary of Bioinformatics and Computational Biology. Edited by Hancock and Zvelebil
ISBN 0-471-43622-4　© 2004 John Wiley & Sons, Inc.